DRUGS and the HUMAN BODY

with Implications for Society

SEVENTH EDITION

DRUGS and the HUMAN BODY

with Implications for Society

SEVENTH EDITION

Ken Liska

Upper Saddle River, New Jersey 07458

Library of Congress Cataloging-in-Publication Data
Liska, Ken,
Drugs and the human body : with implications for society / Ken Liska.— 7th ed.
p. ; cm.
Includes bibliographical references and index.
ISBN 0-13-177321-6
1. Drugs. 2. Drugs—Social aspects. 3. Drugs of abuse.
4. Psychotropic drugs.
[DNLM: 1. Pharmaceutical Preparations. 2. Pharmacology.
3. Sociology, Medical. QV 55 L769d 2004] I. Title.
RM300 .L53 2004
615'.1—dc21 2003054844

Project Manager: Kristen Kaiser
Editor-in-Chief: John Challice
Vice President of Production and Manufacturing: David W. Riccardi
Executive Managing Editor: Kathleen Schiaparelli
Assistant Managing Editor: Beth Sweeten
Production Editor: Robert Runck
Composition: Laserwords, Inc.
Manufacturing Buyer: Alan Fischer
Manufacturing Manager: Trudy Pisciotti
Marketing Manager: Steve Sartori
Managing Editor Audio/Visual Assets: Patty Burns
AV Editor: Jessica Einsig/Abigail Bass
Art Studio: Precision Graphics
Art Director: Jayne Conte
Cover Designer: Bruce Kenselaar

Pearson Prentice Hall
Pearson Education, Inc.
Upper Saddle River, New Jersey 07458

Printed in the United States of America.

10 9 8 7 6 5 4 3 2

ISBN 0-13-177321-6

Pearson Education Ltd., *London*
Pearson Education Australia Pty., Limited, *Sydney*
Pearson Education *Singapore*, pte. Ltd
Pearson Education North Asia Ltd., *Hong Kong*
Pearson Education Canada, Ltd., *Toronto*
Pearson Educatión de Mexico, S.A. de C.V.
Pearson Education–Japan, *Tokyo*
Pearson Education Malaysia, Pte. Ltd.

To Paula Meta

Knowledge is of two kinds. We know
a subject ourselves, or we know
where we can find the information.

Samuel Johnson, 1709–1784

Contents

CHAPTER 3 *Federal Laws: The FDA and Drug Testing—Penalties for Illicit Use* 70

CHAPTER 4 *What Happens to Drugs After We Take Them* 96

CHAPTER 5 *Drugs at the Synapse* 127

CHAPTER 6 *Narcotic Analgesics: Opiates and Opioids* 144

CHAPTER 7 *Cocaine, Amphetamines, Caffeine, Nicotine, and Other Stimulants* 173

Preface

Attention, Students! You have in your hands a book loaded with facts about drugs in our culture. To obtain the most from it, I urge you to make use of the following aides:

- More than 180 carefully selected Web sites (URLs) are included for your browsing. They are found in the text and at the ends of chapters. It is very important you type in the Web address *exactly* as shown. Here are 5 super introductory sites you can browse right away:

  `http://www.a1b2c3.com/drugs/`
  `http://www.health.org` (National Clearinghouse for Alcohol and Drug Information)
  `http://www.usdoj.gov/dea/` (Drug Enforcement Administration)
  `http://www.drugfreeamerica.org/drug_info.html`
  `http://www.nida.nih.gov/DrugPages/`

  Of course, you may use your computer's search engines (such as Google, Mamma, or Lycos) to locate hundreds of additional Internet sites (I have given you many suggestions at the ends of certain chapters). You will soon learn that many of the sites you find are not valuable, and you must be selective.
- **Drugs in Sports** is both the newest chapter (Chapter 17) and among the most pertinent to rampant drug use in America. This topic is now of major importance in America and throughout the world.
- **Women's Health** and the importance of drugs therein is given greatly expanded coverage in this Seventh Edition. See the index for many facts, especially on unique drug responses and to obtain information from government sources.
- For the chemistry-minded, **Structure–Activity Relationships** are presented in Appendix I. Learn about the chemical parts of molecules that engender pharmacological response.
- If you need help with words, a **Glossary** of over 200 terms will prove invaluable. Try it out (Appendix V).
- Appendix II presents a 42-question **self-test on understanding alcoholism**, with answers and comments by experts in the field.
- **DAWN data** summaries show us which drugs are causing problems, where in America they are prevalent, and to whom they represent a threat. See Appendix III and throughout the text.
- Several dozen true **case histories** give us real-life examples of drug abuse. Topics range from an LSD trip to potheads to a failed diet plan.
- **Keyword** listings and **learning objectives** at the beginning of each chapter provide focus, and in-chapter and end-of-chapter **study questions** will move you to reexamine the chapters and learn the material. Some of the study questions will ask you to take

a stand on a drug or its effect on society. I urge you to equip yourself with the knowledge, state your position, and back it up with facts.
- Check out the inside front and rear book covers for more drug data.

To the Professor: This new Seventh Edition edition features many enhancements and additions.

- New is Section 2.7, **The Future of Drug Discovery** or "Drug Discovery 21st-Century Style." Drug discovery has been caught up in the great wave of technological change involving the human genome, the human proteome, the accompanying tools for research, and the support from computer-based Bioinformatics.
- **Hormone replacement therapy (HRT)** is now included with estrogens in Section 2.5. HRT is a topic rife with research—so much so, in fact, that conclusions can be difficult to draw.
- **Complementary and alternative medicine (CAM)** has become an important part of the nation's health system, and is now included in Section 1.3, with definitions and references for additional information.
- **Internet sales of drugs and devices** is very big business, but the FDA and others have serious reservations about this practice, including counterfeit drugs (new in Section 3.9).
- **Drugs and pregnancy** (Section 4.4) is new, as are the URL references.
- **Women's Health** (Section 4.9) is given much greater emphasis in this Seventh Edition, with references to more information on the Internet and from government sources. The topic of Alcohol and Women is now in Section 9.4.
- **Club drugs, raves**, and other current drugs of abuse are now emphasized in Chapter 12 generally and in Section 12.6, specifically.
- **Aspirin in the Prevention of Heart Attack and Stroke** (Section 14.5) has been brought up to date with new research reports.
- **The discussion of COX-2 inhibitors** has been expanded with new reports and Internet references (Section 14.10).
- New is Section 16.10, **Are Teens Too Smart for Drug Testing**?
- Throughout the text, many other topics have been added or updated, including circadian rhythms (Section 4.6), placebo (Section 1.3), drug addiction (1.4), generic drugs (1.10), ephedra (2.3), phytoestrogens (2.5), the search for anti-AIDS drugs (2.8), children's drugs (4.6), drug interactions (4.7), synaptic transmitters (5.7), nicotine (7.17), the chemical treatment of alcoholism (9.10), marijuana and hashish (11.1), marijuana as medicine (11.7), diet aids (15.12), and hydroxyethylstarch in sports (17.2). NIDA data are updated throughout the text. Note that all the URLs in the previous edition were searched, defunct sites discarded, and many new sites added; this applies to all in-text and end-of-chapter URLs.

A Seventh Edition **Instructor's Manual** (IM) is available gratis to adopters of this text. The IM provides a 725-question **Test Bank**, with all answers included; answers to all in-chapter and end-of-chapter questions for all 17 chapters; many more Web sites; a listing of approximately 100 federal and national organizations and resource centers related to drug information (via telephone, e-mail, and URLs); and several dozen newsletters and references. In addition, relevant publications and agencies are supplied specifically for each chapter.

This book can be used in general health and drug education classes, counselor training, parent groups, and allied health fields in general—nursing, dental hygiene, medical technology, physical therapy, occupational therapy, medical records, and industrial hygiene.

As before, most drugs have been selected for inclusion on the basis of their high potential for abuse and misuse or their dramatic effect upon contemporary North American society. I discuss prescription drugs, over-the-counter drugs, street drugs, and recreational drugs.

This book provides facts about drug sources, history, action in the body, side effects, interactions, tolerance, abuse potential, dosage, dependency, drug delivery systems, and alternatives to drugs in use today. Students appear eager to discuss the impact of drugs on their lives, and this book provides a host of golden opportunities. Consider drug dependence (Chapter 1); quackery (Chapter 3); drugs and sex (Chapter 4); drugs and stress (Chapter 5); pain killers (Chapter 6); cocaine and amphetamines (Chapter 7); the fetal alcohol syndrome and lowering the drinking age (Chapter 9); what marijuana does to the body (Chapter 11); the resurgence of LSD and look-alike drugs (Chapter 12); oral contraceptives for men (Chapter 13); personal drug testing (Chapter 16); and drugs in sports (Chapter 17).

I have furnished brief histories of a number of topics, including the development of insulin, the discovery of the ergot alkaloids, and the use of LSD in America. I believe that the history of drugs is fascinating—and, as in the case of cancer quackery and the origin of the food, drug, and cosmetic laws—often highly instructive and personally useful.

Chapters 1–5 address definitions, concepts, theories, laws, and procedures that can be applied generally to many drug categories. For example, in Chapter 4, I discuss how drugs are administered, how they are distributed throughout the body, and what eventually happens to them. In this chapter, the concepts of drug life, drug metabolism, and interactions help explain how drug action can persist in the body and how drugs can have unexpected actions.

Assuming that the reader has little or no background in science, I have used few chemical formulas. For the student who has the necessary background, however, Appendix I provides a challenging study of structure–activity relationships of fourteen pharmacological classes of drugs. A glossary is provided in Appendix V for those who are concerned about nomenclature.

Reader involvement in the text is encouraged by in-text problems and end-of-chapter questions and by means of a 42-question self-test on alcoholism (with answers and discussion in Appendix II). Also, in Chapter 1, students are given the chance to establish a beginning vocabulary of physiology terms and to become familiar with the metric system.

Acknowledgments

I thank reviewers of the Seventh Edition for their many valuable suggestions: Glenn Eberhart, University of California, Santa Cruz; John Lovern, Alan Hancock College; and Peter Malo, Oakton Community College. Special thanks go to Ms. Kristen Kaiser, Project Manager—Chemistry, Prentice Hall. I also give credit to James Cordell, MD, Heinz Hoenecke, MD, Paula Meta Hoenecke Liska, and Mr. Lowell Van Tassel. Finally, I thank Mary-Jane Johnson, PowderJect Pharmaceutical, Oxford, UK, for her kind assistance.

My goal has been to create a useful and informative text, and I welcome feedback and suggestions.

Ken Liska
kenn_liska@ix.netcom.com

DRUGS and the HUMAN BODY

with Implications for Society

SEVENTH EDITION

CHAPTER 1 The Magic Bullets

Key Words in This Chapter

- drug
- placebo
- addiction
- dependence
- complementary/alternative medicine
- chemotherapy
- receptor site
- generic drug
- metric system

Learning Objectives

After you complete your study of this chapter, you should be able to do the following tasks:

- Write the definition of a drug.
- Explain the nature of the placebo effect.
- Understand the nature of addiction (or drug dependence), tolerance, and the withdrawal syndrome.
- Cite the role of chemotherapy in modern medicine.
- Offer an overview of the receptor site theory.
- Know where to get information about drugs.
- Differentiate between brand names and generic names of drugs.
- Understand and use the metric system.

1.1 Introduction

What guides your daily life? securing an education? family responsibilities? America's war on terrorism? Surely there are many compelling factors. But I posit that drugs—whether OTC, prescription, recreational, or street—can also be a major player. Drugs are advertised at our every turn, available at our every glance, as a remedy for some illness, alcohol for a good time, a prophylaxis of disease, the nicotine on which one is dependent, or perhaps a substance the DEA has its eye on. America's drug culture, licit or illicit, affects just about all of us: manufacturer, distributor, prescriber, user, watchdog, or prosecutor. All of this makes your study of *Drugs and the Human Body* appropriate—even necessary.

Just as our drug culture is broad, so is the scope of this book. I shall introduce you to all major and many minor aspects of drugs in America today. As our subtitle, "With Implications for Society" suggests, I will link the drug to its impact on how we live. Major examples of this are alcohol (America's biggest drug problem, agreed?), tobacco, cocaine, marijuana, drug laws, birth-control pills, diet pills, pain pills, drugs in sports, and hallucinogens.

The following facts may help motivate you in your study of drugs and the human body. Your pocketbook is affected by drugs: each year drug abuse and addiction cost the United States $110 billion in law enforcement, medical bills, and lost earnings. Illegal drugs cause thousands of deaths each year and help spread communicable diseases such as AIDS and hepatitis C. In a recent year, more than three-quarters of new heroin users were between ages 12 and 15. An overwhelming 80 percent of adults in jail today have a history of alcohol and drug abuse. A new, large-scale federal study concludes that, in America today, drug abuse and drinking are bundled with high-risk sex. A United Nations report estimates that organized crime syndicates gross about $1.6 trillion a year, with the biggest growth area in drugs—now a larger global industry than manufacturing new cars; altogether, the illegal drug business comprises about 8 percent of world trade. If you and I are drinkers, we stand a chance of joining the more than 1 million Americans arrested each year for driving under the influence. Scholastically, D and F students drink three times as much as others. Government agencies estimate that as many as 1 in 10 of Baltimore's residents are heroin addicts. Today illicit drug use is significant across nearly all age groups (see Figure 1.1).

The sobering statistics go on and on, and we tend to become inured by them. Nonetheless, a knowledge of the drugs we are likely to encounter is essential for making decisions as individuals and members of society. Hundreds of Americans will make that first call to Alcoholics Anonymous today or take that first dose of Valium to get to sleep tonight. A high school football player may see for the first time what an anabolic steroid pill looks like. A 12-year-old may be offered her first toke on a marijuana joint. Many may soon vote on a ballot initiative to legalize the growing of marijuana for personal consumption.

In this book, you will be able to examine the source, dosage, action, and fate in the body of every major high-use, high-abuse drug in use today. You will be able to study prescription drugs, over-the-counter drugs, recreational drugs, and street drugs. The first five chapters in this book are introductory, presenting general information that can be applied to specific drugs. We then examine individual pharmacological[1] categories such as painkillers, stimulants, alcohol, and marijuana.

To help tweak your curiosity and spur you on, several dozen true case histories are included, along with study questions in the chapters and at the end. Many of these questions are controversial and are designed to permit you to take a stand or even debate your classmates. By the way, that is another aspect of the topic of drugs. Legitimate opposing views can be applied to topics throughout this text.

1.2 What Is a Drug?

The word *drug* has different meanings depending on the situation. The expression *street drugs* suggests narcotics, uppers, downers, PCP, hallucinogens, and cocaine. Many people equate street drugs with "dope," regardless of the pharmacology of the street drug (*dope* originally meant opium and its narcotic alkaloids, morphine and

[1]**Pharmacology** is the study of the distribution, actions, and fate of drugs in the body. See the Glossary (Appendix V) for additional definitions and explanations.

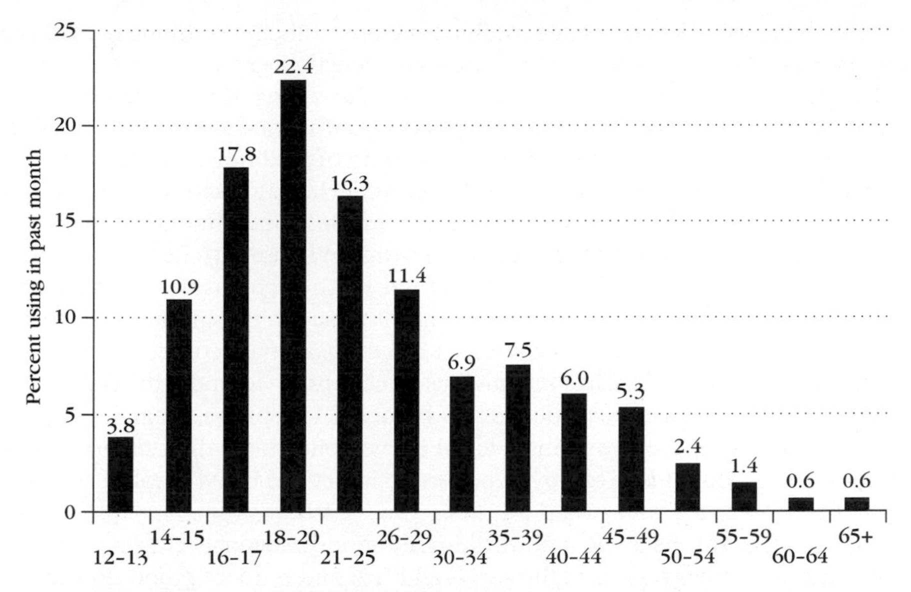

Figure 1.1 Past-Month Illicit Drug Use, by Age: 2001 (*Source*: Results from the 2001 National Household Survey on Drug Abuse. Vol. I. Summary of National Findings. Substance Abuse and Mental Health Services Administration (SAMHSA), Department of Health and Human Services. Printed 2002.)

codeine). Students: please note that terms such as *alkaloid* are defined in the Glossary, Appendix V.

The prescribing physician views drugs in a much more positive light. To him or her, they are the agents that cure or prevent illness by altering or enhancing some physiological process. The physician uses drugs to

- Fight infection (antibiotics, sulfa drugs, antiseptics).
- Reverse a disease process (antihypertensives, heart stimulants, antidepressants).
- Relieve symptoms (morphine, aspirin, steroids, antacids).
- Restore normal function (insulin, iron, gonadotropins, diuretics, hormones, tranquilizers, bronchodilators, anticoagulants).
- Aid in diagnosis (radiopharmaceuticals, skin tests, urine tests, barium enemas).
- Inhibit normal body processes (oral contraceptives, monoamine oxidase inhibitors [MAOI], anesthetics).
- Maintain health (vitamins, minerals, vaccines).

The physician has thousands of modern prescription drugs from which to choose.

But there is another definition we should consider. If our body responds in some way to a substance, either physically, biochemically, or mentally, that substance can be said to have a drug effect. This response leads us to a general and broad definition:

> A **drug** is any absorbed or applied substance that changes or enhances any physical or psychological function in the body.

If we accept this definition and recognize that body functions include mental activity, digestion, metabolism, blood circulation, sexual function, growth, wound repair, and so on, we must conclude that *almost anything can act as a drug*. The *Food and Drug Administration* (FDA), under the law, may regulate as a drug any substance other than a food "intended to affect the structure or any function of the body of man."

Many obvious examples of drug effects exist, including those seen in the treatment of high blood pressure, hay fever, and cancer. But what about the less-obvious cases? Is the dextrose found in candy bars a drug? When dextrose (glucose) is administered in saline to a postoperative patient or to a comatose, severely hypoglycemic individual, it is surely a drug. When the sugar in our diet acts to alter insulin production in the pancreas, it is surely a drug. Ethyl alcohol, which we consume in alcoholic beverages, has the obvious effect observed in the drinker but also has the capacity to cause birth defects, to induce a low blood sugar level (hypoglycemia), and to relieve the painful facial nerve condition called *tic douloureux*, for which ethyl alcohol is given by hypodermic injection. Ethyl alcohol is a multiaction drug.

The nicotine in tobacco is a powerful peripheral vasoconstrictor, which means that it constricts blood vessels in the hands and feet and reduces blood flow in these areas. Nicotine is a powerful drug.

Do you think of the caffeine in coffee, tea, and cola drinks as a drug? You should, because caffeine is a central nervous system stimulant (see Chapter 7, "Cocaine, Amphetamines, Caffeine, Nicotine, and Other Stimulants"). Caffeine can have strong effects on the brain, spinal cord, heart, and kidneys. Brussels sprouts, cabbage, and kale can inhibit normal production of thyroid hormone in certain people, leading to goiter. That result can be considered a drug effect. Monosodium glutamate (MSG), a widely used flavor enhancer, can show startling drug effects in susceptible people—who become acutely ill after consuming the substance (see Section 4.10, "Excretion of Drugs").

To a significant part of our society and to the Chinese, ginseng is considered a valuable drug. This aromatic bitter is believed to act as a tonic and as a restorative. But ginseng has found no value as a drug in modern medicine, is not employed by the prescribing physician, and is not included in the official compendia (see the discussion of the *United States Pharmacopoeia* and the *National Formulary* in Section 1.9, "Where to Get Information About Drugs"). Obviously, whether ginseng is a drug depends on one's training, experience, and prejudices.

If we are not careful when using certain insecticide sprays, we can inhale the chemical or absorb it through our skin. Such ingestion of an insecticide has been shown to result in chemical changes in the liver, such as increased enzyme synthesis. That is a drug effect.

Arsenic, in very small doses, was deliberately ingested by certain women hundreds of years ago to produce the pale complexion considered at that time to be a mark of beauty. Further, in very small doses arsenic or arsenic-based compounds have found medicinal value in the treatment of African sleeping sickness, and more recently in acute promyelocytic leukemia. In large doses, arsenic is clearly a cell poison. The importance of the amount given is reflected in the old Latin phrase, *dosis sola facit*—"the dose alone decides." Recreational drugs comprise yet another category. Surely, tobacco, alcohol, marijuana, and cocaine change or enhance one body function or another. For a valuable Web site on recreational drugs, browse **http://www.a1b2c3.com/drugs/**.

In summary, we conclude that just about everything has the potential of being a drug. Whether a substance is or is not a drug may depend on the patient, the dose given, environmental circumstances, the illness, prior treatment, and the desired effect. Each substance must be evaluated on its own merits under the special conditions of use, and the pharmacological results obtained must be evaluated to see whether they have come from the drug and not from placebo effect.

Problem 1.1 *Is water a drug? Is salt a drug? In what situations might they be considered drugs? In what situation might aspirin not be considered a drug?*

1.3 The Placebo Effect

Drug experts have learned that many people can experience typical drug effects when they *believe* they have taken a drug but actually have not. This effect is called the **placebo effect**. Because the patient believes that he or she is receiving a genuine drug (or believes in the doctor who assures him or her that the treatment will work), the patient expects a positive response and often receives that positive response. The actual placebo (derived from the Latin phrase, "I will please") is a dosage form prepared to look like the real drug but instead contains nothing more active than, say, milk sugar.

Placebos are used intentionally by physicians and researchers in two ways. First, consider a geriatric patient who may be so dependent on her "little red pill" that she thinks she cannot get along without it, although its side effects outweigh its benefits. Her doctor, with the family's knowledge, may have the pharmacist prepare a placebo that the patient cannot distinguish from the real drug. She is placated, and the side effects are eliminated. Second, in investigating the effectiveness of a new drug, the manufacturer will arrange to have half the patients receive the drug and half receive a placebo. The entire evaluation is also conducted double blind; that is, neither the person receiving the drug nor the person administering the drug knows which is the placebo and which is the real drug.[2] The manufacturer knows that inclusion of a placebo is necessary to determine how much of the response is due to the drug and how much is due to the patient's *expectation* (that is, to the power of suggestion).

This expectation is recognized today as formidable. For example, in one study on a new tranquilizer, 40% of the patients who received the placebo said the treatment worked and made them feel better. In another study on pain, it was discovered that the placebo satisfactorily relieved pain of various causes in 35% of patients. By comparison, in the same study, large doses of morphine relieved pain in only 75% of the subjects. This result shows that placebos can have great power in the human and must be taken into account in evaluating a drug's effectiveness. Many other examples of placebo responses have been reported in medical literature, including the successful application of placebos in the treatment of arthritis, hay fever, hypertension, headache, constipation, acne, seasickness, ulcers, and even warts. Placebos can even induce typical drug side effects such as skin rash, nausea, and anaphylactic shock.

---

[2]Today the ultimate protocol would be a multi-center, randomized, double-blind, placebo-controlled, crossover design, followed by an open-label study.

In a British study, aspirin tablets with a brand name marking were found to be more effective in relieving headaches than unbranded aspirin, apparently because the patients believed, from past experience, advertising, and hearsay, that they could trust the branded tablets to be effective.

Case History *Baldness Drug Versus Placebo*

Minoxidil in 2% topical solution (Rogaine) is prescribed for the treatment of baldness. Its effectiveness was tested against a placebo. After four months of use, of 717 patients who received minoxidil, 41% reported no hair growth, 32% reported minimal growth, 25% reported moderate growth, and 0.7% reported dense growth. Of 714 patients who received the placebo, 58% reported no growth, 31% reported minimal growth, 10% reported moderate growth, and 0.4% reported dense growth. On this basis, the placebo appears to work as a hair growth restorer.

Actually, the placebo effect is but one manifestation of the more profound observation that the mind and the body are inseparable, and that the human organism has a great capacity for self-healing. Hence, attempts to treat most bodily diseases as though the mind were in no way involved must be considered archaic. Seventy-five years ago, Sir William Osler, the father of modern medicine, said, "The cure of tuberculosis depends more on what the patient has in his head than what he has in his chest." Research today supports the view that almost every ill that can befall the body—from headache to heart disease—can be influenced in some way by the person's mental state.

Support exists for the belief that a good mental attitude can fight off disease. Certain neurotransmitters can affect the activity of some white blood cells that defend against infection. Also, certain white blood cells (lymphocytes) are known to have receptors that are sensitive to neurotransmitters. Evidence also exists that the immune system may influence the nervous system.

A University of Copenhagen meta-analysis (see Section 5.7) of 114 published clinical trials[3] concluded that the placebo effect is a myth and should be ignored in clinical trials. We note that this study has received criticism.

Problem 1.2 *Measuring the placebo effect is difficult, because research subjects often get better on their own. List three reasons why research subjects get better on their own.*

Holistic medicine (or **holism**), an understanding of the patient's whole situation, is practiced today in both orthodox and unorthodox circles. In the holistic approach, the patient's emotional, spiritual, social, lifestyle, and environmental factors are considered in the diagnosis and treatment of the disease. The

[3]*Scientific American*, October 2001, p. 16.

American Holistic Medical Association (AHMA) was founded in 1978 by physicians who wanted to use nontraditional methods as adjuncts to traditional ones. Most holistic approaches rely heavily on psychology. Holism has been criticized by some because of the emphasis some practitioners place on acupuncture, "psychosynthesis," "homeotherapeutics" (muscular alignment through music), "psychoelectronics," personal awakening, and physical manipulation. Unorthodox approaches aside, no one can deny the importance of the mind in the treatment of physical ailments.

We can distinguish between two types of placebo effects: the clinical and the verbal. In the former, a make-believe drug is given, and the power of suggestion takes over. In the latter, the words spoken by a physician, nurse, or friend have the same effect. In both the clinical and the verbal approaches, it is believed that certain natural body drugs are released, among them endorphins (see Chapter 6, "Narcotic Analgesics: Opiates and Opioids"), adrenal hormones, and neurotransmitters (see Chapter 5, "Drugs at the Synapse"). Generally, people accept today that acupuncture for pain treatment works similarly to a placebo by causing the release of endogenous endorphins.

Complementary and Alternative Medicine (CAM) is now a $35 billion-a-year business in the United States. The National Institutes of Health (NIH) categorizes CAM into:

Homeopathy, in which diseases are treated using drugs that, given to a healthy person in tiny doses, would produce symptoms similar to those of the disease.

Biologically based herbal or nutritional regimens, intended to prevent or treat illness or promote health. Included here are unconventional products such as chelators, shark cartilage, or bee venom.

Naturopathy and folklore, which consider disease to be a disruption in the body's natural healing processes. Techniques are designed to restore health rather than to directly treat the disease. Examples are the Chinese qigong (pronounced "chi-gong"), and the Japanese reiki ("ray-key").

Mind-body systems such as biofeedback, meditation, hypnosis, transpersonal hypnotherapy, and past-life regression.

Bioelectromagnetic approaches, which attempt to apply magnets, electrical currents, or copper bracelets.

Body-based systems such as chiropractic, acupuncture, aromatherapy, cranial manipulation, massage, body and soul conferencing, and psychic healing.

In 1998, the U.S. Congress established The National Center for Complementary and Alternative Medicine (NCCAM) as part of the NIH. NCCAM is "dedicated to exploring complementary and alternative medicine practices in the context of rigorous science, training CAM researchers and disseminating authoritative information." Call toll free 1-888-644-6226, or browse `http://nccam.nih.gov/`.

An estimated 123 million Americans try herbs, vitamins, minerals, enzymes, amino acids and other "natural products" to cure their ills, lose weight, prevent cancer, elevate their mood, build muscles, get to sleep, heal their joints, improve their memory, and tune their libido. But often there is precious little sound scientific evidence behind the hype, and there can be real dangers. Our FDA has over 2900 adverse-event reports on ginkgo, St.-John's-wort, ephedra, zinc, ginseng and many other CAM products. A University of California San Diego report concludes that St.-John's-wort does nothing for people with major depression; in fact, it can adversely interact with Prozac. Garlic products can interfere with AIDS drugs. See Sections 3.2 and 12.11 for more consumer information.

Complementary and alternative medicine has become an important part of the nation's health system. In one recent year, 75 out of 117 U.S. medical schools offered elective courses in CAM or included CAM topics in required courses. The American Medical Association has included articles on CAM in all 10 of its journals. For pro and con views on CAM, browse **http://www.holisticmed.com** and **http://www.quackwatch.com/index.html**. Also see the article in *Newsweek*, December 2, 2002, pp. 45–73.

1.4 Drug Dependence (Addiction)

Especially in the field of drugs, a clear perception of the meaning of a word is a necessity. The words *drug addict* and *drug addiction* have been in use for longer than we can remember, most often to describe a person who has become deeply involved with one of the narcotic analgesics such as morphine or heroin. Today, another term is used to describe people who are deeply involved with drugs. They are said to be **drug-dependent** persons. They may be dependent on a narcotic, alcohol, a barbiturate, cocaine, or one of the many other possible drugs.

Because **addiction** and **addict** are still a large part of our language, let us examine the original meaning of these terms. From a medical-legal point of view, addiction was considered to mean a drug-induced change in the physical state of an individual, in which she or he required the continued presence of the drug to function normally. Further, upon abrupt termination of the drug's use, the addict would suffer through a physical crisis of varying degree known as a **withdrawal syndrome** (also termed *abstinence syndrome*). The withdrawal crisis could be ended at any time by readministering the drug.[4]

According to this definition, then, an addict's body somehow changes physiologically so that it requires the drug for normal existence, much as it requires insulin for normal existence. As a corollary to this definition, it was realized that the addict developed a **tolerance** to the drug, so that ever-increasing doses of the drug had to be taken to achieve the desired effect. This definition of addiction is still widely applied. This definition is useful in describing the narcotic addict, the barbiturate addict, and the alcoholic—all of whom develop a physical reliance on their particular drug, develop a tolerance for it, and suffer withdrawal when the drug is abruptly removed. Incidentally, the symptoms of withdrawal are typically the opposite of the acute effects the drug exerts in the first-time user. For example, heroin—at first use—relieves pain, sedates, and constipates. In heroin withdrawal, we see muscle and bone pain, agitation, insomnia, and diarrhea. Withdrawal from brain-depressant drugs such as alcohol generally involves insomnia, agitation, tremor, and overactivity.

Modern therapists realize, however, that not all addiction results in physical reliance on a drug, and therefore there is not always a physical withdrawal syndrome. Cocaine and nicotine are drugs with which people become overwhelmingly involved.

---

[4]The three classes of drugs that typically produce withdrawal symptoms after heavy use are the depressants (such as alcohol, benzodiazepines, and barbiturates), the opiates (heroin, morphine, methadone, and other synthetics), and the stimulants (cocaine and all amphetamine types).

But cocaine and nicotine do not typically produce withdrawal symptoms like those seen with heroin or the barbiturates. Instead, they usually produce a psychic craving for the drug. Thus, the term **drug dependence** was introduced to apply to all situations in which drug users developed a reliance—either physical or psychological (psychic). Note that tolerance is not a necessary corollary in the concept of drug dependence. An example of this is cocaine, which induces psychic dependence but does not induce the development of tolerance in smaller doses. The concept of drug dependence is broad and is applicable to hallucinogens and minor tranquilizers, as well as hard narcotics.

Yet another approach, based on behavior, has been taken to describe heavy drug involvement. In this context, addiction is seen as a *behavioral pattern* in which the person has an overwhelming compulsion to use the drug, an anxiety about an adequate supply, and a high tendency to return to the drug after withdrawal. Addiction is viewed as an extreme on a continuum of involvement with drug use (see Figure 1.2). Thus, habituation to a drug and preoccupation with it are steps on the road to overwhelming involvement with the drug.

Defined on a behavioral basis, addiction can occur without the person becoming physically dependent or developing a tolerance. (The opposite can also happen, however. Tolerance to drugs can develop without concurrent addiction. The major tranquilizer chlorpromazine is an example.)

The behavioral approach makes it easier to understand why the terminally ill cancer patient taking the Brompton Cocktail (see Glossary) would not be considered a drug addict—nor would the fetus of an addict mother.

In addiction, behavioral pattern reinforcement is of two types. The process by which a repetitive action brings pleasure or reward is termed **positive reinforcement**—as in eating, drinking, or sexual behavior. Drugs can be more powerfully and persistently rewarding than natural reinforcers to which the body is accustomed. In the addict, periods of abstinence are marked by feelings of discomfort and craving, and the addict is motivated to take the drug not because of reward, but to avoid painful stimuli. This type of motivation is termed **negative reinforcement**. Both positive and negative reinforcement are operating in the maintenance of a typical addiction, such as alcoholism.

The most addictive substance Americans encounter is nicotine. Approximately one-third of people who are serious smokers become addicted, compared to about one-quarter of heroin users. For cocaine and alcohol, the figures are around 15–16%. About 11% of amphetamine users become addicted, and about 9% of marijuana users become addicted.

Some 3.2 million Americans and 1.2 million people in Western Europe are addicted to hard drugs, says the United Nations.

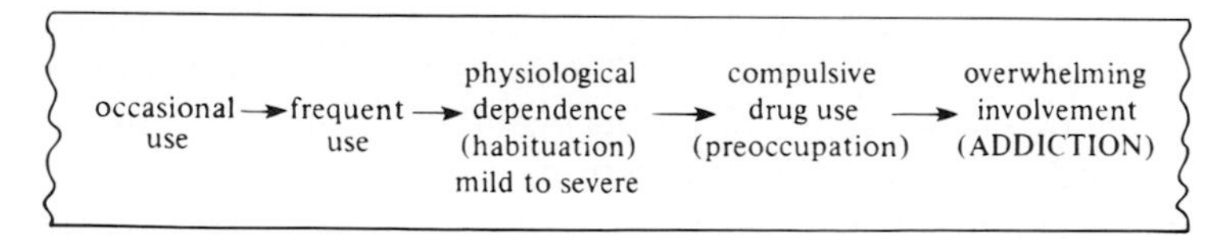

Figure 1.2 Addiction can be viewed as the extreme on a continuum of involvement with drug use.

To summarize, we have listed three definitions of the process in which a person can become deeply involved with a drug:

1. *Addiction*: Physical change and development of tolerance and withdrawal syndrome
2. *Drug dependence*: May be physical or psychological; withdrawal and tolerance are not necessarily seen; applicable to any drug situation
3. *Behavioral pattern of compulsive use*: The extreme on a continuum of drug involvement; high tendency to relapse; physical reliance and withdrawal are not prerequisites

Scientists have been investigating the brain's **reward center**, and possible molecular mechanisms involved in addiction. That is, they are looking for chemicals (such as neurotransmitters, described in Chapter 5, "Drugs at the Synapse") whose presence, overabundance, or deficiency in the brain could explain how addiction arises in the first place. We now understand that the neurotransmitter dopamine (Section 5.3, "What Happens at the Synapse?") is at the heart of the brain's **natural reward system**. Dopamine is so effective at making a person feel good that the brain carefully regulates its release. Dopamine is allowed out of the brain for only brief moments and then is picked up and removed by a natural guard, or transporter molecule, made in the brain. This guard sends the dopamine back to its storage cell. According to The National Institute on Drug Abuse, all drugs of abuse—including cocaine, amphetamines, heroin, alcohol, and nicotine—disrupt the normal flow of dopamine, stimulating its release and increasing brain levels up to 10 times the normal rate. This action is believed to be significantly involved in producing drug-induced feelings of pleasure and reward, and over time, addiction and vulnerability to withdrawal symptoms. George Uhl's research team at the Addiction Research Center at NIDA has found that cocaine "handcuffs" the transporter guards (Section 5.1, "Introduction"), making them unable to bind and remove dopamine. Unbound, the dopamine molecules continue to produce their pleasurable effects. It should be noted that chemist Mark Wightman of the University of North Carolina, Chapel Hill, has evidence that dopamine is not the ultimate reward substance, but rather a messenger molecule, actually one of several factors involved in the pleasure process.

In summary, research shows that the brain's reward system underlies addiction to cocaine, alcohol, heroin, morphine, amphetamines, angel dust, and nicotine. These drugs powerfully activate neuronal circuits in the brain's "reward center" found in the ventral tegmental area, connected to the nucleus accumbens.

Do you have an **addictive personality**? Are you doomed from birth to develop a dependence on opiates, cocaine, alcohol, or nicotine? Not necessarily, but take a look at some statistics. There is now statistical evidence that genes play a role in predisposing people to drug abuse. Large numbers of fraternal and identical twins were studied at the Medical College of Virginia, and researchers concluded that genetic factors in female twins play the major role in the progression of marijuana and cocaine involvement from use to abuse to dependence. Researchers concluded that marijuana and cocaine abuse and dependence are highly heritable and that genetic factors are responsible for 60–80% of the differences in abuse and dependence between fraternal and identical twin pairs.

We point out that family and social environmental factors are also influential in determining predisposition to drug abuse. Researchers at Harvard University have statistical evidence that genetic influences are stronger for abuse of some drugs than for others.

See Section 9.9 for comments on the familial connection to alcoholism, and Section 2.5 for the inherited predisposition to diabetes mellitus. On the Internet see **http://health.bobstx.com/archive/200011-healthy_personality.htm** and **http://www.psychosocial.com/addiction/vulnerab.html**.

Finally, should addiction be considered a disease, much like tuberculosis, or is addiction a moral deficiency? Most therapists today believe that an addict is diseased and cannot control his or her drug use any more than a cancer victim can control the growth of a tumor. They say that trying to quit on your own is like trying to operate on yourself. A few dissenting voices say that the disease theory is poppycock, and that addicts should not be absolved of their responsibilities by convincing them they have a permanent disease that can be managed but never cured.

1.5 Chemotherapy

Chemotherapy is a term frequently used in modern medicine and is defined as the planned attack on a disease using a specific chemical—one that has been designed and prepared for that very purpose. It is expected that the drug will have little or no harmful effect on the patient. This approach is in contrast to the obsolete, hit-or-miss, "shotgun" treatments that employed herbs, extracts, tinctures, or mixtures of drugs in the hope that something would work.

The father of chemotherapy is considered to be Paul Ehrlich (1854–1915), a German chemist who discovered Salvarsan, the first chemotherapeutic agent. Ehrlich planned the synthesis of Salvarsan and its rational use in the treatment of syphilis. Salvarsan killed the syphilis spirochete but managed to kill a few patients as well. Ehrlich continued his search, however, for the "magic bullet"—a drug that he hoped would travel directly to the disease center, do its job, and not harm any healthy tissue.

To some degree, Ehrlich's dream has been realized. *Antibiotics* are a kind of magic bullet. They can kill bacteria or prevent their growth without harming human cells, because they inhibit growth processes special to the bacterial cell. Another example of a magic bullet is radioactive iodine, used to treat thyroid cancer. Almost all iodine, including radioactive iodine, injected into the body ends up in the thyroid gland, the only tissue where its effects are felt.

Another idea for a magic bullet exists, which, if it succeeds, may be the greatest advance yet in treating cancer. The idea is to link a poisonous agent to a *monoclonal antibody* that binds specifically to a tumor antigen—in effect, delivering the poison to the tumor cell and nowhere else. In essence, a monoclonal antibody is made by an immune system in response to an antigen and is highly specific to its antigen. If we can make antibodies to cancer cells and affix cell poisons to the antibodies, the poisons will be carried by the antibodies to the only place in the body where the antibodies can bind—the antigenic cancer cell. Drugs based on the monoclonal antibody concept are termed *immunotoxins*. Rituxan, approved in 1997, was the first monoclonal antibody developed to treat cancer. Zevalin, another monoclonal antibody, is aimed at non-Hodgkins lymphoma, and is a radioimmune agent, inasmuch as it also delivers a dose of radioactivity to the tumor. Humira (adalimumab) is Abbott Laboratories' monoclonal antibody approved for reducing signs and symptoms of rheumatoid arthritis. Made by recombinant DNA technology, Humira blocks the human necrosis factor involved in the body's normal inflammatory response.

In addition to monoclonal antibodies, scientists are planning to attack cancer through the use of interferons, interleukins, colony-stimulating factors, and tumor necrosis factors. Research in cancer chemotherapy is intense, because the benefits to humanity are so great.

A problem that Ehrlich recognized continues to trouble chemotherapy today, viz., the possibility of any of three kinds of complications: drug allergies, side effects, and adverse effects. Although the risk of **allergic reaction** for most drugs is only 1–3%, effects can be severe, including skin rash, hives, allergic liver damage, and life-threatening anaphylaxis. Although allergy to penicillin is well known, other allergies can be totally unanticipated. **Side effects** are undesirable consequences of the way a drug works. For example, aspirin can prevent stroke, but also thins the blood, causing bleeding. A monoamine oxidase inhibitor used to treat depression can also allow the buildup of catecholamines. A drug used to inhibit androgen in prostate cancer therapy can also feminize the patient. And according to an article in *The Canadian Journal of Psychiatry*, a small number of users of the antidepressant Anafranil report that when they yawn they have an orgasm. **Adverse effects** are unwanted and mostly dose related, and can include liver and kidney damage (why could one expect both of these?), nausea or vomiting, diarrhea or constipation, mental effects such as dizziness, confusion, depression, and aggression, insomnia, breathing problems, and heart problems such as tachycardia or bradycardia (see Glossary, Appendix V). See Section 3.3 for additional examples of adverse effects. And remember, although the drug manufacturer tests a product for drug complications, it is impossible to predict every complication in a diverse population of millions of users. This is why some patients will defer using a new drug for up to a year after it reaches the market. Severe adverse effects have led to the withdrawal of potential blockbuster drugs from the market, including Pondimin, Seldane, Redux, and Rezulin. Readers concerned about drug safety are directed to the patient package insert that accompanies a drug, or to browse **www.fda.gov/medwatch/new.htm**. An FDA drug specialist can be reached at 888-463-6332.

Some great milestones have been reached in the history of chemotherapy: discovery of the sulfa drugs, anticancer agents, penicillin and other antibiotics, effective antihistamines, synthetic anti-inflammatory steroids, drugs to treat tuberculosis and malaria, oral anti-diabetics, and drugs to prevent blood clots.

Specifically, chemotherapeutic research has yielded Thorazine (major tranquilizer, 1955); Enovid (contraceptive, 1958); Tofranil (tricyclic antidepressant, 1961); Valium (minor tranquilizer, 1965); Ampicillin (antibiotic, 1965); Capoten (antihypertensive, 1982); Vasotec (antihypertensive, 1987); Interferon Alpha (antiviral, 1987); Retrovir (antiviral, 1988); Prozac (antidepressant, 1989); Taxol (anticancer, 1994); and Interferon Beta (antiviral, 1995).

The effectiveness of chemotherapy in the area of tuberculosis is well known. Furthermore, largely because of advances in drug therapy, four of the 10 leading causes of death in 1900 no longer prevail, and deaths from once-feared diseases have declined dramatically.

The chemotherapeutic route is the one that modern medicine has chosen to take. We see examples of this approach in the treatment of infectious conditions, high blood pressure, diabetes mellitus, allergic conditions, cancer, arthritis, and mental illness. Many drugs are synthetic in origin because the chemist, following the chemotherapeutic approach, sets out to alter one of nature's drugs or an earlier synthetic drug, so as to maximize the therapeutic effects and minimize the side effects.

Chemotherapy implies an attack on the cause of the disease, not just the relief of symptoms. In chemotherapy, the goal is to eliminate pathogenic bacteria, to rectify a hormonal imbalance, or to arrest a cancerous growth. On the other hand, drugs given for symptomatic relief, while greatly appreciated, do little more than relieve a headache, reduce secretions, or treat an upset stomach. In fact, symptomatic relief may work to the detriment of the patient by masking symptoms the physician needs to know about in order to make a proper diagnosis. Should we give painkillers to a child with developing appendicitis? Relief of symptoms does not necessarily remove the cause of the symptoms, as shown by the treatment of migraine headaches with the ergot alkaloids. Relief may be dramatic, but the headaches occur as frequently as ever. Similarly, a drug that successfully lowers high blood pressure may do nothing to correct its cause.

We still have a long way to go before chemotherapy can solve all of the world's major health problems. Today, we have no effective chemotherapeutic agents for curing hardening of the arteries, arthritis of the bones and joints, or respiratory diseases such as emphysema, bronchitis, and asthma. We can treat the symptoms of most mental illnesses, but we cannot correct their underlying causes. Success in treating some cancers has been gratifying, but the fact remains that the great majority of cancer patients will die of their disease. Cardiovascular diseases (heart attack, stroke, and high blood pressure) remain the leading cause of death in the United States. We have no truly effective medicines for herpes infections or for viral infections in general. In fact, there is no drug that will cure or prevent acquired immune deficiency syndrome (AIDS) (see Section 2.8, "The Search for Anti-AIDS Drugs").

In addition to lacking significant success in the areas of cancer and cardiovascular disease, we are actually losing ground in the chemotherapy of certain diseases that respond to antibiotics. Drug-resistant strains of microorganisms that cause gonorrhea, malaria, and dysentery have developed. (See Chapter 15, "Additional Over-the-Counter [OTC] Drugs and Chemicals," for a discussion of how unrestricted OTC distribution of antibiotics and use of suboptimal doses prophylactically can contribute to the development of resistance.) For information on resistance to antibiotics, see the FDA Web site **www.fda.gov/oc/opacom/hottopics/antiresist.html**. Also see **www.cdc.gov/drugresistance/community/**.

Problem 1.3 *A person has diarrhea and takes paregoric. Another has a skin rash and applies an anti-inflammatory cream. Categorize these treatments as chemotherapy or as the treatment of symptoms.*

1.6 Receptor-Site Theory

Throughout this book, we will be discussing the action of drugs on various organs or tissues of the body. We may state, for example, that morphine acts on the brain or that digitalis acts on the heart. But then the question will arise: Just how does a drug such as morphine cause its effects on the brain? Where does the drug go after it gets into the bloodstream, and by what mechanism does it actually produce its effects?

The modern explanation of the details of drug activity is based on the **receptor site theory**. According to this concept, drugs do not act just anywhere in the body; rather, they act at specific locations in tissues or organs. For a drug to act at a specific location, there must be present at that site specialized cells to which the drug molecule can attach by virtue of its size, shape, and chemical-electrical characteristics. Now, these cell tissues—which we term the **receptor site** or **cellular target**—do not exist so that Valium or cocaine could have an action in the human body. These receptor sites exist to receive natural, endogenous (see Glossary, Appendix V) chemicals (such as hormones or adrenaline) that are made and used by our body in carrying out its day-to-day functions. It happens that some chemicals that humans have obtained from nature or have synthesized can also bind to a receptor site. They just happen to have the correct size, shape, and electrical characteristics.

There are many different kinds of receptors in the human body, just as there are many different kinds of drugs. Receptors exist in the heart and blood system, in the lungs and bronchi, and in organs such as the liver, kidney, adrenal cortex, and gonads. Receptors also exist in the various parts of the brain and spinal cord. These receptors differ from each other, so a drug that can fit a receptor in the bronchi and act as an anti-asthmatic would not necessarily fit a receptor in the kidney and act as a diuretic. A typical human cell has perhaps 50 different receptors.

Receptor sites exist in enzymes and at synapses (see Sections 4.5 and 5.2, respectively). Some of the better understood receptors are the opiate type (with six different subtypes, specific for morphine or endorphins, for example); calcium antagonist type; adenosine type; GABA type (Section 5.7); dopamine type; serotonin type; alpha-adrenergic type (Section 5.4); beta-adrenergic type (Section 5.4); and the muscarinic cholinergic type (which has two subtypes). Also, there are receptors for estrogens, androgens, and histamine.

The binding of a drug to a receptor site has been likened to the fit of a key in a lock. The right drug is the "key" that can fit the receptor "lock" and turn on the desired biological response. Sometimes, several keys that are slightly different will fit the same lock and open a door. The same is true of certain drugs. Several slightly different drugs may be able to bind to a receptor site and cause a biological action. If they are given simultaneously, competition for the receptor site may result. Which drug will win out, and therefore which drug action will occur, depends on the number of molecules of each type of drug and on the **affinity** (binding power) of each for the receptor site. Binding power also plays a role in determining how long a drug will remain active in the body. The drug that fits a receptor and causes a response is termed an **agonist**. In the opiate-opioid category, morphine, heroin, and oxycodone are pure agonist opioids (see Section 6.9). The lock-and-key analogy can be carried a step further. Do you have a key that fits your front door but will not turn and open the door? Consider the same situation occurring at a receptor site. A chemical makes a good fit, binds strongly to the receptor, but itself has no drug action; it does not have quite the correct structure to cause any biological response. By occupying a site, it is in effect preventing a drug action that would result from the attachment of other endogenous or exogenous substances. The term **drug antagonist** has been applied to drugs that bind to a receptor, effectively excluding the binding of other substances. Naloxone is an antagonist to the action of opiate narcotics. (See Figure 6.5

for a diagram depicting drugs binding to receptor sites.) One point should be noted here: Our comparison of drugs and receptors to keys and locks is useful only up to a point. Receptor sites are not rigid, inflexible locks. They are protein molecules, and some degree of accommodation is possible.

When a patient swallows a medication, it is absorbed from the *gastrointestinal* (GI) tract and distributed via the blood to all parts of the body. If enough molecules of the drug reach receptor sites, the receptors will be saturated, and drug action will follow. Eventually, the medication taken will be displaced from its binding sites, either by competition with another chemical that has a greater affinity for the site, by enzyme-catalyzed destruction, or because the concentration of the drug in the blood has fallen as the drug is excreted from the body. Another dose of the drug must be taken to reestablish high blood levels and to saturate the receptor sites once again.

Researchers use the receptor site theory to design and understand the action of important drugs such as methadone, antihistamines, and tranquilizers. One corollary of the theory for therapists is that they must urge patients to take all of the medication prescribed over the course of treatment; the binding sites must be occupied continuously (saturated) if results are to be achieved. On a practical level, the receptor site theory has been used in the design of the drug pralidoxime, an antidote for accidental poisonings by insecticides of the cholinesterase type. The idea for pralidoxime came from an examination of the structure of the receptor site for acetylcholinesterase.

Recently, the NMDA receptor, a protein some call "the Holy Grail of neurotransmitter-receptor molecular neurobiology," has been identified and cloned in rats. Named after N-methyl-D-aspartate, one of the agonists that activates it, the NMDA receptor plays an important role in excitatory nerve transmission in the *central nervous system* (CNS) and in the regulation, growth, and degeneration of neurons. Despite its name, NMDA acts primarily as a receptor for glutamate, an excitatory CNS neurotransmitter. Evidence exists that NMDA has a role in the manifestation of the deleterious effects of epilepsy and stroke.

Problem 1.4 *Germicides and antitoxins are agents that work in the human body. When we attempt to apply the receptor site theory to their action, what difficulty do we experience?*

1.7 Some Advanced Concepts

In the preceding discussion of receptor site theory, we did not answer the question, "How does the drug elicit a biological response?" We know that the first step is attachment to the receptor. After that, it is believed that many different mechanisms are used by human tissue to accomplish the end result—drug action. Let us discuss a few of these mechanisms.

In the adenylate cyclase mechanism, it is postulated that the receptor site consists of protein molecules embedded in a two-dimensional lipid layer located in a cell plasma membrane. Two conformationally three-dimensionally distinct types of receptors are postulated. Drug binding results in conversion of the inactive receptor into more of the active form.

$$\text{Receptor}_{\text{inactive}} \rightleftharpoons \text{Receptor}_{\text{active}} + \text{Drug} \rightleftharpoons \text{Active receptor-drug complex}$$

Binding of, say, a hormone molecule affects the adenylate cyclase-catalyzed process that converts *adenosine triphosphate* (ATP) into *cyclic adenosine monophosphate* (cyclic AMP). The cyclic AMP, in turn, regulates biologically active molecules, which complete the drug action.

A second type of mechanism is believed to explain the ability of cardiac glycosides to affect nerve conduction in heart muscle. A drug such as digitalis is believed to bind to a protein receptor on a plasma membrane, resulting in the closure (or near closure) of a channel through which sodium and potassium ions normally migrate. This closure results in a change in the availability of sodium and other ions and a change in depolarization and conduction of nerve tissue. The end result is a more forceful and more efficient contraction of the heart. This mechanism is also believed to be the basis for opiate action in the brain. It is believed that the receptor for opiates consists of protein molecules embedded in a cell membrane and existing in two interconvertible conformations. One conformation preferentially binds agonists; the other preferentially binds antagonists. Opiate pharmacological activity occurs when the opiate binds to the active (or agonist) conformation of the receptor. This binding results in a change in permeability of the cell membrane to the passage of sodium ions, causing blockage of the excitatory effects of acetylcholine and glutamic acid. Thus, the narcotic effects of opiates appear to be based on their action at synaptic, sodium-permeable channels.

The protein in the cell plasma membrane to which the drug or agonist initially binds is typically complexed with a G protein (so named because it binds to a guanine nucleotide). Researchers have found that more than 100 receptors convey messages through G proteins (of which at least 20 distinct forms have been isolated). These receptors are linked to G protein-dependent effector mechanisms, including adenylate cyclase, cyclic GMP phosphodiesterase, certain enzymes, and membrane channels regulating the flow of inorganic ions. A few of the pharmacological effects mediated by G proteins include estrogen synthesis, glycogenolysis, fat metabolism, conservation of water, slowing of the heart, muscle contraction, elevation of blood pressure, and changes in electrical activity of neurons.[5]

In a third type of receptor site mechanism, the binding of the drug to the receptor is believed to occur inside the cell, in the cytoplasm. A steroid drug, for example, binds to a protein that is soluble in the cytoplasm and that can, through an intermediate form, migrate to the cell nucleus and act on *deoxyribonucleic acid* (DNA). The DNA, thus modified, acts to synthesize new protein. This process constitutes the drug action.

Finally, let us take a look at the acetylcholine receptor (See Chapter 5, "Drugs at the Synapse," for a discussion of the biological role of acetylcholine). These receptors are distributed throughout the body in various parts of the autonomic nervous system. We limit this discussion to the acetylcholine receptors at the myoneural junction. As in the adenylate cyclase mechanism, the receptors are protein molecules embedded in a two-dimensional lipid layer in a cell membrane. The receptor molecules span the membrane from inside to outside. They are believed to consist of an anionic functional group, to which the quarternary nitrogen of acetylcholine is attracted, and the positive pole of a dipolar group, to which the carbonyl oxygen atom

---

[5]For a valuable reference on G proteins, see **http://www.nobel.se/medicine/laureates/1994/illpres/index.html**.

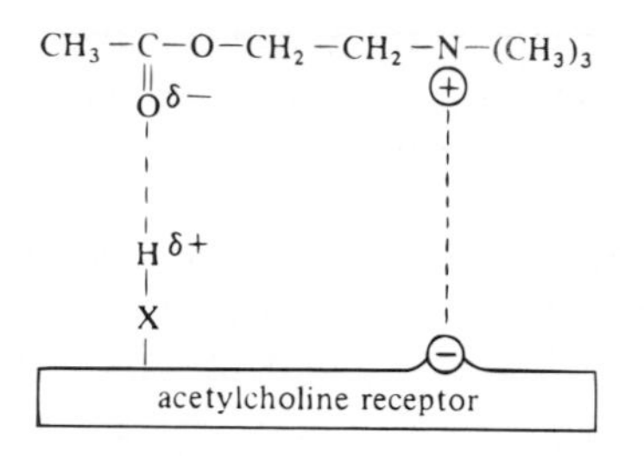

Figure 1.3 A possible mechanism by which acetylcholine might bind to its receptor site.

binds through dipole-dipole interaction (see Figure 1.3). Upon binding of acetylcholine (or its active congeners) to the receptor site, conformational changes are believed to occur in the membrane proteins, opening up channels for the passage of ions such as sodium. Because migration of ions affects nerve depolarization, nerve transmission is thus affected. The channels can close again in a millisecond, so drug action is the summation of the opening and closing of thousands of such channels. Note that receptors involved in the so-called muscarinic actions of acetylcholine function by a mechanism different from the one described here.

We should ask one final question before leaving the subject of receptor site theory. How does the theory explain the action of a drug antagonist such as naloxone? Recall that opiate receptors are believed to exist in two interconvertible conformations, one active and the other inactive. If the antagonist drug were to bind preferentially to the inactive conformer, the equilibrium between the two forms would swing toward the inactive.

$$\text{Inactive receptor-drug complex} \rightleftharpoons \text{Drug} + \text{Receptor}_{\text{inactive}} \rightleftharpoons \text{Receptor}_{\text{active}}$$

The consequence of this reaction would be that the possibility of active conformation would be greatly diminished. There would be no place for agonist molecules to bind.

1.8 A New Vocabulary

A stimulating new vocabulary awaits the student of drugs and drug actions. In this section, a few of the most important words, prefixes, and suffixes encountered in discussions of drugs are presented. These constitute only a brief introduction; therefore, the reader is urged to expand on them by consulting a textbook of medical terminology.

To begin, we must give special attention to the word *pharmacology*. In pharmacology, we are dealing with the study of drugs—their sources, how they enter the body, what effects they have, where and how they act, what dose to give, their possible interaction with other drugs, and how the body acts to change and eliminate them. In fact, pharmacology is the subject of this book, whether we are discussing aspirin, marijuana, or morphine.

Here are some additional words that can act as the beginning of your word list:

Word	*Meaning*
acute	brief, intense, or severe (as opposed to chronic)
chronic	not severe, but continuing for a long time
carcinogenic	capable of causing cancer
teratogenic	damaging to the embryo, especially in the first trimester of pregnancy
calculi	stones (What are renal calculi?)
euphoria	a state of well-being, especially an exaggerated one having no basis in reality
peripheral	at the extremities; away from the center (What is peripheral circulation in the human?)
peristalsis	rhythmic motion of the stomach and intestines associated with digestion of food
syndrome	all of the signs and symptoms associated with a disease
synergistic	working together to give superadditive affects
tachycardia	an abnormally fast heart rate
bradycardia	an abnormally slow heart rate
renal	pertaining to the kidneys
visceral	referring to the internal organs of the body
febrile	feverish
emetic	a drug that induces vomiting
antipyretic	a drug that reduces fever

Prefix	*Meaning*	*Example*
hyper-	more than the normal or customary	hyperventilation
hypo-	less than the normal or customary	hypoxia
cardio-	referring to the heart	cardiovascular
gastri-	referring to the stomach	gastritis
myo-	muscle	myocardium
post-	after or behind	postpartum
anti-	against	anticonvulsant
intra-	within	intravenous
poly-	many	polynuclear
dys-	painful, bad, difficult	dysmenorrhea
osteo-	pertaining to bone	osteoarthritis
a-	without	asymptomatic

Suffix	*Meaning*	*Example*
-emia	in or of the blood	glycemia
-itis	inflammation of	appendicitis
-ectomy	incision or removal of a part	appendectomy
-algia	pain in a part	neuralgia
-uria	in or of the urine	polyuria

1.9 Where to Get Information About Drugs

We now have more sources of information on drugs than ever before. Reference books are available that give us names, product descriptions, constituents, and pharmacological actions for a myriad of drugs. You may use the following to contact some of the important drug abuse-oriented associations via telephone, e-mail, or the Internet.

African American Family Services
Tel. (612) 871-7878

African American Parents for Drug Prevention
Tel. (513) 961-4158
http://www.emory.edu/NFIA/

Al-Anon Family Group
Tel. (800) 356-9996
http://www.al-anon.alateen.org

Alcoholics Anonymous (AA)
Tel. (212) 870-3400
http://www.alcoholics-anonymous.org

Alcohol Hotline (24-hour, 7 days)
Tel. (800) 252-6465

American Council on Drug Education
Tel. (800) 488-DRUG
http://www.ACDE.org

Center for Substance Abuse—Prevention Workplace Helpline
Tel. (800) 843-4971
e-mail: **helpline@samhsa.gov**

Cocaine Hotline
Tel. (800) 262-2463
http://www.drughelp.org

Do It Now Foundation Publications Hub
http://www.doitnow.org/pages/nowhome2.html

Drug Abuse Resistance Education (DARE)
Tel. (800) 223-DARE
e-mail: **Webmasters@dareamerica.com**
http://www.dare-america.com

Drug Enforcement Administration (DEA)
Tel. (202) 307-7936
http://www.usdoj.gov/dea/

Drug Policy Information Clearinghouse
Tel. (800) 666-3332
e-mail: **askncjrs@ncjrs.org**

Food and Drug Administration (FDA)
http://www.fda.gov/

Mothers Against Drunk Driving (MADD)
Tel. (800) 438-6233
http://www.madd.org

Narcotics Anonymous (NA)
Tel. (818) 773-9999
http://www.wsoinc.com

National Clearinghouse for Alcohol and Drug Information
http://www.health.org

National Coalition of Hispanic Health and Human Services Organization
Tel. (202) 387-5000
e-mail: **cossmho@cossmho.org**
http://www.cossmho.org

National Center for Health Statistics
http://www.cdc.gov/nchswww/default.htm

National Council on Alcoholism and Drug Dependence, Inc.
Tel. (800) 622-2255
e-mail: **national@ncadd.org**
http://www.ncadd.org

National Families in Action
http://www.emory.edu/NFIA/

National Institute on Alcohol Abuse and Alcoholism (NIAAA)
Tel. (301) 443-1677
http://www.niaaa.nih.gov/

National Institute on Drug Abuse (NIDA)
http://www.nida.nih.gov

NIDAInfofax (information on drug abuse and addiction, in English and Spanish)
Tel. (888) 644-6432

Office of Minority Health Resources
Tel. (800) 444-6472
http://www.omhrc.gov

ParentsPlace.com(parenting home page)
http://www.parentsplace.com

Pavnet Online, sponsored by seven federal agencies
http://www.pavnet.org

Most libraries have a copy of the useful book, *Physicians' Desk Reference*, familiarly known as the PDR. Revised annually, the PDR lists drugs alphabetically by brand name, by manufacturer, and by generic and chemical name. A product classification index and a manufacturer index are useful, as is a color photo product identification section. The heart of this 2,000-page book is the product information section, wherein composition, contraindications, actions, warnings, precautions, adverse reactions, dosage, and administration are detailed. As a reference book on approximately 2,500 drug products, the PDR is unexcelled. Also available is the *Physicians' Desk Reference for Nonprescription Drugs*. Its format is identical to that of the original PDR, but the book covers only OTC drugs sold in the United States. It also provides a valuable guide to herbal medicines plus lists of sugar-free, alcohol-free, lactose-free, and galactose-free products.

Drug data in the PDR are similar to those printed on the official package circulars that are inserted by the manufacturer in the carton sent to hospitals and local pharmacies. **Package inserts** constitute another good source of product information. A court decision requires that in the case of prescribed estrogens, all the information on the package insert should be made available to the patient. This policy is likely to become a trend as more emphasis is placed on the side effects of drugs—and as we recognize more drugs with a high risk/benefit ratio.

In the United States, two official compendia of drugs are published. They are *The United States Pharmacopoeia* (USP) and *The National Formulary* (NF). These official compendia set forth standards of identity, strength, purity, packaging, storage, and labeling of drugs and related articles describing therapeutic value. Currently, the NF is incorporated into the USP and is published as one compendium, *The United States Pharmacopoeia*. A primary objective of USP standards is to ensure uniformity of drug products from lot to lot and from manufacturer to manufacturer. Hence, the consumer is confident that when he or she buys brand A of aspirin USP in one locality, it is the same as brand B of aspirin USP purchased somewhere else. Standards in the USP are intended to protect drug products from deterioration, from contamination, from failure to dissolve in the digestive tract, and from content inconsistency. The wisdom of publishing such official compendia for the protection of the public is apparent. Most of us will have little opportunity to utilize information given in the USP because the bulk of the information is intended for the manufacturer or pharmacist. We recognize, however, that the designation *USP* for preparations such as aspirin, peppermint spirit, pine tar, kaolin, isopropyl rubbing alcohol, mineral oil emulsion, penicillin, or ephedrine helps ensure uniformity, potency, and efficacy in these products at the time of purchase and use. A Web site for more information on the USP is: **http://www.usp.org/pubs/review/rev_049a.htm**.

An additional source of drug information for the laity is USP DI *Advice for the Patient*, available from the United States Pharmacopoeia Convention, Inc., 5645 Fishers Ln., Rockville, MD 20857.

Local bookstores and libraries offer many books and paperbacks dealing with such aspects of drugs as street names, licit and illicit use, product identification, drugs and sex, alcoholism, and national policy. Our national concerns about drug abuse are reflected in the books dealing with heroin, morphine, methadone, marijuana, and central nervous system stimulants.

Authoritative publications can be obtained from the *Drug Enforcement Administration* (DEA), a division of the United States Department of Justice, Washington, DC 20537. DEA publications cover the science of drug identification, smuggling, foreign connections, cost to society, educational policies, and so on. DEA's homepage is **http://www.usdoj.gov/dea/index.htm**.

For a catalog of FDA "Publications and Audiovisual Materials for Consumers," write to the United States Department of Health and Human Services, Food and Drug Administration, Office of Public Affairs, 5600 Fishers Ln., Rockville, MD 20857.

Drug Education for Youth (DEFY) typifies the local organization offering drug education pamphlets to the public.

A fine source of information about OTC drugs is the *APhA Handbook of Nonprescription Drugs*, published by the American Pharmaceutical Association, 2215 Constitution Ave. NW, Washington, DC 20037. The handbook is updated yearly.

A national drug information center has been established by Families in Action, Suite 204, 2296 Henderson Mill Rd., Atlanta, GA 30345. See **http://www.emory.edu/NFIA/**.

A plethora of information on alcohol and drugs can be obtained from the National Clearinghouse for Alcohol and Drug Information (NCADI), P.O. Box 2345, Rockville, MD 20852. The clearinghouse, operated by the National Institute of Mental Health, is the focal point for federal information on drug abuse, distributes publications, and refers specialized and technical inquiries to federal, state, local, and private information resources.

1.10 Brand Names and Generic Drugs

A typical drug, such as the minor tranquilizer Halcion, may be known by as many as four names (excluding street names):

1. The generic name (USAN)
2. The USP or NF name
3. The chemical name
4. The brand (proprietary) name

Generic names are established by the *United States Adopted Name Council* (USAN Council), composed of representatives from the American Medical Association (AMA), the American Pharmaceutical Association, the United States Pharmacopoeia Convention, and the FDA. The substance named must be of demonstrated therapeutic usefulness. Once assigned, the name is never given to another drug. Except in cases of older drugs, the USAN is identical to the USP or NF name. The generic name has nothing to do with the trade name or the drug's chemistry. The generic name is given for general recognition purposes, not necessarily for commercial purposes—no matter who makes the drug.

Of course, each manufacturer establishes the name under which the product is to be traded (subject to approval by the FDA and the United States Patent and Trademark office). Brand-name drugs are patented under an exclusive trade name for a period of 20 years. Trade names are jealously guarded and are actively promoted because they are the identity keys through which the product is often prescribed by the physician, dispensed by the pharmacist, recognized by the nurse, and taken by the patient. The manufacturer of a drug may sell the right to make and distribute the drug to another company, which in turn may market the drug under the generic name.

Thus, a system has developed in this country whereby a drug may appear on the market and on prescriptions under both its trade name and its generic name. While many brand-name drugs are still under patent protection[6] and are not available from other manufacturers by generic name, it is true that patents are expiring or have expired on products that account for billions of dollars in annual sales. These products include chlorpromazine, hydrochlorothiazide, propoxyphene hydrochloride, ampicillin, erythromycin, and chlordiazepoxide. When a patent expires, other companies are free to manufacture and sell generic versions of the drug, with FDA approval. Note that generic names are usually not capitalized, whereas trade names are. Some of the more popular generic drugs are listed in Table 1.1. More than 8,000 generics are currently made and sold with FDA approval.

The claim is not true that brand-name drugs are produced only by large, well-known firms, while generics are made by small, unknown companies. Generic diazepam (brand name Valium) is now made by a dozen companies, including the two giants American Cyanamid and Warner-Lambert Company. Actually, many large firms advertise and sell a drug carrying their brand name that has been manufactured by a company specializing in generic drugs. A striking case in point is the widely used antibiotic ampicillin. Ampicillin is available under 224 product labels but is produced by only 24 formulators. Also, Glaxo SmithKline's SK-Hydrochlorothiazide is manufactured by Roxane Laboratories, Inc., a generic firm. No matter who manufactures the ampicillin, FDA requirements demand standards of strength, purity, and effectiveness. Furthermore, under the FDA's monitoring and testing program, drug samples are periodically collected from all manufacturers, large and small, brand and generic. The FDA requires that generic drugs must be used to treat the same illnesses and include the same precautions, warnings, and label instructions that apply to trade-name drugs.

Table 1.1 Some of the More Popular Generic Products

Generic Product	*Drug by Brand Name*	*Category*
Acetaminophen w/Codeine	Many brands	Pain reliever
Albuterol	Proventil, Ventolin	Bronchodilator
Alprazolam	Xanax	Minor tranquilizer
Amitriptyline	Elavil	Antidepressant
Atenolol	Tenormin	Antihypertensive
Cycrin	Depro-Provera	Contraceptive
Glyburide	Micronase	Oral antidiabetic
Hydrocodone/APAP	Hycodan	Antitussive, analgesic
Ibuprofen	Motrin	Analgesic
Levoxyl	Synthroid	Hypothyroid and goiter drug
Naproxen	Naprosyn	NSAID (see Section 14.1)
Prednisone Oral	Deltasone	Glucocorticoid
Propoxyphene-N/APAP	Darvocet-N-100	Pain killer
Trimethoprim/sulfa	Polytrim combtn	Antibacterial
Veetids	Various brands	Antibiotic
Verapamil	Verelan	Antihypertensive

[6]The 20-year clock on a patent starts when the inventor makes the application, not when it is granted. Processing and approval may deplete many of the years.

Formerly, most physicians and patients were reluctant to permit substitution of the equivalent generic-name drug for the brand-name drug that had been prescribed. Because of agitation by consumer groups and others, however, laws in all 50 states have now been passed to permit the pharmacist to substitute the usually less expensive generic drug, provided that the physician has not prohibited the drug.

The trend is toward more liberal substitution laws, and cost appears to be the major factor. Surveys have shown that some generic-name drugs cost only one-fourth as much as their brand-name equivalents. According to the University of California at Berkeley Wellness Letter, generics retail for 30–80% less than their brand-name equivalents. A recent study by the *Los Angeles Times* discovered that the average cost of generic drugs in one recent year was $18.45 versus $61.33 for the comparable brand-name drug. The *Times* found found that 30 Keflex capsules can cost $49.71, but the generic cephalexin only $7.46; similarly, Tenormin cost $32.66, but atenolol only $4.29. However, the generic-drug market is still much smaller, in dollars, than the brand-name market: in one recent nine-month period Americans spent $19.4 billion on generics and $98.6 billion on brand-name drugs.

Are generic drugs equivalent in effectiveness as well as in name? This question is the crux of the matter. Special ads are prepared for doctors, pharmacists, and the public that purport to show differences in **bioequivalence** between brand-name and generic drugs—that is, differences in the absorption of the drug into the bloodstream (termed **bioavailability**) and subsequent pharmacological action. Supporters of generic drugs, however, cite the fact that all generic manufacturers must perform bioequivalency studies on their drugs and report the results to the FDA. If the generic drug's bioequivalence does not closely match that of the brand-name drug, the FDA will not approve the drug. (Thus, generic conjugated estrogens were taken off the market in 1991 when they did not produce the same blood levels of estrogen as Premarin.)

Actually, bioequivalence differences between brand and generic drugs are small. An FDA review of new generic drugs approved since 1984 showed an average difference of only 3.5%, which was no greater than the difference between one batch of a brand-name drug and another batch in the same manufacturing process. FDA scientists tested several thousand samples from generic drug companies. Only 27 samples, about 1% of those tested, did not comply with standards of potency, dissolution, content uniformity, product identification, moisture determination, or purity. The FDA also tested 429 samples representing at least three different batches of all 24 *narrow-therapeutic-index* (NTI) drugs then on the market (drugs in which the difference between the therapeutic dose and the toxic dose is small). Both generic and brand-name drugs were included. Only five of the samples—all aminophylline tablets, a bronchodilator—failed to meet USP standards. None of the defects in the generic drugs posed a public health hazard. Some other NTI drugs are Coumadin, Dilantin, Lanoxin, Theo-dur, and Tegretol.

Based on these results and the fact that brand-name products show similar failures at a similar rate, the FDA recommends that doctors continue to consider prescribing generic drugs when appropriate to offer lower-cost products to consumers. The director of FDA's Office of Generic Drugs says, "Generic drugs contain exactly the same active ingredients as the brand-name drug and are just as safe and effective."

A point of caution should be noted when we are considering NTI drugs such as the heart drug digitoxin, the anticoagulant warfarin, the thyroid drug levothyroxine, and other drugs for contraception, high blood pressure, and asthma. *If you are taking one of these drugs, either brand or generic*, it is best to stick with that type. If you switch, your blood levels must be monitored because the small possible differences in bioequivalence might become critical.

Generic drugs continue to gain in popularity and now comprise nearly half of the 2.4 billion prescriptions written in the United States each year. Part of the popularity of generics lies with cost-conscious health maintenance organizations. Generics also have great popularity among union members, consumer advocate groups, insurance companies, and retired people—again, mainly because of the lower cost. The Generic Pharmaceutical Industry Association estimates annual savings of over $1 billion to the American public.

For the cost conscious, remember, too, that OTC drugs can be just as effective as more costly prescription alternatives. For example, OTC Zantac can work as well as prescription Prilosec at as little as one-seventh the cost.

For nine FAQs on generics, see the September/October 2002 issue of *FDA Consumer* magazine, page 24, or FDA's website **www.fda.gov/cder/ogd/**. For a pro-generic viewpoint, see **http://www.healtheon.thehealthpages.com/articles/ar-gener.html**.

1.11 Examples of Naming Drugs

	First Example	*Second Example*
Generic name (USAN):	dextroamphetamine sulfate	acetaminophen
USP name:	dextroamphetamine sulfate	acetaminophen
Proprietary (trade) name:	Dexedrine	Tylenol, Tempra
Chemical name:	alpha-methylphenethylamine sulfate	*N*-methyl-*p*-aminophenol
Street name:	dexies	

A minor point, but one that is possibly vexing for the reader, is the inclusion of names of *salts* in official descriptions of drugs. What is meant by *sulfate, nitrate*, and *hydrochloride* in such names as *morphine sulfate* USP or *vincristine hydrochloride*? These suffixes identify the acid from which the water-soluble form of the drug was made. Morphine is an alkaline molecule; it forms a salt when it reacts chemically with an acid.

$$\underset{\text{(the free base)}}{\text{Morphine}} + \text{Sulfuric acid} \longrightarrow \underset{\text{(a salt)}}{\text{Morphine sulfate}} + \text{Water}$$

Drug manufacturers know that the salt form is usually water-soluble and is often more stable toward heat, moisture, and light. Many different acids can be used to form a salt: sulfuric, hydrochloric, nitric, phosphoric, citric, tartaric, lactic, maleic, benzoic, and salicylic.

The base from which the salt is made can be referred to as the **free base** (because it is free of its salt form). You probably have heard of cocaine "free-basing."

Chemically, cocaine is an alkaline, alkaloidal base. It is smuggled into this country as the nonvolatile hydrochloride salt form, but it is converted back to the free base by some users who want to inhale the fumes of the more easily volatilized free-base form (see Section 7.2, "History and Pharmacology of Cocaine"). One gram of cocaine hydrochloride dissolves in 0.4 mL of water, but 1 gram of free base requires 600 mL of water to dissolve.

1.12 The Most Widely Used Drugs

To treat or prevent illness, the American physician has more than 5,000 drugs available, 90% of which have been developed since 1938. American patients currently spend yearly over $154 billion on almost 3 billion prescriptions. That cost is increasing by $22 billion a year. Females obtain about 50% more prescriptions than males, and white persons account for 60% more prescriptions than nonwhites. General practitioners prescribe fewer drugs than do cardiologists, dermatologists, allergists, and internists.

Geriatric medicine is big business and is getting bigger. More than 27 million Americans (11% of our population) are 65 years or older, and it is projected that this age group will nearly double within the next 50 years. Older adults spend more than $3 billion yearly on prescription and OTC medicines. The average geriatric patient purchases more than 13 prescriptions a year, including refills. In one survey, 87% of geriatric patients 75 years and older were taking prescription drugs. As many as 95% of nursing home patients may be taking drugs, and overprescribing is common. (See Section 4.6, "How People Can Respond Differently to Drugs," for a discussion of effects of age on drug action.)

What are the most popular drug categories and individual drug preparations in use in the United States? Cardiovascular and central nervous system products are the leading categories; abroad, anti-infectives are the leading sellers. Amoxicillin, atenolol, furosemide, Lipitor, Premarin, Symthroid, and Zithromax are among the top U.S. drugs.

1.13 Getting Used to Dosages in the Metric System

We can forget about the old apothecary system of weights and measures, with its grains, drams, scruples, minims, and fluid ounces. Prescription writing today is done exclusively in the metric system. And while the United States is the only major nation in the world that has not adopted the metric system generally, it is only a matter of time until all aspects of our lives will have become metricized.

Quantities of drugs are measured in grams, milligrams, and micrograms. Volumes of drugs are measured in liters, milliliters, or cubic centimeters.

1. The gram (g) is the basic unit of weight in the metric system. There are 454 grams in 1 pound. One one-thousandth of a gram is 1 milligram (mg). One one-thousandth of a milligram constitutes 1 microgram (μg). One thousand grams is 1 kilogram (kg). Some examples of use are indicated in the following example:

Drug	*Usual Dose*
A sulfa drug for urinary infection	1–2 g
Typical barbiturate	50 mg
Vitamin B_{12}	100 μg

Problem 1.5 *(a) Aspirin tablets have long been sold as containing 5 grains of aspirin. If 1 grain equals 65 mg, how many milligrams of aspirin does each tablet contain? How many micrograms of aspirin does each tablet contain? (b) How many pounds are equivalent to 1 kg? (c) If you weigh 140 pounds, what is your weight in kilograms? (d) A metric ton is 1000 kg. Convert 1 metric ton to pounds.*

2. The liter is the basic unit of volume in the metric system. We more commonly encounter the milliliter (mL), which is one one-thousandth of a liter. The unit *cubic centimeter*, or *cc*, is used synonymously with *mL*. A pint contains 473 mL. A gallon is equal to 3.784 liters. A teaspoonful is generally considered to be 5 mL (although we find wide variations). Some examples of metric system doses are as follows:

Drug	*Usual Strength or Dose*
Insulin	100 units/cc
Antibiotic eyedrops	5 mg/mL
Cough syrup	2 teaspoonfuls (10 mL)
Worm medicine	5 mL per 10 kg body weight

Problem 1.6 *(a) A fluid ounce is approximately 30 mL. Using the equivalents just given, how many teaspoonfuls are contained in 1 fluid ounce? (b) A patient has a daily urinary output of 1.25 L. How many pints is this? How many cubic centimeters is this? (c) How many teaspoonful doses are there in 1 L? in 1 pint? (d) Convert 150 mL to liters. (e) Convert 10 L to gallons.*

To familiarize yourself with doses of powerful drugs, consult Table 1.2.

Table 1.2 The Ten Most Powerful Drugs Used in Medicine*

	Based Solely on Dosage	
Drug	*Usual Dose*	*Category*
ethinyl estradiol	0.02 mg	contraceptive
fentanyl	0.1 mg	opioid analgesic
reserpine	0.1 – 0.25 mg	antihypertensive
digoxin	0.125 – 0.5 mg	cardiac glycoside
nitroglycerin, sublingual	0.15 – 1.2 mg	vasodilator
alprazolam	0.25 – 0.5 mg	minor tranquilizer
conj. estrogen (Premarin)	0.3 – 1.25 mg	estrogen
atropine	0.4 mg	anticholinergic

Table 1.2 continued

	Based Solely on Dosage	
Drug	*Usual Dose*	*Category*
scopolamine	0.4 mg	antimuscarinic
ergotamine	0.5 – 1.0 mg	vasoconstrictor, uterine stimulant

*Excludes vitamins such as B12 (dose: 3–6 mcg), and LSD (dose 50 mcg). Students: can you find even more potent drugs? Send your candidates to the author at **<kenn_liska@ix.netcom.com>**.

Web Sites You Can Browse on Related Topics

ATTENTION INTERNET USERS: In addition to the URLs listed below, you can search for topics as suggested. A search can yield numerous sites, but careful evaluation is necessary.

SAMHSA National Household Survey on Drug Abuse
http://www.samhsa.gov
http://www.health.org
Also: Search for "SAMHSA"

Addiction Theories
http://www.intervention.com/defns.html
Also: Search for "drug addiction"

Generic and Trade Names
Search "generic drug names"
Answers to FAQ (Frequently Asked Questions)
http://www.druginfonet.com/pharmfaq.htm

General Information
http://www.drugfreeamerica.org/drug_info.html

Receptor Site Theory
http://www.csuchico/psy/BioPsych/neurotransmission.html
See diagram of:
http://www.sdsc.edu/10tw/week19.96/10tw.html
Also search for "receptor site"

Study Questions

1. Name six ways in which the physician uses drugs; cite one drug example in each category.
2. Euphoriants such as marijuana are mood-altering substances. Do you consider mood alteration a drug effect? Explain.
3. After reading the section on placebos, can you see another role for the physician and the therapist in addition to *physical* treatment of the patient?
4. Define (**a**) drug, (**b**) placebo, (**c**) chemotherapy, (**d**) addiction, (**e**) tolerance, (**f**) generic name, (**g**) bioavailability, (**h**) peristalsis.
5. What are the aspects of modern chemotherapy that distinguish it so clearly from the herbal or plant extract treatment of past years?
6. Using any of the definitions given in this chapter, is it possible that a human could ever become dependent on aspirin? What criteria would have to be met before you would classify this person as an aspirin addict?
7. **a.** Consider a heavy cigarette smoker that you know. From your observations, is this smoker showing habituation, preoccupation, or addiction to cigarettes?

   b. An ad in our local newspaper boasted, "STOP SMOKING–In only two hours. Only $39.99 complete." What is your evaluation of this claim?
8. Assume that a drug induces psychic, but not physical, dependence. What kind of withdrawal syndrome can we expect to observe when the drug dependence is abruptly

terminated? (See Chapter 6. "Narcotic Analgesics: Opiates and Opioids," for a description of the withdrawal from heroin.)

9. A classroom has seats for 25 students, and they are all filled. Standing in the classroom are 30 husky football players who would like to sit down. How is this situation like that at the receptor site? (*Hint*: Consider fit, binding, affinity, saturation, competition, and concentration.)

10. Once epinephrine binds to its receptor site in an artery, why doesn't it stay there forever?

11. Refer to the meaning of words given in this chapter, and then provide definitions for the following: (**a**) *hyper*critical, (**b**) *hypo*ventilation, (**c**) *post*nasal drip, (**d**) *myo*neural junction, (**e**) tonsill*ectomy*, (**f**) *intra*cranial, (**g**) sub*acute*, (**h**) *gastrectomy*, (**i**) pericard*itis*, (**j**) *osteo*path, (**k**) my*algia*, (**l**) protein*emia*, (**m**) *dys*function, (**n**) *afebrile*, (**o**) *pyr*etic, (**p**) albumin*uria*.

12. What is the "USP" and for what purpose is it published?

13. Here are four names for a common drug. Identify the generic name, the street name, the chemical name, and the brand name: blue birds, sodium pentobarbital, sodium 5-ethyl-5-(1-methylbutyl)barbiturate, Nembutal.

14. A college student devised a test demonstrating the power of the placebo effect. He gave a group of fellow students coffee that he said was decaffeinated (it was) and then had them perform a manual task. The time for completion of the task was carefully noted for each student. A day later, he gave half of the original group of students coffee that he said was decaffeinated (it was); the other half of the group received coffee he said was fortified with extra caffeine (he lied; it was the same decaffeinated coffee). All students were again carefully timed in their performance of the same manual task. None were permitted the use of coffee, caffeine, or other stimulants during the days of the test. The test was quite successful in demonstrating the placebo effect. What were the results?

15. Using any of the definitions of a drug given in this chapter, decide whether or not the Pill (the oral contraceptive) is a drug. Are vitamins drugs? the fluoride in toothpaste? insulin? water?

16. After reading this chapter, devise an explanation for the apparent success enjoyed by African witch doctors.

17. In a study of 288 cancer patients, researchers identified "placebo responders"—those who obtained pain relief 50% or more of the time from a placebo. The responders included more professionals than unskilled workers, more women with children than without, and more persons who had a traumatic interruption of marriage through death, separation, or divorce. What one experience in common do placebo responders generally appear to have that nonresponders do not?

18. Girls and young women get hooked on cigarettes, drugs, and alcohol more quickly than boys, and they suffer the consequences faster and more severely, says the National Center on Addiction and Substance Abuse at Columbia University. If the study is valid, give at least three reasons for these gender differences.

19. A patient brings home a prescription consisting of 200 mL of a liquid. The doctor has directed the patient to take 1 teaspoonful four times a day. How long will the prescription last (i.e., until it is all used up)? See the text for needed equivalents.

20. If the dose of a drug is 25 mg/kg body weight and a patient weighs 176 pounds, how many *grams* of drug should he be given?

21. **a.** Which is the better buy, 1 avoirdupois pound of soybeans for $1.00 or 30 g of soybeans for $0.09?

b. One gallon of gasoline for $1.50, or one liter for $0.45?

22. Beer is approximately 4% alcohol by volume. Calculate how many milliliters of pure alcohol are contained (total) in a typical six-pack.
23. Which has a greater volume, 1 quart of milk or 1 liter of milk? Prove your answer mathematically.
24. In America, chewing gum has been marketed in tobacco pouches, fruit juice in flasks, and cola drinks in amber-colored, beer bottle–shaped containers with the word "draft" in large letters. TV ads for the cola include neon signs and other bar-like imagery. Considering what we have read in this chapter about the power of suggestion, debate the proposition that our children are being culturally seduced.
25. *Advanced study question*: Select one teaspoon and one tablespoon from your kitchen and take them to your chemistry department or pharmacist for the determination of their capacity in milliliters. How do your spoons compare to the theoretical capacities of 5 mL and 15 mL, respectively?
26. *Advanced study question*. You know that there are receptor sites in the human body for barbiturates because you know that barbiturates act in the human to induce the sleepy state. Obviously, the receptor sites aren't there so that Eli Lilly and Co. can make a profit selling Seconal. Why are they there? Speculate on their purpose and on what we might start looking for in the brain. Do the same for morphine. (*Hint*: See Chapter 6, "Narcotic Analgesics: Opiates and opioids.")

CHAPTER 2 Where Drugs Come From

Key Words in This Chapter

- Plant drugs
- Alkaloid
- Opium
- Morphine
- Atropine
- Cocaine
- Ergot
- Marijuana
- Mescaline
- Insulin
- Growth hormone
- hCG
- Estrogen
- Gene splicing
- Anti-AIDS drugs

Learning Objectives

After you complete your study of this chapter, you should be able to do the following tasks:

- Define active ingredient and alkaloid.
- Give the source of morphine and codeine.
- Explain atropine's source and actions in the body.
- Give cocaine's source and history of use.
- Understand how digitalis is used as a heart stimulant.
- Relate some of the history of marijuana.
- Understand that mescaline is the psychedelic active ingredient in peyote.
- Name the psychedelics in Mexican mushrooms.
- Explain what is meant by gene-spliced insulin.
- Give some of the medical uses of estrogen.
- Cite some of the risks associated with estrogen use.
- Discuss how growth hormone is applied in dwarfism.

2.1 Introduction

Thousands of prescription and OTC drugs are available for use today. While some of these drugs have been known for hundreds of years, others have been on the market for only a few years. Today, we obtain drugs from many sources, as shown in Table 2.1.

Table 2.1 Sources of Modern Drugs

Source	*Examples of Drugs*
Plants*	Morphine, digitalis, cocaine, oncovin, atropine
Animals*	Insulin, estrogen, tetanus antitoxin
Laboratory synthesis	Diazepam, aspirin, antihistamines, barbiturates
Microorganisms	Penicillins, bacitracin, tetracyclines
Gene splicing (recombinant DNA technology)	Insulin, growth hormone, interferon, Factor VIII

*A significant portion of modern drugs have their roots in folklore (e.g., curare, reserpine). People have used plants and animals to cure diseases for thousands of years.

2.2 Drugs from Plants

Our first drugs came from plants, which continue to be a very important source. In this section, we discuss plants that yield drugs of significant licit use, in addition to other plants that are the sources of drugs of abuse.

The discovery of a potent drug in a plant—for example, cocaine in *Erythroxylon coca*—was made unknowingly by someone who either tasted the plant or observed its effects on an animal that consumed the plant. The stories of these discoveries are lost in antiquity.

Plants were used by early humans in two ways:

1. A plant or its extract was used in a religious ritual as a divinatory agent. This use was especially true of plants such as peyote and the psilocybe mushrooms, which contain hallucinogenic active ingredients. People believed that by consuming the plant, they could communicate with the gods and discern their will.
2. A plant or its extract was used medicinally. The wide range of illnesses believed to be treatable included mental illness, diarrhea, worm infestations, lung diseases, and diabetes mellitus. Plants, once established in the medical folklore of a people, were used for centuries.

Both of these uses of plants continue in various parts of the world today.

One other notorious use for plants should be mentioned. Some famous poisonings have been carried out with plants—a remarkable example being the "cup of hemlock" that Socrates drank in 399 B.C. Socrates died because he ingested the alkaloid coniine, the active ingredient of a poisonous plant known botanically as *Conium maculatum*. The witches' brews of the Middle Ages were concoctions prepared from several famous plants, including belladonna (known as *Atropa belladonna*, or deadly nightshade) and stramonium (known as *Datura stramonium*, jimsonweed, and thorn apple). Atropine and scopolamine, two of the active ingredients in these plants, are capable of producing hallucinations in humans. The "flying about on broomsticks" that was supposed to have occurred during the witches' rites was, in reality, a drug-induced hallucinatory experience. Scopolamine was formerly used as the hypnotic (sleep-inducing) ingredient in some OTC sleep aids.

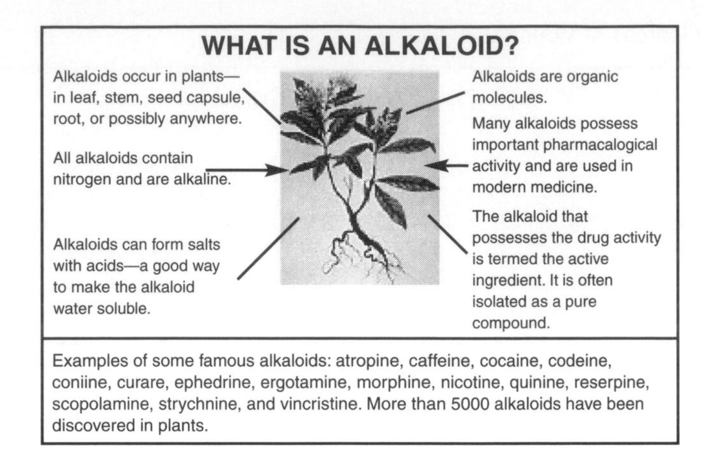

The substance in a plant that is responsible for its pharmacological effects is termed the **active ingredient**. Sometimes this ingredient is an **alkaloid**.

Formerly, the entire plant, or perhaps only the leaves, was dried and ground up, and this process constituted the drug preparation. In modern medicine, the plant part is dried, and the active ingredient is extracted and purified, yielding a much more potent, reliable, and reproducible agent.

The question is often asked, "Could I grow a certain plant in my backyard, cultivate it, and obtain the drug?" The answer is that a given species of plant will probably grow in your area, but there is no guarantee that the soil and climate there will promote the same plant constituents in the same quantities as those found in the plant's natural habitat. Some notable failures have occurred when plants were brought to the United States and were grown for their drug content. The time of harvest may be important, too. In the ephedra plant, the alkaloid content is greatest just before the first fall frost.

Modern plant scientists (called *pharmacognosists*) visit native peoples to learn of their folklore medicine and to practice the ethnobotanical approach to drug discovery. Once a promising plant is identified, it is collected, dried, and shipped to the laboratory for botanical identification and isolation and identification of the active ingredient.

Reserpine, digitalis, curare, vincristine, and other important drugs have come to us through folklore medicine, and this route continues to be an important source of drugs.

Drugs from plants continue to be big business in the United States, with annual sales now at $10 billion. More than 200 organizations worldwide are investigating new uses of plant-derived drugs, especially in the fight against AIDS, cancer, diabetes, and cardiovascular disease. The United States National Cancer Institute is spending $8 million in a 5-year program to check 10,000 drugs each year for efficacy against cancer and AIDS.

A word of caution to the homeowner: Many household plants are poisonous, and such plants are among the nation's leading sources of poisoning in children. At least 80 common plants are known to be toxic, including the buttercup, calla lily, daffodil, elderberry, hyacinth, hydrangea, iris, jonquil, larkspur, mistletoe, morning glory, narcissus, oleander, philodendron, potato sprouts, ranunculus, rhubarb-blade, sweet pea, tomato vines, tulip, and wisteria. Symptoms of poisoning can vary from a mild stomachache, skin rash, or swelling of the throat to involvement of the heart, kidney, and other organs. Know the names of the plants in your yard. Call your poison control center if you suspect a poisoning.

Compounds from the sea are being investigated for possible anticancer, antifungal or other activity. Sponges, sea squirts, and other organisms have yielded anticancer drugs now in clinical trial.

2.3 Examples of Plant-Source Drugs in Use Today

Ipecac is a drug extracted from the dried root of a small shrub native to Brazil. It is widely available from poison control centers as syrup of ipecac (see Figure 2.1) and is used to induce vomiting in people (often children) who have swallowed a toxic substance. Some parents keep a bottle of syrup of ipecac on hand for emergency use, along with the telephone number of their nearest poison control center. As with all drugs, label directions should be followed scrupulously.

Emetine, the active ingredient in syrup of ipecac, stimulates the reflex center in the brain that controls emesis. Because emetine is potentially a very toxic substance, great care must be used in determining the total amount of syrup of ipecac that is administered to a child or even to an adult. Fluid extract of ipecac, which is 14 times more concentrated than ipecac syrup, is still for sale in this country. Deaths have occurred when the fluid extract was used by mistake. In other countries, emetine is valued for its effect in the treatment of amebic dysentery.

Case History *Ipecac Poisoning*

A 26-year-old, previously healthy woman was admitted to a hospital complaining of chest tightness, palpitations, extreme fatigue, and shortness of breath upon exertion. She admitted to drinking three or four 1-ounce bottles of ipecac syrup daily for 3 months after meals to induce vomiting in order to lose weight,[1] but she said she had taken no other drugs. Her pulse was 150 beats/min; her electrocardiogram showed inverted T waves. Despite intensive medical support, she developed progressive hypotension, ventricular tachycardia, and fibrillation, and she died. Of the two main alkaloids in ipecac—emetine and cephalin—emetine is the more cardiotoxic, and its accumulation in the body with prolonged use is especially dangerous. Cephalin has primarily a nauseant and vomiting action.

The stems of the **ephedra** plant (*Ephedra sinica*, and other species), a shrub listed in the Chinese dispensatory as early as 1569 under the name **Ma huang**, are the source of the alkaloid **ephedrine**. Since 1924, ephedra has been used as a bronchodilator and decongestant in the treatment of asthma. Today ephedra is widely

[1]**Bulimia** is an eating disorder characterized by episodic binge eating typically followed by purging behavior, such as self-induced vomiting or laxative abuse.

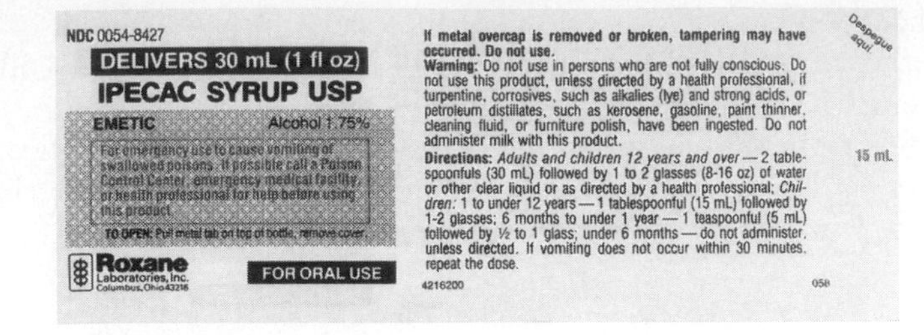

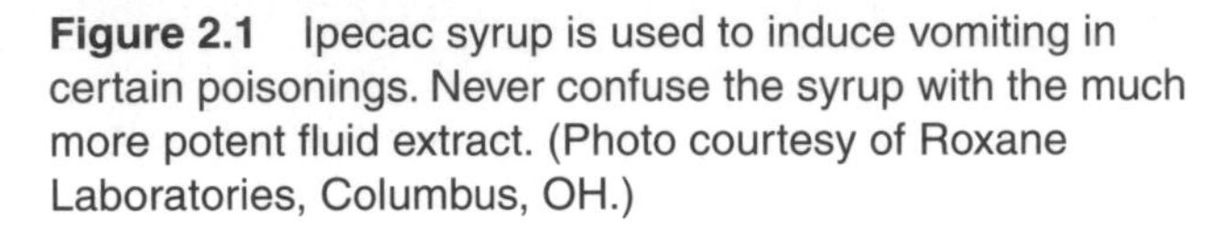

Figure 2.1 Ipecac syrup is used to induce vomiting in certain poisonings. Never confuse the syrup with the much more potent fluid extract. (Photo courtesy of Roxane Laboratories, Columbus, OH.)

used in nutritional supplements by professional athletes, teenage dieters, gym rats, and others interested in sports nutrition. The supplement market is replete with dozens of heavily advertised products containing ephedra extract. But user beware: ephedra has been linked to many heart attacks, strokes, and over 50 deaths, primarily owing to its strong heart-stimulatory action, especially when combined with caffeine, or guarana, kola nut, or exercise. An article in the *New England Journal of Medicine* considers ephedra-containing supplements unreasonably hazardous to health. The National Football League and the U.S. military have banned ephedra use. Yearly sales of ephedra products are in the billions of dollars, and the industry vigorously opposes regulation. For more pharmacology, see the next paragraph on ephedrine.

Ephedrine, like epinephrine, is a sympathomimetic drug, meaning that it mimics stimulation of the nervous system that governs our response to stress or "fight-or-flight" situations (see Chapter 5, "Drugs at the Synapse"). Ephedrine stimulates the heart and CNS, and these actions can be unwanted—even dangerous—side effects when ephedrine is used as a bronchodilator in asthma.[2] As a strong vasoconstrictor, ephedrine is used to treat hypotension. A typical oral dose of ephedrine is 15–50 mg. Pseudoephedrine and ephedrine are stereoisomers as found in the plant. As an alpha-adrenergic agonist (Section 5.4), pseudoephedrine acts as a nasal decongestant (Section 15.3, "Cold Medicines and Decongestant Preparations"), but critics advise caution in its use. Its use should be avoided in children under 12.

Opium is the dried exudate of the incised seed capsule of the opium poppy plant, *Papaver somniferum* (see Figure 2.2). The active ingredients in opium are valued for their analgesic (pain-killing) effects. **Morphine** and **codeine**, the two most important analgesic alkaloids of opium, are termed *narcotic analgesics*. Paregoric, also known as *camphorated tincture of opium*, is used by the laity to control diarrhea. The "opium eaters" of hundreds of years ago in Europe consumed laudanum, or tincture of opium. Refer to Chapter 6 for the complete story of opium alkaloids, their pharmacology, and their licit and illicit uses.

---

[2]Recently, the FDA warned about an OTC product that contained Ma huang and kola nut. The combination of the stimulant ephedrine from the Ma huang and caffeine from the kola nut caused life-threatening incidents in more than 100 people. Also see Section 7.12 for a description of the lethal drug combination of Ma huang and caffeine.

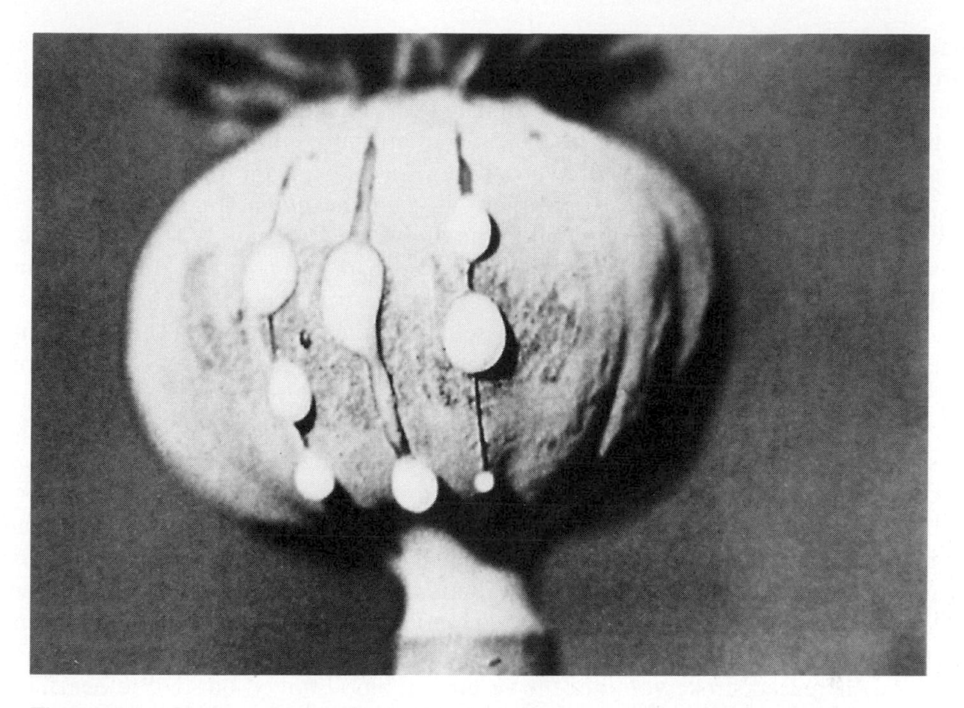

Figure 2.2 Crude opium is obtained by incising the unripe seed capsule of the opium poppy and collecting the milky exudate. Upon exposure to the air, the exudate turns brown. It is extracted for its content of morphine and codeine.

Atropine, from the plant *Atropa belladonna* (deadly nightshade), is an important drug in modern medicine. Atropine has been used in cold preparations to dry secretions of the mouth, nose, and bronchi; it has been used to reduce secretions and mobility of the GI tract in, for example, the treatment of peptic ulcers. Atropine has been employed in conditions of the eye that require dilation of the pupil (indeed, the name *belladonna* is a reference to the "beautiful woman" of Roman and Egyptian times who instilled a water extract of this plant into her eyes to produce dilated pupils, a sign of beauty). Atropine's effects on the heart are dose related: In very small doses, atropine can slow a too rapidly beating heart, but in larger doses it can increase the rate of a too slowly beating heart. Atropine acts as an antispasmodic to the intestines and colon. Because atropine is a general inhibitor of the effects of acetylcholine (see Section 5.4), it is used to treat cases of accidental poisoning by insecticides of the organic phosphate and carbamate types. These insecticides inhibit acetylcholinesterase, the enzyme our bodies make to destroy acetylcholine after its action at a synapse is complete. In carbamate poisoning, acetylcholinesterase is inhibited, permitting acetylcholine to function long after it should have been destroyed.

Mushroom poisoning can be of various types, but in the case of the fungus *Amanita muscaria*, the toxic alkaloid muscarine is produced. This alkaloid mimics the effects of acetylcholine; hence, this particular type of mushroom poisoning is treatable with atropine.

Atropine is a highly potent and potentially dangerous drug, as shown by the tiny quantity of it employed in a typical dose—1 mg. Doses of 10 mg or more can induce hallucinations, delirium, and coma. Diphenoxylate (Lomotil) is a currently popular drug for the management of diarrhea associated with the use of strange water supplies. Travelers to foreign countries often take the drug along and use it prophylactically. Each tablet, however, contains 0.025 mg of atropine sulfate, and one should be acutely aware of the possibility of overdosage, especially in children.

Digitalis, from the leaves of the plant *Digitalis purpurea* (foxglove), has had great usefulness as a cardiac stimulant in the treatment of certain types of heart disease. The active ingredients in digitalis are digitoxin, digoxin, and gitalin. In certain individuals, the heart loses its ability to beat with efficient contractions that force the blood effectively through the cardiovascular system. As a result of this heart failure, tissue fluid (water) tends to accumulate in the lungs or extremities—a condition laymen refer to as *dropsy* and medical practitioners as **edema**. The principal symptoms and signs of congestive heart failure are cardiac enlargement, edema, weakness, a prolonged circulation time, shortness of breath, and, in advanced cases, accumulation of fluid in the lung. This condition was recognized in 1785 by the English physician William Withering, who also learned of a folk remedy used to treat the condition. Withering successfully used extracts of the leaves of *Digitalis purpurea* to treat congestive heart failure. He also evaluated the drug, noting cautions in its use.

Termed *cardiac glycosides*, the active ingredients in the digitalis plant are potent chemicals. A dose of less than 1 mg can result in a much more forceful heart contraction, increased stroke volume (the volume of blood forced out in one beat), reduced heart rate, and a striking reduction in fluid accumulation in the body. In other words, digitalis makes the heart pump more efficiently. The toxic (poisonous) dose of digitalis is close to the therapeutic dose (there is a low margin of safety). No one should ever attempt self-medication with this drug, because only a physician knows the precautions necessary in establishing individual digitalization and maintenance dosage schedules. Other, safer drugs are supplanting digitalis today.

A word of caution to the home gardener: Digitalis is an attractive ornamental plant, and its seeds are commercially available. But one should never plant digitalis in areas frequented by children or pets, because the leaves and seeds are highly poisonous. Ingestion of only a few bitter leaves has resulted in death.

Reserpine is another example of folklore as a rich source of modern medicines. Reserpine's source is the root of the plant *Rauwolfia serpentina*, known to the ancient Hindus, who used it as a treatment for insanity and hypertension. Native to India, the plant was rediscovered in the 1930s and was brought to the attention of Western medicine in 1955.

Reserpine is useful in the treatment of mild essential hypertension (high blood pressure). Although now generally supplanted by other drugs, reserpine was a profoundly important discovery in the treatment of psychotic persons. Before about 1950, schizophrenics and other psychotics were "treated" mainly by placing them in an institution. Large mental hospitals were crowded with all types of patients, ranging from the mute catatonic to the raging psychotic. About the only kind of drug treatment for them was transient sedation.

Reserpine was one of two drug discoveries made in the mid-1950s that revolutionized the area of chemotherapy of the psychotic patient. Reserpine was found to quiet rage and excitement in the psychotic. With reserpine, patients became indifferent to stimuli around them, and thus they became much more manageable. The word *tranquilizer* was coined and became part of the English language.

Reserpine accomplishes its tranquilizing and blood pressure-lowering effects by depleting the body of its stores of catecholamines and serotonin (see Section 5.4). These compounds are responsible, in part, for the degree of excitement experienced at any given time. Reducing their levels will result in a generalized depression of many body functions. Thus, the explanation of the success of reserpine in treating schizophrenics becomes obvious.

Unfortunately, undesirable side effects can accompany the use of reserpine—sleepiness, nightmares, slowing of the heart, nausea, and diarrhea, among others. At present, the rauwolfia alkaloids have generally been replaced by other, more efficacious agents.

Cocaine, a local anesthetic and CNS stimulant alkaloid, is obtained from the leaves of the South American shrub *Erythroxylon coca* (Figure 2.3), where it occurs typically in 0.5–1.0% concentrations. Leaves of the coca bush can be harvested by the third or fourth year after planting, and they continue to produce for 20 years. Grown in Peru, Bolivia, Colombia, and other South American countries, cocaine smuggled into the United States comes mainly from Colombia. While cocaine is notorious for its abuse potential, one must remember that it also has legitimate medical applications. Physicians use it in certain surgical procedures in which excessive bleeding must be controlled (as in surgery of the nose). It is also used as an anesthetic in intubation procedures. Cocaine has been used in the Brompton Cocktail as a CNS stimulant for countering the narcotic effects of morphine. Use of this "cocktail" originated in England as part of the hospice movement, emphasizing humanitarian aspects of terminal cancer care. Relief of pain in terminally ill patients is afforded by the opiate in the cocktail. For further pharmacology of cocaine, see the discussion of CNS stimulants in Chapter 7.

Figure 2.3 The leaves of this South American shrub, *Erythroxylon coca,* are the only source of the alkaloid cocaine.

The coca plant is a woody shrub with bright green leaves and red berries, providing three to four harvests per year. In the period 1200–1300, the Incas venerated the plant as an agent for communicating with the gods; they considered its leaves sacred. After the Spaniards destroyed the Incan society, the use of coca spread to the common people. Later, the plant came to be chewed by Andean porters and guides, who found that it gave them boundless energy and freedom from hunger and fatigue. Today, many mestizo people in the area from northern Argentina to Colombia are coca chewers (coqueros), and they still regard the plant as sacred. Both men and women use coca at social rituals such as birthdays, weddings, and funerals, healing ceremonies, and magic rites. The pleasure it gives is apparently as much social as it is pharmacological. "Chewing" consists of either packing leaves between the teeth and the cheek (the buccal cavity), where saliva makes a natural extract, or placing crushed and roasted leaves (mixed with the alkaline ashes of a jungle tree) into the same cavity. Coca chewing appears not to result in drug abuse or dependency: in the approximate 30–60 g (1–2 ounces) of leaves chewed daily, only about 0.025–0.045 g of cocaine is absorbed into the bloodstream.

Cocaine was an ingredient in the first Coca-Cola product, which was introduced by an American pharmacist named Pemberton in 1886 and was advertised as a remedy for depression. By 1906, the distributor of Coca-Cola had eliminated the cocaine and included caffeine in the formula. Caffeine continues today to be a significant drug ingredient in cola soft drinks. (See Chapter 7 for much more information on cocaine.)

The **vinca alkaloids**, known as vincristine and vinblastine, are obtained from the leaves of a common household plant, *Vinca rosea* (periwinkle). This plant is also named *Catharanthus rosea*. This plant is another example of drugs that came to the attention of scientists because they had been mentioned in folklore (although for conditions other than cancer). Vinca is a rapidly growing plant that is cultivated for its beauty. Nature has placed in it several potent alkaloids that have become highly useful in the treatment of Hodgkin's disease (a primary lymph node neoplastic disorder) and acute lymphocytic leukemia in children. The introduction of vincristine therapy for childhood leukemia was hailed by the American Medical Association as one of the three outstanding advances in drug progress for 1963. Both vinblastine sulfate (Velban) and vincristine sulfate (Oncovin) are official in the USP.

Any drug that is effective against leukemia is welcome, especially when it is effective against leukemias of childhood. Some leukemias are termed *acute*, because they develop quickly and soon jeopardize life. Others are termed *lymphocytic* or *lymphoblastic*, depending on the type of white blood cell involved. Hodgkin's disease is a chronic form of leukemia, producing enlargement of the lymph tissue, spleen, and liver. Today, vincristine (Oncovin), in combination with mechlorethamine, prednisone, and procarbazine (the MOPP treatment), is the drug of choice for advanced stages of Hodgkin's disease. Vincristine, given with prednisone, is also the drug of choice in childhood acute lymphoblastic leukemia, in which it produces complete remission (i.e., return of the white cell count to normal) in about 90% of children on their first course of treatment. When some of these children relapse (i.e., when the disease strikes again) and are given a second course of vincristine therapy, 70–80% experience remission.

The vinca alkaloids are not without serious side effects. Nervous tissue is adversely affected, resulting in muscle weakness, headache, and double vision. Alopecia (falling out of hair) occurs in about 20% of patients receiving Oncovin.

Curare, from the vines of *Chondodendron tomentosum*, has long been known to the Indians of Peru and Ecuador who placed it on the tips of their arrows to kill or paralyze animals or enemies by producing paralysis—not the rigid kind that results from strychnine poisoning, but the flaccid relaxation of muscles attached to the bony skeleton. Curare progressively blocks nerve conduction to the muscles, first in the arms and legs, then finally in the rib cage, which the animal uses for breathing. The brain does not appear to be affected, because victims remain lucid until the end. The natives have a unique quality control system. If a monkey hit by a dart is only able to get from one tree to the next before he falls paralyzed, this is superior-grade, "one-tree" curare. "Two-tree" curare is less satisfactory, and the "three-tree" grade is definitely inferior.[3]

Today, the chief therapeutic application of tubocurarine chloride USP (the purified, active ingredient of curare) is to induce muscular relaxation in surgical patients under general anesthesia. Use of tubocurarine chloride permits the anesthesiologist to use less anesthetic and permits easier insertion of an endotracheal tube. Of course, if enough curare is given, the patient will stop breathing. Respirators must be on hand to take over if this situation happens.

Tubocurarine chloride is considered to be a competitive antagonist to acetylcholine (see Section 5.4, "What Are the Transmitters?"), occupying the receptor site in place of the acetylcholine. (Acetylcholine is the agonist at the neuromuscular junction, the anatomical endplate at which motor nerves are connected to skeletal muscle.) Typically, 6–9 mg of tubocurarine is injected to produce muscle relaxation, an indication of its great potency.

From the chemical structure of tubocurarine chloride, chemists have gotten ideas for synthetic compounds that have similar pharmacological actions. Thus, we have today the synthetic neuromuscular blocking agents decamethonium bromide NF and succinylcholine chloride USP, and we have eliminated total reliance on a foreign plant source.

Ergot has a phenomenal and exciting history. The ergot alkaloids (ergonovine, ergotamine, and four others) are obtained from a fungus, *Claviceps purpurea*, that infects rye and other grains growing in Europe and North America. The parasitic growth on the rye grain, termed a *sclerotium*, is about 2 cm in length and is colored purple (see Figure 2.4). Ergot is a veritable treasure house of potent pharmacological constituents. Careful inspections of our grains prevent ergot from getting into our bread, but in the Middle Ages and for hundreds of years thereafter, epidemic-level contamination of rye bread by the ergot fungus was common. Peasants ate the bread and received heavy doses of the ergot alkaloids, and the results were shocking. Pregnant women aborted. Convulsions occurred. Circulation of the blood in the extremities was so severely restricted that fingers, toes, and limbs became gangrenous and dropped off. Skin became black and was believed to have been consumed by the holy fire, or St. Anthony's fire—to whose shrine the peasants flocked for relief (and got it because they did not have contaminated bread to eat while on the journey to the shrine). The last great epidemic in Europe occurred in southern Russia in 1926–1927. An outbreak of ergotism occurred in Ethiopia in 1978 in connection with famine conditions in that country.

Midwives knew that the powerful ingredients of ergot could be put to obstetrical use. They used ergot to stimulate the uterus (womb) to contract in order to hasten

[3]M. Krieg, *Green Medicine*, Rand-McNally, Skokie, IL, 1964, p. 225.

Figure 2.4 Ergot is a fungal parasite found growing as a curved sclerotium on rye and other grains.

the birth of the infant. Any drug that has this action is termed an **oxytocic**. Midwives also used ergot to contract the muscles of the uterus in order to stop postpartum hemorrhage. These uses continue today, but physicians know they must exercise great care in administering the dose of ergot alkaloid because it is easy to overstimulate the uterus (with serious consequences). Ergonovine maleate NF (Ergotrate) can be administered intramuscularly, intravenously, and orally. Side effects can include nausea, vomiting, hypertension, and cramping.

On the subject of oxytocics, it is instructive to note the presence of a pituitary gland hormone that is secreted naturally at full term to initiate labor. Termed *oxytocin*, the hormone can be obtained from natural sources but is now made synthetically. It is available under the trade name Pitocin. It is administered by injection or by buccal tablets and is used for the induction of labor and the control of postpartum hemorrhage. Oxytocin, as secreted naturally, also stimulates the lactating mammary gland to make milk available to the suckling infant (milk ejection). Oxytocin, too, is a powerful drug, not without its side effects.

Ergot alkaloids also are employed in the treatment of the symptoms of migraine headache. Ergotamine tartrate USP (Gynergen) is effective in controlling 90% of acute migraine episodes and has been the drug of choice for this condition. Cafergot is another commercial preparation that adds caffeine to supplement the vasoconstrictor effect of the ergotamine. One theory to explain the occurrence of migraine headaches suggests that arteries in the brain become dilated (expanded) and exert pressure on the brain tissue, resulting in pain. If this suggestion is correct, the relief given by the ergot alkaloids is understandable, for they are vasoconstrictor (artery-shrinking) drugs. Ergotamine tartrate relieves the symptoms of migraine but does not remove the cause, because the frequency of migraine attacks is not diminished. The ergot alkaloids are alpha-adrenergic blocking agents (see Section 5.4). They also antagonize serotonin. Newer antimigraine drugs, the so-called *triptan* drugs, target the nerves and blood vessels associated with migraine headaches. Examples are Amerge, Maxalt, Relpax, and Zomig.

Ergot alkaloids contain the chemical grouping known as *lysergic acid.* In 1943, the Swiss chemist Albert Hofmann handled the *N,N*-diethylamide derivative of lysergic acid, accidentally got some of it on his skin, and experienced the first LSD "trip." The story of the fascinating discovery of LSD and its far-reaching implications for society is told in Chapter 12, "Hallucinogens, Street Drugs, Designer Drugs, and Some Observations."

2.4 Plant Sources of Illicit Drugs

Marijuana (or marihuana), a preparation of the leaves and flowering tops of the plant *Cannabis sativa*, has been presenting challenges to various cultures and societies for 5,000 years. In the pre-Christian era, the Chinese referred to marijuana as the "liberator of sin," and in a later period they named the plant "delight giver." Hindus have long called it the "heavenly guide" and the "soother of grief."

Cannabis is a tall, annual weed, sometimes reaching a height of 4 meters (13 feet), as seen in Figure 2.5. The male and female plants will grow in almost any waste or fertile area. Scientific cultivation of cannabis plants has revealed remarkable variations in plant characteristics, size, and concentration of the active ingredient. One researcher commented that cannabis is either a genus composed of more than 100 species or a single species that has many variations. Plant scientists recognize the Indian, Mexican, Thai, Korean, Iowan, and Russian variants. They may differ as much as 50-fold in their tetrahydrocannabinol concentration (0.07–3.7%). The predominant wild-growing U.S. form of marijuana is *C. sativa*; a second form, *C. indica*, is grown as a source of the resin and may contain 2–5% of the psychoactive principle.

C. sativa is the source of hemp fiber. Its seeds (hemp seed) are rich in oil and serve as a valuable source of food. (Section 16.3 has a case history on another hemp product.) Today hemp oil is also sold as an ingredient in cosmetic products, including lotions, shower gels, and lip balm. As a medicine, the cannabis plant was important not only to peoples of ancient China and India but also to physicians in Europe

Figure 2.5 Marijuana plant (*Cannabis sativa*). Users and growers typically learn how to distinguish male from female plants.

and the United States in the late 19th and early 20th centuries. Marijuana was official in the USP in the early 1930s, and a fluid extract of the plant was made and sold by such major U.S. drug companies as Eli Lilly and Company, Bristol-Myers, and Parke-Davis as a treatment for neuralgia, mental depression, rheumatism, and gout. In the 18th and 19th centuries, great plantations of marijuana were established in Missouri, Mississippi, and Kentucky and provided large quantities of hemp fiber for the manufacture of rope and cloth.

Concurrent with these licit uses through the centuries was the illicit consumption of the drug for its euphoriant and intoxicant effects. Chemists have identified 421 different compounds in marijuana; the heat produced by smoking marijuana causes chemical changes that increase this number to more than 2,000. Sixty-one different lipid-soluble cannabinoids exist in marijuana, and 11 of these are specifically termed *tetrahydrocannabinols*. The major active euphoric principle is **Δ^9-tetrahydrocannabinol (THC)**.[4] **Hashish**, which mainly comes from the Middle East, consists of the drug-rich resinous secretions of the cannabis plant, which are collected, dried, and compressed into a variety of forms, such as cakes, balls, or

[4]The distinction between Δ^9-THC, and Δ^8-THC is based on the position of the carbon-to-carbon double bond in the THC molecule. A newer system of numbering designates Δ^9 as Δ^1-*THC*. See Figure 13 in Appendix I.

cookie-like sheets. Historically, "hasheesh" has been smoked, eaten, or drunk by millions of peoples, especially in Muslim areas of North Africa and the Near East. Hashish in the United States varies in potency as in appearance (see Figure 11.1), ranging in THC content from trace amounts up to 13%. *Hashish oil* (actually not related to hashish) is made by repeated extraction and concentration of cannabis to yield a dark, viscous liquid—current samples of which average about 20% THC. In terms of its psychoactive effect, a drop or two of hashish oil on a tobacco cigarette is equal to a single joint of marijuana.

Less THC exists in cannabis leaves, and practically none is present in the stems, roots, or seeds. Somewhat less-potent preparations are *bhang* and *ganja*. The term *sinsemilla* (meaning "without seeds") refers to the product made from specially cultivated flowering tops (known as colas or buds) of unfertilized female plants. Much sinsemilla is grown illicitly in California and is prized for its high THC content (up to 8%).

Thai sticks (from Thailand) are six 12-inch-long compressed bundles of buds and stems of high-THC-content marijuana. The sticks are the form in which the drug is transported; for use, however, the bundle is broken up and is smoked in a joint or pipe. Recently, Thai sticks have become less important because home-grown pot is rich in THC and is readily available.

Officially, all of these preparations are grouped together under the term *cannabis*. The "pot" or "grass" available to most American users today is of increasingly higher THC concentration. In the 1960s, THC averaged 0.1–2%; in the 1970s, it averaged 1%; today, it averages 6%. In smoking a typical "joint," the user will receive a dose of 5–25 mg of THC, assuming that he or she is smoking one of the more potent batches.

In 1937, marijuana became an outlawed drug. The U.S. Congress passed the Marijuana Tax Act, partly in response to many sensational news stories, spread, some say, deliberately and falsely to hasten the illegalization of the drug. By 1937, all 48 states and the District of Columbia had anti-marijuana laws in their codes. The federal act legally defined marijuana as a narcotic drug, in the same drug category as the opium alkaloids (which, of course, it is not). Marijuana dropped from medical use. So great was society's abhorrence of marijuana that in Missouri, Utah, and Illinois, possession of a single marijuana cigarette could have resulted in a life sentence.

Since the 1970s, authorities have sprayed the herbicide **paraquat** on fields of growing marijuana in attempts to destroy the plant at its source. The first domestic use of paraquat spray occurred at Red Bay, Florida, in August 1982, when 80 acres of harvest-ready marijuana were destroyed. Spraying with this herbicide is controversial because of the possible toxic effect of paraquat residues in the pot that gets to the illicit market. Paraquat is now banned from use on national forests, but the DEA has announced that paraquat and two other herbicides will be used to spray marijuana plants growing on private lands.

For the remainder of the story on marijuana, the extent of its present usage, and the controversy surrounding the drug, see Chapter 11, "Marijuana."

Peyote (mescal button) is obtained from an unimpressive, golf ball-sized, gray-green cactus (see Figure 2.6) that grows in the Rio Grande valley from Mexico City to the Texas border, and in some areas of Texas, Arizona, and California. Peyote was known to the Aztecs, who considered this hallucinogen a divine messenger. In 1692, Indians established the mission El Santo de Jesus Peyotes, which showed the early influence of this plant. The mescal button results from slicing off

Figure 2.6 The peyote cactus, shown here in flowering stage, is the source of the hallucinogenic alkaloid mescaline.

the top of the cactus and drying it. When used in a religious ritual, some 4–12 buttons are ingested by each member of the group, always sitting around a fire. A period of meditation follows, during which visual aberrations, nausea, and vomiting may occur. Natives report seeing visions of vast fields of golden jewels, changing kaleidoscopically. They feel removed from earthly cares. A peyote trip can last for 6–10 hours.

A group of about 250,000 North American Indians, comprising the Native American Church, won a battle in 1965 in the U.S. Supreme Court for the right to use peyote legally as a sacrament in their modified Christian ritual. Their religion is called the *peyote religion*, and their services are termed *peyote meetings*. However, in 1991 the U.S. Supreme Court ruled that state drug laws could prohibit the use of peyote by American Indians for religious reasons. Currently, 15 states with almost 65% of the American Indian population and the federal government allow the religious use of peyote. Actually, only a small, very religious minority of American Indians practices peyote use. The Indians have had success in using peyote in the treatment of widespread alcoholism.

In 1966, Arthur Kleps founded a psychedelic religion, the Neo-American Boohoo Church, and claimed that the use of LSD was sacramental, similar to the peyote ritual. Kleps's appeal to the courts to protect his use of LSD was rejected, the judge ruling that an organization with "Row, Row, Row Your Boat" as its theme song was not serious enough to qualify as a church.

Mescaline, named after the Mescaleros Apaches, is the active alkaloidal hallucinogen in peyote and has a simple chemical structure closely related to that of the amphetamines. The average hallucinogenic dose of mescaline is 5 mg per kilogram of body weight. Under the 1970 Controlled Substances Act, mescaline is classified as a schedule I drug, and its mere possession, whether pure or as part of the plant, is punishable. (Drug schedules are described in Chapter 3.)

Mescal buttons can be found in the illegal drug market, but mescaline is rare. Apparently, the operators of clandestine laboratories do not find it worthwhile to synthesize it. According to the *Do It Now Foundation*, only about four out of every 10 street samples offered as mescaline actually turn out to be mescaline (most are LSD or LSD-PCP mixtures). The foundation bases its estimate on chemical analyses of actual street purchases. Because of the great number of forays made by drug seekers into the Western deserts to pick mescal buttons, there is some danger of the plant becoming extinct.

Psilocin and **psilocybin** are hallucinogenic alkaloids obtained from "sacred" mushrooms of Central America (Figure 2.7). The most important of the sacred mushrooms, long used by the mushroom-worshipping Indians of Mexico, is *Psilocybe mexicana* (other genera are *Conocybe* and *Stropharia*). Plant scientists believe that mushroom eating and worship was widespread, because stone images of mushrooms carved into the shape of a god date back to 1000 B.C. in Guatemala.

Figure 2.7 The hallucinogenic psilocybe mushroom. See text for discussion.

Frescoes depicting the use of mushrooms have been found dating to 300 A.D. History's first recorded use occurred during the coronation feast of Montezuma in 1502.

Once again, religious ritual provided the setting for the use of the hallucinogenic principles. Divination, prophecy, and worship followed ingestion of *teonanacatl*—"flesh of the gods." Hofmann, the discoverer of LSD, also investigated these Mexican mushrooms, determined the chemical structure of their active ingredients, and prepared the alkaloids synthetically. Psychotomimetics such as psilocin and psilocybin were often used in the 1950s and 1960s by writers, painters, and certain entertainers to achieve creativity, to release their "hidden potential," or just for thrills.

The hallucinatory experience produced by psilocybe alkaloids is roughly comparable to that produced by LSD, except that the LSD trip lasts about twice as long. Dangers involved stem not from physical harm but from the potential for inducing psychotic states that persist long after the expected end of the psychedelic experience. Tolerance to these hallucinogenic mushrooms (indeed, to most hallucinogens) quickly develops, and a period of about five days must elapse before the user can "get high" again.

On the street, the rip-off rate on supposed hallucinogenic "shrooms" is strikingly high. Psilocybin (along with mescaline) heads the list of most often misrepresented hallucinogens, with fewer than 20% of the street samples actually containing psilocybin. Many of the mushrooms sold on the street are of the grocery store variety, frozen and spiked with LSD or PCP. On a weight basis, LSD is about 50–100 times as potent as psilocybin.

2.5 Drugs from Animal Sources

2.5.1 Insulin

In the summer of 1921, the team of Charles Best and Fred Banting, working at the University of Toronto School of Medicine, proved that the pancreas was the source of a hormone that could "cure" diabetes mellitus, a disease recognized by the ancient Greeks. They named the hormone *insulin*.

Best and Banting discovered that they could make dogs diabetic by surgically removing the pancreas, and then they could make the dogs well again by injecting pancreatic extracts obtained from another animal source. After they successfully treated 80 diabetic dogs with their new extract, Banting and Best's report of their work spread to the newspapers, and soon telegrams, letters, and phone calls came from diabetics all over the world appealing for this new "cure." A then small pharmaceutical firm named Eli Lilly and Company won the right to manufacture and supply insulin. They continue today to be the major supplier to the approximately 1.5 million American diabetics who take insulin injections.

Beef and pork pancreas are the source of most of the insulin sold today, but they present problems. First, almost 8,000 pounds of the gland is required to produce 1 pound of insulin. Second, almost all humans who have received commercial beef insulin for longer than two months have developed antibodies to the foreign proteins in this product. In cases where the patient's antibody level is so high that he or she is resistant to the beneficial effects of the insulin, pork insulin may be substituted for beef. Fortunately, we have a variety of alternatives to beef: pork, goat, sheep, horse, and sei whale. All of these animals produce insulins that do not have exactly the same chemical structure as human insulin but are close enough to substitute very well in the human. Of course, the ready availability of some animals at the slaughterhouse plays a major role in the selection of the animal.

Since 1978, scientists have known that human insulin can be produced by bacteria through the process of recombinant DNA technology (gene splicing; see Section 2.6). Briefly, in this remarkable technique, the human gene that codes for the synthesis of insulin is incorporated into the genetic material of a bacterium such as *Escherichia coli* or a yeast such as *Saccharomyces cerevisiae* (bakers' yeast). The genetically manipulated microorganism is then grown in huge tanks, and as it grows and multiplies, the microorganism synthesizes insulin along with all of its own natural protein molecules. The insulin is then isolated, purified, and bottled in vials for injection. It is identical to human insulin. Gene-spliced insulin, when highly purified, can induce no antibody production, because it is not a foreign protein—although it is made by a foreign organism. Its supply is limitless, in contrast to the limited quantities of slaughterhouse insulin. Eli Lilly and Company markets its gene-spliced insulin as Humulin (see Figure 2.8). Novo Nordisk markets its gene-spliced insulin (made using bakers' yeast as the microorganism) as Novolin, available as R (regular), L (lente), N (NPH), 70/30, and other forms including PenFill injection cartridges.

Insulin promotes the uptake and metabolism of glucose (blood sugar) into skeletal, cardiac, and smooth muscle tissues of the body. It also promotes glycogen storage, fatty acid synthesis, amino acid uptake, and protein storage, thus playing an important anabolic (tissue-building) role in liver, fat, and muscle. In the nondiabetic, insulin is released only during mealtimes in response to ingested carbohydrates. Typically, the diabetic musters too little insulin, which can cause the blood glucose level to rise dangerously. This condition is referred to as *hyperglycemia*. Administration of insulin from an exogenous source lowers a too high blood glucose level and prevents sugar from spilling over into the urine. If, by accident, too much insulin is taken, the level of glucose in the blood may fall far below its normal range of 70–110 mg/100 mL of

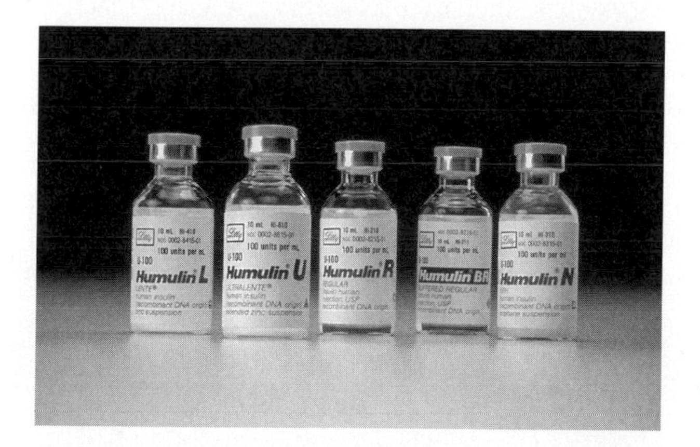

Figure 2.8 Gene-spliced insulin is marketed by Eli Lilly and Company as Humulin®. (Photo courtesy of Eli Lilly and Company.)

blood, and the individual is then said to be *hypoglycemic*. (We should recognize that hypoglycemia also exists in otherwise perfectly healthy people and that it can make the person physically weak and at times quite irritable.)

Diabetes also can occur in the juvenile form, which is "brittle" and difficult to treat (type 1). Diabetes can also manifest itself in later years (maturity-onset, or type 2 diabetes). About 1 in 20 Americans will show diabetic symptoms sometime during life. As long as they follow the treatment regimen, diabetics can lead essentially normal lives. Untreated diabetes, however, can have serious consequences: damage to the retina of the eye, disrupted fat metabolism, cuts that do not heal, and damage to other body organs. Recently, diabetes was cited as the second highest cause of blindness and the fifth leading cause of death by disease in this country. Some 325,000 new cases of diabetes are diagnosed in the United States each year. The risks of developing diabetes based on familial disposition are given in Table 2.2.

Altogether, three approaches to the treatment of diabetes mellitus, used singly or in combination, are available to the physician.

1. Control of diet
2. Injection of insulin
3. Oral antidiabetic drugs (also termed *oral hypoglycemics*)

Clinicians have attempted to administer insulin by oral, rectal, vaginal, intranasal, intravenous, and skin routes, but the only practical method is subcutaneous injection. Insulin, indeed any protein, taken orally is inactivated by the enzymes and juices of the digestive tract.

Because diabetics vary so greatly in their need for injected insulin, a variety of dosage forms are available; they differ in strength, onset time, and length of action. Strength of insulin is measured in USP units. Insulin is available from both gene-spliced (recombinant DNA) and from animal sources. Lilly's Humulin BR is gene-spliced and is a special form intended for external insulin pump use only. Lilly's regular Iletin I is of beef-pork pancreas origin.

Table 2.2 Genetic Predispositions to Diabetes[a]

Relatives with Diabetes	*Maximum Risk (percent)*
Parent plus grandparent and aunt or uncle on other side of family	85
Parent plus grandparent or aunt or uncle on other side of family	60
Parent plus first cousin on other side of family	60
Parent plus first cousin on other side of family	40
Parent	22
Grandparent	14
First cousin	9

[a] A person's chances for developing diabetes at some point in life are predictable from the table. Thus a person with a high risk factor for development of diabetes may be referred to as a *prediabetic* during any period in life before symptoms appear.

Diabetes patients should know that over the past several years, thousands of insulin pumps have been sold to diabetics who wear the device strapped to their bodies. The small, battery-driven pump uses a computer to deliver measured amounts of insulin through a needle inserted under the skin, but the computer is not able to sense blood sugar levels. Dose adjustments are made by the patient on the basis of capillary blood samples obtained by finger pricking several times a day. Insulin pumps cost between $1,000 and $2,500. Needle-free insulin injection systems are also commercially available. Vitajet and Medi-Jector EZ are two products that deliver the insulin in an ultrafine stream—about one-third the size of the smallest needle. No needles mean no discomfort. The newest way to deliver insulin is up the nose with a spray pump. California Biotechnology, Mountain View, manufactures Nazlin, an insulin nasal spray.

Oral antidiabetic drugs are not insulin, nor are they insulin substitutes. They are synthetic hypoglycemic agents, taken orally, that stimulate a sluggish pancreas to release more endogenous insulin. Thus, they are useful only in diabetics whose pancreases still produce some insulin. Used especially in maturity-onset patients, oral antidiabetic drugs have been known and used since 1955. Chemically, they are sulfonylureas. Examples of oral antidiabetic drugs are chlorpropamide (Diabinese), tolazamide (Tolinase), tolbutamide (Orinase), glipizide (Glucotrol), and glyburide (Micronase, Diabeta). Glimepiride is not yet approved in the United States.

Diabetics can subscribe to two magazines: *Diabetes Forecast*, American Diabetes Association, National Service Center, 1600 Duke St., Alexandria, VA 22314, and *Diabetes Countdown*, Juvenile Diabetes Foundation, 60 Madison Ave., New York, NY 10010.

2.5.2 Thyroxine

Gently place two of your fingers at the front center base of your neck, just above the bony structure, and swallow. You may then feel your thyroid gland, a two-lobed structure that secretes the hormone thyroxine. We need thyroxine in order to grow to normal physical and mental maturity and to maintain normal body metabolism. Diseases associated with the thyroid gland include the following:

1. *Myxedema*—Adult hypothyroidism, manifested by coarse hair, hoarse voice, dry skin, and slow mentation
2. *Cretinism*—Childhood hypothyroidism, manifested by mental retardation, dwarfed size, protruding tongue, and potbelly
3. *Hyperthyroidism*—Characterized by nervousness, weight loss, and a too-high metabolic rate

For use in hypothyroid conditions, exogenous thyroxine can be obtained from the thyroid glands of slaughtered animals, usually pigs. It is also obtained synthetically. Thyroid USP, a fine powder marketed in tablet form, is given as a substitute drug to patients who have diminished or absent thyroid function resulting from surgical removal of the gland, radiation therapy, or failure of the gland to function adequately. In children, adequate amounts of this hormone are essential if normal bone growth and brain development are to be achieved.

Thyroxine is a potent drug. It controls the rate at which body tissues metabolize nutrients. Effects of overdosage with thyroxine include nervousness, insomnia, headache, a too-fast heart rate (tachycardia), weight loss, and sweating. Some people take thyroid preparations to lose weight. But if they have normal thyroid function to begin with, they run the risk of overdosing.

2.5.3 Growth Hormone

The tallest adult human on record is 8 feet 11 inches tall, and the shortest is a tiny 23 inches. What made the difference, of course, was how much growth (somatotropic) hormone these two individuals received from their pituitary glands. **Human growth hormone** (hGH) is a large molecule composed of 191 amino acids joined together in a specific, precise way so that when it circulates in the blood system, the ends of the long bones (called the *epiphyses*) are stimulated to develop and grow. Fortunately, for most of us, the pituitary gland secretes just about the correct quantity of growth hormone, and we attain a reasonable height (5 feet 10 inches for the average U.S. male and 5 feet 5 inches for the average U.S. female). Actually, we now know that growth hormone works by means of intermediary compounds termed *somatomedins*. Growth hormone stimulates the production of somatomedins in the liver, brain, and kidneys.

Until recently, if a child was far below normal height and the physician decided that hGH would help, human cadavers would be used as the source. This is because growth hormone from nonhuman sources is not sufficiently similar chemically to do the job. Cadaver hormone, however, has given way to synthetic hGH obtained by recombinant DNA technology (gene splicing.) As with human insulin, *Escherichia coli* was the host organism whose DNA was altered to accept the fragment of human genetic material that codes for hGH production. Clinical trials with genetically engineered hGH were carried out successfully in the early 1980s in 10 medical centers across the country with children whose bodies were unable to produce enough of the hormone to sustain normal growth. The gene-spliced hormone, given to the children three times a week for a year, is easier and less expensive to make than the hormone extracted from cadavers. The Genentech Company has received FDA approval to market its Protropin brand of gene-spliced hGH. Cost for a year of treatment is estimated to be $8,000–10,000. Eli Lilly and Company markets its Humatrope brand of recombinant hGH. Would-be basketball players should be aware that there is no evidence that hGH will make normal children taller. We know that an excess of hGH can result in the condition known as acromegaly, which affects the hands, feet, face, and head. Genentech's recombinant form of hGH is slightly different from the natural form in that it contains an extra methionine residue.

Table 2.3 gives an example of the effectiveness of exogenous growth hormone in promoting growth in an abnormally short individual. In another case, a 10-year-old grew five inches in a year after taking recombinant hGH. Male patients continue to receive growth hormone injections until they reach a height of 5 feet 6 inches (5 feet 3 inches for females). On the other hand, a Stanford University study concluded that an average of only 2 inches was gained by short but otherwise healthy children who were given growth hormone (*New England Journal of Medicine*, February, 1999). Care must be taken, because too much growth hormone has been implicated in diabetes mellitus among men taking the hormone.

After the epiphyses close, no further growth is possible, and no amount of growth hormone will be effective in stimulating growth. Some individuals, however, have informed me that they are still growing in their early twenties. An Internet reference for

Table 2.3 Year-to-Year Effects of Growth Hormone Treatment (2–5 mg Given Three Times Weekly)[a]

Year	*1957*	*1958*	*1959*	*1962*	*1964*
Age	17	18	19	22	24
Height	4'3"	4'6"	4'8"	5'1"	5'4"
Weight	68	71	82	100	95

[a] Note that without treatment, expected height at 24 years would have been 4 feet 5 inches (a gain of 11 inches).

quality information on hGH is **http://www.aace.com/clin/guidelines/hgh.pdf**.

2.5.4 Gonadotropins

Women ovulate because of the physiological influence of two gonadotropic hormones released by the pituitary gland: *follicle-stimulating hormone* (FSH) and *luteinizing hormone* (LH). The target organ for these two hormones is the ovary; there they help stimulate the development, maturation, and release of one or more egg follicles each month. Some women are infertile because their pituitary gland does not secrete gonadotropin, and consequently they do not ovulate. In this circumstance, it is possible for the physician to administer gonadotropic hormones obtained from an animal source—the urine of either postmenopausal women or pregnant mares. Large quantities of such urine samples are collected and extracted in a manner that isolates the active hormone, leaving behind the impurities. Menotropins for injection (Pergonal) is a common preparation of FSH and LH obtained from human urine.

About three-fourths of infertile women treated with exogenous gonadotropins ovulate. One of the complications of this therapy is possible over-stimulation of ovarian follicular development, with subsequent multiple pregnancy. In a few cases, seven or more eggs matured all at once, were fertilized simultaneously, and developed into a multiple pregnancy that did not proceed to term. Cases of stillborn sextuplets have also occurred. However, when the treatment is successful and a reasonable number of eggs mature, they develop into normal offspring.

This discussion emphasizes the fascinating biological control mechanisms based on highly specific hormonal action on target organs. Ovarian tissue responds to the proper hormones, whether of endogenous or exogenous origin. The hormones, in turn, act only upon certain specific tissues.

Clomiphene citrate USP (Clomid), a synthetic substance, is also used in cases of infertility due to lack of ovulation. Oral administration of clomiphene results in increased pituitary release of FSH and LH. Apparently, the drug acts by disrupting the feedback inhibition of gonadotropin release that estrogens exert on the pituitary gland. About 30% of selected infertile patients treated with Clomid conceive. The offspring appear to be normal. Clomid is also prescribed for men who are producing an insufficient concentration of sperm.

2.5.5 Human Chorionic Gonadotropin (hCG)

Weight reduction clinics seem like an unlikely place to encounter hCG, but that is exactly where many women today use this substance as a drug. hCG is promoted in such clinics as a substance that, when injected, hastens the loss of body fat (used, of course, as part of an ongoing overall treatment regimen). Women who have taken hCG for weight reduction confirm the clinics' claims. Representatives of the FDA, however, state that there is absolutely no rational application of this hormone of pregnancy in weight reduction. Medical authorities state that hCG has no known effect on fat mobilization, on appetite or sense of hunger, or on body fat distribution. They believe hCG is ineffective in causing a more attractive or normal distribution of fat.

hCG is indeed a vital hormone of pregnancy. This glycoprotein, secreted by the placenta soon after conception, stimulates the corpus luteum to enlarge and secrete progesterone for maintenance of the pregnancy. Gynecologists administer hCG to infertile women, in conjunction with FSH and LH, to induce ovulation. Chorionic Gonadotropins for Injection USP is made from the urine of pregnant women and is marketed as A.P.L., Pregnyl, Profasi HP, and others.

hCG detection in the urine is the basis for inexpensive ($9–$13) tests for pregnancy sold OTC. The FDA has now approved over 100 different pregnancy tests, some examples of which are Advance, Answer, Conceive, Confirm, Clear Blue Easy, e.p.t., Fact Plus, and Precise. Manufacturers claim that these products can be used at home to detect pregnancy as early as within minutes on the first day of a missed menstrual period and that the tests are accurate 97–99% of the time when a positive result indicates pregnancy. hCG has been made by recombinant DNA technology at Integrated Genetics, Inc., Framingham, Massachusetts. Ovulation prediction kits, now sold OTC, detect the spike in the level of luteinizing hormone that occurs in the middle of a menstrual cycle.

Because hCG is a hormone of pregnancy, it is not found in men—except for the rare event of (testicular) teratomatous tumor containing chorionic (embryonic) elements. Injected into men, hCG stimulates the interstitial cells of the testes to secrete male sex hormone. Perhaps it is this function that has motivated some athletes to take hCG to try to build bigger bodies. (See Chapter 17 for more information about drugs in sports.) The FDA has approved menotropins for injection (Humegon), a form of gonadotropin obtained from the urine of postmenopausal women.

2.5.6 Estrogens

Simply put, the estrogenic hormones are what make the difference between a prepubertal girl and a postadolescent woman (Figure 2.9). All of the secondary sex characteristics that develop in females are due, in some measure, to the secretion and action of the natural estrogens (estradiol, estriol, and estrone). Indirectly, estrogens contribute to the shaping of the skeleton, tone and elasticity of urogenital structures, and growth of the long bones at puberty. While the ovaries are clearly the main site of estrogen production, other tissues, such as the adrenal glands, liver, skeletal muscles, and fat tissue, contribute significantly. So does the placenta during pregnancy. In males, the testes produce small amounts of estrogens.

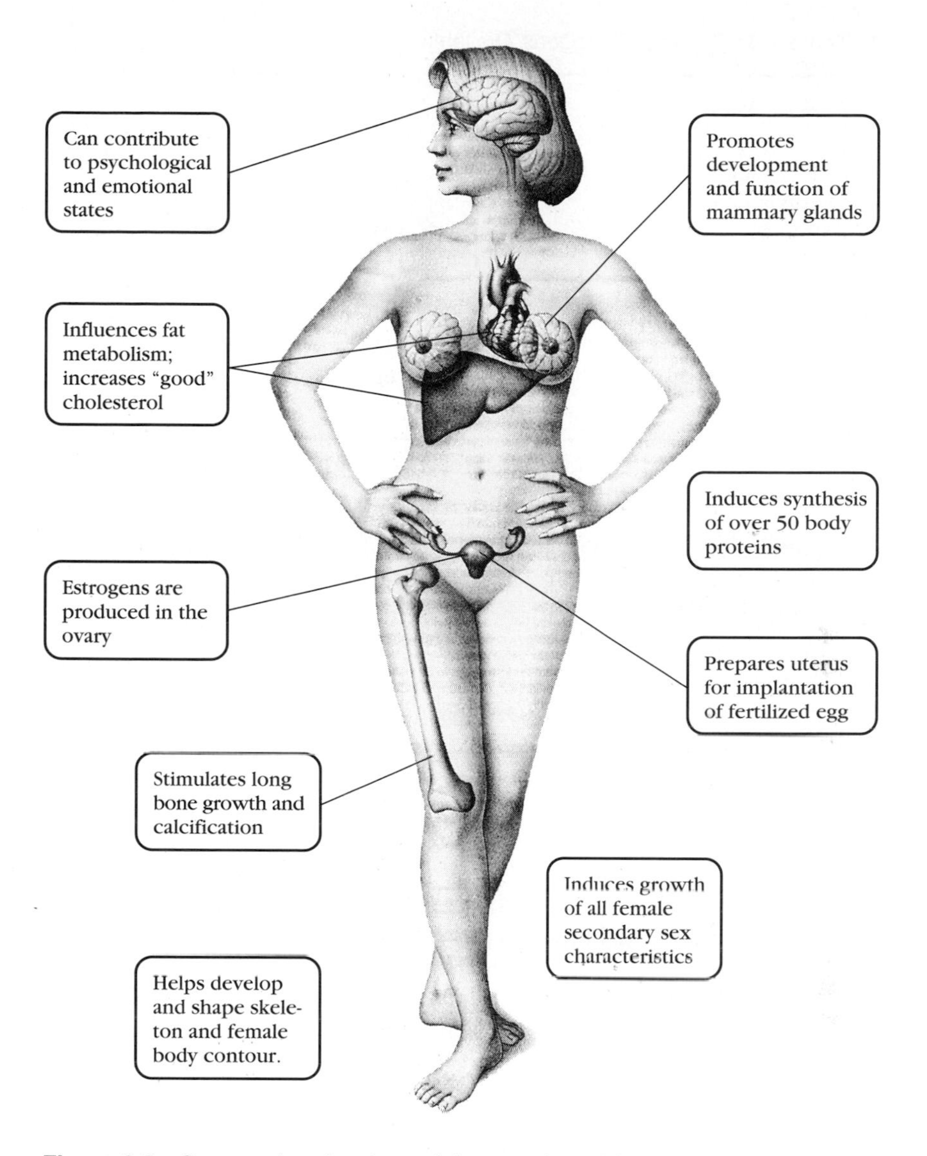

Figure 2.9 Source of and actions of the amazing, ubiquitous estrogens.

Table 2.4 shows the types of estrogen available, both from natural sources and by synthesis.

With the estrogens, the physician has long had highly potent substances for the treatment of a wide variety of female hormone-deficiency conditions and other morbid states. These conditions include the menopausal syndrome, female hypogonadism, abnormal uterine bleeding, failure of ovarian development, atrophic vaginitis, female castration, and osteoporosis. Estrogens such as Premarin are heavily prescribed in **Hormone Replacement Therapy** (HRT). About 25 percent of the nation's 40 million women older than 50 take estrogen to make up for the loss of the hormone after menopause, many in the belief that it will help maintain a youthful appearance. Women in the United States receive about 20 million prescriptions yearly for estrogen to replace the natural hormone. **Prempro** (conjugated

Table 2.4 Types of Estrogens Available[a]

Generic Name	*Trade Name*
Chlorotrianisene	TACE
Diethylstilbestrol	Stilphostrol
Estradiol	Estrace, Estraderm (transdermal)
Conjugated estrogens	Premarin
Esterified estrogens	Estratab (from Mexican yam plants)
Estrone	Ogen
Ethinyl estradiol	Many trade names (used in the Pill)
Phytoestrogens	Various vegetable products
Quinestrol	Estrovis

[a] Progestins (Section 13.2), such as the steroid contained in Prempro, are not estrogens, and are not included here.

estrogens/medroxyprogesterone acetate tablets), a heavily prescribed drug that combines natural estrogens with a progestin, is indicated for the prevention and treatment of osteoporosis. This combination is for women who still have their uteruses, as the progestin protects against uterine cancer. Do not take Prempro if you have had abnormal vaginal bleeding, abnormal blood clotting, or are pregnant.

Hormone-replacement therapy has been intensively practiced for decades, not only to treat the immediate symptoms of the menopause, but to prevent osteoporosis, lower the incidence of fractures, reduce the risk of heart attack and stroke, lower bad LDL cholesterol, and raise the good HDL. But there is now compelling information from randomized controlled trials that casts doubt on some aspects of HRT's effectiveness and even safety. A new report, the *International Position Paper on Women's Health and Menopause*, funded by the NIH and the Giovanni Loren Zini Medical Science Foundation of Italy, concludes that rather than protecting women from heart attack and stroke, HRT *increases* their risks. The lengthy report states that no large randomized, controlled studies have determined whether HRT reduces fractures. Two new trials show that it does not improve urinary incontinence in older women. These new reports are backed up by the landmark Heart and Estrogen-Replacement Study (HERS), which led the American Heart Association to reverse its recommendation that estrogen be prescribed solely to prevent heart disease and strokes. And the five-year National Health Initiative study of Prempro (note that this is a combination of estrogen and a progestin) concluded that women who took the combination drug had a higher risk of invasive breast cancer, coronary heart disease, stroke, and pulmonary embolism than those who took a placebo. We note that the estrogen skin patch and the widely used Premarin (Wyeth-Ayerst) contain estrogen only, with no progestin. To add to the confusion surrounding this topic, the Post-menopausal Estrogen/Progestin Intervention (PEPI) Trial had earlier concluded that estrogen-replacement therapy has positive effects on HDL-cholesterol levels in postmenopausal women. Further, the Brigham and Women's Hospital Nurses Health Study of 121,700 women found that estrogen dramatically reduces older women's risks of dying from heart attack, but it also increases the risk of breast cancer. The FDA has approved HRT to prevent osteoporosis but not to treat it. A published National Cancer Institute study concluded that women who took estrogen were significantly more likely to develop ovarian cancer than women not on estrogen. See Section 13.4 for a discussion of cancer risks associated with the use of birth-control drugs. Regarding **Testing for Risks and Benefits**, we again caution the reader: some clinical trials of estrogen products are performed *retrospectively* by interviewing women who have used estrogen products in the past. As broad and well designed as the trials appear to be, they prove nothing. They do offer statistical evidence on which risk is based. *Prospective* studies, which are planned to follow people currently taking drugs, require many years of record keeping but offer more substantive information, albeit still statistical.

The spate of HRT research studies, sometimes contradictory, has left menopausal and postmenopausal women concerned and undecided about the hormones they are taking. Many may need to confer with their physician and join in the decision making. As time passes, the choices may become clearer. Right now, estrogen alone appears to be a reasonably wise choice for women who have no uterus but who need relief from menopausal symptoms. Also, the use of estrogen plus progestin to prevent osteoporosis may be wise if the patient believes the benefits outweigh the risks. Women realize that hormones are not the only option. Being physically active, getting adequate amounts of calcium, and using local lubricants and topical estrogen for vaginal dryness are viable alternatives.

For more on HRT see: **`http://www.4woman.gov.HRT/index.htm`**.

HRT deals with powerful drugs. It is wise not to switch between products, change doses, or swap with friends. Potential users are thus well advised to consult with their physician to review conditions, products, and risks.

Following the bad news about HRT, the dietary-supplement industry aggressively pushed products such as Promensil, Remifemin, and Pro-Gest as nontraditional menopause remedies, in addition to a dizzying array of such herbals as red clover, soy, and dong quai.

CAUTION: *unlike pharmaceuticals, dietary supplements are subject to few government regulations; they can be marketed without proof of safety, effectiveness, or rigorous scientific research.*

By the way, male readers, the use of testosterone as a form of HRT in aging males is rising. Presumably capable of stopping or reversing diminished libido and muscle and bone loss, testosterone replacement therapy (by injection) and other uses, has soared to 1.5 million prescriptions yearly.

CAUTION: *despite the hype, the fact remains that no large-scale clinical trial has ever demonstrated that testosterone helps counter the effects of aging in males over 30 years of age. Furthermore, in prostate cancer patients, extra testosterone is anathema.*

Estrogen is believed to avert thinning of the bones (osteoporosis), a serious disease in the elderly. Some women who choose not to nurse their babies also take estrogen to reduce painful engorgement of the breasts after pregnancy.

Estrogens are indicated in the treatment of prostatic cancer in elderly males. The basis for this use is the feminizing effect estrogens have on all male reproductive system tissue, including the prostate gland. Normal and cancerous prostatic cells cease to grow under the influence of female hormones. Using the same therapeutic concept, male sex hormones (termed *androgens*) are administered for the control of inoperable breast cancer in women.

Most of the estrogenic drugs used today are prepared synthetically. However, conjugated estrogens USP (Premarin), an important product that has been widely prescribed for the menopausal syndrome, is obtained from pregnant mare urine. Although it contains several estrogens that are identical to those found in humans, its primary estrogen is estrone, which is the least dominant estrogen in the human female. Equine estrogens substitute in the human in replacement therapy for naturally occurring or surgically induced estrogen deficiency states, but there is controversy about their safety. Estrogens and endometrial cancer have been linked in various research studies (the endometrium is the inner lining of the uterus). In each of these studies, women with endometrial cancer were compared with equal numbers of controls matched for age, area of residence, intact uterus, and so on. Each study showed that the women with endometrial cancer had a greater history of estrogen use than the controls. For example, in a retrospective Seattle study, nearly half of the cancer patients were estrogen users, whereas only about one-sixth of the controls were estrogen users. This finding is not proof, but it is accepted as strong epidemiologic evidence upon which the *risk* of estrogen use is based. Note that most recently, epidemiologists have concluded that this increased risk of endometrial cancer is eliminated if the dose of estrogen is lowered to 0.625 mg daily. Also, it appears that combining progestin with the estrogen greatly reduces the risk of uterine cancer.

Wyeth-Ayerst Laboratories, the manufacturer of Premarin, publishes prescribing information that should be read by prospective users. Wyeth-Ayerst says that independent case studies show a 4.5–13.9 times greater risk of endometrial cancer in postmenopausal women exposed to exogenous estrogen for more than a year. This risk is further substantiated by the discovery that incidence rates of endometrial cancer have risen sharply since 1969 in eight areas of the United States that have population-based cancer-reporting systems. The increased incidence could be related to the growing use of estrogen in past decades. Using exogenous estrogen increases a woman's risk of gall bladder disease 2.5-fold.

The FDA, upon review of studies of estrogen use, has concluded that menopausal and postmenopausal women who take estrogens have an increased risk of endometrial (uterine) cancer. This risk is proportional to the duration of estrogen use and is particularly high with use of five years or longer. The FDA has published the following advice concerning the risks and benefits of estrogen therapy:

> "The usefulness of estrogens in treating certain symptoms of the menopause ... is well established. In most women undergoing menopause, however, if psychosomatic and anxiety symptoms predominate, these can often be managed with reassurance and, if necessary, with anti-anxiety mediations .... Estrogens are obviously used to a far greater extent and for a longer time, however, than can be accounted for by the incidence or duration of acute menopausal symptoms ... estrogen use appears to exceed by far that required for short-term management of the menopausal syndrome."

Estrogen-containing creams are available on prescription for direct vaginal application in menopausal women. Transdermal estradiol (i.e., skin patch application), approved by the FDA in the 1980s, offers advantages over the oral route, because as much as 30% of oral estrogen is destroyed in the GI tract. Also, oral estrogen is more likely to induce enzyme systems in the liver and to increase the risk of blood clots and hot flashes. Oral estrogens are taken once daily; patches are changed twice weekly. See Section 4.2, "Routes of Drug Administration," for more on skin patches.

Estrogen in skin creams and cosmetics is widely promoted as an agent to delay or correct the effects of aging, but there is no scientific evidence that you are getting your money's worth. No cosmetic on the market can reverse the effects of aging. Beware of misleading advertising in this area.

A danger exists in applying estrogen to the skin; the possibility exists that estrogen will be absorbed into the bloodstream and will be circulated to all body parts, where it can have significant side effects. For example, breast enlargement was seen in a 10-month-old girl and in a 5-year-old boy on whom estrogen hair creams had been used. An 82-year-old woman and a 54-year-old man had side effects after using a hair lotion containing estrogen, and an 8-month-old girl began to develop breasts after her diaper rash was treated with an estrogen product. Remember: Estrogens are hormones, and they are powerfully active in milligram doses. If you are using a cosmetic containing a "hormone," check it out with a pharmacist.

A new group of drugs is being developed, dubbed "**designer estrogens**," that attempts to maximize estrogen's good effects while reducing the bad. Scientifically, these drugs are *selective estrogen receptor modulators* (SERMS). One that has been approved by the FDA is Evista (raloxifene, Eli Lilly). Evista is in the same category as **tamoxifen**; both act as estrogens in some tissue, but block it in others—possibly a great boon for women prone to breast cancer.

A new dosage form for estrogen use in vaginal dryness is the **estrogen ring** (Estring). Similar to a diaphragm in size, the ring, placed in the vagina, releases estogen at low dose levels for 90 days. Little is known about long-term effects. Some women may prefer the ring to the cream.

CAUTION: *Never take any estrogen product (including an oral contraceptive) if you are pregnant, or even if you suspect you are pregnant. Sufficient evidence exists to conclude that estrogen causes birth defects. Even a few doses could be disastrous for your fetus. (For more on the contraindications of estrogen use in pregnancy, see Chapter 13, "The Pill: An Update.")*

Phytoestrogens are chemicals produced by many different plants (herbs, grains, seasonings, fruits, others) that, when ingested, mimic or interact with estrogenic hormones in animals. Phytoestrogens are not potent female sex hormones, and they are usually quickly metabolized in the body and excreted. There appears to be little good evidence that plant sources of phytoestrogens pose a health risk to humans who consume them; in fact, according to some animal studies, eating soy-based phytoestogens may confer some protection against breast, colon, prostate, or other cancers. However, it appears that soy products are no better than placebo in reducing hot flashes in menopause, and other factors must be considered before conclusions can be drawn here. Similarly, black cohosh, flaxseed oil, evening-primrose oil, and vitamin E have failed to show convincing evidence of usefulness in menopause.

I shall include at this point a discussion of an estrogen that is obtained synthetically and has developed into a controversial drug. *DES* stands for **diethylstilbestrol** USP, a product of the chemistry laboratory used in estrogen therapy for more than 40 years. In the 1940s, 1950s, and 1960s, DES was given to estrogen-deficient women in the first trimester of pregnancy and to women threatened with miscarriage or premature labor. DES was considered a miracle drug that could even promote "bigger and stronger babies," according to one drug company's advertisement. Approximately 5 million women were treated with DES. DES is now *never* used in pregnancy since the discovery in 1971 that 1 in 1,000 daughters of women who took the drug will develop cervical or vaginal cancer, and some 40–60% will develop some form of benign abnormality in their reproductive organs. The cancer typically shows up when the daughters are in their late teens or early twenties. No clear scientific evidence exists that sons of mothers who took DES during their pregnancy suffer an increased risk of cancer.[5] Medical examinations for DES-exposed sons include the requirement of checking for small or undescended testicles, conditions that predispose a male to genital cancer. Medical experts are calling for more research on possible extended effects of DES, and two consumer groups—DES Action and DES Cancer Network—offer support and information to victims.

DES is administered to cattle to promote growth, thus facilitating faster arrival on the market. Today, FDA regulations require termination of DES use long enough before slaughter that no significant quantities of DES remain in the meat sold to the consumer. It is likely that ultimately the use of DES in cattle will be totally banned.

DES has medical use today in prostate cancer and in hormone therapy. DES was formerly used as a "morning-after" pill, that is, as an abortifacient taken the morning after sexual intercourse. A large dose was given all at once, and any fertilized ovum failed to implant and develop. This use of DES is now strongly discouraged by the FDA, which has removed from the market the 25-mg tablets formerly used in the morning-after treatment. See Section 13.6 for a discussion of the use of the oral contraceptive Ovral as a morning-after pill. Prostaglandins are now also used as a morning-after drug. Preven is the first emergency contraceptive kit approved by the FDA. The kit includes high doses of two birth control pills plus a home pregnancy test.

Finally, DES is used in cases of rape or incest. The FDA approves of DES as a postcoital contraceptive in emergency situations such as rape or incest, provided that the manufacturer supplies necessary labeling information for the patient.

Chlorotrianisene (TACE) is a synthetic estrogen chemically related to DES. Chlorotrianisene does not occur naturally and is never to be used during pregnancy because of possible teratogenic effects.

Unquestionably, the prominence estrogens once enjoyed is being replaced by caution in their use. Evidence that estrogens can be carcinogenic is statistical, based on retrospective studies. Nonetheless, judicious use of estrogens is clearly indicated.

1. Menopausal women can benefit from estrogen therapy, but the physician should prescribe as little as needed for as short a period as necessary.
2. The use of estrogen in pregnancy is contraindicated.
3. The use of DES as a morning-after pill is inadvisable.
4. Women who do take estrogen for any reason should be apprised of the risks, as well as the benefits.
5. Federal law requires that the information on the package insert be made available to the patient receiving estrogen.

One type of person can take estrogen and never worry about subsequent cancer of the female reproductive organs. Transsexuals of the male-to-female type are given estrogen to develop breasts, round out their figure, grow softer hair, and so on. Combined with the necessary surgical procedures, this therapy can work well,

[5]However, in 1985 a U.S. District Court jury made the first award to a male who claimed genital damage and emotional trauma because his mother used DES while she was pregnant with him.

as indicated by statements made by Christine Jorgensen in an interview some years ago. The use of estrogens and progestins as contraceptives in the Pill is of great importance. Chapter 13, "The Pill: An Update," is devoted to a discussion of the Pill and of the risks and benefits associated with its use.

2.5.7 Miscellaneous Animal-Source Drugs

Some drugs bear a unique relationship to their animal source. Duck and chick embryos, for example, are growth media for biological preparations (viral antisera, antivenins, and vaccines). Horse blood is a source of antibodies produced when the immunologic response system of the horse is challenged by an antigen. Tetanus toxoid is prepared this way, as are equine antivenins.

Human blood fractions give us the gamma globulins, which are used as injections when immediate emergency protection is required against the hepatitis virus or German measles. In the latter case, if a pregnant woman who has never had German measles, and therefore has no antibody titer against the virus, is exposed to the disease (especially in her first trimester), it is imperative that she be given gamma globulin. In this way, she obtains a blood level of antibodies to the measles virus and is protected against the known teratogen.

Human blood is also collected and treated to isolate factors involved in blood coagulation. Hemophiliacs depend on exogenous blood coagulation factors obtained in this manner. Note that a quantity of blood can be removed from a person's system, the clotting factors taken out, and then the red cells and other formed elements *returned* to the vascular system of the donor within an hour. Under this system, individuals can donate blood once a week.

Liver injection is a water-soluble extract of mammalian livers that contains vitamin B_{12}. This injection is useful in cases of pernicious anemia. A protein called *intrinsic factor* must be present if the anemic patient is to properly utilize vitamin B_{12} administered by the oral route. Intrinsic factor from dried hog stomach and duodenum can be taken if the human supply is deficient.

Parathyroid Injection USP is obtained from cows' parathyroid glands. Parathormone, the active hormonal agent in this preparation, is vital to the control of calcium levels in the body.

Poisonous plants and animals are the source of useful drugs in modern medicine. Curare, the South American arrow poison (Section 2.3), is the classical example of a muscle relaxant, and viper venom contains an anticoagulant that is the prototype for Aggrastat. The skin of an Ecuadorian tree frog yields a potent pain killer.

Boston's Biopure Corporation makes Hemopure, a hemoglobin product from cattle. Human and cattle hemoglobin are remarkably similar.

2.6 Drugs from Gene Splicing

In Section 2.5, I described the process in which insulin is obtained by gene-splicing techniques (recombinant DNA technology). It was mentioned that a portion of human DNA that codes for the production of a certain protein substance can be taken from a human cell and incorporated into the genetic makeup of a host bacterial cell.

From that time on, the host cell will synthesize the human protein substance as though it were its own (which it now is). Hence, we have a new method for the unlimited production of certain hard-to-get drugs.

Insulin is only one of many protein drugs that can be obtained by genetic engineering. hGH, interferon, a vaccine for hoof-and-mouth disease, hCG, a vaccine against hepatitis B, lipocortin (a human anti-inflammatory protein), interleukin-2, tumor necrosis factor, erythropoietin, renin (an enzyme in cow's stomach used in cheese making), *tissue plasminogen activator* (TPA), and somatostatin have all been made by gene-splicing techniques. Factor VIII, a protein required in the blood-clotting process, has also been made by gene splicing; this is of great importance to hemophiliacs. Kogenate FS is a recombinant factor VIII product distributed directly to hemophilic patients. Another trade name for this product is Helixate FS.

The FDA has approved Monsanto's genetically engineered *bovine somatotropin* (BST), a hormone that increases milk production in cows. Genentech has approval to market gene-spliced Pulmozyme, a form of human DNase for treating cystic fibrosis. At present, more than 100 genetically engineered drugs are in clinical trial; another 20 are ready for FDA approval and a dozen more are already on the market.

Only protein drugs, that is, molecules composed of amino acids joined in peptide bonds, are obtained from gene splicing. This statement has to be true, because the genetic material (DNA) that we incorporate into the host organism can only direct the arrangement of amino acids into protein. To make epinephrine by gene splicing, for example, we would have to provide the host organism with the human DNA that codes for the synthesis of an adrenal gland. (This process we cannot do, at least not yet. However, a patent has been issued for a genetically engineered mouse.)

The potential for recombinant DNA technology is staggering. We can anticipate great advances in drug synthesis, weapons against disease, food crops that will grow anywhere, unlimited hormone supplies, and animals modified to optimal capacity. Disadvantages, while few, might be significant: Recombinant insulin (Humulin-R) costs about one-third more than porcine insulin, and great care must be taken to ensure the purity of the product—because bacterial proteins carried over to the drug product could induce antibody formation in the patient.

2.7 The Future of Drug Discovery

The way new drugs are discovered is rapidly and dramatically changing. Traditionally, chemists would synthesize many drugs, some planned for a disease, some not, and pharmacologists would test these compounds in **drug screens**, using microorganisms or animals. This is slow, expensive, and usually hit or miss. More recently, computer-aided molecular design (CAMD) was developed to analyze variables and ideas, and suggest structures likely to be active in the particular chemotherapeutic problem under study. Another recent concept is **combinatorial chemistry**, in which chemists synthesize thousands of closely related drug compounds, the ideas for which have come from computer programs. A high-throughput technique is then employed to rapidly assay thousands of compounds to find the top candidates for further testing. With combinatorial chemistry, researchers can efficiently search a "haystack" of compounds to find the "needles' (promising candidates) to be tested further. Pharmacognosists have developed ideas for synthetic drugs by examining the structures of **naturally occurring agents** in plants and sea life. For example, the story of the natural drug quinine gave impetus to the successful search for similar synthetic analogs. Additional examples are:

Reserpine and the synthetic antipsychotic drugs
Morphine and synthetic analogs
Curare and synthetic neuromuscular blockers
Progesterone and the anovulatory progestins
Estrogens and DES
Adrenal cortex steroids and synthetic anti-inflammatory drugs
Adrenaline and the amphetamines
Belladonna alkaloids and synthetic analogs

Today the future of drug discovery is quite different. It now lies in computers, the human genome, bioinformatics, pharmacogenomics, and proteomics. Many large pharmaceutical companies are investing multimillions of dollars in these disciplines. Harvard University has developed an institute of proteomics.

Some needed background: The Human Genome Project (HGP)—now essentially complete—mapped, sequenced, and annotated the human genome. A genome, whether it be human, mouse, fruit fly, worm, or yeast, constitutes all the twisted, double-stranded DNA that is responsible for making every cell, and for building and operating a fully functional organism. In mapping the human genome, scientists determined the actual sequence of the 3.1 billion base pairs in human DNA. The grand sequence of these bases—adenine (A), thymine (T), guanine (G), and cytosine (C)—spells out our estimated 35,000 genes. The genes, strung out along our 46 chromosomes, carry instructions for making all 30,000 body proteins—organs, enzymes, hormones like insulin, endorphins, and tissues, among many others. Proteins in human cells are exceedingly important, because they conduct a cell's biochemical reactions and act as messengers between cells. The genome mapped by teams of researchers from across the world is a "consensus" genome from a number of human samples. Scientists believe that 99.9 percent of human genes perfectly match us all, but the remaining 0.1 vary from one human to another. You may have a T in one of your sequences in a position where I have a C. Variations can be benign or disastrous, as in the case of Tay-Sachs disease, Down syndrome, hemophilia, or PKU. Because of these tiny genetic variations, many drugs work on only 30–50% of the human population. Because of genetic variations, some people can be "fast" metabolizers of a drug, others "slow" metabolizers; the difference can be as much as twenty-fold. Further, a drug that is beneficial in one person may actually poison another. Scientists at Iceland's deCODE genetics, Inc., have assembled a highly detailed reference map of human genetic variations, based on 146 Icelandic families. They plan to hunt down genes that account for inherited diseases. French families have also been studied for genetic variations linked to familial diseases. Beyond these physical databases, millions of human cell cultures or "living" genome repositories preserve unique genotypes, allowing for the study of planned drug design.

Following is a discussion of the concepts that will drive the future of drug discovery. **Bioinformatics** is the application of computer technology to the management of information in molecular biology. It is a fast-growing area of computer science applied to prediction of protein function and form, and to the design of drugs. Bioinformatics creates computer-generated, simulated protein structures that are like those from a given human gene. This helps the researcher learn about the protein the human gene encodes, and—if it is a disease protein—search for drugs to block it. In other words, if we can understand the genetic basis of a disease, then we can predict what protein it produces and look for appropriate drugs. Bioinformatics is being used to study the MLHI gene involved in human colon cancer. A new challenge facing bioinformatics is automating not just molecular-biology data, but also that of gene and enzyme names, structural proteins, diseases, and physiological pathways.

The discipline of **proteomics** was given impetus by completion of the HGP. Now the emphasis has shifted from DNA to the protein complement of the human body. Thus proteomics is the study and cataloguing of all human proteins. One researcher defined proteomics as "the analysis of complete complements of proteins present in defined cell or tissue environments and their variations in time and space." The goal for scientists is to be able to

- identify all body proteins (some 30–50% of our proteins remain unknown).
- identify proteins present in cancer cells but not in normal cells.
- determine a protein's three-dimensional structure, with the goal of locating sites at which the protein is most vulnerable to drugs.
- determine how proteins interact (network) with each other to accomplish special cell tasks.

A consortium of research companies plans to elucidate the complete human proteome in three years. Actually, however, the human proteome is indefinite. It differs slightly from person to person (because DNA can differ) and at times even in the same person. For example, researchers are now able to compare the patterns of proteins present in the blood serum of patients *with* and *without* ovarian cancer. The National Institutes of Health is developing a database to catalogue and predict structures of proteins.

Proteomics has a long way to go. The drugs we have presently target only about 400 proteins, representing only a fraction of the human genome. Receptor sites, nuclear receptors, and metabolic enzymes constitute the great majority of present drug targets. Research is now under way to use proteomics to find targets for potential cardiovascular disease drugs, and for ovarian and breast cancer.

Pharmacogenomics (or pharmacogenetics), as the name implies, is the application of genetics to pharmacology. A new field of study, pharmacogenomics is designed to help physicians determine drug regimens for individual patients by identifying genes that control drug reactions. Researchers study how genetic variations affect the way patients respond to drugs—that is, how genetic differences are manifested in drug targets (receptor sites) or in the 400 enzymes that catalyze human drug metabolism. If you and I have inherited different drug targets, we might differ in how we respond to a drug. If we have inherited different metabolizing enzymes, we may experience different drug efficacy or toxicity. Ultimately, the goal of pharmacogenomics is to tailor pharmaceutical treatments to an individual's specific genetic profile ("personal genomics"). The basis is the person-to-person variation that could be used to "personalize" chemotherapy. Each of us might learn from the string of 3 billion base pairs in our DNA what diseases we are susceptible to. Can you imagine your physician someday having your personal gene map on file in his office, or you carrying a genetic profile card in your wallet?

Pharmacogenomics is still in its infancy, but it holds promise as one way to get the right medicine to the right patient at the right time. Here is a hypothetical example: A person is diagnosed with squamous cell cancer. Scientists recognize that the squamous-cell-cancer group can consist of five or six different genomic (genetic) classifications of disease, each with a different outcome and each with a different treatment. For pharmacogenomics to work in this example, DNA "markers" must be found that indicate the connection between the patient's genetic makeup and the drug to be used. A marker is an inherited or acquired variation in DNA at a single base (A, T, G, or C), Right now, genomics is being applied to study patients' responses to statin drugs for lowering cholesterol and to predict asthma patients' responses to albuterol. We note that genomics will not replace physician judgment but can make a difference in the trial-and-error process. Pharmaceutical companies are now developing gene tests to be prescribed with their new drugs when these drugs come on the market (in about three years). Physicians would use similar gene tests before prescribing existing drugs to their patients. Roche Pharmaceuticals is working on a genetic test for rheumatoid arthritis patients who should not use the oft-prescribed methotrexate as a treatment but instead select a more aggressive therapy.

To help you better understand new approaches to drug discovery, here are a few **helpful internet Web sites**. For help in defining terms and and understanding procedures, consult these glossaries:

http://www.genomicglossaries.com/content.omes.asp
http://genomics.pharma.org/lexicon/index.html
http://www.nhgri.nih.gov/DIR/VIP/glossary.

A general reference for bioinformatics is

http://www.bioplanet.com/whatis.html.

For proteomics, browse **http://www.e-proteomics.net/**.
For the Human Genome Project, see **http://www.ornl.gov/hgmis/**.

Today, we hear about **natural, organic**, and **synthetic drugs** and the possible differences among them. If a drug is obtained from a plant or an animal, it can be termed *natural*. If a plant is grown without the use of synthetic chemical fertilizers or synthetic chemical pesticides, its products can be said to be *organic*. Although advertising copywriters have used the terms *natural* and *organic* to help sell a lot of vitamins, dietary aids, and drugs, it must be recognized that every substance has a fixed chemical structure and many substances have been duplicated by synthesis in the laboratory. What is more, there is no chemical variation or pharmacological inconsistency between a drug obtained naturally and the same drug obtained synthetically. Vitamin C from rose hips is the same as vitamin C obtained by laboratory synthesis from sorbitol. Choline from natural sources is identical to synthetic choline. Some people may find it important that no fertilizer or pesticide was used in the cultivation of the plant that supplied their drug, but the natural drug itself cannot be inherently better than the synthetic version. And the synthetic version is likely to be far less expensive.

2.8 The Search for Anti-AIDS Drugs

The devastating epidemic we call acquired immune deficiency syndrome (AIDS) began in 1981 when the U.S. Center for Disease Control and Prevention reported on an unusual viral disease in gay men. Now, some two decades later, 58 million people worldwide have been infected with human immunodeficiency virus (HIV), and 25 million have died after their immune system was attacked by the virus, exposing them to some opportunistic infection or disease, including cancer, tuberculosis, and pneumonia. Up to 900,000 Americans (and 40 million worldwide) are infected by HIV and have AIDS or AIDS-related complex (ARC). An estimated 350,000 Americans have died of the disease.

As with the common cold virus, HIV is a submicroscopic parasite consisting of a core of RNA wrapped in a protein coat. It replicates by entering and attacking a human cell (Figure 2.10). Because they enter the cell, viruses are much more difficult to treat with chemotherapy than bacteria, which replicate outside the human cell.

HIV attacks the human immune system by destroying CD4 lymphocytes—white cells that are essential to our immune system response to invading pathogens. HIV spreads rapidly in the body: up to 10 billion new HIV virus particles can be replicated daily in an infected human.

The AIDS virus is a sneaky, insidious agent. It produces strains that are resistant to chemotherapy, seriously complicating treatment of the disease. A recent national study found that more than half of the people receiving HIV/AIDS care are infected with strains of virus that resist one or more antiviral drugs. The study found that three-fourths of the 1647 persons sampled (78 percent) had resistant strains. Of these, 70 percent were resistant to the older reverse transcriptase inhibitor drugs such as AZT, and over one-third were resistant to the newer class of protease inhibitor drugs. Resistance also developed quickly to the newest nonnucleoside reverse transcriptase inhibitors. Many clinicians now require that virologic resistance testing be done so that they can tailor drug regimens to fit lifestyles of their patients.

Today chemotherapy of HIV/AIDS is based on three types of antiviral drugs: the **nucleoside reverse transcriptase inhibitors (NRTIs)**, the **nonnucleoside reverse transcriptase inhibitors (NNRTIs);** and the **protease inhibitors (PIs)**. Note that the "ase" suffix means that we are dealing here with enzymes—specifically enzymes that are crucial to cell replication in AIDS. Around 1996, use of **combination 'cocktail'** drugs introduced a new treatment era in the battle against AIDS. The triple-punch (or "three-drug cocktail") combination of indinavir, AZT, and 3TC has been among the most potent treatments of HIV infection. Trizivir is also a three-in-one combination pill designed to cut down on dosing. At first very successful, "mix and match" drugs are becoming less effective as viral drug resistance develops. Chemotherapy for AIDS can cost $10,000–$12,000 a year for drugs alone.

NRTI drugs such as zidovudine (Retrovir) are synthetic nucleoside analogs—that is, bogus molecules that are so similar to critically needed viral substrates that the key viral enzyme (reverse transcriptase) mistakes it for the true substrate and, as a result, is inhibited by it. Think of it this way: A car is built with an engine that fits perfectly under the hood but has no cylinders. The car is not going anywhere; neither is the virus. See Section 4.5, "Metabolism of Drugs" for more information on competitive inhibition. Unfortunately, Retrovir can have serious side effects. Other NRTI drugs are lamivudine (Epivir, 3TC), didanosine (Videx, ddl), zalcitabine (ddC), stavudine (Zerit, d4t), and abacavir (Ziagen).

Tenofovir (Viread) is a nucleotide RTI, very similar to the nucleoside drugs. However, nucleotide compounds differ in that they are chemically preactivated and thus require less biochemical body processing to become biologically active. Viread

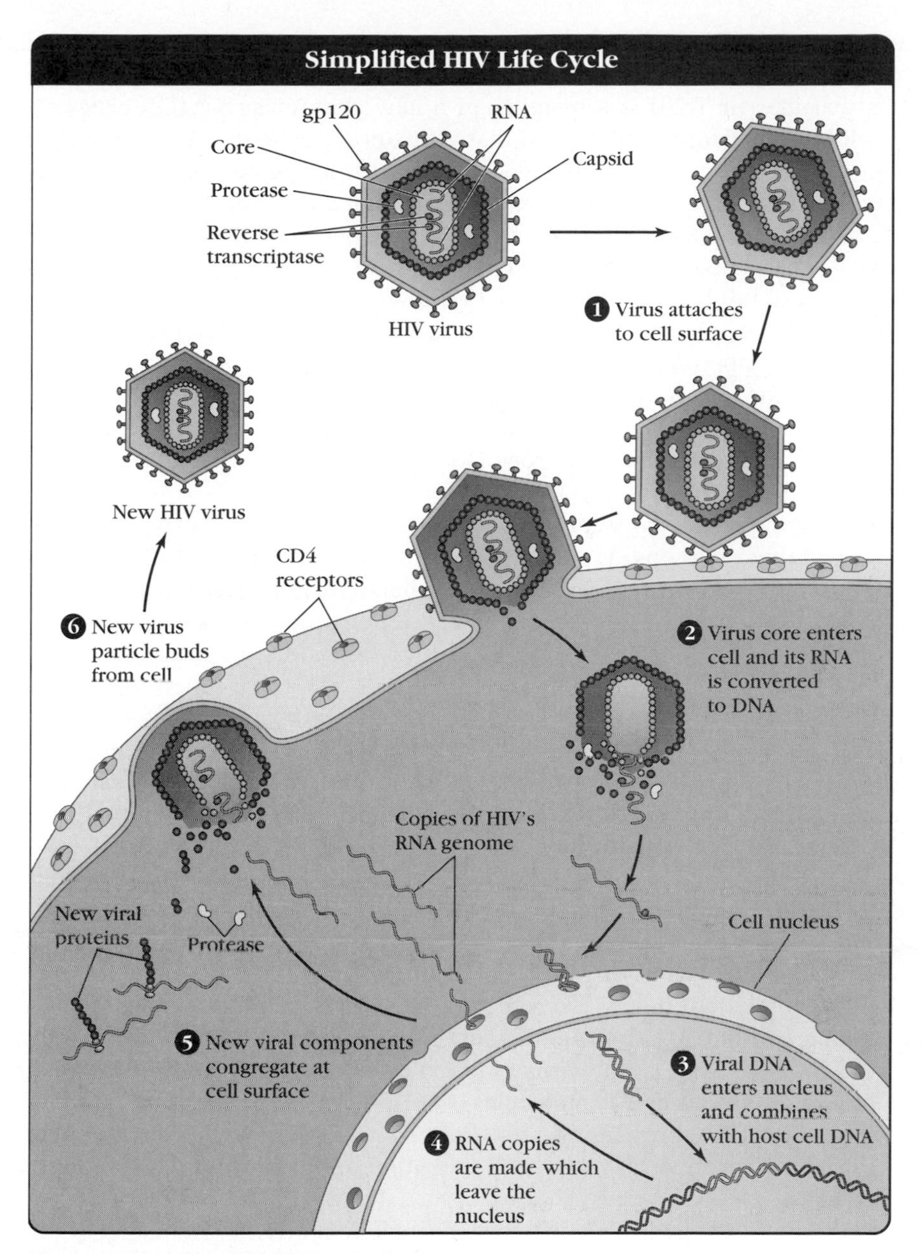

Figure 2.10 Simplified HIV Life Cycle.

must be used in combination with other drugs including NRTIs and at least one PI or NNRTI. For more, see **`http://www.aidsmeds.com/drugs/`**.

Approved by the FDA in only 119 days, nevirapine (Viramune) was the first nonnucleoside RTI drug; it is used in combination with other AIDS drugs. Among the 16 drugs in this category are Dupont's efavirenz (Sustiva) and delavirdine (Rescriptor). Serious drug side effects are seen.

Protease inhibitor drugs such as indinavir (Crixivan), ritonavir (Norvir), and saquinavir (Invirase) act to disrupt HIV's life cycle by blocking the production of protease, an enzyme essential to the virus's survival. This in turn blocks the escape route of the virus; it cannot receive a protein coat and leave the host cell. (See Figure 2.10.) Studies have shown that St.-John's-wort interferes with the action of protease inhibitors.

Enfuvirtide (or T-20) is a member of a new class of anti-AIDS drugs called **entry inhibitors** (or fusion inhibitors) that act by preventing the HIV from entering the blood cells that it kills. T-20, given FDA fast-track review, is administered in combination with standard anti-AIDs drugs. It is not a cure for AIDS.

The AIDS virus hibernates in the body; the vast majority of people infected show no clinical signs of disease. Thus, there is a need to develop drugs that will prevent the development of full-blown AIDS or ARC. Such drugs are not now available. We do not know if they would have to be taken for the life of the infected person.

Yet another approach to treatment of AIDS deals with restoring the patient's damaged immune system. Scientists are now investigating immune modulators—drugs that can reinvigorate the weakened immune defenses. Ampligen and interleukin-2 control the body's production of interferons. Isoprinosine and alpha-interferon have been investigated for their immuno-stimulatory effects.

Drugs are available to treat the opportunistic infections that beset AIDS patients. Injectable pentamidine and trimethoprim-sulfamethoxazole (Bactrim) are standard treatments for *Pneumocystis carnii* pneumonia, the leading cause of death in AIDS patients. Trimetrexate, originally developed as an anti-cancer agent, is also used to treat this pneumonia, although the FDA considers the drug highly experimental and has tried to restrict its use to patients who have failed to respond to more proven drugs. Trimetrexate shows serious adverse reactions. Ganciclovir (Cytovene) treats an eye infection from cytomegalovirus, a kind of herpes virus. Foscarnet (Foscavir) is an FDA-approved alternative to ganciclovir. Recombinant human alpha-interferon (brand names Roferon-A and Intron-A) is useful in treating Kaposi's sarcoma, a skin cancer often contracted by AIDS patients. Fluconazole (Diflucan) is useful in treating candidiasis, a yeast infection of the mouth, and cryptococcal meningitis, a bacterial brain infection. Erythropoietin (Procrit) alleviates some of the anemia resulting from AZT therapy.

A different approach to AIDS is the drug CD4, a synthetic version of the CD4 glycoprotein that normally occurs on the surface of T4 helper cells, a type of white blood cell that is especially vulnerable to infection by the AIDS virus. To enter the immune system, the virus binds to the CD4 glycoprotein. The idea behind the use of synthetic CD4 molecules is to flood the body with decoys to which the virus will attach, preventing the virus from attacking the CD4 "doors" on cells. Synthetic CD4 does not kill the virus; rather, it only inhibits its spread to uninfected cells.

Since chemotherapy has not solved the problem of AIDS, vaccine development is being given high priority. But here again, there are formidable obstacles. The AIDS virus shows great genetic variability. Many strains are known, each with a different outer protein coat, which is the part usually recognized and attacked by the antibody in the vaccine. Nevertheless, animal studies show that a vaccine is at least theoretically possible in humans.

Because we have no cure for AIDS, either with a chemical or with a vaccine, prevention is the only effective attack on this disease. Prevention includes educating intravenous drug users not to share their hypodermic syringes, because this practice appears to be strongly associated with the spread of the AIDS virus. Drug abusers who frequent "shooting galleries" are at high risk of AIDS exposure. Also, more than 75% of "secondary" AIDS cases involve people who have had sexual contact with an

IV drug user and children whose parents use drugs intravenously. IV drug users are strongly urged to soak their syringes in household bleach or alcohol or to boil the entire syringe in water for 15 minutes. Procedures such as these are believed to kill the AIDS virus, as well as any other microorganisms. Because marijuana has been implicated in a lowered immune response, it is possible that its use predisposes a person to infection with AIDS. Other drugs similarly implicated (but not proven) are amyl and butyl nitrite and alcohol.

The National Institute of Allergy and Infectious Diseases (NIAID) conducted a national, multicenter study on the use of illicit drugs by HIV-infected pregnant women. The study, known as the Women and Infants Transmission Study (WITS), followed 530 HIV-infected women, and found a significantly higher risk of transmitting HIV to infants than in the case of HIV-infected women who did *not* use drugs while pregnant. The most commonly used drug was cocaine, used singly or in combination with other drugs.

CAUTION: *The threat of AIDS and its consequences can motivate the patient to try any drug or substance purported to be a cure. Unscrupulous dealers have treated AIDS victims with snake venom, enemas, umbilical cord extract, thymus extract, acupuncture, garlic, herbs, megadose vitamins, laetrile, amino acids, zinc, selenium, meditation, and colostrum. One AIDS patient is taking 100 pills a day; another spends $600 a month for food supplements, $300 for acupuncture sessions, $250 for colonics, and $500 for sessions with a holistic doctor. Many patients travel to Mexico to purchase investigational drugs not sold in the United States. Although no one can advise a patient whose life is threatened to avoid a purported cure, one can advise that unscrupulous dealers are all too ready to fleece the easily deceived person with bogus alternative therapies. Furthermore, we must consider the power of the placebo effect when we take any drug (see Section 1.3, "The Placebo Effect").*

Web Sites You Can Browse on Related Topics

ATTENTION INTERNET USERS: In addition to the URLs listed below, you can search for topics as suggested. A search can yield numerous sites, but careful evaluation may be necessary.

General Information About Many Drugs
http://www.ncjrs.org/htm/chapter2.htm
http://www.drugfreeamerica.org/

AIDS-Related Drugs
http://www.fda.gov/bbs/topics/answers/ANS00800.html

Also Search for "AIDS drugs"

Ma Huang
http://www.ephedra.net/

Diabetes-Insulin
Search for "Eli Lilly Company"

Magic Mushrooms, Psilocybin
Search for "hallucinogens"

Study Questions

1. In what ways did early humans use plant drugs?
2. Name four drugs that were discovered by modern medicine because they had achieved notoriety in folk medicine.

3. Define (**a**) active ingredient, (**b**) pharmacology, (**c**) CNS, (**d**) hypoglycemic, (**e**) edema, (**f**) toxicity, (**g**) schizophrenia, (**h**) analgesic, (**i**) uterus, (**j**) vasoconstrictor drug, (**k**) cretinism, (**l**) gonadotropin, (**m**) menopause.
4. Match the drug in the left-hand column with its pharmacological action in the right-hand column:

a. Digitalis	**1.** Narcotic analgesic
b. Atropine	**2.** Bronchodilator in asthma
c. Reserpine	**3.** Inhibitor of acetylcholine
d. Ipecac	**4.** Emetic
e. Cocaine	**5.** Heart stimulant
f. Opium	**6.** Anti-cancer drugs
g. Curare	**7.** Migraine headache treatment
h. Vinca alkaloids	**8.** Antihypertensive
i. Ergot	**9.** Induces paralysis
j. Ephedrine	**10.** Local anesthetic and CNS stimulant

5. Medical records show a dramatic drop in the number of institutionalized mental patients, beginning in the mid-1950s. What drug helped accomplish this, and how did it affect the mental patients?
6. What is meant by the term *alkaloid*? Where do we obtain alkaloids? Give three examples of alkaloids.
7. Identify the plant source and name two alkaloids used successfully in the treatment of leukemias.
8. Through the centuries, *Cannabis sativa*, the plant source of marijuana, has been used commercially as well as abused as a euphoriant. Cite two commercial or practical uses for the cannabis plant.
9. Some states are decriminalizing the use of marijuana, but not of hashish, bhang, or ganja. Why not?
10. Table 2.5 lists various estrogens marketed today. Identify those that (**a**) are synthetic; (**b**) occur as such in the human; (**c**) are isolated from mares' urine; (**d**) are used in a commercial skin patch.
11. In your opinion, is it proper to state that insulin from beef, pork, or gene splicing is a cure for human diabetes mellitus? Is Nicorette® gum a cure for nicotine addiction? How could a true cure for diabetes be achieved?
12. Human hormones circulate everywhere in the bloodstream, but they produce their effects only on specialized body tissue that is capable of responding to them. We term this tissue the *target organ*. What is the target organ for hGH, FSH, LH, and hCG?
13. State the relationship by matching the item in the left-hand column with the name in the right-hand column. Items in the left column may have more than one match:

a. Peyote	**1.** THC
b. Cannabis	**2.** Psilocin
c. Mexican mushrooms	**3.** Mescal button
d. Cocaine	**4.** Brompton Cocktail
e. Digitalis	**5.** Cardiac glycoside
	6. Hashish
	7. Mescaline
	8. Sinsemilla

14. **a.** Define estrogen.
    b. Name the three naturally occurring estrogens.
    c. Why could there be a need for estrogen in menopause?
    d. Explain the controversy about the safety of estrogens such as Premarin.

15. Briefly, and in simple language, explain what is meant by gene splicing. List four drugs that have been prepared by this technique.

16. *Student project*: Prepare a table of all the plant-source drugs mentioned in this chapter. List their botanical source, plant part used, active ingredient, medicinal use, cautions, and possible abuses.

17. True or false:

a. The agent that causes AIDS is a bacterium.
b. Medically speaking, it is easier to cure bacterial infections than viral infections.
c. At present, there is no cure for AIDS, but an effective vaccine has been developed.
d. Retrovir (AZT) is an anti-AIDS drug that controls the replication of the virus but does not cure.
e. AZT works by fooling the virus into accepting it as a genuine substrate needed for viral growth and replication.
f. One of the adverse effects in the use of Retrovir is anemia.
g. ARC is an acronym for "adiabatic respiratory control," a new method of treatment of AIDS.

18. Syrup of ipecac is widely used to induce vomiting in children who have swallowed a poison. However, the induction of vomiting is clearly contraindicated with certain poisons. Name some of these poisons.

19. Human growth hormone is used to stimulate growth in children who are deficient in the hormone. Now, proposals are made to use it also in short children who are not deficient in the hormone, just to make them more socially acceptable. The argument is that short children (especially boys) are stigmatized by deviant appearance and thus have a low self-esteem. Would you request hGH for your child, just because he or she is short? Why or why not?

20. Do you agree it is fair and legal for a research company to patent a human gene—a gene that may be found only in your body, or only in a small number of humans? Should a company claim rights to human DNA? Who "owns" the particular gene? Who spent the research dollars to elucidate it?

21. After reading Section 2.7, offer an explanation for the fact that a given drug can be toxic in one person but completely safe in another.

CHAPTER 3

Federal Laws: The FDA and Drug Testing—Penalties for Illicit Use

Key Words in This Chapter

- Food, Drug, and Cosmetic Act
- Carcinogen
- Quackery
- Federal and state laws
- Teratogen
- NDA
- Controlled drug
- FDA

Learning Objectives

After you complete your study of this chapter, you should be able to do the following tasks:

- Give the origin and need for the Food, Drug, and Cosmetic Act and the FDA.
- Explain the terms LD_{50} and drug lag.
- Define carcinogen and teratogen.
- Explain the five schedules of drugs under the Controlled Substances Act.
- Tell where to find penalties for illicit use of controlled substances.

3.1 The "Good Old Days"

Mail-order morphine. OTC coca leaves. Cocaine nasal sprays. Homegrown opium poppies. These items were routine occurrences in America at the turn of the century and were as legal then as selling aspirin is today.

In most states in 1900, no laws existed to control the sale of narcotics, to govern claims made by the purveyors of patent medicines, or to regulate devices advertised and sold as cures for human illnesses. Hoax, fraud, and deception ran rampant in America. More than 50,000 different patent medicines were sold in 1906, many of which were advertised as cures for tuberculosis, mental illness, dysmenorrhea, and even cancer (see Figure 3.1). Others claimed that they would "renovate" the stomach,

Figure 3.1 These "patent medicines" were a constant danger to millions who believed their false promises but had little else to rely on for serious ailments. Typical products claimed to "renovate" the stomach, liver, and kidneys, and to cure cancer, diabetes, gallstones, and weak hearts.

liver, and kidneys. Teething babies were given "Mrs. Winslow's Soothing Syrup." The babies became manageable, indeed, as would any child given daily doses of opium.[1] This era was the era of "Kick-A-Poo Indian Sagwa" and "Warner's Safe Cure for Diabetes."

Physicians commonly prescribed opium for menstrual or menopausal complaints or "female troubles." No wonder that two out of three morphine addicts in the 19th century were women. In those "good old days," no controls or restrictions existed on claims for curative powers, and no efforts were made by any agency to force the manufacturers to prove the safety of their products.

Phony devices were sold door-to-door, and the naive public fell prey. One claim was that your arthritis would be cured if you purchased a black box with two external wires. One wire was to be wrapped around the arthritic limb, and the other wire was to be dipped into a pail of water. Thus, the energy in the water was transferred to the limb, and a cure was achieved. Inspection revealed that the box contained nothing but air.

Correction of these deceptive practices did not come quickly. The first major step occurred in 1906 when Congress passed the first Food and Drug Act. This act dealt more with crude and impure food additives than with drugs, and shady practices in

[1]In 1886, one physician stated, "There is a triad of infant murderers, and their names are Godfrey's Cordial, Paregoric, and Mrs. Winslow's Soothing Syrup."

the area of medications continued. In the mid-1930s, a weight-reduction product was marketed that, unbelievable as it seems today, contained nitrophenol as the active ingredient. People who used it lost weight, and some lost their lives. Today, we recognize nitrophenol to be a highly toxic substance. In 1937, a small drug company marketed a solution of a sulfa drug for oral treatment of bacterial infections. Diethylene glycol was used as a solvent in this product (the same substance we use today as automobile antifreeze). A total of 107 patients died from the poisonous effects of ethylene glycol, many of them children being treated for sore throats. In the fury of criticism that followed, the firm's chemist (who formulated the elixir) committed suicide, but the firm was not prosecuted because before 1938, there was no law prohibiting the sale of dangerous, untested, or poisonous drugs. This fiasco and others, such as a "new and improved eyelash dye" that caused blindness in one user, prompted Congress to amend the 1906 Food and Drug Act.

3.2 The Federal Food, Drug, and Cosmetic Act

The 1938 **Food, Drug, and Cosmetic Act** was a major attack on unsafe and deceptive drug promotion in America. The act created the modern Food and Drug Administration (FDA) and required manufacturers to furnish the FDA with proof of the drug's safety in humans and to submit a *New Drug Application* (NDA) for approval by that agency. It authorized factory inspections. For the first time, sale of devices was regulated by a federal law. Drug effectiveness, however, was less of a concern than drug safety, and only occasionally was effectiveness questioned or an exaggerated claim challenged. Another weakness in the 1938 act lay in the provision that if the FDA took no action on an NDA within 180 days, approval became automatic.

A horrific discovery in 1961 helped motivate Congress to tighten existing drug laws. Thalidomide was introduced in 1958 by the German firm Gruenenthal Chemie, Ltd., as one of the safest sedatives and sleep inducers, producing normal, refreshing sleep with no morning-after effects. It was sold to prevent nausea in pregnancy, and was advertised as suicide-proof and "completely harmless." What is more, it was sold in Europe without a prescription. Thousands of pregnant women took thalidomide in Europe, Australia, and Japan. But in the next 3 years, there came from these countries reports of rare birth defects in the form of seal-like flippers in place of arms and legs. There were 12 cases in 1959, 83 in 1960, and 302 in 1961. The array of unusual deformities was called **phocomelia**. Working independently, physicians in Germany, Australia, and Scotland showed the link between the phocomelia and the use of thalidomide in the first trimester of pregnancy. At the end, some 8,000 babies in 20 countries were afflicted, some born with flipper-like limbs, others without ears or eyes, and some deaf and blind, to live their lives in hopeless, dependent isolation. Some thalidomide-damaged babies were born years after the drug was banned; their mothers had saved the drug in the medicine cabinet long after the drug was unavailable in pharmacies.[2]

[2]Ironically, the dextrorotatory enantiomer of the drug is effective against morning sickness and is safe to use. The pharmacologically inactive levoratory form is the teratogen. As sold, thalidomide was the racemate, containing both forms.

Tragedy was narrowly averted in America when the FDA refused to approve the request to market thalidomide here. Dr. Frances O. Kelsey, the FDA medical officer who reviewed the safety data, delayed approval on grounds of disturbing suspicions about the drug's toxicity. Her actions averted untold calamity, a fact duly noted later by the President of the United States. Ironically, a large proportion of phocomelia babies were born to wives of American physicians whose husbands had received samples of the drug from the American firm trying to get approval to market it in the United States. Other Americans obtained the drug while traveling abroad.

That teratogenic substances such as thalidomide damage fetal anatomy but not genetic makeup is apparent from the fact that numerous thalidomide-damaged women have given birth to relatively normal babies. In 1986, for example, an armless, legless woman (thalidomide baby) gave birth to a normal 7 1/2-pound girl. Today, thalidomide continues to be used to treat leprosy, to help heal severe mouth ulcers in AIDS patients, and as an immunosuppressive agent in organ transplant rejection cases.

Although thalidomide was never marketed in the United States, the narrow escape stimulated Congress to amend the 1938 act. The Kefauver-Harris Drug Amendments of 1962 strengthened new-drug clearance procedures and ensured a greater degree of safety. Also, for the first time, drug manufacturers were required to prove to the FDA the effectiveness of their products before marketing them.

Nearly 4 decades after thalidomide was banned in ignominy, the FDA approved it for use in leprosy patients. Other possible uses include treatment for brain cancer, lupus, Kaposi's sarcoma, and certain forms of blindness. However, the thalidomide Victims' Association of Canada and others are strongly opposed to any "rehabilitation" of thalidomide.

Over the years, Congress continued to amend and strengthen the 1938 act. A 1965 amendment gave the FDA power to seize illegal supplies of controlled drugs, to serve warrants, to arrest violators, and to require all legal handlers of controlled drugs to keep records of their supplies and sales.

Devices were now scrutinized as the FDA accumulated knowledge of extensive deception of the public. The following items were a few of the gadgets:

- The Zerret Applicator was said to produce "Z-rays" that would expand the atoms of the body, thereby curing all disease. Users were warned not to cross their legs while using the device, because that action would cause a "short circuit" in it. More than 5,000 were sold at $50 each.
- The Spectrochrome, hailed by its inventor as a device that would cure all diseases, projected colored light that had been passed through water. The inventor wrote a textbook on spectrochrometry and explained how it would restore "radio-emanative equilibrium" in the body if the time of treatment was selected based on the rules of astrology. A cult of believers 9,000 strong paid dues to the inventor and took "treatments" from him.
- The Chiropra Therapeutic Comb, imported from Germany, was said to treat diseases if it was used to scratch the skin in a certain area for a certain disease.
- In the 1920s, a system of disease diagnoses called Radionics utilized dried blood specimens from patients. Send in your dried blood, and back would come your diagnosis on a postcard. When authorities sent in the dried blood of a rooster, back came the diagnosis of sinus infection and bad teeth.

- The Radiotherapeutic Instrument, invented by Ruth Drown, a Los Angeles chiropractor, was a device that would treat a patient by remote control anywhere in the world. Drown was eventually taken to court, where a defense witness testified that she had been cured of pneumonia from California while attending a convention in Atlantic City. The witness was later identified as the chairman of a board of education of a large school district, responsible for hiring school science teachers and organizing health education programs. Thousands of Californians, including many notables, patronized the Drown establishment before the state closed it down.
- The Uranium Wonderglove, the Vrilium Tube (at $306 each for a few grams of a cheap chemical), the Orgone Accumulators, the Magic Spike, and the anti-arthritis copper bracelet were a few of the many other science fiction gadgets used to fleece thousands of Americans before state and federal authorities cracked down. (Copper bracelets, however, are still being advertised. One bracelet sold for $100, and the user was instructed to wear one of them on the left wrist and the other on the right ankle! Another ad warned that your old bracelet was becoming obsolete, and you needed a new one!) In one recent year, an Illinois company sold $10 million worth of "IONIZED" bracelets that could "ease pain and increase strength."

Today, the FDA continues to ban worthless drugs and devices from the market. Diet aids (see Figure 3.2) and arthritis "cures" are swindles that are repeatedly offered to an all-too-believing public, as are the brain fuels, sauna corsets, hair growers, miracle cosmetics, full-stop slim cubes, total-man capsules, electric bed boards, electromagnetic field generators, weight-loss earrings, snake doctor electroshock devices (including one that would work on your photo!), ozone generators, and the perennial breast developers and penis enlargers. A recent FDA consumer magazine reports on the seizure of fraudulent medical devices including Silky Touch, a radio-wave device for removing unwanted hair without electrolysis; Hemorr-Ice, an insertable rod for shrinking hemorrhoids; The Brain Tuner, an electromagnetic contraption for improving I.Q. scores and treating drug addiction; and Lotus Pond Healthy Balls and Centi-Cure for treating arthritis, stiff neck, and rheumatism.

Two salesmen landed in prison for claiming in 1996 that their "REM Super Pro Generator" would "shatter cancer cells and AIDS cells in people," and for saying, "Your kids get chicken pox, use the Super Pro Generator immediately, and it will knock it out." Clearly, worthless drugs and devices haven't disappeared; all the hoaxes are still here. Be sure to visit this health quackery site: **http://www.nia.nih.gov/health/agepages/healthfy.htm**.

For an article about outrageous claims made for "emu oil" and other worthless products, see *FDA Consumer* magazine, Nov/Dec 1999, pp. 23–25.

The FDA is of tremendous importance to Americans. It oversees more than $1 trillion worth of our products, including pills (human and animal), our food supply, cosmetics, blood supply, vaccines, infant formulas, and product labeling. It regulates medical tests and devices, and the facilities that apply them. On the Internet, browse **WellnessLetter.com**, a Web site that has articles on over 75 dietary supplements.

Consumer beware: A century after the patent medicine rage, we must still daily evaluate claims made in magazines and the supermarket press for commercial products, especially nutritional and OTC products. The ads claim "boost your memory, exclusive better nutrition, the natural choice, backed by 40 years of research, latent

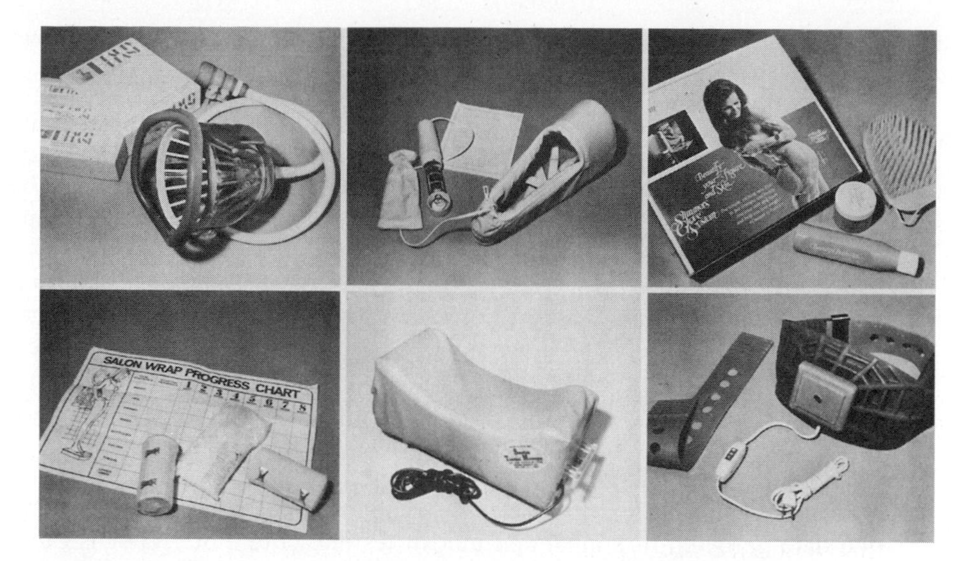

Figure 3.2 The postal service took legal action to halt the sale of these six products through the mail. One product (top row, left), a water-sprinkling device accompanied by three kinds of creams, was sold as a bust developer. Another (top row, center), a battery-powered vibrating face mask, was supposed to keep wrinkles away. The other four were flagrantly promoted as weight-reducing aids. All these types of hoaxes are back on the market, under different names, soon after their legal removal. (Source: *FDA Consumer* magazine.)

weapon, in the battle against, bone protection, recommended by the German Kommission, dramatic benefits, standardized extract, revolutionary dietary ingredient, great news, compelling new studies, totally new, 20% more calcium, beyond calcium, high-quality European, original clinical studies, found safe and beneficial, enzymatic therapy, rely on" (actual quotes from just the first 3 pages of a major nutrition magazine). This type of approach has been used for a long time and has sold a vast amount of merchandise. It would seem fair to expect more substantial data and compelling reasons to buy.

In 1968, the FDA won a ruling that the commercial success of a drug does not constitute substantial evidence of its safety and efficacy. In 1976, the Medical Device Amendments Act was passed; it requires that manufacturers ensure the safety and effectiveness of medical devices, including certain diagnostic and laboratory products.

Although it took almost 70 years, the pendulum finally swung from a no-control situation to a "prove-it-safe-and-effective-or-you-can't-market-it" position. But had the regulations become too stringent? Were they retarding the introduction of desirable, useful new drugs? Had the protector become a hindrance to progress?

3.3 The FDA and the NDA

The FDA, responsible for approving the marketing of all new drugs and generics that are sold in the United States and for monitoring their use after approval, has established a rigorous program of evaluation for safety and effectiveness through which every new drug must pass. The program of evaluation consists of the following stages:

1. *Investigational New Drug (IND) stage*. The sponsor of a new chemical entity (NCE) must submit evidence to the FDA that the new drug has therapeutic activity in animals, that it shows promise in humans, and that it appears to be safe for human testing. The sponsor must also submit a protocol describing the planned testing in humans.
2. *Human Testing stage*. Here, the drug passes through three phases: Determination of the dosage and absorption characteristics in a few human volunteers, human testing in a limited number of sick patients, and human clinical testing on large numbers (hundreds or thousands) of patients to determine its safety and effectiveness (see "Understanding Clinical Trials," Justin A. Zivin, *Scientific American*, April 2000, pp. 69–75.).
3. *The New Drug Application (NDA)*. At this point, the sponsor applies for approval to market the drug and submits a voluminous report containing all of the data gathered in animal and human testing. Often, the NDA is thousands of pages long. In its crucial review of the NDA, the FDA examines all evidence of safety and effectiveness, of risks versus benefits, and then decides to approve marketing, to disallow marketing, or to defer its decision until additional data are developed. Since 1992, the FDA has had the authority to collect *user fees* from manufacturers seeking marketing approval.

The FDA requires that drug manufacturers perform two important animal tests: the chronic (long-term) toxicity studies for carcinogenicity and teratogenicity.

1. **Carcinogenicity**. Is the drug capable of inducing cancer if ingested over a long period of time? Some patients must take their medication for months or even years. Hence, long-term testing for possible induction of cancer is a necessity.
2. **Teratogenicity**. Is the drug capable of causing thalidomide-like damage to the unborn child, especially in the first trimester of pregnancy? Given our memory of babies born with stumps for arms, this test is considered to be one of the most important of all.

For each new drug examined, these two tests cost the drug manufacturer a total of almost $1 million. About 500 animals are needed for each test (mice, rats, hamsters, or guinea pigs). Expensive tests such as these contribute to the time and cost of developing a new product, which is now estimated at up to 15 years and more than $500 million per product. *Note:* The Pharmaceutical Research and Manufacturers of America claims the costs of developing and marketing new drugs averages $800 million, a figure disputed by the watchdog group Public Citizen, which pegs the average cost at $240 million.

In addition to cost, *delay* of the introduction of the new drug is a consequence of such extensive testing. Consider, for example, the test for possible carcinogenicity. The rodents (usually mice) used in this test are divided into a control group and an experimental group. Both groups are given the same food and water, breathe the same air, live in identical cages, and are handled by the same technicians. In fact, *everything* is identical except that the experimental group of mice are fed daily a fixed amount of the chemical (drug) being investigated. After many months of this procedure, all of the mice are killed, and their organs are examined for signs of a significantly greater incidence of cancer among the mice that received the drug than in mice in the control group.

In the tests for teratogenicity, pregnant mice are fed the drug. The just-born offspring of these mice are then examined for evidence of teratogenicity (cleft palate, fused bones, abnormal limbs, and so on).

Such tests require months to carry out, plus additional time for interpretation and written summation of results. Added to this process are the necessary tests for possible acute toxicity (i.e., whether the drug is poisonous when given in a single, large dose). Here, the results are expressed in terms of an **LD_{50}**, which is defined as the lethal dose that results in the death of 50% of the test animals used. For example, the LD_{50} of chloroform is 3,283 mg/kg given by subcutaneous injection in the mouse. The LD_{50} of DDT is 113 mg/kg given orally in the rat. The lower the LD_{50} value, the more acutely toxic (poisonous) the drug. Researchers like to see large LD_{50} values, which means the drugs are safer to use. The LD_{50} for the same substance can vary widely in different species. For example, dioxin (a contaminant found in certain herbicides) has an LD_{50} of 1μg/kg body weight in the guinea pig but 22μg/kg in the male rat and 114μg/kg in the mouse. (These values make dioxin one of the most potent poisons ever synthesized.) The route of administration (oral, dermal, parenteral) can also influence the value of the LD_{50}.

The period of time required for FDA approval of a new drug for prescription sale is termed the *drug lag*. The drug lag for approving new molecular entities (as opposed to new formulations) has been as long as 3 years, but in 1996—with the use of computer-assisted NDAs—the FDA lag time averages 15 months. Priority medicines were evaluated in an average 7.8 months. Critics have claimed that the FDA's drug-approval program is too slow compared to other countries. But a recent analysis published in the *Journal of the American Medical Association* (JAMA) shows that the United States far outpaces Germany and Japan in new-drug approval and is close to that of England. These four countries account for 60% of global pharmaceutical sales.

The FDA says that its insistence on proof of a drug's safety and usefulness has spared the American public a number of drug disasters. They cite drugs that over the past 25 years were banned in the United States but were sold in foreign countries where they caused serious side effects or death. The FDA is proud of its ban on Aminorex, an amphetamine-like appetite suppressant sold in Germany, Switzerland, and Austria in the 1960s that caused hundreds of cases of pulmonary hypertension and the death of several users. The agency, however, has made mistakes. It approved the NSAID Duract, but a year later the manufacturer voluntarily withdrew the drug after it was associated with 4 deaths and 8 liver transplants. The blood pressure drug Posicor resulted in 24 deaths and 400 injuries due to interactions with other drugs; the diet drugs fenfluramine and Redux caused heart damage; and the anti-histamine Seldane triggered deaths when used with other drugs. The FDA says that when these drugs were approved, the full scope of adverse effects had not shown up.

The FDA has established a "fast track" approval process for drugs it ranks as having important therapeutic value. These include drugs to fight AIDS, cancer, and Alzheimer's disease.

The FDA has also proposed streamlining the paperwork required of a pharmaceutical company filing an NDA. Presently, a typical NDA averages 100,000 pages; this could be cut considerably by allowing manufacturers to submit summaries instead of detailed reports. Foreign countries demand much less red tape and paperwork than does the United States for the same drugs. A new ruling by the Department of Health and Human Services will cut NDA time by allowing more use of clinical studies done in foreign countries. Actually, it is now routine for American companies to seek approval for their new molecular entity products, first in foreign countries and then domestically. In recent years, 80% of our new drug entities were marketed abroad first.

One positive result of FDA policies has been the decline in "me-too" drugs—preparations that are rushed in to capture some of the market of a highly successful drug but differ only slightly from the drug in chemical design, strength, or ingredients. Such imitative drugs offer no new benefits or advantages.

3.4 Drugs Are Recalled, Too

More than 700 firms manufacture and distribute drug preparations for use in America. They sell everything from morphine and heart medicines to aspirin and corn removers. With this great number of companies and products, it is important to have an agency such as the FDA to check their safety and effectiveness. The FDA conducts a great variety of laboratory tests on drug products. Unfortunately, it finds many mistakes and asks that the offending products be recalled voluntarily by the manufacturer. If the company refuses, the FDA can initiate legal action; for example, it can ask a court to order a recall or to seize the product to prevent further distribution. The FDA has ordered product recalls for the following reasons: an injectable product was not sterile, or the vials leaked; the product contained less drug than it was labeled to contain (subpotent); it contained more drug than it should have (superpotent); vials were unlabeled, or labeling was incorrect; tablets did not dissolve or disintegrate in the proper manner; the preparation contained the wrong product; the formulation was ineffective; there was bacterial contamination or a high lead level; an NDA was not obtained before the product was marketed; or use of the product resulted in too many cases of serious side effects. In one recent year, over 1,050 products was recalled in the United States for safety reasons.

Sometimes the efforts of the FDA to challenge a GRAS substance (one in use for a long time and Generally Regarded As Safe) stir up a great controversy. Such was the case with the cyclamates (now banned and awaiting the results of more tests) and more recently with saccharin. We think of saccharin more as a food additive than a drug, but either way the FDA is responsible for deciding its safety. Saccharin has not been banned, but a warning statement must appear on the label, much like the one on a pack of cigarettes. One of the reasons for the difficulty in making a clear-cut, final decision about saccharin is that the disturbing information we have about it has come from tests in animals (rats), and some find it hard to apply these results to humans. If it causes cancer in rats, will it necessarily cause cancer in humans? The uncertainty in making this kind of decision goes far beyond saccharin. It influences research in carcinogenicity, cancer chemotherapy, and all research situations where animal data must be interpreted with humans in mind.

Problem 3.1 *(a) If you reject the rat as a suitable test animal for drugs that are to be used in humans, what other method for testing safety and effectiveness would you propose? What are the disadvantages of your alternative method? Is it legal to use humans as test animals? (b) If you discovered that every substance known to cause cancer in humans has also been proven to cause cancer in rats, would you change your mind about not applying rat results to humans?*

3.5 Teratogens

Earlier in this chapter, I described the test for teratogenicity. A **teratogen** is any drug, chemical, infectious agent, or form of radiation that can cause damage to a developing embryo, with resulting birth defects. The defects can include an abnormal joint, a growth deficiency, a too-small head, a genital or heart defect, cleft palate, or the catastrophic thalidomide-like baby born with stumps for arms or legs. Teratology is the science of drug-linked birth defects. Alcohol is the most common teratogen (see Chapter 9), and thalidomide the most famous. One of our newest teratogenic health hazards is the anti-acne drug isotretinoin (Accutane), discussed in Section 15.9. Because the estrogen in birth control pills can act teratogenically, it is important that birth control pills **not** be taken for even a day if pregnancy is suspected. The list of possible teratogens is so great that the pregnant woman is well advised to expose her fetus to no drug, no chemical, no alcohol, and no radiation, unless her physician approves. Remember, the first trimester of gestation is most critical, although a teratogen can cause damage at any time.

The FDA has established five pregnancy categories to indicate a drug's teratogenic potential. For drugs that might be given during pregnancy, the label must indicate the category to which the drug belongs. The categories are as follows:

- *Category A*. Studies indicate no teratogenicity; risk to the fetus is remote.
- *Category B*. Either animal studies are negative but there have been no studies in women, or animal studies are positive but studies in women are negative for teratogenicity. Hence, risk to the fetus is low.
- *Category C*. Either animal studies are positive for teratogenicity but studies in women have not been reported, or there is no information from either animal or human studies. The decision to use or not is based on risk/benefit considerations.
- *Category D*. Enough evidence has been gathered from human studies to indicate a clear risk of teratogenicity in pregnant women. The drug is contraindicated in all but the most serious or life-threatening situations.
- *Category X*. There is little or no doubt that the drug is a teratogen, based upon animal and/or human studies. The drug should not be used by pregnant women.

According to the FDA, the letters A through X only imply a gradation of risk that may or may not exist, and that a drug labeled X must be judged in terms of the benefits and risks. *Note to women*: if you are using a minor tranquilizer and planning to become pregnant, be sure to read the manufacturer's warning on the patient package insert or in the PDR.

Unfortunately, most drugs a pregnant woman ingests readily pass the placental barrier (see Figure 4.3), and the embryo (first trimester) and the fetus (second and third trimesters) are fully exposed to their possible teratogenic effects. X-rays are also capable of inducing birth defects. There must be a good reason for a woman to be X-rayed while she is pregnant. In spite of ubiquitous warnings, however, hundreds of thousands of women across the United States ingest dangerous drugs while they are pregnant, as shown in Figure 3.3.

The virus that causes German measles (rubella) grows easily in (and can devastate) embryonic tissue, and pregnant women who have had German measles have subsequently given birth to defective babies. Fortunately, vaccines such as Biavax (Merck, Sharp & Dohme) are available for immunization against rubella and other diseases. Gamma globulin injections are prescribed for pregnant women who have been exposed to German measles and do not have natural immunity. Gamma globulin is the fraction of blood that contains most of the known antibodies; it is obtained from the general population through blood bank donations. Gamma globulin can confer a passive immunity that lasts for several weeks.

Can the father's drug exposure or lifestyle affect the newborn? Yes, say researchers in medical science, but the evidence is sketchy. If the father is a heavy smoker or is exposed to vinyl chloride, arsenic, or the anesthetic nitrous oxide, his wife has an increased risk of miscarriage or stillbirth. A statistical relationship exists

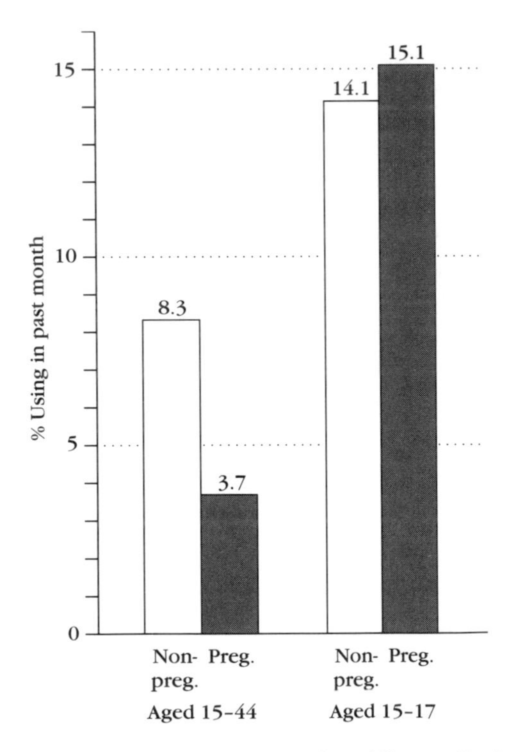

Figure 3.3 Illicit Drug Use Among Pregnant American Women During the Past Month, by Age, 2000 + 2001 Samples. Drugs include marijuana, hashish, cocaine, PCP, LSD, heroin, and alcohol. (Source: SAMHSA Office of Applied Studies, 2001 National Household Survey on Drug Abuse: Vol I. Summary of National Findings. Printed 2002.)

between the father's exposure to ionizing radiation, hydrocarbons, certain solvents, and paints and an increased risk of leukemia in the children he subsequently sires. Among known sperm-damaging substances are lead, solvents such as benzene and toluene, paint thinners, carbon disulfide, certain pesticides, Kepone, marijuana smoke, large quantities of alcohol, and ionizing radiation. The prospective father would be wise to strictly limit his exposure to potential toxins for at least 3 months before insemination. (Sperm require about 3 months to mature to the point of ejaculation.)

3.6 Quackery

According to the FDA, quackery, the promotion and sale of useless remedies promising relief from health conditions, is a $27 billion-a-year business in America. Furthermore, about 38 million Americans use "quack" health products or treatments each year, despite the fact that the remedies being hawked are generally ineffective and may, in fact, be harmful. Three areas in which quackery flourishes are cancer, arthritis, and aging. The modern quack's prime target is the senior citizen. The approaches include the illustrated brochure, the supermarket tabloids, TV commercials, testimonial ads, storefront clinics, and phony foundations. Eighty percent of older people have at least one chronic health condition. The quack feeds on ignorance. An FDA study found that three-fourths of people in a survey believed that extra vitamins provide more pep and energy; one-fifth thought that diseases such as cancer and arthritis are caused by vitamin and mineral deficiencies; and 12% surveyed reported self-diagnosed arthritis, rheumatism, or heart trouble.

Hundreds of phony diets, drugs, devices, and therapies have been promoted as cures for cancer, including jojoba oil, goat serum, Hett cancer serum, antineol, Bamfolin, CH-23, H.11, Polonine, the grape diet, the Gerson diet, the macrobiotic diet, the Rand treatment, and the Chase dietary method (including daily enemas with lemon juice, grapefruit juice, or coffee: a "poison by mouth but a stimulant rectally").

A list of the top 10 health frauds, as compiled by the FDA, follows:[3]

1. *Fraudulent arthritis products*. Copper bracelets, Chinese herbal medicines, megadoses of vitamins, and snake or bee venoms find a big market among the 40 million people who have arthritis—many of whom self-treat, even when under a physician's care. These treatments may be credited with what are actually spontaneous remissions.
2. *Spurious cancer clinics*. Often located in Mexico, these clinics promote the use of Laetrile, vitamins, minerals, etc. and often induce people to abandon legitimate cancer therapy that may be curative.
3. *Phony AIDS cures*. "Guerrilla" clinics in the United States, the Caribbean, and Europe claim cures with large doses of antibiotics, typhus vaccine, and herbal teas—all of which are unproven. For information on approved AIDS treatment studies, call the National Institute of Allergy and Infectious Disease (1-800-TRIALS-A).

[3] *FDA Consumer*, Vol. 23, No. 8, 1989, p. 29.

4. *Instant weight-loss schemes*. These schemes often use toll-free telephone numbers to gain credit-card sales and get around mail-fraud regulations.
5. *Fraudulent sexual aids*. Spanish fly, strychnine, mandrake, yohimbine (in unapproved dosage forms), licorice, zinc, herbs, and Chinese "crocodile penis pills" are all ineffective as aphrodisiacs, and some may be dangerous as well.
6. *Baldness cures and other appearance aids*. So far, only minoxidil is approved for treating certain cases of baldness. Other purported baldness cures, as well as bust enlargers for women, are frauds.
7. *False nutritional schemes*. Bee pollen, wheat germ capsules, and the like are harmless but ineffective—a waste of money!
8. *Chelation therapy*. Administration of the amino acid EDTA with vitamins and minerals has not been proven to clean out the circulatory system, and this treatment can be very expensive.
9. *Unproven use of muscle stimulators*. Though approved for certain medical purposes, these devices won't get rid of your beer belly, wrinkles, cellulite, or other flaws.
10. *Candidiasis*. Though *Candida albicans* can cause health problems, some promoters claim that 30% of the population is hypersensitive to this organism and attribute a wide range of maladies to the hypersensitivity. The claim is false, and the touted treatments are ineffective.

Cancer quackery has fleeced thousands of Americans of millions of dollars. At the peak of its business in the 1950s, the Hoxsey Cancer Clinic of Dallas, Texas, was treating more than 10,000 persons. Each had paid $400 for the Hoxsey cancer treatment—a total income of $4 million. Actually, the FDA estimated a total of $50 million was paid for the Hoxsey treatment. Such potential for income indicates how vulnerable the cancer patient is and how fear will promote trust in almost any "cure" that is proffered by even the most unscrupulous dealer. President William Howard Taft once said, "There are none so credulous as sufferers from disease," and this statement is especially true of frightened victims of cancer. Harry Hoxsey's treatment for internal cancer consisted of a "pink medicine" and a "black medicine," the formulas for which Hoxsey obtained from his great-grandfather, whose horse had been cured of a leg cancer after eating certain pasture plants. The $400 each patient paid included a lifetime supply of the two medicines. As the FDA inspectors prepared to take Hoxsey to court, they discovered that, of the 400 persons Hoxsey claimed to have cured, some had diagnosed themselves as having had cancer; some had cancer but had been treated by accepted methods before going to Hoxsey; and the rest were Hoxsey patients who had died of cancer or still had the disease. Not even one true cure could be verified. Ten years of litigation were required before the Hoxsey hoax was finally stopped. At one point, a "crusade of prayer" and a write-your-congressman program were begun in support of Hoxsey's treatment.

Quackery and deception have not disappeared. Recently, three men were arrested for operating a "Fountain of Youth" scheme. Their drugs, imported from Germany (foreign drugs are a widely used gimmick), were said to cure ills ranging from depression to diabetes, rheumatoid arthritis, radiation damage, aging, and psychological problems. "Cell therapies" were claimed. Their concoction contained procaine (a local anesthetic) plus "sexual tonics." The men were convicted and sentenced. In another recent case, a man and his wife collected $500,000 by claiming to cure AIDS, cancer, multiple sclerosis, arthritis, and other diseases using ozone treatments, European enzymes, hydrogen peroxide intravenous drips, and a "diapluse" machine. The ozone, which has no known therapeutic value, was introduced rectally. One patient from Australia paid $30,428 for 36 treatments. The pair was convicted and sentenced.

Laetrile (amygdalin, "vitamin B-17"), utterly discredited as a cancer cure in the 1980s, has reappeared in cyberspace. Despite FDA warnings, a man in Queens, New York, working out of his basement, used the Internet to promote and dispense Laetrile, using false promises that the drug would prevent and even cure cancer. Be it known that the National Cancer Institute reported in the *New England Journal of Medicine* in 1982 that Laetrile is not an effective cancer cure, and can even be harmful.

The FDA has a Web site for protection against health fraud. See **http// vm.cfsan.fda.gov/~dms/wh-fraud.html**. See also **http://www.quackwatch.com**.

3.7 Current Cancer Therapy

Appendix I discusses the chemistry of a few of the chemotherapeutic agents used to treat various cancers. These synthetic chemicals are examples of one of three currently accepted methods used for the treatment of cancer:

1. Chemotherapy
2. Surgery
3. Radiation

Research is also under way to develop an immunological approach to cancer therapy, that is, to induce the cancer patient's body to manufacture antibodies that will destroy the cancer cells but not harm normal cells.

Everyone admits that we have a long way to go in the development of cancer chemotherapy, but as I pointed out in Chapter 2 in the discussion of vinca alkaloids, some noteworthy successes have been achieved. Some patients, especially children, appear to have been cured of their cancer for as many years as we have followed them since chemotherapy was first begun. Scientists recognize that some types of cancer, such as leukemia, can be treated successfully; other types, however, such as one form of lung cancer in smokers, are utterly resistant to any chemotherapeutic approach.

The chemotherapeutic anticancer drugs suffer from lack of specificity and from debilitating side effects. Nausea, weakness, and loss of hair may accompany their use. This is a situation in which we are still seeking the magic bullet. Early detection of cancer offers the best prognosis for cure. Learn cancer's early warning signs, and heed them.

3.8 How Federal Law Affects You

By what right can a Drug Enforcement Administration (DEA) agent detain a citizen and conduct a search for drugs? What drugs and substances are within his or her jurisdiction? Is morphine legal? Is LSD illegal? What penalties face a U.S. citizen for trafficking in drugs? These questions can be of the utmost importance to us or to members of our family. The answers are obtained by considering one very important federal law.

In 1970, Congress passed Public Law 91–513, the **Comprehensive Drug Abuse Prevention and Control Act**, more familiarly known as the Controlled Substances Act. The act repealed and superseded more than 50 patched-together pieces of drug legislation, including the Old Harrison Narcotic Act, the federal Marijuana Regulations, and the Food, Drug, and Cosmetic Act insofar as it applied to drugs. Substances and drugs that are regulated by the 1970 act are termed *controlled substances*. Distilled spirits, wine, malt beverages, and tobacco are not controlled by this act; rather, they are covered by other federal laws.

In 1986, Congress passed the Anti-Drug Abuse Act of 1986, also known as the Drug-Free America Act (Public Law 99–570). Its 15 titles supersede much of the 1970 Controlled Substances Act and include greatly expanded penalties for the sale, manufacture, possession, or trafficking in controlled substances, especially where large quantities or repeat offenses are involved. (See Section 3.9 for the actual penalties.) Designer drugs are included with wording that anticipates future analogs or modifications so that they are controlled even before they are synthesized. The manufacture, sale, or trafficking in drug paraphernalia is controlled, but only in interstate shipment. Included in the definition of paraphernalia are cocaine freebase kits; wired cigarette papers; seven different kinds of pipes with or without screens, hashish heads, or punctured bowls; roach clips; miniature spoons; bongs; chillums; and chamber or carburetor pipes. This act also addresses the problem of money laundering.

The 100th Congress passed yet another drug bill, the Anti-Drug Abuse Amendment Act of 1988. This massive drug bill approved the death penalty for drug bosses and their hit men, for cop killers who murder in the act of a drug-related crime, and for others convicted of "serious drug felonies." (A serious drug felony is defined as one that under law carries a prison sentence of at least 10 years.) The act denies or withholds certain federal benefits—such as farm or student loans—from persons convicted of drug offenses. Mere possession of crack can now bring a severe jail sentence.

The Controlled Substances Act, although modified by the 1986 and 1988 amendments, is still the basic law and continues to control the distribution of all depressant and stimulant drugs and other drugs with abuse potential such as hallucinogens. It established five classifications, or schedules, of controlled substances.

Schedule I drugs are those drugs or substances that have a high potential for abuse and *no* currently accepted medical use in the United States. Furthermore, with schedule I drugs, there is a lack of accepted safety under medical supervision. Some of the better known schedule I drugs are listed here:

Amphetamine variants (DMA, PMA, STP, MDA, MDMA, TMA, DOM, DOB)
Heroin
Thebacon
Bufotenine
LSD
Ketobemidone
Mescaline
Methaqualone
Peyote
Phencyclidine (PCP) and analogs (PCE, PHP, PCC)
Psilocybin
Tetrahydrocannabinols (as found in "grass," sinsemilla, hash oil, Thai sticks)

Marijuana remains in schedule I even though it has been used in the treatment of glaucoma and of nausea resulting from cancer chemotherapy. Schedule I drugs are not legally available to the general public anywhere in the United States. They are outlaw drugs.

Schedule II drugs are those drugs and substances that have a high potential for abuse but also have a currently accepted medical use in treatment, with or without severe restrictions. Abuse of schedule II drugs may lead to severe psychological or physical dependence.

Schedule II drugs are legally available in the United States by prescription. Examples of schedule II drugs are the following:

Opium	Preludin	Morphine
Benzedrine	Cocaine	Dexedrine
Dolophine	Methamphetamine	Pantopon
Ritalin	Methadone	Rapidly acting barbiturates
Percodan	Demerol	Hydromorphone
Dilaudid		

Note that many of the drugs listed in schedule II are widely prescribed in the United States and are considered to be highly useful substances in treatment.

Schedule III drugs are considered to have a lower abuse potential than those in schedules I and II. They have a currently accepted medical use in the United States. Their abuse may lead to moderate or low physical dependence or high psychological dependence. Examples of schedule III drugs are the following:

Paregoric (and other preparations with limited amounts of narcotics)	Doriden	Chlorphenteramine
	Nalorphine	Testosterone products
	Noludar	Hydrocodone
	Anabolic steroids	Codeine
Vicodin	Benzphetamine	GHB
Aspirin	Acetaminophen	

Schedule IV drugs have a low potential for abuse compared to those in schedule III. They have a currently accepted medical use in the United States. Their abuse may lead to limited physical or psychological dependence compared to schedule III drugs. Tranquilizers and some of the less addicting barbiturates are listed in schedule IV. Some examples include the following:

Ativan	Chloral hydrate	Valmid
Benzodiazepine-type minor tranquilizers (e.g., Valium, Librium, Serax, Halcion)	Paraldehyde	Darvon
	Phenobarbital	Placidyl
	Dalmane	Equanil, Miltown
	Equanil	Talwin Rohypnol

Schedule V drugs have an even lower potential for abuse than schedule IV drugs. They have a currently accepted medical use in the United States. Their abuse may lead to limited physical or psychological dependence. Most of the items in schedule V consist of preparations containing limited quantities of narcotic drugs often used in cough medicines or as antidiarrheals. They correspond to the old "exempt narcotic" classification. Examples:

Terpin hydrate and codeine
Cough medicines with codeine phosphate
Cheracol
Phenergan
Robitussin A–C
Brown Mixture
Lomotil

The DEA publishes a Controlled Substances Inventory list that catalogs about 1,800 schedule II drugs and about 3,500 schedule III, IV, and V drugs. As a substance becomes more widely abused, the substance may be elevated in schedule. An example is PCP, which has now been raised to schedule I. You can browse the list of all scheduled drugs at **http://www.deadiversion.usdoj.gov/schedules/**.

Manufacturers and distributors of drugs are required by the Controlled Substances Act to register with the U.S. Attorney General, who in turn determines production quotas for schedule I and schedule II drugs. Complete, accurate records of all stocks of controlled drugs are mandated by the act. Even labeling and packaging requirements are spelled out. The act requires a prescription for a controlled substance, forbids refilling of a prescription for any schedule II drug, and prohibits prescription refills of schedule III or IV drugs more than five times after the initial date of the prescription. A prescription for a schedule III or IV drug may not be filled or refilled more than 6 months after the date it was written. A prescription for a schedule II drug must be typewritten or written in ink or indelible pencil and must show the signature of the doctor. Prescriptions for schedule III, IV, or V drugs may be issued orally (as over the telephone), as well as in writing.

The Durham-Humphrey Amendment to the Controlled Substances Act has given rise to the term *legend drugs*. This term refers to the legend "Caution: Federal law prohibits dispensing without a prescription." The Durham-Humphrey Amendment reflects the necessity of reducing indiscriminate use of potentially dangerous drugs. Some drugs are quite dangerous (the heart stimulant digitalis, for instance). Other legend drugs are safer but could be habit-forming or used indiscriminately to the detriment of the patient (taking too much of certain diuretics results in potassium loss from the body, and antihistamines can cause sleepiness). Of course, not every drug for which a prescription is written is a legend drug. OTC drugs may be dispensed or sold legally without a prescription. The FDA decides which drugs are safe enough to be included in the OTC category (see Chapter 15).

Some drugs are marketed for both prescription-only and OTC sales. Examples are the 3 heartburn drugs, Pepcid, Tagamet, and Zantac. The OTC versions have lower doses and easy-to-understand labeling.

About 2,800 agents are currently employed by the DEA, an arm of the Justice Department. These "narc" agents have jurisdiction when any of the controlled substances or drugs mentioned in the preceding discussion are transported across state lines and thus enter interstate commerce. All 50 states have also passed drug control laws, and an offender can be subject to state as well as federal prosecution. DEA agents are assigned to the United States and to 31 foreign countries. Their efforts are concentrated on drugs of high abuse and high addiction potential, such as heroin, the opiates, LSD, cocaine, PCP, and hallucinogens.

The most interesting drugs are the schedule I drugs (heroin, LSD, peyote, psilocybin, etc.). Because there is no legitimate medical use for these substances, they are not available legally for distribution (except in some cases for scientific research purposes). This fact means that almost all of the use of these substances in America today is illicit use, and mere possession of any of these schedule I drugs makes one liable to prosecution. Morphine, a schedule II drug, has almost as high an abuse potential as heroin but is medically useful in the United States. In terms of illegal possession, manufacture, or distribution, the 1970 act does not distinguish between any of the schedules of drugs. Theoretically, one can be prosecuted for illegal possession, manufacture, or distribution of a schedule V drug just as for a schedule I drug.

The Controlled Substances Act covers 61 pages, and it is not possible to discuss all aspects of it here. However, other topics covered in the law include counterfeiting drugs, continuing criminal enterprise, definition of *drug-dependent person*, and definition of *addict*.

3.9 Straight Talk About the Internet Sale of Drugs and Devices

The Internet has become an advertising medium for a vast number and variety of products, and health concerns are now the sixth most common reason people go online. One study showed that 70% of people look on the Internet before visiting a doctor. Unfortunately, it is easy to bypass the few controls that exist for the Internet, and anybody can claim just about anything. As a result, "rogue sites" offer products or practices that are already illegal in the offline world. Examples: A California company used the Internet to sell a home kit advertised as a blood test for the AIDS virus; the kit was unapproved, and the maker fabricated test results. A breast-enlargement device is sold (see also Section 3.2) that promises a 55 percent enlargement in 10 weeks. The seller claims FDA approval, but in fact the FDA has not reviewed it for effectiveness but allows it on the market. The herb catspaw is sold as an immune-system stimulant, but no good scientific studies have been conducted on its effects in humans. Remember, there is a large scientific jump from an animal test to the effect in humans. Also, testimonials can be elicited from anyone; they have been a favorite con device for hundreds of years. As for beta-carotene (vitamin A), one authority does not recommend this supplement for *anyone*. In fact, one scientific study found that beta-carotene actually increases the risk of lung cancer in smokers. The FDA decries as fraudulent and illegal Internet claims such as "cures Alzheimer's and AIDS" and "proven effective in treating over 650 infectious diseases." The sales pitches go on and on.

One more reason to view drugs advertised on the Internet with suspicion is the uncertainty regarding side effects. Many herbs have as many potential risks as possible benefits. If you have high blood pressure, diabetes, a bleeding tendency, liver or kidney disease, are currently taking other prescription or OTC drugs, or are pregnant, think carefully about dosing yourself with an product bought over the Internet. Some sites on the Internet only require customers to fill out a questionnaire before ordering prescription drugs. On the Internet, there is no direct doctor-to-patient interaction, no physical examination, and no evaluation of safety, drug interactions, or appropriateness. FDA officials say that consumers seeking health products online can find dozens of sites that are legally questionable. Some of the Web sites are operated from outside the United States.

Warning people to be wary, the FDA and the Federal Trade Commission announced joint efforts to fight deceptive health claims for products advertised on the Internet through the formation of "Operation Cure All." Internet users can complain by telephoning toll free 1-887-382-4357.

On the Internet you can read preposterous claims such as a cure for arthritis made from beef tallow and a magnetic device to treat high blood pressure and liver disease. "The Internet is a veritable minefield of false and misleading claims," says Bruce Silverglade of the Center for Science in the Public Interest, **`http://www.cspinet.org/`**. Caveat emptor.

A word about the **legality of imported drugs** (Canada, Mexico, or other foreign countries). Newspaper and magazine ads claim it is now legal to import drugs into the United States for personal drug use. Not true, says FDA spokesman Ray Formanek, Jr. By law, unapproved, misbranded, or adulterated drugs are prohibited from importation, and usually all drugs imported by individuals fall into one or more of these categories. This includes foreign versions of FDA-approved medications and drugs dispensed without a prescription. Beware of **counterfeit drugs**. Our FDA is battling a global problem of fake, substandard pharmaceuticals in the United States. The World Health Organization estimates that 5–7 percent of drugs sold in the United States are fake (contrasted to 25 percent of Mexican drugs and over 50 percent in parts of Africa). Drug counterfeiters prepare copies of the real item—right down to color, scoring, company markings, and packaging (see Figure 3.4).

However, counterfeit drugs often show poor bioavailability and are subpotent, out of date, or made in substandard facilities. They may be manufactured in one foreign country and packaged and shipped in another. Profits from their manufacture and sale can be huge.

Phony drugs can enter the United States in the possession of travelers who cross the border, buy the drug, and reenter. Travelers are allowed to reenter with a 90-day supply, and often make several border crossings in one day. There are 1,700 pharmacies in Tijuana, Mexico alone!

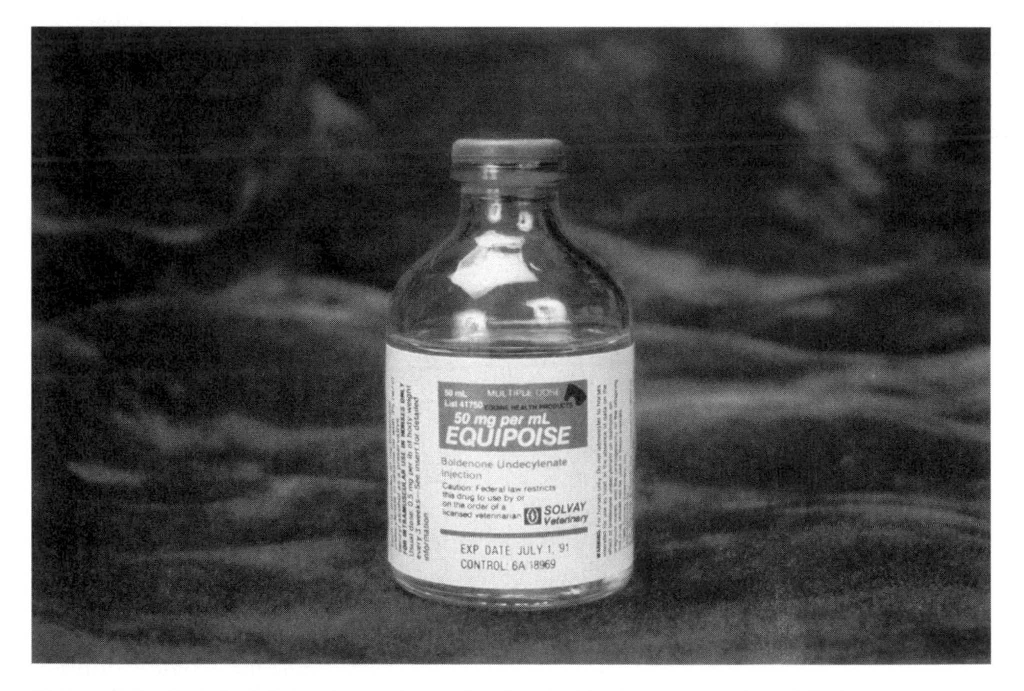

Figure 3.4 Counterfeiters duplicate packaging for black-market sales. (Photo courtesy United States Department of Justice, Drug Enforcement Administration.)

Fake drugs also enter via mail deliveries from the more than 1,000 Internet pharmacies, many of which deal with drug sources in Mexico, Switzerland, India, and Thailand. Another source of counterfeit drugs is bulk import of raw material or pills from abroad, as in the case of an antibiotic made by a Chinese bulk-drug manufacturer.

Some examples of fake drugs are Anavar (see Figure 17.1); counterfeit Viagra from Malaysia; bogus Ponstan (an anti-inflammatory drug); and the antibiotic Cipro, purchased from one of the 3,400 Internet Web sites offering it. Internet Cipro pills made in Mexico, India, or Thailand have a high probability of being fake, and may even be nothing more than cornstarch! For more see "Imported Drugs Raise Safety Concerns," Michael Meadows, *FDA Consumer*, September/October 2002. For buying on-line see **`www.fda.gov/cder/drug/consumer/buyonline/guide.htm`**.

Ten particularly risky drugs are now monitored by the FDA, especially in Internet sales, and at import. The 10 drugs are:

Drug Trade Name	*Generic Name*	*Use*
Accutane	isotretinoin	intractable acne
Actiq	fentanyl	pain
Clozaril	clozapine	severe schizophrenia
Lotronex	alosetron	irritable bowel syndrome
Mifeprex	mifepristone (RU486)	abortion
Thalomid	thalidomide	skin sores
Tikosyn	dofetilide	heart conditions
Tracleer	bosentan	hypertension
Trovan	trovafloxacin	antibiotic
Xyrem	sodium oxybate	narcolepsy

Of course, the potentially acute problem here, especially with Internet sales, is failure to observe risk management procedures. With Accutane (Section 15.9), this failure could be disastrous. With other risky drugs, bypassing important information about side effects or not taking necessary tests could result in serious problems. The FDA is warning Americans not to buy these drugs on the Internet. See the useful article "Imported Drugs Raise Safety Concerns" in *FDA Consumer* magazine, Sept./Oct., 2002, Vol. 36, Nos. pp. 18–23.

3.10 Penalties for Illicit Use of Controlled Drugs

Under the Controlled Substances Act, it is unlawful for any person knowingly or intentionally, except as authorized,

- **a.** to manufacture, distribute, or dispense, or possess with intent to manufacture, distribute, or dispense, a controlled substance; or,
- **b.** to create, distribute, or dispense, or possess with intent to distribute or dispense, a counterfeit substance.

The *maximum* penalties provided by the 1986 Amendments for violation of (a) or (b) are as follows:

For a first offense involving

1. 1 kg or more of heroin or a mixture containing detectable amounts of heroin,
2. 5 kg or more of cocaine or a mixture containing detectable amounts of coca leaves or cocaine or ecgonine,
3. 50 g or more of cocaine free base (crack),
4. 100 g or more of phencyclidine (PCP) or 1 kg or more of a mixture containing a detectable amount of PCP,
5. 10 g or more of LSD or a mixture containing detectable amounts of LSD,
6. 400 g or more of fentanyl,
7. 1,000 kg or more of a mixture or substance containing a detectable amount of marijuana,

the sentence is imprisonment for not less than 10 years or more than life, except that if serious bodily injury results from the use of the controlled substance, not less than 20 years or more than life, and a fine of as much as $4,000,000 if the defendant is an individual or $10,000,000 if the defendant is other than an individual, or both. For repeat offenders, the penalties are doubled. For first-time offenders convicted under this law, a minimum 5 years of supervised release is imposed after the term of imprisonment; for repeat offenders, a minimum 10 years of supervised release after imprisonment is imposed. Probation or suspension of sentencing of any person sentenced under this law is forbidden.

For a first offense involving

1. 100 g or more of a mixture or substance containing heroin,
2. 500 g or more of a mixture or substance containing coca leaves, cocaine, or ecgonine,
3. 5 g or more of cocaine free base,
4. 10 g or more of PCP or 100 g or more of a mixture or substance containing PCP,
5. 1 g or more of a mixture or substance containing LSD,
6. 40 g or more of a mixture or substance containing fentanyl,
7. 100 kg or more of a mixture or substance containing marijuana,

the sentence is imprisonment for not less than 5 years or more than 40 years, except that if death or serious bodily injury results from the use of the controlled substance, not less than 20 years or more than life, and a fine of as much as $2,000,000 if the defendant is an individual or $5,000,000 if the defendant is other than an individual, or both. For repeat offenders, the penalties are doubled. For first-time offenders convicted under this law, a minimum 4 years of supervised release is imposed after the terms of imprisonment; for repeat offenders, a minimum of 8 years. Probation or suspension of sentences is forbidden.

The law distinguishes between possession with intent to distribute and simple possession. In the latter case, maximum penalties are less severe. For more on cocaine, see Section 7.5.

Under the Controlled Substances Act, it is unlawful for any person knowingly or intentionally to possess a controlled substance unless such substance was obtained directly with a prescription or order, from a practitioner while acting in the course of his professional practice.

The *maximum* penalties provided for simple possession violation of (a) above are as follows: imprisonment of not more than 1 year and a fine of at least $1,000 but not more than $5,000, or both, except that if he or she is a second-time offender, imprisonment must be a minimum 15 days but not more than 2 years, and the fine a minimum of $2,500 but not more than $10,000. A third-time or more offender must serve a minimum 90 days in jail but not more than 3 years and is fined a minimum $5,000 but not more than $25,000. Minimum sentences under this law may not be suspended or deferred. Further, conviction under this section requires that the offender pay the costs of the investigation and prosecution of the offense, except if the court determines that the defendant lacks the ability to pay.

Under this act, a judge may grant probation to a first-time offender convicted of simple possession. Probation may not exceed 1 year, and if the offender was under 21 years of age, he or she may apply to the court for an order to expunge the proceedings from all official, public records. After the record is expunged, the offender need never admit or acknowledge the arrest or indictment in response to any inquiry made of him or her for any reason.

Case History *LSD Conviction*

L., a teenager, was on his way to a rock concert when he was stopped by a federal undercover agent at the local airport. Three grams of LSD was found in his bag. L. was arrested, tried, convicted, and sentenced to 15 years in a federal correctional institution. The unique aspect of his sentencing was that he received 5 years for actual chemical LSD and 10 years for the 440 g of paper on which the drug was absorbed. The Supreme Court has now upheld the claim of federal prosecutors that the paper (or sugar cube, gelatin, orange juice, or whatever) in which the LSD is carried *must be included in the weight of the illegal substance* when determining how long to imprison an LSD offender. In other words, prison time for drug felons can now be based on the weight of the mixture containing a detectable amount of illegal drug, not just on the pure chemical itself. This concept also holds for cocaine traffickers who "cut" their drug with sugar (or whatever) before distributing it. If L. had diluted his LSD in a glass quart container of orange juice, it is possible he could have been sentenced to life without parole in a federal prison. Note in this case that LSD is a drug sold in microgram quantities, and that apparently there was intent to distribute the drug, possibly across state lines.

3.11 State Laws

Space limitations prevent me from summarizing the penalties imposed by all states for the illegal possession or sale of drugs; such rules differ from state to state. I will, however, list the penalties assessed in the state of California as an example, Table 3.1 summarizes California's penalties for illegal sale or possession of drugs.

A **Uniform Controlled Substances Act** has been passed in the state of California. In a manner similar to the federal Controlled Substances Act, the California law

Table 3.1 Penalties Under California's Uniform Controlled Substances Act of 1977

Provisions of the Health and Safety Code	*Sentence*	*Fine*
California Law—Narcotics Except Marijuana		
Section 11350—Possession of narcotics except marijuana	Imprisonment in state prison	A fine of up to $20,000 may also be imposed for each offense.
Section 11351—Possession for sale of narcotics except marijuana	2, 3, or 4 years in state prison	A fine of up to $20,000 may also be imposed for each offense.
Section 11352—Transporting, importing, furnishing, gift, sale, etc., of a nonmarijuana narcotic (e.g., heroin)	3, 4, or 5 years in state prison	A fine of up to $50,000 may be imposed for sale of 0.5 ounces or more of a substance containing heroin, or upon the second or subsequent conviction for violating Sections 11351 or 11352.
Section 11353—Anyone 18 years or over who solicits, induces, encourages, or intimidates any minor to violate narcotic laws, other than those dealing with marijuana	3, 4, or 5 years in state prison	Not to exceed $20,000 for each offense.
Section 11354—Anyone under 18 who solicits, induces, encourages, or intimidates any minor to violate narcotic laws, other than those dealing with marijuana	Imprisonment in state prison	
California Law—Marijuana		
Section 11357—a. Possession of concentrated cannabis (e.g., hashish)	Maximum of 1 year in county jail	And/or $500 fine.
b. Possession of not more than 1 avoirdupois ounce (28.35 g) of marijuana, other than concentrated cannabis (misdemeanor)[a]	—	Not more than $100 for each offense. Not subject to booking. Three or more convictions, education required.
c. Possession of more than 1 avoirdupois ounce (28.35 g) of marijuana, other than concentrated cannabis[a]	Maximum of 6 months in county jail	And/or $500 fine.

Table 3.1 continued

Provisions of the Health and Safety Code	*Sentence*	*Fine*
Section 11358—Planting, cultivating, harvesting, drying, or processing marijuana	Imprisonment in state prison	—
Section 11359—Possession for sale of any marijuana	Imprisonment in state prison	Not to exceed $20,000 for each offense.
Section 11360—a. Transporting, importing, selling, furnishing, giving marijuana b. Gift, transport of less than 1 avoirdupois ounce of marijuana, other than concentrated cannabis (misdemeanor)	—	Not to exceed $100. Not subject to booking.
Section 11361—Use of a minor to sell, or sale to a minor of, marijuana	3, 4, or 5 years in state prison	Not to exceed $20,000 for each offense.

[a]Records of convictions under these subsections are to be destroyed 2 years after the date of conviction.

establishes five schedules of controlled substances. Heroin, LSD, peyote, marijuana, and psilocybin are classified as schedule I substances. Morphine, methadone, cocaine, and amphetamines are classified as schedule II substances. In California, no person other than a physician, dentist, podiatrist, or veterinarian can legally write a prescription. Pharmacists must maintain prescriptions for controlled substances on file for a period of 3 years. Veterinarians must not prescribe, administer, or furnish a controlled substance for themselves or for anyone else. Recognizing the potential for abuse of PCP and the danger in its use, California law forbids even the possession of the chemical ingredients necessary to synthesize PCP if there is intent to manufacture the drug. The same holds true for the chemical precursors of methamphetamine.

Web Sites You Can Browse on Related Topics

FDA on the Internet
http://www.fda.gov/

Quackery
http://www.mtn.org/quack/devices/devindx.htm
http://www.nih.gov/nia/health/pubpub/health.htm
http://www.quackwatch.com/index.html

To Subscribe to NIDA Notes:
e-mail **nidanotes@hq.row.com**

Partnership for a Drug-Free America
http://www.drugfreeamerica.org

Women and Drug Use
http://www.lindesmith.org/library/focal_women.html

Office of Women's Health
http://www.fda.gov/womens/default.htm

Teratogens
http://www.ctispregnancy.org/ctis.html

Hoaxes
Search for "medical hoaxes"

Study Questions

1. Drug X claims to be a cure for cancer; drug Y claims to be a cure for diabetes. Would it be wise to permit unrestricted distribution of these drugs on the American market, so that their effectiveness can be proved by *actual use*?
2. Define (**a**) carcinogen, (**b**) teratogen, (**c**) LD_{50}, (**d**) chronic toxicity, (**e**) mutagen, (**f**) subpotent preparation, (**g**) abuse potential, (**h**) NDA, (**i**) phocomelia.
3. **a.** Identify the federal laws that were passed in 1906, 1938, and 1962 that dealt with drugs.
   b. What did each law specify or establish?
4. Teratogens are capable of damaging the developing fetus. Name one drug, one chemical solvent, one disease virus, and one type of radiation that are teratogenic.
5. If you accept the statement made in this chapter that no drug is always 100% safe in all patients, then where would you draw the line in regard to the drug's risk/benefit ratio? If the drug were very useful (as in treating older diabetics) but caused two deaths per year (out of 230 million Americans), would you permit its continued use or ban it?
6. Users of marijuana are concerned about ingesting residues of Paraquat, a defoliant sprayed on marijuana plants in Mexico. Paraquat has an LD_{50} of 57 mg/kg (orally in the rat) and 80 mg/kg (in the skin of the rat).
   a. Is Paraquat more or less acutely poisonous than DDT? (See the text for DDT values.)
   b. Give a reason why the LD_{50} of Paraquat given by the skin route in the rat is higher than it is when given by the oral route.
7. The Germans, English, and Mexicans all require less paperwork, less testing, and less proof of the safety of new drugs than the United States does. Does this mean that they care less for the safety of their citizens? If you answer yes, give your reasons.
8. Cancer cure hoaxes have appeared in the United States since the 1800s. Reread the descriptions of the hoaxes in this chapter and list the methods of operation that almost invariably characterize such schemes. (In other words, identify the standard techniques employed.)
9. When is morphine a licit drug, and when is it illicit? Is heroin ever a licit drug today?
10. What distinguishes a schedule II drug such as Seconal from a schedule I drug such as LSD?
11. OTC drugs such as aspirin can be sold legally without a prescription. Antibiotics such as penicillin, however, are legend drugs. Why shouldn't antibiotics such as penicillin be sold over the counter just as aspirin is?
12. Some DEA officials are concerned about the provision of the 1970 Controlled Substances Act that allows a 1-year probationary sentence to be applied to first-time offenders caught possessing a schedule I controlled substance. After much hard work, narcotics agents see offenders go free on probation. Would you do away with this provision?
13. Into which of the five pregnancy categories (Section 3.5, "Teratogens") would thalidomide fall if categorized today?
14. True or false:
    a. Another name for the Comprehensive Drug Abuse Prevention and Control Act of 1970 is the "Controlled Substances Act."
    b. Another name for the Anti-Drug Abuse Act of 1986 is the "Drug-Free America Act."

c. Penalties under the 1986 Drug-Free America Act are more severe for the first-time offender than for the repeat offender.

d. A repeat offender under the 1986 act who was convicted of an offense involving more than 50 g of cocaine free base could possibly be imprisoned for as long as 20 years and fined as much as $8 million.

e. The 1988 Anti-Drug Abuse Act (Section 3.8, "How Federal Law Affects You") provides for the possibility of the death penalty for a person convicted of a first offense involving more than 50 g of cocaine free base.

15. In February 1991, testosterone was declared a controlled substance under federal law. Does this now make all adult males lawbreakers?

16. Oklahoma was the first state to pass a computerized prescription drug tracking system. Every prescription for Demerol, codeine, and a dozen other potentially addictive drugs is tracked by computer along with the patient's and doctor's name. Critics claim violation of doctor-patient privacy and thus a threat to health care personnel and unintimidated drug prescribing. Take a stand and explain your point of view.

17. In California, people convicted of possessing any amount of any illegal drug will automatically lose their driver's license for 6 months. This law applies to someone convicted of smoking one marijuana joint while walking in the park. Minors who are convicted have their driver's license application delayed for 6 months. Do you support the provisions of this law? Why or why not?

18. *Advanced study question*: From advertisements in magazines, books, and newspapers, compile a list of questionable products or outright hoaxes that are currently being offered to the American public. Your nearest FDA office will help you. Examine each product from the standpoint of claims made, language used to motivate buyers, ingredients, and purported therapeutic action. Include devices and appliances, as well as drugs, in your list.

19. *Advanced study question*: An early patent medicine, Curfurhcdake Brane Fude, was advertised as giving harmless relief of headache and containing no poisonous ingredient. The medicine contained acetanilide, antipyrine, caffeine, sodium bromide, potassium bromide, and 24% alcohol by volume. Using an older pharmacology text, check on the safety of each of these ingredients.

20. Brochures in the mail claim "Gerovita" will prevent aging and senility, stating that the "famous Romanian anti-aging formula will make you look younger overnight," plus cure arthritis, heart disease, depression, and diabetes. In the light of what you have read in this chapter, analyze these claims.

CHAPTER

What Happens to Drugs After We Take Them

Key Words in This Chapter

- Route of administration
- Parenteral
- Skin patch
- Implants
- Fat-soluble
- Transdermal
- Blood-brain barrier
- Biotransformation
- Enzymes
- Drug interactions

Learning Objectives

After you complete your study of this chapter, you should be able to do the following tasks:

- Give the names and features of seven routes of administration of drugs.
- Explain the nature of the blood-brain barrier.
- Explain the nature of the placental barrier.
- Cite the role of liver enzymes in drug metabolism and development of tolerance.
- Define the half-life of a drug.
- Cite five factors that can affect drug response.
- Define and give examples of drug interactions.
- Define synergism.
- Relate how drugs can be excreted from the body.

4.1 Introduction

We swallow a tablet of a certain medication and then forget about it, assuming it will get where it is supposed to, and do some good. But the process of ingestion, absorption from the gastrointestinal (GI) tract, distribution via the blood, pharmacological action, possible storage, and, ultimately, metabolism or excretion can be quite complicated.

In this chapter, we study the ways in which medicinal agents can be administered, their dosage forms, how they can get into the bloodstream, their distribution to tissues of the body, and their ultimate fate. We also examine the possibility of drug interactions.

When you have completed this chapter, you will know a great deal more about your body and how it can absorb, store, utilize, change, reject, and excrete drugs.

4.2 Routes of Drug Administration

A medicinal agent can be administered by one or more of the following routes.

4.2.1 Oral Ingestion

Most drugs are taken orally. They are expected to disintegrate in the acidic stomach contents and enter the blood by absorption from the stomach or small intestine. A few oral drugs are deliberately designed to be insoluble in the fluids of the GI tract so that they will remain inside the intestinal lumen. Povan, a drug for treating pinworm, is an example of such a substance. A physician may direct that an oral drug be taken with meals in order to minimize any irritant effect the drug may have on the stomach lining. The great advantage of the oral route is the ease of administration and the expected effectiveness of this route. There are also disadvantages, however. Food taken concurrently may sometimes interfere with the drug's action; certain drugs are destroyed by digestive juices if taken orally (e.g., insulin); a few people simply cannot swallow even a medium-sized tablet; the taste of the medication may be obnoxious; and once a patient swallows a drug, its subsequent absorption and action usually cannot be stopped (short of inducing vomiting). Unconscious persons, of course, must not be given drugs by the oral route.

For the first time, the powerful narcotic fentanyl comes in a lollipop form. Oralet is a red lozenge on a stick, intended for children and adults as an anesthetic premedication and for anesthesia itself. Sucking on the lollipop produces drowsiness in 10 minutes.

4.2.2 Inhalation

Inhaling a substance usually leads to rapid absorption into the bloodstream because of the rich supply of blood vessels in the lungs and nasopharyngeal region. With cocaine, the nasal route actually gives faster access to the brain (7 seconds) than the parenteral route (14 seconds). Nasal decongestants work locally by shrinking the nasal mucous membrane. Decongestant sprays and inhalants have the disadvantage that their heavy use may result in a worse congestion state than the one for which they were used. Substances sold in inhalers must be volatile and easily able to pass into the gaseous state. But even some nonvolatile drugs—including proteins used in therapy—can be inhaled in the form of a fine aerosol mist or tiny-particle spray. Inhalant devices are available for treating asthma. A drawback to inhalation therapy is the possibility of drug-caused irritation of mucous membranes.

4.2.3 Parenteral Administration (by Injection)

Drugs given by a route that does not involve the GI tract or body orifices are said to be given parenterally. We can classify parenteral administration, that is, hypodermic injection, into IM (intramuscular), IV (intravenous), IP (intraperitoneal), subcutaneous (beneath the skin), and intradermal (within the skin). The great advantages of parenteral administration are that it affords direct (or nearly direct) access to the bloodstream, rapid drug action, and can be instantaneously stopped if necessary. Giving a drug by injection eliminates concern about its taste or its susceptibility to destruction by digestive juices. The disadvantages include the pain of injection, as well as the need for sterility and the necessity of putting the drug into solution or a very fine suspension. Injectables usually cost more than other drug forms. For a discussion of needle-free injection systems, see the discussion of insulin administration in Section 2.5.1.

Hypodermic needles, of course, can be involved in the transmission of disease, such as HIV infection. But drug-use paraphernalia other than needles and syringes can also infect. University of Miami researchers found evidence that using contaminated water for rinsing needles, cotton swabs for filtering drug solutions, and "cookers" such as spoons or bottle caps for dissolving drugs can be responsible for transmitting the AIDS virus.

4.2.4 Body Orifices

Medical substances can be administered in the form of eye, nose, or ear drops; eye inserts; sublingual (under the tongue) tablets; vaginal inserts; and rectal suppositories or retention enemas.

Andean porters obtain their cocaine by stuffing the coca leaves into their buccal cavity (the space between the teeth and the cheek). Some tobacco users get their nicotine the same way. The FDA has approved buccal nitroglycerin for anginal attacks.

Drug absorption from body orifices may not be as rapid as absorption from other routes, and part of the dose may be lost by drainage out of the orifice. A severely nauseated person who cannot tolerate aspirin by mouth can be given aspirin by rectal suppository.

4.2.5 Skin Application

Application of a medicinal substance to treat a local skin condition is widely practiced. The skin is the largest organ of the body, and while it is remarkably impervious to many organisms and substances, it sometimes can act as a route for getting drugs into the bloodstream. Thus, when we apply a drug to the skin to treat a local condition, we must recognize that a whole-body action may sometimes also result. Salicylic acid and salicylates (oil of wintergreen) are rapidly absorbed even through intact skin, and there have been cases of systemic poisoning after application of these substances to large areas of skin.

Ointments fall into four categories, depending upon the base or vehicle used to carry the active ingredient:

1. Hydrocarbon-based ointments, or oleaginous ointments, often contain petrolatum as the base.
2. Absorption bases are oil-in-water emulsions that permit the incorporation of additional water solutions.
3. Water-removable bases can be readily washed from skin or clothing.

4. Water-soluble bases, or creams, are greaseless and water-soluble.

Medications can also be applied to the skin in aerosol spray form.

4.2.6 Transdermal (Skin Patch) Systems

One of the newest means of administering drugs is the **skin patch** which is worn on a convenient area of the body and from which a drug is slowly absorbed into the bloodstream. An example is the prescription product Transderm SCOP for motion sickness. Scopolamine is one of the best anti-motion sickness drugs, but too large a dose all at once is toxic. Hence, scopolamine is ideally suited to administration via skin patch.

Nitroglycerin, testosterone, and clonidine skin patches are available by prescription, as is Estraderm, a source of estrogen worn on the abdomen. Nitroglycerin patches cost about $1 apiece but work for up to 24 hours (sublingual tablets remain effective for less than 30 minutes). Clonidine is an antihypertensive that appears on the market as Catapres-TTS. Duragesic skin patches contain the pain reliever fentanyl. For a discussion of nicotine skin patches, see Section 7.20. The FDA has approved an estrogen/progestin skin patch.

There are limits on what kinds of drugs can be administered via skin patches. First, the drugs must be potent. Second, they must have significant solubility in both oil and water in order to pass the semipermeable membrane in the patch and the skin. Also, these drugs typically are the type that are taken over long periods of time.

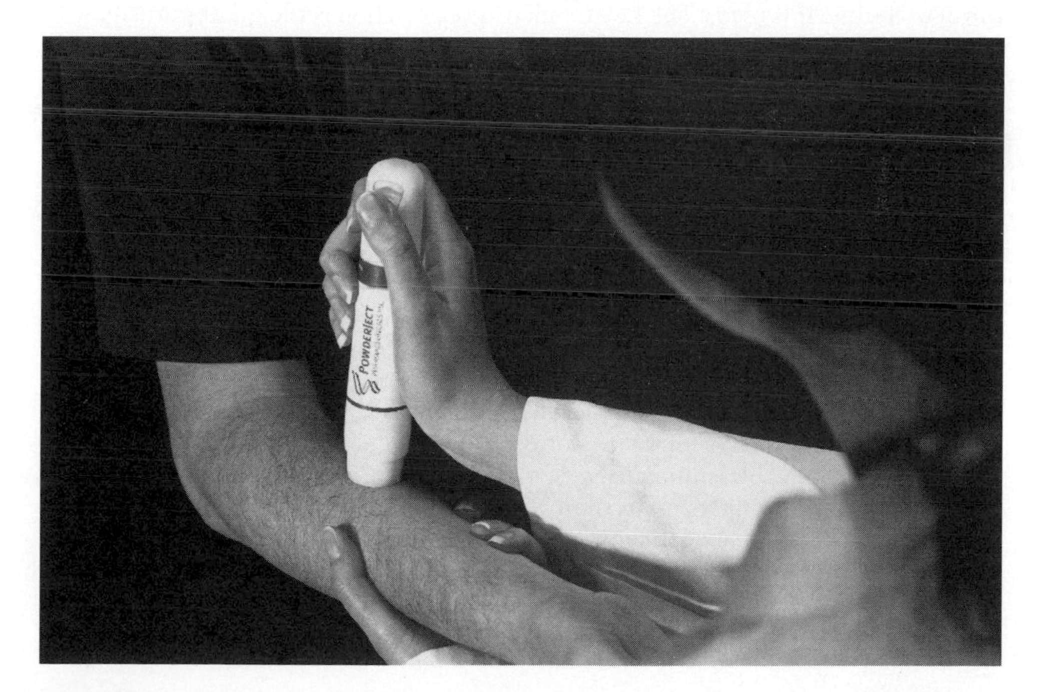

Figure 4.1 The PowderJect device painlessly delivers minute amounts of powdered vaccine. (Photo courtesy PowderJect Pharmaceuticals, The Oxford Science Park, Oxford, OX4 4GA UK.)

Powder injection uses a special device to administer tiny amounts (1–2 mg) of powder-form vaccine without the use of a needle. The hand-held PowderJect System (see Figure 4.1) is pressed against the skin, releasing a jet of helium gas, accelerating particles of powdered vaccine to a very high velocity and injecting them into the outer layer of skin (epidermis). The process is painless.

4.2.7 Implantable Drug-Delivery Systems

Research has shown that drug delivery can be accomplished by the surgical implantation of a drug pellet, reservoir, or pump. The FDA has also approved a new, concentrated form of morphine for use in implantable pumps. Doctors program the pump to release the proper dose, and patients return once a week or monthly for refills or dose adjustment. One patient has worn a pump for 8 years. (See also insulin pumps, Section 2.5.1.) Research has shown the feasibility of implanting a contraceptive drug (Progestasert) into the uterus, from which it slowly dissolves into the bloodstream for up to 1 year. Contraceptives can also be implanted under a woman's skin, in thin rubber capsules from which the drug is slowly released over a period of months or even years. Fertility is restored soon after implant removal.

4.2.8 Microsponges

Microsponges are tiny, synthetic spheres that can be "programmed" to release drugs or cosmetics in response to pressure, time, or temperature changes. For example, a foot powder would release more antifungal drug with each step (or a wet-wipe towel each time it is used). The key to microsponge action is the great porosity with consequent huge surface area that can be filled with the drug or cosmetic in fluid form. A system sold OTC in the United States is the benzoyl peroxide microsponge that permits controlled release of an acne medication.

A summary of routes of drug administration and the roles they can play is given in Figure 4.2.

4.3 Dosage Forms for Drugs

Tablets are the most common dosage form for drugs. They are swallowed, chewed, or held sublingually. Scoring of some tablets facilitates breaking them in half if a smaller dose is needed. A tablet's color or shape can help identify it, as can special markings used by the manufacturer. Some tablets are wrapped in foil to keep out moisture until used. Others are coated with a substance that keeps the tablet intact until it gets to the small intestine. Some tablet coatings simply make the tablet easier to swallow or more palatable.

> ***Problem 4.1*** *Why may it be necessary to delay tablet disintegration until it reaches the small intestine? (Hint: What is one chemical difference between stomach juices and intestinal juices?)*

Almost all tablets have binders added to the active ingredient to hold the tablet together. Thus, an aspirin tablet labeled as containing 325 mg of aspirin weighs much

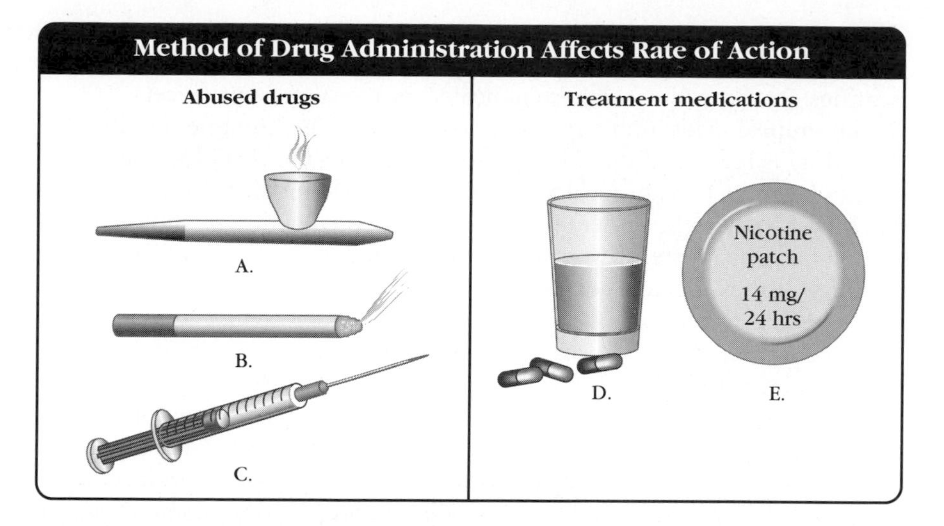

Figure 4.2 The method of administering a drug can play a major role in the drug's delivery and rate of action, which in turn can affect its abuse liability or potential as a treatment medication. Smoked drugs that are delivered through a "crack" pipe (a) or a cigarette (b), and injected drugs that are administered through a hypodermic needle (c) reach the brain rapidly — an action that is an important factor in the strong effect and abuse potential of such drugs. Drug abuse treatment medications given orally (d) or through a skin patch (e) take longer to reach the brain. Such slower delivery and rate of action are important factors in the milder effect and lower abuse liability of treatment medications. Source: *NIDA NOTES*, March/April 1997, p. 9. National Institute on Drug Abuse.

more than 325 mg. If a tablet is compressed too hard during its manufacture, it may fail to disintegrate in the patient's GI tract. The manufacturer carries out disintegration tests to ensure bioavailability; so does the FDA.

Capsules made of gelatin are widely used dosage forms. Capsules can mask the irritant nature of a drug or its bad taste; they also help keep moisture away from the active ingredients. Use of colored capsules aids in drug identification. Liquid drugs can be sealed inside capsules. Sustained-release ingredients may be incorporated into capsules or tablets in order to provide active medication over a longer period of time. This task is accomplished by using different sizes of drug particles that have different solubility rates in the GI tract, or by using an erosion-controlled polymer-drug complex that disintegrates over a long period. Contac's "tiny time pill" is an example of an erosion system. The active ingredient is thus released more slowly, and high blood levels of drug are maintained for a longer time.

Liquid medications are popular because of their ease of use. They include syrups, elixirs, spirits, suspensions, and liquids for external use. A physician who directs that "1 teaspoonful" of a liquid be taken really means 5 mL (or 5 cc). We know from measurements that some teaspoons hold more than 5 mL and some hold less, so it is best to obtain a special container that measures in the metric system. Your local pharmacist can probably supply one. (For more information on measurements in the metric system, see Section 1.13.)

Putting a drug into a liquid state often means dissolving it in water. The presence of water often means a shorter shelf life for the drug, because water, along with heat or sunlight, can help cause chemical changes that make the drug worthless. Watch the expiration date for any prescription, but especially so for water-based preparations. Refrigeration, when so indicated, is a very good practice.

Rectal suppositories offer a means of local or systemic treatment of illness. Provided the drug is released from the suppository vehicle, the rich blood supply of the rectum can be the means of local absorption and systemic distribution. Suppositories can be used when the patient is unable to take medication by mouth, when the medication has a bad taste, or when the patient is unconscious. Disadvantages to the use of suppositories include possible uncertain, slow drug absorption, local irritation, and possible accidental loss of the medication.

Some additional dosage forms for drugs are powders; eye, ear, and nose drops; sublingual tablets; effervescent tablets; and retention enemas.

The "unit dose," a single dose of a tablet, capsule, or other drug form packaged by the manufacturer, has become popular for use in hospitals and nursing homes. It is more expensive to package single doses this way, but waste is reduced because there is no leftover medication. Many hypodermic preparations come in unit doses with disposable syringes, which is highly desirable because dirty needles can transmit infectious diseases.

4.4 How Drugs Get to Their Target

The process by which drugs enter the body, move through it, interact with their target tissues, and ultimately undergo elimination is called *drug distribution* (technically, **pharmacokinetics**). The unloading area in this process is the capillary bed, the rich network of tiny blood vessels in the lung, stomach, or walls of the small intestine. Drugs can enter the bloodstream as well as leave it at the capillary bed. Once in the blood, a drug may be delivered to many other body fluids, compartments, or tissues. Some of these target fluids and tissues are as follows:

Body Fluids

1. Intracellular fluid—inside the cell
2. Extracellular fluid—surrounding and bathing the cell
3. *Cerebrospinal fluid* (CSF)—cushioning the spinal cord and brain
4. Placental blood supply—the unborn baby's blood; ordinarily, it never mixes with its mother's blood
5. Intraocular fluid—inside the eye
6. Synovial fluid—inside the bony joints
7. Urine

Body Tissues

1. Muscle tissue
2. Organs of the body—liver, spleen, brain, kidney, heart, etc.
3. Fat tissue, comprising 20–50 percent of body mass
4. Albumins and other blood proteins
5. Bone

Fat (adipose) tissue dissolves and stores substances that are fat soluble. In fact, we can consider body fat to be a kind of reservoir for fat-soluble drugs, from which the drug is released over a period of time. Solubility in body fat accounts for the unusually long time drugs such as the benzodiazepines (e.g., Valium), tetrahydrocannabinol, PCP, and methaqualone remain in the body or are detectable in the urine (see Chapter 16). Vitamins A, D, E, and K are fat-soluble and therefore are found in fatty tissue. Vitamin C and the B vitamins, on the other hand, are water-soluble. They are not stored in fat tissue. In fact, because they are highly water-soluble, they are not stored in the body and must be ingested more frequently than the fat-soluble vitamins. The parent who has proudly given a growing infant plenty of vitamin B supplement soon learns how water-soluble riboflavin is. In a short time, the infant's urine is bright yellow as the unutilized water-soluble vitamin B_2 is excreted. It has been said that Americans have the most heavily vitaminized sewer systems in the world.

Albumins and other proteins in the blood (i.e., plasma) can attract and reversibly bind drugs by means of noncovalent bonds. Significant quantities of drug can be bound to plasma protein. In fact, in laboratory experiments, as much as 94.5 percent of Prozac (see Section 5.7 for a description of Prozac) was found bound to human serum protein, including albumin. Thus, this binding can be considered a drug reservoir and can prolong the life of the drug in the body. Drugs can compete with each other for plasma binding sites.

Thyroid gland tissue has a very high affinity for iodine, and almost all of the iodine we ingest accumulates in the thyroid gland by way of the blood supply.

Bone marrow is the site of red and white blood cell synthesis. Marrow tissue collects from the blood all of the ingredients it needs to manufacture red blood cells and leukocytes. Bone acts to store heavy metals, such as lead.

We have said that once in the bloodstream, a drug may have access to all body fluids and compartments. We know, however, that there is often unequal distribution of a drug in the body. Part of this inequality is explained by considering two barriers to distribution: the blood-brain barrier and the placental barrier.

4.4.1 The Blood-Brain Barrier

With its rich blood supply, the brain would be expected to absorb quickly almost any drug we ingest. Not so. Hardly any penicillin, for example, given by mouth or by injection, reaches the brain or CSF. The general rule is that fat-*in*soluble drugs have a difficult time penetrating the CNS. They are said to be stopped at the **blood-brain barrier**, the fat-covered membranes that surround capillaries in the brain. Fat-soluble drugs can better penetrate these membranes (like dissolves like). Contributing to the barrier effect is the fact that capillaries in the brain have almost no pores (whereas in the rest of the body, capillary pores are present). We know that drugs such as the barbiturates, tranquilizers, LSD, narcotics, marijuana, and cocaine strongly affect the CNS. They must therefore have enough fat-like character to pass the blood-brain barrier quickly. To the question, "How then can glucose, a water-soluble substance, enter the brain?," scientists answer, "Only by mediated transport, a process by which carrier molecules and channel-forming chemicals shepherd desired substances across cell membranes."

4.4.2 The Placental Barrier

I hesitate to perpetuate a misconception by using the word *barrier* to describe the situation that exists at the placenta (see Figure 4.3). In actuality, almost all of the

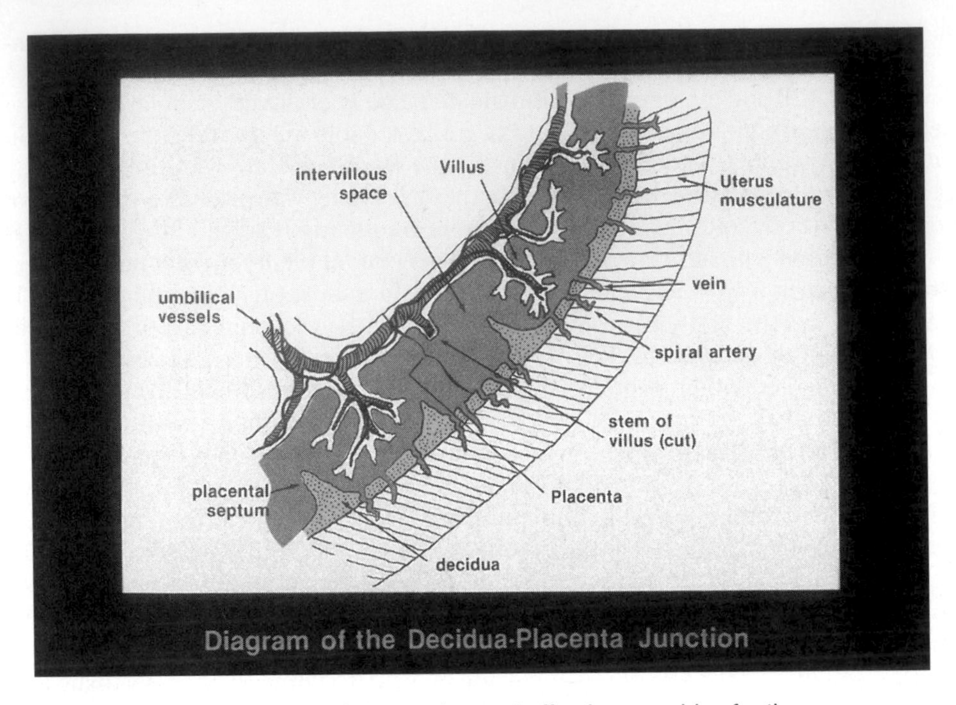

Figure 4.3 The placenta, joining mother and offspring, provides for the nourishment of the fetus. Fetal blood and maternal blood are separated by a semipermeable membrane that allows passage of blood-borne soluble substances but blocks high-molecular-weight proteins and particles.

drugs that enter the mother's blood supply will cross over into the unborn's circulation. A fetus is exposed to essentially all of the substances taken by the mother. If a pregnant woman smokes tobacco, her unborn baby, in effect, smokes too. If she consumes alcoholic or caffeinated beverages, so does her baby (see the discussion of the fetal alcohol syndrome in Chapter 9, "Alcohol and 100 Million Americans"). If she takes a drug such as thalidomide, her fetus unfortunately gets a dose of this teratogen as well. And, if, at delivery, the mother is given a pain-reducing drug such as Demerol, the newborn child will also be under its influence. On the other hand, the placenta is an effective barrier to the passage of bacteria, to high-molecular-weight proteins, and to certain drugs such as the anticoagulant heparin.

The key to understanding the placental barrier, such as it is, is the placenta—the spongy network of blood vessels and structures in the uterus through which the fetus derives its nourishment. Capillaries in the placenta are more easily penetrated by fat-soluble substances, but the barrier to fat-*in*soluble drugs is not nearly as effective as it is in the brain. Ordinarily, maternal blood never mixes with fetal blood, but substances in the maternal blood are just a cell membrane away from the fetal capillary circulation.

Drugs and Pregnancy

Certain drugs can have acute effects if the mother uses them prenatally—i.e., during gestation, especially the first trimester—and postnatally if she breast feeds her newborn.

We learned in Section 3.5 that the placental barrier isn't much of a barrier at all, and we also know that most drugs a lactating mother ingests will be present in her breast milk. A good rule for pregnant women to follow is to take only those drugs and substances that are absolutely necessary or prescribed during gestation, and avoid all others. Chemicals in cosmetics, aerosols, household products, and smog should also be avoided. OTC products should be *assiduously* avoided. A study of pregnant women showed that during their pregnancy, all of them had taken two or more OTC or prescription drugs, with the average exposure being eleven drugs per woman. What the women probably did not know is that 80 percent of the OTC and prescription drugs currently on the market have never been tested for safety during pregnancy. Certain prescription drugs can pose real threats to the embryo and fetus. However, if a pregnant woman gets sick, or if she comes to the pregnancy with a chronic condition that requires treatment, a clinical decision must be made based upon the potential toxicity of the prescription drug and the mother's physical and mental health. Because medication use in pregnancy is common, a mother-to-be needs to know which drugs to avoid. Some of the drugs known to be potentially harmful to the fetus are: aspirin and most other NSAIDs (especially near full term); tetracycline; sulfa drugs; the anticonvulsant Dilantin; antimigraine drugs; caffeine; the anticoagulant warfarin; the anti–high-blood-pressure drugs thiazide and reserpine; iodine (in vaginal douches); excessive amounts of vitamins C, D, and K; sex hormones; oral contraceptives; nicotine; and the tranquilizer Librium. Women should also be sure to read the critically important cautions on the use of the anti-acne drugs Accutane and Tegison (Section 15.9). Two good Web sites on the topic "Drugs in Pregnancy" are **http://www.drughelp.org/** (then click on "research section" under the Research button) and **http://www.perinatology.com/exposures/druglist.htm**.

Alcohol's potentially tragic effects during pregnancy are discussed in Section 9.5. **Cocaine** use during pregnancy can damage the fetus anatomically and developmentally, including cardiovascular, neurological, and excretory systems. Cocaine babies may suffer from low birth weight. News Item: A 24-year-old South Carolina woman was sentenced to twelve years in prison for "killing her unborn fetus by smoking crack cocaine." **Heroin** babies tend to be low in birth weight, but there is no evidence that heroin is a teratogen as dangerous as alcohol. **Tobacco** use in pregnancy can result in low-birth-weight babies who later score low in standard tests of physical function—and who are more likely to die in the first year of life. Far worse, a recent German study found direct evidence of a link between a mother's tobacco smoking and the presence of a powerful cancer-causing drug in the urine of newborn babies of these mothers. The carcinogen, NNK, is linked to cancers in humans and animals. NNK can pass the placental barrier and presumably will continue to be found in the babies as they grow up passively inhaling their mother's cigarette smoke. Other drugs such as PCP, barbiturates, and certain prescription medications can also increase risks in the neonate. It is generally accepted that pregnant women addicted to narcotics or cocaine will give birth to babies similarly addicted and who must experience the withdrawal syndrome (see Sections 6.11, 8.5, 10.3 for more information). See Section 3.5 for teratogen drug pregnancy categories.

It is of interest that the **placenta** itself, formerly discarded, is now often saved for its content of enzymes, proteins, and hormones. Placenta now has about 135 uses in medical science and commerce. It appears in "protein-rich" face creams, shampoos, and body lotions. Its extract is being tested in dentistry as an agent to treat some gum diseases. Placenta also is used to treat arthritis, skin irritations, and eye problems. Gamma globulin can be obtained from the placenta.

In summary, this section has discussed the distribution of drugs—how they get into the bloodstream, their transport to body compartments and fluids, and the possibility of barriers to their passage to organs such as the brain.

4.5 The Metabolism of Drugs

Some may find it strange to read about the "metabolism" of drugs, as though they were digested in the same way as foods are. But there is an analogy here, for drugs are chemicals, and the body has developed enzymes and processes to handle drugs much as it handles foods.

Almost all of the medicinal agents we take are foreign to our body. Consider aspirin, penicillin, barbiturates, Valium, and the sulfa drugs. These chemicals are unique chemicals, not at all like the substances we encounter in foods or as normal body constituents. In a sense, it is remarkable that our bodies are able to accommodate and ultimately get rid of these strange substances. The process by which the human body accepts a drug, alters it chemically to eliminate the drug action, and then prepares it for excretion is termed **metabolism** (also known as **biotransformation**). The products of chemical breakdown of drugs are termed **metabolites**. See Table 16.1 for some typical metabolites.

For metabolism, the body's most important organ is the liver, with the kidney, the GI tract, and the blood plasma also contributing. In the liver, a drug may be handled in the following ways:

1. It may be changed chemically, as by hydrolysis. A different inactive product can result, although sometimes the metabolites themselves are pharmacologically active.
2. It may be conjugated, or chemically linked to a normal body substance such as glycine. The conjugated product is usually pharmacologically inactive and more water-soluble, thus being more easily eliminated by the kidneys.
3. It may be altered by oxidation or salt formation to become more soluble in the urine, and thus more easily excreted. In this way, a fat-soluble drug becomes water-soluble.

In the liver are structures called *microsomes*, which contain naturally occurring proteins called *enzymes*. These microsomal enzymes hook on to drugs and foreign substances and catalyze their chemical breakdown into different, pharmacologically inactive products. Indeed, the liver's **cytochrome P-450 mixed-function oxidase** enzyme system has come to be recognized as playing a key role in the body's defense against endogenous and exogenous chemicals. These cytochrome enzymes (whose name comes from their red color) are extremely important in determining the body's ability to handle drugs and chemicals in the environment, because they can speed up the metabolism of thousands of different substances. They can catalyze the metabolism of alcohol, barbiturates, prostaglandins, fatty acids, benzene, carbon tetrachloride, and other halogenated hydrocarbons, certain insecticides, certain air pollutants, chemicals in soaps, deodorants, cosmetics, and perfumes, dyes, antioxidants, thyroxine, and endogenous steroids, to name just a few. Two additional drug-metabolizing systems are the glutathione S-transferases and UDP-glucuronosyl transferase.

Because of their genetic makeup, people can differ significantly in the quantity and quality of cytochrome P-450 enzymes in their liver. Some researchers believe that this variation explains how people can react differently to the same drug—that is, why a certain dose of a drug may be too little for one patient but too much for another. One person can be a "fast metabolizer" and another a "slow metabolizer." The difference between them can be as high as a factor of 20 (see also Section 2.7).

Enzymes are proteins, and they possess receptor (or active) sites exactly analogous to those we discussed in Section 1.6. When an enzyme acts as biological catalyst, speeding up the metabolism of an ingested drug or chemical (called the substrate), it first binds the drug or chemical to its active site. While in this bound state, the substrate is altered chemically, and the breakdown products (the metabolites) are released to the bloodstream for final excretion in the urine. The enzymes themselves are not consumed in the process; rather, they are free to hook up to additional substrate molecules. Thus, in a few seconds, one enzyme molecule can "turn over" or metabolize tens of thousands of substrate molecules.

Enzyme action can be inhibited competitively or noncompetitively. In **competitive inhibition**, the active site on the enzyme can be occupied by some antagonist so closely related in size, shape, and chemistry to the natural substrate that the natural substrate is effectively blocked out (examples are shown diagrammatically in Figure 6.5 and in the discussion of anti-AIDS drugs, Section 2.8). Sheer numbers plus natural affinity will dictate which molecule, substrate, or antagonist will most effectively compete for the active site on the enzyme. (See also albumins in Section 4.4 and naloxone in Section 6.9.) In **noncompetitive inhibition**, enzyme activity is impeded in some cases by agents that combine with the enzyme on a site other than that used by the substrate. If the conformation of the enzyme is changed, the inhibition can be irreversible. Cyanide is fatal in the human because of its irreversible binding to and disruption of respiratory enzymes.

In a few, rare cases, an enzyme can biotransform an ingested substance into a poisonous material or even into a carcinogenic compound. Enzymes do not always work to our advantage. It is known that Valium is enzymatically broken down into products that are more active than Valium itself.

The liver contains the enzyme necessary to metabolize barbiturates. When we swallow a barbiturate sleeping capsule, it disintegrates, and the drug is absorbed into the bloodstream. The liver has an especially rich blood supply and is therefore able to begin immediately the process of enzyme-catalyzed metabolism of the barbiturate. If the person takes heavy doses of barbiturates over many days, the liver's enzyme concentration may be insufficient to handle its detoxification work. But a fascinating event occurs. The liver is stimulated to manufacture more enzymes in order to cope with the challenge. This process is termed **induction of microsomal enzymes.** It can be understood by considering this useful analogy: A fortified city is attacked by 1,000 enemy troops. It withstands the attack but realizes it must marshal more defenders. The next attack, by 10,000 enemy soldiers, is again barely survived, and more defenders are gathered. Thus the very process of attack engenders the means of defense. The liver is the fort; it produces enzymes required for the metabolism and removal of the foreign barbiturate molecules. To some degree, the more of a drug we take, the more we induce enzyme synthesis in the liver and the better we are able to manage and control the drugs we ingest.

The person taking the heavy dose of barbiturates discovers that his body is becoming used to them. He must take a larger and larger dose to get the same hypnotic effect. We term this process the development of **tolerance** to a drug. It can occur with many substances other than the barbiturates. There is a limit to tolerance. Given enough troops, an enemy can overwhelm a fortified city. Similarly, challenged with molecules of barbiturate, our enzyme response system can be overwhelmed, and overdosage occurs. When the liver is no longer stimulated with an ingested drug, induction of microsomal enzyme synthesis is reduced, and liver enzyme levels fall back to normal.

Occasionally, we discover a drug that does not appear to be metabolized extensively or at all in the body. Such drugs are more slowly excreted, some without chemical change. From our study of drug distribution and metabolism, we can see that the fate of a drug is fairly well established at the time of ingestion. A healthy body will absorb the drug and distribute it to various fluids or tissues, depending upon how fat-soluble the drug is. The liver receives a rich supply of the drug immediately and begins the process of metabolism, or removal of the drug from the bloodstream. An equilibrium is established between liver, blood, and tissue (see Figure 4.4). As the liver metabolizes the drug, its level in the blood will fall; this leads to a fall in the concentration of the drug in the tissue. To maintain an effective level of drug at its receptor site in the tissue, another dose of drug must be taken.

Chemicals that are held in reservoirs in the body can be more difficult to remove. Lead in bone, and DDT and tetrahydrocannabinol in fatty tissue, are bound so effectively that they can be retained long after ingestion. One researcher found that 1% of a

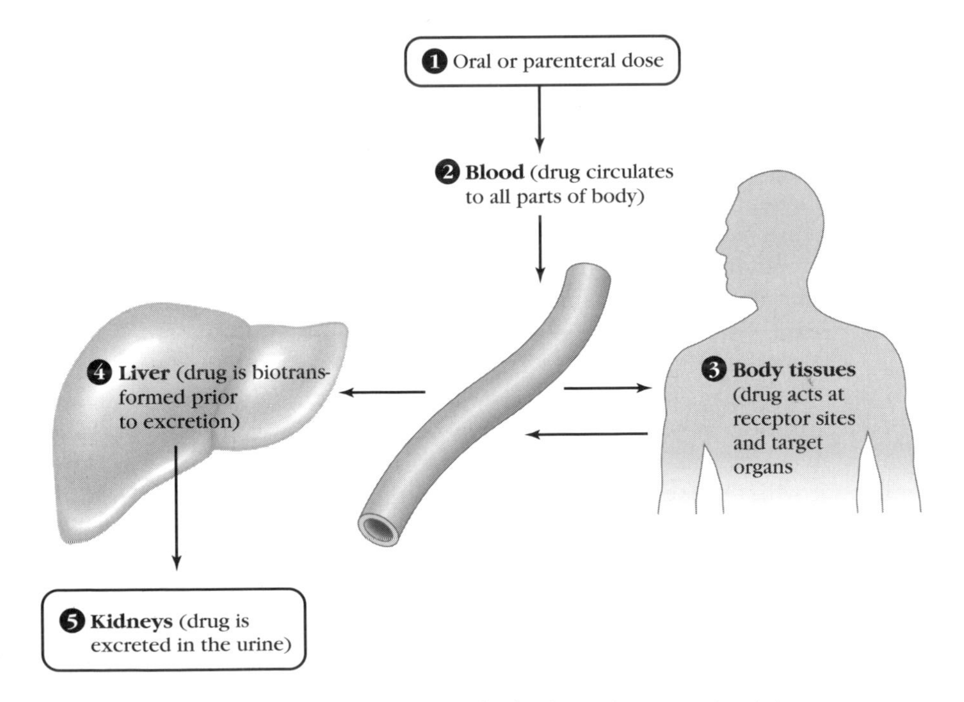

Figure 4.4 The fate of a typical drug in the body, oral or parenteral dose.

dose of tetrahydrocannabinol (marijuana's active ingredient) was retained in the body 4.5 months after ingestion.

The **half-life** ($t_{1/2}$) of a drug is the time it takes for the body to get rid of 50% of an ingested dose. (Stated in more scientific terms, the half-life is the period of time required for the initial blood level to fall by one-half.) The half-life of Valium is 20–50 hours; of methadone, 24 hours. Small amounts of Valium may be present in the body 2–8 days after a single dose is taken. The half-life of secobarbital (Seconal) is in excess of 15 hours, and that of phenobarbital is in excess of 80 hours. The half-life of chloral hydrate falls in the range of 4–10 hours. The half-life for the intravenously administered propofol (Diprivan) is less than five minutes.

Half-lives depend on the chemical nature of the drug taken, the individual, and other factors. This fact leads us to a consideration of physical and behavioral aspects that have a general influence on the action of a drug in the body.

4.6 How People Can Respond Differently to Drugs

Your neighbor can drink strong coffee at bedtime and yet sleep like a baby. You try it, and you are up all night. One woman tolerates an oral contraceptive very well; another suffers so many side effects that she must switch to a different drug. The kitchen ingredient nutmeg is just a spice to some; to others, it is a hallucinogen. There is no doubt that people can react differently to the same drug, both qualitatively and quantitatively. In this section, we discuss how and why such differences occur.

Even in the same person, a drug can have different pharmacological effects depending on how much is given. This concept is termed the **dose-related effect.** Atropine in small doses slows the heart rate; in larger doses, it speeds it up. Alcohol in moderate amounts may either stimulate or decrease respiration. Some of these effects are explained by recognizing that some nerves controlling organs can be stimulated by a low dose of a drug but blocked by a high dose.

Food in the stomach at the time a drug is swallowed can modify the activity of the drug. Partygoers know the difference between drinking alcohol on an empty stomach and drinking after some food has been eaten. Milk can neutralize acidic foods, but it should not be taken with a tetracycline antibiotic, because the calcium in milk is known to interfere with absorption of this antibiotic. Oral forms of tetracyclines should be given 1 hour before or 2 hours after meals.

Biological variability can be an important factor in people's reactions to drugs. Differences in age, sex, weight, physical condition, presence of pregnancy, and mental attitude can play a role. For example, the placebo effect may become important in the patient who identifies with his or her physician and who *expects* the physician to prescribe a helpful drug. **Age** can also play a major role in drug action through five mechanisms:

1. Generally, as we age we significantly increase body fat, and alcohol and fat-soluble drugs remain in the body longer.
2. Liver metabolism of alcohol and other drugs diminishes with age as liver enzymes diminish.
3. Kidney excretion of drugs or their metabolites falls with age by as much as one-third between ages 30 and 90.

4. Aging can mean lowered blood protein, resulting in higher blood levels of the free, active drug, possibly increasing the risk of acute toxicity.
5. With age, we show an enhanced brain and organ sensitivity to drugs.

Women's hormone fluctuations during the 28-day menstrual cycle can affect drug action or metabolism. Examples are epileptic women taking medication reporting more seizures starting a few days before the menstrual period, apparently because their body temporarily eliminates the antiepileptic drug faster, and women with asthma apparently metabolizing theophylline much more slowly during ovulation. However, researchers are reluctant to test experimental drugs on pregnant women.

Overmedication can occur in anyone, but older people are especially vulnerable. Overmedication in persons over 60 years old is responsible for hip fractures, memory loss, auto accidents, and large numbers of hospitalizations. In one real case, a women in her seventies was daily taking eight prescribed drugs: she had a 100 percent chance that at least one drug-drug interaction would occur. For a list of drugs that should *always* be avoided or are generally inappropriate in people age 65 and over, see Sue Ellen Browder's article in *Woman's Day* magazine for May 14, 2002, pp. 76–80.

Circadian rhythms may play a role in how we respond to a drug. Circadian rhythms are daily, predictable, physiological changes driven by a kind of inner "biological clock." They are observed in respiration, temperature, and urine excretion. The most well known rhythm is that of body temperature, which varies by 2–4° C throughout the day, peaking in the late afternoon and troughing in the early morning—even if the person is confined to bed. Pulse rate and blood pressure peak around the same time as temperature. Body levels of glycogen are lowest at 3–6 A.M. Regarding the mechanisms that control or reset the biological clock, scientists at Novartis Research Foundation and Scripps Research Institute in San Diego have discovered the first mammalian photoreceptor, a gene in the inner-layer cells of the retina that senses light and then signals the master circadian oscillator in the hypothalamus. It may do this by directing the synthesis of a protein named **melanopsin**, which may be involved in the production of **melatonin**[1], and thus in the regulation of circadian rhythms. Other mechanisms may also be operative. For more, browse for melanopsin or see the valuable clinical review at **http://www.bmj.com/egi/content/full/317/7174/1704**.

For most medicines, circadian rhythms are not a factor in dosage or treatment. However, the metabolism of a few drugs can be affected. Heparin's effects as an anticoagulant apparently can be affected. In treating asthma, the circadian rhythm of the lung's airways is now considered. The bronchodilator theophylline might work best at night, when the airways are narrowest. The dose of an anticancer drug might be based on rhythms; some tumors have daily temperature cycles, and breast tumors grow most actively around midnight. Knowing that the bone marrow's production of new blood cells peaks around midnight, doctors can best program the release of the anticancer drug 5-fluorouracil, an agent that has the adverse effect of suppressing bone marrow activity.

[1]Melatonin is not to be confused with the pigment melanin.

Chronotherapy recognizes that biological rhythms can affect body function and medical treatment. It attempts to optimize drug response and minimize side effects by coordinating chronobiology with medical treatments. Chronotherapy does not involve new drugs but rather uses old drugs differently.

As for melatonin, it is of interest that this human hormone, made in the pineal gland, is now widely sold in health food stores without governmental control, because it is promoted as a food rather than a drug. It is touted for treatment of jet lag and insomnia, even though it has no observable hypnotic effects. Further, in a rebuff to the ads, a Harvard Medical School study showed that it is not true that older people have lower body levels of melatonin. The potential adverse or long-term effects of melatonin are not known.

As discussed earlier, **genetic factors** are believed to account for some of the differences in our reaction to drugs. If, because of an inborn error of metabolism, a person lacks a certain liver enzyme, he or she may react differently to a drug. This factor may account for drug-induced anemias in nonwhites and possibly for the often discussed susceptibility of the American Indian to ethyl alcohol. A gene-based predisposition to alcoholism is now accepted by many authorities (see Chapter 9). A widespread inherited inability to produce a certain body enzyme may explain why about 10% of whites in Europe and North America metabolize the antihypertensive drug debrisoquine only 1/200 as effectively as people who do not have the genetic defect.

Obesity can affect drug distribution and metabolism. Consider, for example, the benzodiazepines (such as Valium), which are used to treat anxieties. In one study of six obese subjects, it was found that while it took longer than usual to build up a body concentration of benzodiazepine, its half-life was prolonged from the usual 41 hours to 98 hours. The half-life of a single 10-mg IV dose of Valium was more than doubled in obese subjects. Thus, doctors and nurses treating obese patients must learn that it can take much longer for a drug to take effect, and the effects can last much longer than in the nonobese. Benzodiazepines such as Valium are oil-like compounds and preferentially dissolve in body fat. Body fat, in effect, therefore acts as a drug reservoir.

4.6.1 Children's Drugs

There is a paucity of information regarding children's drugs, particularly on dosage levels for the pediatric population and on whether certain adult drugs are safe in children. Drug companies are averse to undertaking expensive pediatric studies for a small market. There is also a real concern for using a child in a clinical study for safety and effectiveness. (In one adult study, vital patient-safety rules were broken and an eighteen-year-old died.) Thus the majority of our drugs have never been tested on the pediatric population.

The prescribing of drugs newly on the market for adults and children is criticized in a study reported in the *New England Journal of Medicine*. The authors concluded that clinicians should avoid prescribing new drugs when older, similarly effective drugs are available. The reason is that safety records for new drugs are limited, and some potentially harmful side effects may be discovered only after a drug is approved and used. The FDA disputes this philosophy, but others note that of the 548 new drug entities approved by the FDA between 1975 and 1999, 56 were eventually withdrawn or given new safety warnings.

Prescribers of drugs know that the age and weight of the patient must be considered in determining the proper dose. An elderly patient may be given the same amount of an antibiotic as a younger adult but may not require the usual dose of a sedative or hypnotic. An old rule called *Clarke's rule* uses body weight to calculate the presumably proper dose for a child:

$$\text{Dose to be given to child} = \frac{\text{weight of child (lb)} \times \text{adult dose}}{150 \text{ lb}}$$

In the metric system, with 2.2 lb equal to 1 kg, the rule is

$$\text{Dose to be given to child} = \frac{\text{weight of child (kg)} \times \text{adult dose}}{70 \text{ kg}}$$

Problem 4.2 *If the adult dose of a drug is 140 mg and a child weighs 100 lb, how many milligrams should be administered to the child?*

Note that Clarke's rule gives us a pediatric dose that is generally safe under usual conditions. In special cases or with very potent drugs, the physician will consider additional factors in determining the dose for a child. Giving medicine to children requires special care. Children are more sensitive than adults to many drugs. For example, alcohol and antihistamines, common ingredients in cold formulas, can excessively excite a child or cause drowsiness. Aspirin can be dangerously toxic if given to a child suffering from chicken pox or flu, or recovering from these or other viral infections (see Reye syndrome, Section 14.4). Barbiturates, which typically sedate adults, can make a child hyperactive; however, amphetamines can calm children.

Many young children are over-medicated. A 1994 study reported in the *Journal of the American Medical Association* found that more than one-half of all mothers surveyed had given their 3-year-olds an OTC drug in the previous month. Remember, there is no drug anywhere that will cure the common cold or speed up recovery. Antibiotics, available on prescription, do not work at all on cold viruses. Paula Botstein, M.D., says, "Just because your child is miserable and your heart aches to see her that way, doesn't mean she needs drugs." To give an accurate pediatric dose, make use of the special plastic medicine cups, oral syringes or droppers, or cylindrical dosing spoons that are available from your pharmacist.

To reach the American Academy of Pediatrics online, browse **`http://www.aap.org/default.htm`**

Parents, make certain you know the difference between regular, extra-strength, children's, and infant's drug formulas, and to prevent possible overdosages, be sure to inform each other if and when you have given your young child a dose.

4.6.2 Drug Tolerance

Tolerance can profoundly affect the way we respond to a drug. If you are a cigarette smoker or a heavy drinker, you may well remember that first smoke or that first drink, and verify how your response has changed! Tolerance—the need to ever increase the dose of a drug to maintain its effectiveness—can develop with alcohol, barbiturates, amphpetamines (central effects), benzodiazepine tranquilizers, opiates and opioids, nicotine (certain effects), and with other drugs. The mechanism by which tolerance develops can be of three types:

1. In **metabolic tolerance**, the liver responds to continued doses of the drug by making more enzymes to destroy (biotransform) the drug. Thus, greater doses of the drug are necessary to elicit the desired response. Days, weeks, or months may be required before liver enzyme levels are high enough to produce tolerance. And when the drug is discontinued, days or weeks may be necessary for enzyme levels to fall to normal.
2. In **pharmacodynamic** (cellular—adaptive) tolerance, receptors in the various tissues affected by the drug lose sensitivity to it. This loss results from physiological adjustments, such as a change in the number of receptor sites. In simple terms, the tissues just get used to the drug, and don't respond as much to its presence.
3. In **behavioral tolerance**, the drug user is repeatedly conditioned by his environment to compensate for the effects of the drug, or he learns how to handle the drug in a way the nonuser cannot.

No matter by which mechanism or combination thereof tolerance develops, it can have a profound effect upon the drug user.

Finally, a person may react unexpectedly to a drug because of the presence of a drug previously or concurrently ingested. We term this phenomenon *drug interaction*.

4.7 Drug Interactions Can Be Dangerous to Your Health

Retail pharmacists fill about 1.66 billion prescriptions a year, 54% of which are new and the remainder refills. Each of us probably has had at least one prescription filled in the last month. Some individuals may use as many as five prescriptions at one time. Others may be self-medicating, with OTC preparations such as sleeping tablets, antihistamines, cold and cough syrups, laxatives, or antacids.

When we use two or more drugs simultaneously, the possibility arises that these drugs will interact with each other in our body. Such **drug interactions** can sometimes do a lot of harm, and physicians and pharmacists are trained to anticipate them and prevent them from happening. However, since the patient is the only one who *really* knows what is being taken, he or she must know enough to play a role in prevention, too.

Drug interactions are of two major types:

1. The drugs may act on each other to potentiate (enhance) each other's effects. For example, barbiturates and alcohol are both CNS depressants when taken singly. When taken concurrently, they potentiate each other's CNS depressant effects and result in a superdepressant, dangerous combination. Such a potentiation is termed **synergism**. It is a 2 + 2 = 5 situation (see Figure 4.5). The particular combination of barbiturates and alcohol is so dangerous that death by respiratory depression is not uncommon. The rule "If you drink, don't take sleeping capsules, and vice versa" is a good one. Alcohol is also notorious for its ability to potentiate the effects of tranquilizers, antihistamines, and sedatives, resulting in drowsiness that can make driving a car or operating machinery dangerous. A description of the potentially

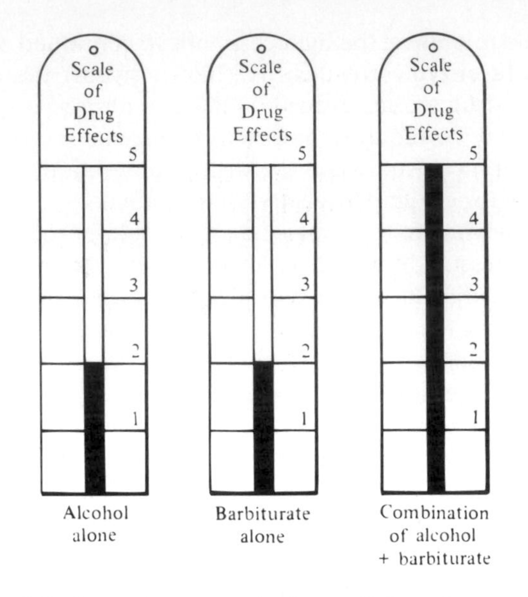

Figure 4.5 Synergism, or mutual potentiation pharmacological effect, between alcohol and barbiturates in the human.

dangerous combination of ephedrine and kola nut is given in Section 2.3. Other kinds of potentiation result when one drug inhibits the metabolism of another or displaces another from its plasma binding site (allowing a greater amount to get to its receptor site). A somewhat different kind of potentiation is seen when aspirin or salicylates are given concurrently with a blood anticoagulant such as dicumarol. Aspirin reduces the number of platelets, which are needed for normal blood clotting; this results in potentiation of dicumarol's own anti-clotting action. Abnormal bleeding can follow from this type of drug interaction. For a number of reasons, aspirin is not the innocuous drug that most people think it is. Certain combinations of antibiotics act synergistically to combat infections more effectively than either antibiotic used alone. Nonsteroidal anti-inflammatory drugs (NSAIDs, described in Chapter 14) can increase the activity of methotrexate, sometimes to poisonous levels.

2. One drug may act as a therapeutic antagonist to another, nullifying the effect of the other drug. A perfect example of this is caffeine (a CNS stimulant) nullifying the effect of sedatives (CNS depressants). Another example is the administration of an oral antidiabetic drug such as Orinase, given to reduce blood sugar, with the concurrent use of an adrenaline-like compound, which acts to increase blood sugar. Anticonvulsant drugs, used to control epilepsy, can increase the body's use of vitamin D and folic acid and cause a deficiency in these nutritional factors. Barbiturates, when combined with a major tranquilizer such as Thorazine, reduce the effectiveness of the Thorazine. A chronic alcohol user may discover that no tranquilizing effect is obtained from Valium, Librium, or Miltown. This is because chronic alcohol use causes induction of the liver enzymes that help metabolize minor tranquilizers. Smokers may discover that they need larger and more frequent doses of analgesics such as Talwin and Darvon to obtain pain relief.

> This need arises because the nicotine in cigarettes acts to enhance the metabolism of these two analgesics. High sodium intake can cause loss of control of manic-depressives treated with lithium.

Drug-condition interactions may occur when an existing medical condition makes certain drugs potentially harmful. For example, people with high blood pressure could experience an unwanted reaction if they ingest a nasal decongestant such as pseudoephedrine (Section 5.6).

Food-drug interactions are well known today, and in some cases they may be of critical importance. Consider the patient who has been depressed and for whom a *monoamine oxidase* (MAO) inhibitor has been prescribed. An MAO inhibitor allows the accumulation of body amines (like epinephrine), with the result that the brain and spinal cord are stimulated and the depression is overcome. MAO inhibitors work by inhibiting the body enzyme that normally helps metabolize such amines. The patient then eats some aged cheese, or Chianti wine, or chicken livers and suddenly gets a strong reaction (severe headache, high blood pressure, brain hemorrhage, even death). It turns out that these foods all contain large amounts of another amine called *tyramine*, which is similar to epinephrine. The MAO inhibitor had made it impossible for the patient to metabolize CNS-stimulating amines, and he or she is now getting an extra dose from an outside source. The combination is too much to handle, and a hypertensive crisis results. Anyone taking an MAO inhibitor should avoid aged and fermented foods, including pickled herring, fermented sausages such as salami and pepperoni, sharp cheese, yogurt, sour cream, beef and chicken livers, fava beans, canned figs, beer, and Chianti.

One glass of **grapefruit juice** or half a grapefruit can acutely increase the blood levels of many commonly used oral drugs. The "grapefruit juice effect" is seen with some high-blood-pressure drugs such as the calcium channel blockers Procardia, Plendil, and Sular, as well as with some antihistamines, immunosuppressants, cholesterol-lowering statins, oral contraceptives, and protease inhibitors. Grapefruit juice apparently interferes by both increasing absorption of the drug and decreasing its metabolism.

St. John's wort is used by millions of Americans for insomnia, anxiety, or depression, but not all users realize that this herb has powerful properties with the possibility of dangerous drug interactions. St. John's wort can interfere with the actions of cyclosporin and with some oral contraceptives, reducing their effectiveness. It can dramatically interfere with the action of the anti-HIV drug indivir. It diminishes the effects of warfarin (a blood thinner) and digoxin (a heart stimulant). Chemicals found in St. John's wort can affect liver function, resulting in either too high or too low blood levels of certain prescription drugs.

Oral contraceptives are known to lower blood levels of folic acid and vitamin B_6. Usually this depletion is not serious, but in pill users who have a pre-existing diet-related deficiency of folic acid or vitamin B_6, oral contraceptive use could be critical. Such women should eat extra green leafy vegetables, which are good sources of folic acid.

Other food-drug interactions include the following:

1. Large quantities of bananas or potassium supplements can cause a harmful buildup of potassium when combined with ACE inhibitors, such as Capten or Vasotec.

2. The ability of vitamin K from liver or green leafy vegetables to hinder the effectiveness of anticoagulants.
3. Soda pop or acid fruit or vegetable juices producing enough acidity to cause a drug to dissolve in the stomach instead of in the intestine. Poorer drug absorption into the bloodstream can result.
4. Soybeans, brussels sprouts, cabbage, or kale inhibiting normal production of thyroid hormone in susceptible individuals. These vegetables contain substances identified as *goitrogens*.
5. Mineral oil, a laxative still popular with some people, interfering with the body's absorption of fat-soluble vitamins (D and K). The vitamins dissolve in the mineral oil, which is itself excreted. As little as 4 teaspoonfuls of mineral oil daily can cause vitamin loss.
6. Psyllium gum (as in Metamucil) can prevent absorption of riboflavin.
7. Phenytoin can cause vitamin D deficiency.
8. Chlorpropamide taken concurrently with alcohol can cause the flush reaction.
9. When alcohol (beer, wine) is taken concurrently with certain cephalosporin-type antibiotics (e.g., Cefotan), an Antabuse-like reaction can occur, manifested by flushing, headache, sweating, and tachycardia.
10. Two drugs that block the production of acid in the stomach (Zantac and Tagamet) also block the action of an enzyme that catalyzes the breakdown of alcohol, thus increasing the level of alcohol in the blood. Research has shown that patients on these drugs can get legally drunk on less than two glasses of wine.

Armed with all of this knowledge, the alert consumer can do the following actions to prevent drug interactions:

1. Take only those drugs that are absolutely necessary; avoid all others.
2. Unless directed to do so, try to avoid taking more than one drug at a time, and strictly follow the doctor's directions on when and how to take the drug and what foods and liquids to avoid.
3. Read the labels on OTC drugs, and ask your pharmacist about possible drug interactions.
4. Eat a variety of foods. This will help minimize the effects of drug-food interactions, should they occur.

Table 4.1 summarizes some of the more common and important drug interactions.

Need help evaluating potential drug interactions? Here are some Internet Web sites you can browse:

www.drugchecker.drkoop.com
http://ohioline.osu.edu/hyg-fact/5000/5406.html
http://www.personalhealthzone.com
http://vm.cfsan.fda.gov/~ird/fdinter.html
http://www.projinf.org/fs/drugin.html

Women, for dozens of drug interactions of special interest to you, see

http://www.4woman.gov/FAQ/drug.htm (or click "drug interactions" on the site's homepage).

Table 4.1 Some of the Adverse Interactions of Commonly Used Drugs

When a Person Is Taking This Drug	*And Combines It With This Drug*	*This Adverse Effect May Occur*
Heart Drug		
Digitalis	Thiazide diuretics	Increased digitalis toxicity
Digitalis	Reserpine	Increased digitalis toxicity
Digitalis	Barbiturates	Enhanced digitalis metabolism
Sedatives		
Barbiturates	Alcohol	Greatly increased sedation (synergism)
Chloral hydrate	Alcohol	Greatly increased sedation (synergism)
Barbiturates	Oral anticoagulants	Diminished anticoagulant effect
Barbiturates	MAO inhibitors	Increased CNS depression
Barbiturates	Male sex hormones	Diminished activity of sex hormones
Barbiturates	Oral contraceptives	Inhibition of contraceptive action
Barbiturates	Oral antidiabetic drugs	Enhancement of barbiturate activity
Minor tranquilizers		
Valium, Librium, Serax, etc.	Alcohol	Increased CNS depression (additive effect)
Valium, Librium, etc.	MAO inhibitors such as Parnate, Eutonyl, Marplan	Oversedation
Major tranquilizers		
Phenothiazines (Thorazine, Mellaril, Compazine, etc.)	Alcohol	Oversedation (additive effect)
Major tranquilizers, as above	Thiazide diuretics	Can result in shock
Major tranquilizers, as above	Antihistamines	Additive effect (CNS depression)
Major tranquilizers, as above	Morphine	Enhanced sedation
Haldol and Innovar	Lithium	Increased tranquilizer toxicity
Drugs for treating parkinsonism		
1-Dopa	MAO inhibitors	Hypertensive crisis
Artane, Pagitane	Phenothiazines	Lowered blood levels of the phenothiazines
Oral contraceptives	Tegretol, Dilantin, rifampin, certain antifungals	Possibly diminished contraceptive action

Table 4.1 continued

When a Person Is Taking This Drug	*And Combines It With This Drug*	*This Adverse Effect May Occur*
Antidepressives		
Prozac, Paxil, Zoloft	MAO inhibitors	Nausea, shivering, confusion, muscle contractions
Bronchodilator		
Primatene (for asthma)	Tagamet or antibiotics	Potential life-threatening crisis
Painkillers		
Aspirin	Anticoagulants	Hemorrhage
Aspirin	Alcohol	Increased incidence of GI bleeding
Aspirin	Probenecid (Bememid)	Inhibition of probenecid
Demerol and narcotic analgesics generally	MAO inhibitors	Increased CNS depression and respiratory depression
Narcotic analgesics generally	Sedatives	Increased sedation effect
Butazolidine	Oral antidiabetics	Increased effect of oral antidiabetic drug
Antiepilepsy drugs		
Dilantin	Barbiturates	Diminished anticonvulsant effect due to enzyme induction
Dilantin	Oral contraceptives	Increased seizure tendency; fluid retention
Dilantin	Chronic alcohol use	Diminished anticonvulsant effect
Antibiotics		
Tetracyclines and penicillin G	Antacid or milk	Reduced effectiveness of the antibiotic
Penicillin	Tetracycline	Diminished activity of penicillin
Tetracycline	Oral iron preparations	Inhibited absorption of the iron preparation
Steroids		
Estrogen, progesterone	Phenobarbital	Diminished steroid activity in test animals
Oral contraceptives	Anticoagulants	Diminished effect of anticoagulant
Hydrocortisone	Antihistamines	Diminished hydrocortisone activity

Table 4.1 continued

When a Person Is Taking This Drug	*And Combines It With This Drug*	*This Adverse Effect May Occur*
Antihistamines		
Benadryl, Chlortrimeton	Alcohol	Increased sedation
Benadryl	LUVOX and certain blood pressure drugs	Magnified effects of LUVOX
Antihistamines generally	Barbiturates	At first, severe CNS depression; then they nullify each other
Seldane, Hismanal	Certain antibiotics and antifungals	Rare cases of heart attack
Drugs for high blood pressure		
Ismelin	Alcohol	Extra fall in blood pressure
Ismelin	Amphetamines	Inhibition of Ismelin
Apresoline	Thiazide diuretics	Enhancement of Apresoline's effect
Aldomet	Methamphetamine	Inhibition of Aldomet
Miscellaneous		
Ginkgo biloba	ASA, anticoagulants	Hemorrhage
Insulin	Thyroid preparations	Possible increase in insulin requirements
Insulin	Terramycin	Enhancement of insulin activity
Insulin	Inderal	Marked hypoglycemic effect
Viagra	Nitroglycerin	Acute blood pressure drop
Melatonin	Serotonin reuptake inhibitors such as Prozac	Serotonin syndrome (stroke, heart attack)

4.8 Drugs and Sex

It is possible that in some people certain drugs[2] can affect sexual drive (libido) and sexual performance in either a positive or negative way. We know, however, that the reported incidence of sexual disturbances is very low and should not be considered a common side effect of most drugs.

In Chapter 9, "Alcohol and 100 Million Americans," we will see that alcohol increases sexual desire but depresses performance. Impotence is also sometimes reported in persons taking the anti-high blood pressure drugs clonidine, prazosin, and chlorthalidone with amphetamines, MAO inhibitors, haloperidol (Haldol), or the anti-alcohol drug Antabuse.

All phenothiazine-type antipsychotics are capable of inhibiting ejaculation, as is the antihypertensive drug guanethidine (Ismelin). Reduction of libido is seen with certain minor tranquilizers of the benzodiazepine type, reserpine, methadone, and opiates (generally).

All of these drug effects on sex are encountered infrequently and are highly variable. Sometimes the same drug (e.g., Valium and Elavil) can decrease libido in one person but increase it in another. Cholestyramine, a cholesterol-lowering drug, is reported to increase libido. There have been some reports of a near-aphrodisiac effect of L-dopa, used to treat Parkinson's disease.

The antidepressant drug trazodone (Desyrel) is known to cause priapism (abnormal erection of the penis) in a significant number of users; in some, surgery was required to terminate the priapism. Yohimbine, from the bark of a tree native to Africa, has been sold illegally as an aphrodisiac ("passion pill"). Promoters cite a single Canadian hospital study in which 23 sexually unresponsive people were given yohimbine for 10 weeks. Six of the 23 responded with arousal. The prescription drugs Yocon and Yohimex contain yohimbine. Medical researchers have identified yohimbine as an alpha$_2$-adrenergic blocking agent (see Section 5.4) useful in the treatment of some types of male erectile impotence. It can also be used as a mydriatic (agent that dilates the pupil of the eye). Be aware, however, that yohimbine was identified 25 years ago as a hallucinogen capable of activating a schizophrenic psychosis. This drug does not appear to be safe for casual use.

Viagra (sildenafil, Pfizer), a phenomenal drug for treating impotence, acts by binding to (and inhibiting) one of the phosphodiesterases that catalyzes the breakdown of *cyclic guanosine monophosphate* (cGMP). Through a complicated biochemical pathway, this results in smooth muscle relaxations and penile erection, with a success rate of about 50 percent.

Because orgasm is mediated by both the sympathetic and parasympathetic divisions of the autonomic nervous system, it is possible that some drugs act on sexual performance by blocking or inhibiting these nerves. Drugs may also act through mood-altering effects.

---

[2]We exclude the sex hormones from this discussion, because they have obvious effects on libido. Spanish fly is discussed in Section 9.4.

4.9 Women's Health

Did you know that since 1994 the FDA has had an Office of Women's Health (OWH), and that it sponsors public health fairs, workshops, and seminars? Among its goals are the advancement of drug therapies in cardiovascular disease, osteoporosis, menopause, cancer, autoimmune diseases, and contraception. OWH is especially anxious to rectify past neglect of women in clinical trials of new drugs, and it stresses that since 1993 drug makers are required to include women in clinical testing, including women of child-bearing age.

The National Institute on Drug Abuse (NIDA) offers information on *Advances in Research on Women's Health and Gender Differences* at **http://www.nida.nih.gov/WHGD/WHGDAdvance.html**. Topics include epidemiology, gender differences in drug effects, pregnancy and drugs, gender differences in drug dependency, comorbidity, and drug-treatment outcomes. Women and gender research is receiving much more attention that it did only a decade ago.

The FDA's *Center for Food Safety and Applied Nutrition* offers bundles of information for women regarding pregnancy, AIDS, cancer, diabetes, eating disorders, heart disease, estrogens, supplements, allergies, health fraud, diets, and nutrition. Its Web site is **http://vm.cfsan.fda.gov/~dms/wh-toc.html**.

A *Women and Clinical Trials Fact Sheet* is published on the Internet at **http://www.womens-health.org/clinfact.html**. Traditionally, researchers have carried out tests on new drugs using male subjects. They cite reasons such as the need to protect a potential fetus (even if the woman is not pregnant), avoidance of legal responsibility in case of teratogenic effects of the drug testing, and possible interference resulting from hormonal changes in the woman. However, the discovery of differences between male and female responses to drug therapy can be observed only if *both* sexes are tested. Further, the fact that a woman's physiology does change during her menstrual cycle makes the study of the effects of these drugs imperative. If only males are used to study drugs, then we must infer that the results of these studies are applicable to females, which is not always the case. Federal law now mandates inclusion of women in federally funded clinical drug trials.

Problem 4.3 *Is it reasonable for women to remember thalidomide and the Tuskegee experiments and thus refuse to be "guinea pigs" in the clinical testing of a new drug?*

Evidence of the increased development of drugs for women is the introduction of "designer estrogens" (Section 2.5), exemplified by Eli Lilly's Evista, plus two dozen additional drugs in clinical trials. Thirteen new contraceptives are in the works. Heart disease—women's No. 1 killer—arthritis, breast cancer, and drugs for stroke are areas of intense pharmaceutical research.

Women account for 75 percent of nursing home residents, 61 percent of annual doctor visits, and 59 percent of prescription drug purchases. Women outlive men by about 6 or 7 years. They have more buying power. The nation's best-selling drug is the 50-year-old estrogen product, Premarin.

The National Council on Alcohol and Drug Dependence (NCADD) supports efforts to educate women about drug use, including alcohol, tobacco, prescription, and OTC drugs, as well as illegal drugs.

The National Women's Health Information Center (NWHIC), managed by the U.S. Public Health Service, offers a Web site, **http://www.4woman.org**, and a toll-free number, 1-800-994-WOMAN. This Web site offers links to more than 1,000 other federal and private women's health Web sites. Just click on the site index. Also check out **http://www.ama-assn.org/special** (then click on "Women's Health Information Center"). Facilities offering special services for women are found at **http://www.SAMHSA.gov/oas/facts.cfm.htm** (then click on "Womens").

4.10 Excretion of Drugs

The final step in the fate of drugs is their excretion from the body. This elimination can proceed along five possible paths:

1. Through the urine
2. In the feces (via the bile)
3. On the breath
4. In perspiration, saliva, or milk
5. As vomit

To be excreted in the urine, the most important of the five pathways, a drug must be water-soluble or chemically alterable into a water-soluble form. As we discussed earlier in this chapter, the process of altering a drug to prepare it for excretion is termed **metabolism** (or *biotransformation*). Drug metabolism, catalyzed by liver enzymes, converts a fat-soluble drug such as a barbiturate into a urine-soluble oxidation product. It converts an amine such as adrenaline into a pharmacologically inactive, urine-soluble oxidative deamination product. It also causes hydrolysis of an ester into smaller, water-soluble fragments. These and other transformations are designed to mobilize the drug out of the bloodstream and therefore out of body fat deposits and off protein binding sites—in short, to get rid of foreign chemicals. Average daily adult urine output is about 1000 mL (946 mL equals 1 quart), and this route of excretion can handle a considerable amount of drug.

That we can eliminate drugs by exhaling them is apparent from the odor of an alcoholic beverage on the breath of a drinker or onions or garlic after a meal. Elimination of drugs on the breath is a significant route only for volatile drugs that are mostly fat-soluble. While it is proceeding, the liver can metabolize the substance into a water-soluble, inactive form. General inhalation anesthetics, dissolved in body fat tissues, are in equilibrium with the same anesthetic dissolved in the blood. The blood passes through the lungs, where it gives up some of its content of dissolved gases into the air we exhale.

Excretion of drugs through perspiration, saliva, or milk is of lesser importance. It should be mentioned, however, that maternal milk can contain drugs and chemicals present in the mother's blood system. Tranquilizers, oral contraceptives, laxatives, and salicylates ingested by the mother can be transmitted through her breast milk. Nicotine is also found in the milk of lactating women who smoke, as much as 0.5 mg/L in heavy smokers. According to *Drug Abuse Update* magazine, researchers have found exceptionally high urine levels of nicotine in infants who were breast-fed by mothers who smoked. The researchers concluded that mothers who smoke and breast feed can pass on the equivalent of 20 cigarettes a day to their babies.

Metronidazole (Flagyl), used in the treatment of trichomoniasis ("trich"), is also secreted in breast milk. During treatment with Flagyl, egotamine, or lithium, women should discontinue breast feeding. In fact, the use of any drug during lactation should be seriously questioned.

Case Histories *Drugs in Breast Milk Kill Infants*

A 3-week-old infant was found dead in her mother's filthy, garbage-strewn apartment. The Corona (California) coroner said the infant had been breast fed and had a blood concentration of 0.266 μ/mL of methamphetamine, enough to kill her. (Adult toxic doses range from 0.6 to 5.0 μg/mL.) The 25-year old mother used methamphetamine regularly; she was sentenced to 5 years in prison on a charge of child endangerment.

Authorities in Tucscon, Arizona, charged a young mother with first-degree murder, alleging she killed her 7-week-old daughter by feeding her drug-filled breast milk. Police found syringes and drug paraphernalia in her home, and she admitted using heroin after her daughter's birth. Heroin can cause the oxygen deprivation that killed the baby. Because the father did nothing to stop the child abuse, he was also charged with murder.

Chemicals such as mercury, lead, and potassium iodide, when introduced into the human body, are excreted in part in the saliva. Excretion of ethyl alcohol in the saliva is well established. Indeed, it has been discovered that the alcohol content of saliva can be made the basis of a test for sobriety (see Section 9.7).

4.11 Drugs in Our Diet

It is important to know that some readily available foods and beverages are the source of drugs and chemicals and that, unknowingly, many of us may be consuming agents that have a significant pharmacological effect on our bodies.

The common kitchen spice **nutmeg**, for example, has a long history of use as a hallucinogen. Reactions to it, however, are diverse, with many people experiencing no mind-altering effects and others experiencing profound distortion of time and space. Headache, cramps, and nausea are also observed after nutmeg ingestion. Prisons dropped nutmeg from their list of kitchen spices after abuse of it was discovered.

As mentioned earlier, many foods are rich in **tyramine**, a chemical that has the beta-phenethylamine chemical structure (see Figure 4a in Appendix I) found in norepinephrine and the amphetamines. Tyramine is capable of producing strong excitatory actions in the body, especially in the presence of MAO inhibitors. Some foods that contain tyramine are aged cheese, avocados, pickled herring, liverwurst, soy sauce, Chianti wine, and bananas.

Monosodium glutamate (MSG) is a flavor enhancer found in almost all canned meats. Its drug effects became well known after it was discovered to be the causative agent in the "Chinese restaurant syndrome." After consuming Oriental-style food flavored with MSG, some people became dizzy and nauseated and experienced palpitations. While MSG has this effect in some people, most users do not react at all.

Caffeine (found in coffee, tea, and cola drinks) is suspected of being teratogenic (see Chapter 7).

Sugar (sucrose) has become a tool with which food manufacturers manipulate our food-buying habits. Sugar occurs in heavy amounts in some breakfast cereals, salad dressings, shake-and-bake flavorings, and formerly in baby foods. Ingestion of large amounts of sugar can alter the production of insulin in the pancreas.

We must also consider the importance of a diet adequate in the essential amino acids and fatty acids. There are nine amino acids found in protein that cannot be synthesized in the body, or at least not rapidly enough to meet the body's needs. These essential factors, then, must be included in our diet. One of them is tryptophan. Tryptophan is the most likely precursor to serotonin, a vital neurotransmitter (see Section 5.7). It is conceivable that if our diet lacks tryptophan, we will experience the problems associated with insufficient serotonin (serotonin is implicated in the establishment of sleep-wake patterns and in our ability to handle stressful stimuli).

Similarly, we must obtain essential fatty acids (linoleic and linolenic acids) from which our body makes prostaglandins. A diet deficient in these fatty acids could result in inadequate production of prostaglandins and consequent interference with normal body functioning.

Web Sites You Can Browse on Related Topics

Pregnancy - Drug Effect
http://www.perinatology.com/exposures/druglist.htm
http://www.efn.org/~djz/birth/complications.html#drug

Drug Interactions
http://www.mhc.com/drug.html
http://www.cfhinofo.org

National Clearinghouse for Alcohol and Drug Information. A must-browse site is:
http://www.health(then click on "drugs")

Pharmacology
Search for "pharmacology"

Scopolamine
Search for "scopolamine"

Yohimbine
www.nim.nih.gov/medlineplus/druginfo/uspdi/202639.html
Also search for "yohimbine"

Circadian Rhythms
http://www.sleepnet.com/narco5/messages/80b.html
www.chronotherapy-circadian-rhythms.com/

Study Questions

1. Which of these three routes of drug administration—skin, inhalation, or oral—would ordinarily provide the fastest access to the bloodstream? Explain your reasoning.
2. To achieve the fastest drug action, should one generally take medication with meals or before eating? Explain.

3. List three general advantages of the oral route of drug administration over the injection route, and vice versa.
4. All other things being equal, which route offers the best chance of getting *all* of a given dose of a drug into your bloodstream—skin, oral, inhalation, or body cavity?
5. Which of the following factors would tend to hasten deterioration of a drug and thus shorten its shelf life—heat, moisture, refrigeration, sunlight, dryness, presence of bacteria, exposure to air, or storage in an amber bottle?
6. **a.** Name two drugs currently available in skin patch (transdermal) dosage forms.
   b. For a drug to be suitable for skin patch application, what requirements must it meet?
7. Define (**a**) parenteral, (**b**) IV, (**c**) IM, (**d**) buccal cavity, (**e**) adipose tissue, (**f**) blood-brain barrier, (**g**) placental barrier, (**h**) cytochrome P-450 enzymes, (**i**) tolerance (to drugs), (**j**) drug half-life, (**k**) dose-related effect.
8. If the adult dose of a tranquilizer is 15 mg, what would be the dose for a 50-lb child? Show your calculations.
9. Of the five possible routes for the body's excretion of drugs, which is the most important? Why is this true?
10. Assume that drug X has a chemical makeup that results in its being water-soluble and that of drug Y results in its being fat-soluble. Which drug will pass the blood-brain barrier more readily? Which will deposit more extensively in fat tissue? Which will tend to be excreted faster in the urine? Which would probably have a greater half-life in the body?
11. Our text discusses body reservoirs for drugs (i.e., tissues that can bind a drug and then slowly release it into the bloodstream). Name three such reservoirs.
12. In both a qualitative and a quantitative sense, people can respond differently to the same drug. Consider two 75-kg men, one 20 years old and the other 75 years old. Predict their responses to a 50-mg dose of a sleeping tablet; a dose of alcohol; the caffeine in two cups of coffee.
13. Could a woman react differently to a drug than a man just because of her sex?
14. Explain the drug interaction known as *synergism*. How does it differ from the simple additive effect of two drugs?
15. Using Table 4.1, predict the adverse effect that might occur if (**a**) the minor tranquilizer Atarax is taken in combination with alcohol; (**b**) an oral contraceptive is taken concurrently with a barbiturate; (**c**) aspirin is taken during a course of anticoagulant therapy.
16. What is meant by circadian rhythms? Give three examples.
17. Examine the accompanying graph and indicate which curve (A, B, or C) would best represent a typical dose response to oral administration of a drug, which to sublingual, and which to transdermal skin patch.

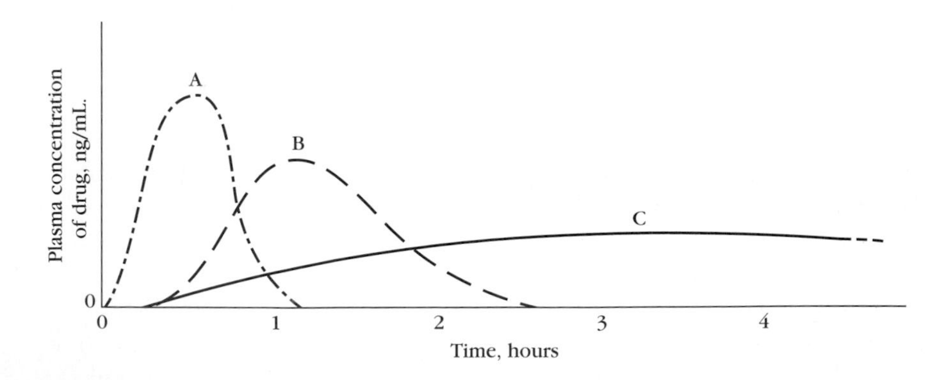

18. *Advanced study question*: Cigarette smoking, prolonged exposure to insecticides, and chronic alcohol use can affect a person's response to prescription drugs. Explain how this change can occur.
19. *Advanced study question*: Some families are now keeping records of prescription drugs and other substances taken by family members over a period of many years. What useful purpose could such records serve? At what times during the family's history would it be especially wise to keep records? Besides drugs, what other factors, techniques, or types of exposure should be recorded?
20. *Advanced study question*: Doctors have been known to overprescribe drugs, sometimes with deleterious results. In one real case, an 84-year-old woman was receiving all of the following drugs, most of them two or three times a day: isorbide dinitrate, Lanoxin, quinidine, dipyridamole, Lasix, meclizine, Phazyme, Tagamet, DSS, Valium, chlorpheniramine, KCl, Mylanta, and Mylicon. Using the PDR, list the possible side effects (if any) of each drug.
21. *Advanced study question*: Upon retiring for the night, a 230-lb male, aged 70, rubs one-quarter tube of salicylate-based ointment onto his sore back muscles in an attempt to relieve pain. Identify six factors that could possibly affect how much of the salicylate enters his bloodstream and how much pain relief he receives.
22. The half-lives of many benzodiazepine and barbiturate drugs *increase* 50–150 percent between age 30 and age 70. Explain.

CHAPTER 5 Drugs at the Synapse

Key Words in This Chapter

- Synapse
- Neurotransmitter
- Acetylcholine
- Norepinephrine
- Dopamine
- Sympathomimetic
- GABA
- Serotonin

Learning Objectives

After you complete your study of this chapter, you should be able to do the following tasks:

- Explain the nature and function of a synapse.
- Discuss the role of neurotransmitters at the synapse.
- Define and contrast cholinergic and adrenergic.
- Explain how enzymes can work to deactivate neurotransmitters.
- Cite the two great divisions of the autonomic nervous system.
- Explain what it means to mimic sympathetic discharge.
- Discuss serotonin's role in the body.

5.1 Introduction

In a moment, you will be reading about the synapse, the critically important space between nerve cells where transmission of a nerve impulse occurs. You will learn about three of the most important neurotransmitter molecules that can act at the billions of synapses throughout the human body, influencing many body functions—such as pain perception, brain activity, mood, blood pressure, heart rate, digestion, elimination, and respiration. Then you will learn that exogenous chemicals (drugs) can act by interfering with the concentration of neurotransmitters in the synapse, or by themselves acting the way neurotransmitters do. All of this information will help you better to understand the pharmacology of various drug classes.

For example, cocaine acts as a CNS stimulant by changing the availability of two important neurotransmitters in the brain. The antidepressant drug Elavil acts by increasing the availability of a brain synaptic transmitter. We could say much the same about the basis of action of mood stabilizers, minor tranquilizers, antihypertensives, sedatives, and psychedelics.

It may help you to know one more fact. Each neurotransmitter in the synapse has a **transporter molecule**—a protein that assists it in getting where it is to go. This statement is true for the catecholamines (dopamine and norepinephrine), 5-HT, glutamate, aspartate, GABA, glycine, adenosine, choline, and many others. Transporters stand ready for action in the synaptic membranes of nerve cells, but they are susceptible to interference by drugs, too.

Also in this chapter, we learn about mental and physical stress, and the nerves, transmitters, and synthetic drugs that can influence our physiological responses. You will learn that an understanding of this topic can be important in understanding asthma treatment, weight loss, treatment of shock, and some people's use of nasal decongestant and other cold medications.

The drugs we examine in this chapter act at the synapse, and this is where our discussion begins.

5.2 What Is the Synapse?

The human nervous system contains some 100 billion neurons, or nerve cells. Some of these neurons are information-gathering, or **sensory**, in nature, while others are involved directly in movement or muscle activity and are termed **motor nerves**.

Neurons carry impulses from sensory organs such as the skin, eye, ear, or nose to the brain. Neurons also carry motor impulses from the CNS to organs of the body, such as the heart, GI tract, pancreas, and gonads and to skeletal muscles of the leg, hand, neck, and eye. Typical motor nerve cells are represented diagrammatically in Figure 5.1. The cell body, containing the nucleus, is the center of cell activity. From it project two kinds of cell processes, **dendrites** and the **axon**. In motor neurons of the type shown here, the axon process can be remarkably long (60 cm or more), as in the case of a spinal nerve activating the muscles of the foot. Conduction of impulses down an axon is extremely fast—about 200 miles per hour. In a nerve cell, dendrites

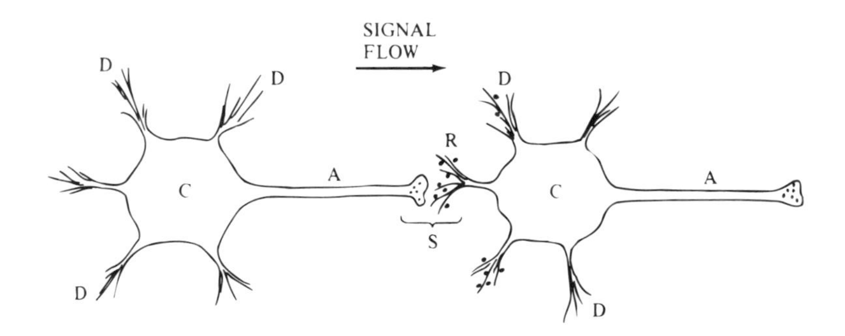

Figure 5.1 Diagrammatic representation of two neurons (nerve cells). Chemical input received through dendrites (D) is changed to an electrical signal that travels down the axon (A). From the ends of the axons, neurotransmitter chemicals are released that continue to carry the signal across a microscopic space called the synapse (S) to receptors (R) on the next neuron. The process is usually repeated many times to create signal paths throughout the nervous system. Cell bodies are labeled C. Cell nuclei are not shown.

function to gather information from other cells or sources; axons serve to transmit impulses to other cells. Now, the description of a synapse becomes important.

Consider two nerve cells that are part of a chain transmitting impulses to some organ or muscle (as in Figure 5.1). The two neurons do not actually touch each other. A space or gap exists of some 20–50 nm between them, called the **synapse** or **synaptic junction**. Figure 5.2 shows an electron micrograph of an actual synaptic junction. Note that in this figure, the synaptic cleft (or gap) is filled with a protein-rich fluid. The b cell represents the end of an axon; the c cell is a dendrite on another neuron or part of a cell in muscle or gland tissue. It is the c cell that contains the receptor sites (see the next section). The nerve before the synapse is termed the **presynaptic nerve**; the one after it is the **postsynaptic nerve**.

There are billions of such synapses established throughout the body, in the brain, in the visceral organs, in the spinal cord, in the musculature—wherever an impulse must be conducted from one neuron to the next or from a neuron to an effector organ. A typical neuron may have more than 1,000 synapses and may receive information from about 1,000 other neurons. All that we do every day, from walking, talking, playing, and eating to thinking, digesting, and healing involves impulses transmitted across synaptic junctions. As a familiar example, consider the involuntary blinking of an eye when an insect flies too close. Sensory neurons detect the presence of the insect and transmit the knowledge by way of a reflex arc to the portion of the brain regulating the motor activity of the eye. The impulse is transmitted across the appropriate synapse in the brain to one of the three cranial nerves that control eye muscle movement. Down the motor axon comes the impulse, at 200

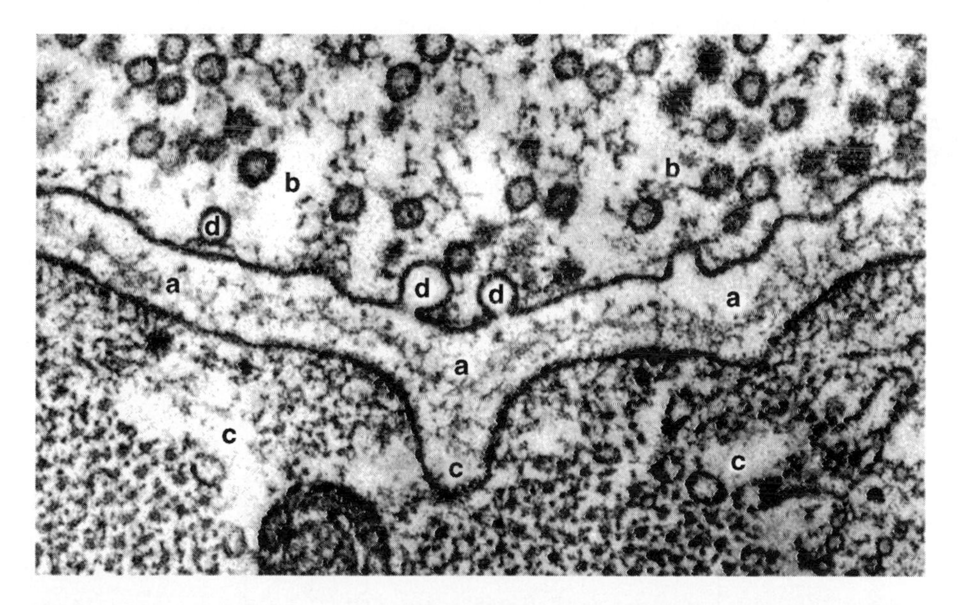

Figure 5.2 A synapse. (a) Fluid-filled synaptic cleft, or gap between the two cells (b and c); (d) the synaptic vesicles (see text), some of which have opened, spilling their contents into the cleft; (e) junctional fold in receptor cell membrane. The impulse is traveling from top to bottom. (Electron micrograph. Henry A. Lester, "The Response to Acetylcholine," *Scientific American,* February 1977, p. 107. Used by permission of John E. Heuser, M.D.)

miles per hour. The impulse is transmitted across another synapse to the cells of the eye muscle, and the eye blinks. All of this action takes place within a fraction of a second. The human body contains some 60 trillion synapses. In the brain itself, an estimated 100 million billion synapses fire and pass messages every day.

5.2.1 Summary

Neurons consist of three general parts: the dendrites, the cell body, and the axon. The dendrites collect information, the cell body processes it, and the axons conduct the impulse very rapidly to the next cell. Somehow, the impulse must get across the synapse, which is the gap between the cells.

5.3 What Happens at the Synapse?

Transmission of nerve impulses from one cell to another is so fast that investigators long believed that it was purely an electrical phenomenon. We now know, however, that chemical **neurotransmitters** are released from the end of one neuron into the synaptic cleft. They diffuse across the tiny space (only a few millionths of a millimeter wide) and bind to the receptor sites on the next neuron, for which they have great affinity and high specificity. Thus, the signal continues on its way, and ultimately the desired physiological result is achieved (see Figure 5.3). In effect, then, the transmitters carry the message across the gap. The process depicted in Figure 5.3 thus represents the transmission of a nerve impulse across a synapse. Molecules of transmitter substance are made and stored in large supply in sacs called *synaptic vesicles* (there may be as many as 10,000 molecules per sac).

When an impulse arrives at the synapse, the vesicles open and release their contents into the cleft within a thousandth of a second. Within another ten-thousandth of a second, these molecules have diffused across the cleft and have bound themselves to receptor sites in the effector cell; transmission is complete.

Thus, neurotransmitters are the chemical signals of the nervous system, and when released, they move across the synapse to other neurons or to certain secretory cells. They also function at the myoneural junction—the connection of a nerve to

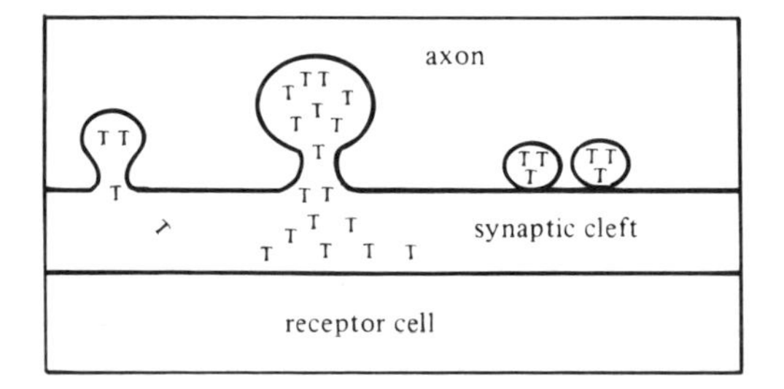

Figure 5.3 Release of neurotransmitter molecules from synaptic vesicles (see (d) in Figure 5.2) with subsequent diffusion into the synaptic cleft and binding to the membrane of the receptor cell (junctional folds in receptor cell are not shown).

a muscle fiber. By the way, Nobel laureate Eric Kandel has concluded that the synapses are the actual storage sites for memory.

What terminates the transmission across a synapse? That is, what stops the neuron from continuously firing its signal across the gap? Transmitter molecules are inactivated in two general ways:

1. **Enzymes** present in the immediate area catalyze the breakdown (catabolism) of the transmitter substance. This process occurs within a thousandth of a second after they have done their job.
2. Energy is supplied by the cells for **active reuptake** of the transmitter molecules. They are reabsorbed back into the vesicles from which they came, thus diminishing their concentration in the synaptic cleft. As much as a third of released norepinephrine, for example, is reabsorbed.

We shall see how drugs can alter both types of transmitter inactivation, thus giving rise to therapeutically useful or sometimes dangerous medicinal agents.

5.4 What Are the Transmitters?

Now we come to the part of our discussion that has application to drugs and the human body, for knowledge of the identity of the neurotransmitters has led to the discovery of new drugs and new therapeutic applications.

There are several important neurotransmitters:

1. **Acetylcholine**. Proof of the existence and action of this neurotransmitter was obtained by Otto Loewi in 1921. Acetylcholine is the chemical mediator at many synapses in vertebrate and invertebrate animals. In humans, acetylcholine is the transmitter in nerves that slows the heart, controls the use of voluntary muscle, constricts involuntary muscle, and keeps the CNS functioning as an integrated unit. Acetylcholine is the transmitter at all preganglionic nerve endings in both the parasympathetic and sympathetic divisions of the autonomic nervous system (see the following discussion) and at postganglionic endings in the parasympathetic division. Acetylcholine is a local phenomenon; it is discharged and deactivated at discrete local points and generally does not appear in the circulating blood. Acetylcholine is distributed throughout the CNS. Nerve endings and effects involving acetylcholine are called *cholinergic*.
2. **Norepinephrine** (NE, norepi). In 1946, von Euler in Sweden obtained proof that norepinephrine (pronounced *nor-epee-NEFF-rin*) is the major transmitter substance at postganglionic nerve endings in the sympathetic nervous system (see the following discussion). Like acetylcholine, NE is stored in vesicles adjacent to the synaptic cleft. Stimulation of the presynaptic nerve results in release of NE into the cleft and thus synaptic conduction.

We should distinguish between norepinephrine (noradrenaline) and epinephrine (adrenaline). The latter has a methyl group on its nitrogen atom; the former does not (see Figure 4a in Appendix I for its structure). Both are catecholamines (pronounced *cat-eh-CALL-ameens*), so named because they are amino derivatives of an organic compound known as catechol. Both are secreted by the adrenal gland (actually, the adrenal medulla). However, NE plays a more important role in synaptic transmission, for it is now considered to be a hormone released by the adrenal gland to stimulate heart contractions, dilate the bronchial tree, and increase contractile strength of arm and leg muscles. A typical blood plasma level of NE is 0.8μg/L, and that of epinephrine is 0.1μg/L, showing that these catecholamines can get into the bloodstream by way of adrenal gland secretion and apparently also via liberation from nerve endings. Epinephrine is a powerful heart stimulant, increasing heart rate and blood output. It is more effective than NE in elevating blood sugar levels, and it is active in raising blood concentrations of free fatty acids. Epinephrine is inactivated by the same enzymes that inactivate NE.

Some advanced information: The adjective **adrenergic** generally refers to nerves in the heart, bronchi, and intestines that release or use catecholamines as their transmitter substances. Adrenergic receptors are further classified as **alpha** or **beta**. Alpha-adrenergic receptors operate through calcium channels and respond to epinephrine, norepinephrine, and other drugs; their stimulation results in contraction of smooth muscle. Two well-known alpha-adrenergic receptor–blocking agents are phenoxybenzamine and phentolamine. Beta-adrenergic receptors respond especially to epinephrine; their stimulation results in relaxation of smooth muscle and stimulation of cardiac muscle. A famous beta-adrenergic receptor–blocking drug (*beta-blocker*) is propranolol (Inderal). In order to explain different actions of various agonists and blockers, it has been necessary to further subclassify adrenergic receptors into alpha$_1$, alpha$_2$, beta$_1$, beta$_2$ and beta$_3$. Beta$_1$ agonist drugs stimulate the heart and promote lipolysis. Beta$_2$ agonist drugs dilate the bronchi and cause vasodilation.

3. **Dopamine**. A close chemical relative of NE is dopamine (dihydroxyphenethylamine). This catecholamine is important primarily in the brain, but there are also receptors for it in the heart and kidney. Dopamine is the neurotransmitter for (1) the brain's extrapyramidal system (coordination and integration of fine muscular movement, such as picking up small objects); (2) the brain's mesolimbic system (memory and emotions); and (3) the hypothalamic-pituitary axis (release of the hormone for lactation). In Parkinson disease, the dopamine content of the caudate nucleus portion of the brain is found to be far below normal. Parkinsonism is treated by giving the patient L-dopa, the biogenetic precursor to dopamine (L-dopa readily passes the blood-brain barrier; dopamine does not).[1] L-Dopa is biotransformed in situ to dopamine.

Problem 5.1 *In a medical textbook, look up the symptoms of Parkinson disease and determine if they correspond to a disruption of any of the three systems listed earlier in this paragraph.*

Dopamine is excitatory to the brain. The antipsychotic major tranquilizers work by blocking dopamine receptors in the brain. Dopamine (Intropin) has been used as a pressor agent in the treatment of shock. It has the typical catecholamine action of

[1]Only 25% of an oral dose of dopamine enters the brain.

increasing cardiac output and raising blood pressure. Intravenous administration of Intropin has little effect on the CNS because the drug does not readily pass the blood-brain barrier. Some researchers have proposed a dopamine hypothesis of mental illness. They believe that excessive production of dopamine in the brain results in the signs of schizophrenia. Support for this theory is found in the fact that the major tranquilizers (the antipsychotics) work by blocking dopamine receptors in the brain. As discussed in Section 1.4, dopamine is believed to be a key chemical in producing drug-induced feelings of pleasure and reward and, over time, addiction and withdrawal symptoms.

Additional human neurotransmitters are GABA (Section 5.7), glutamate/aspartate, serotonin (Section 5.7), histamine, and the tachykinins. Researchers at Johns Hopkins University School of Medicine have concluded that D-serine is made in mammals' brains and acts as a neurotransmitter there, activating the N-methyl-D-aspartate receptor. The D-series of amino acids is a rarity in humans.

Next, we must consider the enzyme-catalyzed deactivation of neurotransmitters in order to understand the action of certain drugs, such as the antidepressant monoamine oxidase inhibitors.

Acetylcholine, once released into the synaptic cleft, is deactivated by hydrolysis into acetate and choline fragments:

$$\underset{\text{biologically active}}{\text{Acetylcholine}} \xrightarrow[\textit{hydrolysis}]{\textit{enzymatic}} \underset{\text{biologically inactive}}{\text{Acetate + Choline}}$$

This reaction is catalyzed by the enzyme **acetylcholinesterase**.

Norepinephrine is deactivated by a chemical change involving oxidation and loss of the nitrogen atom:

$$\underset{\text{biologically active}}{\text{Norepinephrine}} \xrightarrow{\textit{MAO}} \underset{\text{biologically inactive}}{\text{Oxidative deamination products}}$$

This deactivation is catalyzed primarily by a mixture of enzymes known collectively as **monoamine oxidase (MAO)**.

Now, it stands to reason that if we can inhibit the action of the enzyme that catalyzes the destruction of the transmitter, there will be more transmitter substance left in the synaptic cleft. We would expect this reaction to lead to increased conduction across the synapse and more generalized activity of the nervous system employing that transmitter. This is precisely what has been accomplished in both the cholinergic and adrenergic nervous systems. For example, the first listing in Table 5.1 is for tranylcypromine (Parnate), an MAO inhibitor that causes elevated NE brain levels and consequent CNS stimulation. This drug has found application as a mood elevator in mentally depressed patients, although its use is limited because of potentially serious side effects.

Other listings in Table 5.1 are for drugs that inhibit acetylcholinesterase, causing a great buildup of acetylcholine in the synapse with resultant superfiring of cholinergic neurons. The carbamate insecticides are highly efficient examples of this enzyme inhibitor approach. Unfortunately, making use of the same principle, war research has devised nerve gases for the potential destruction of human beings. Table 5.1 also lists drugs that have different mechanisms of action at the synapse. Some (cocaine, for example) can block the active reuptake of NE from the synaptic cleft, thus increasing NE activity. Conversely, some (reserpine, for example) can

Table 5.1 Drugs Acting at the Synapse[a]

Drug	*Action of Drug*	*Pharmacological Result*
Tranylcypromine (Parnate)	Inhibits action of MAO enzymes (MAO inhibitor)	Antidepressant in depressed patients; mood elevator
Imipramine (Tofranil), Elavil	Inhibit mechanism for reuptake of NE into adrenergic neurons	Increased availability of NE; antidepressant, antianxiety agent
Benzodiazepines	Promote the actions of GABA	Antianxiety and anticonvulsant effects
Cocaine	Blocks reuptake of dopamine and NE from synaptic junction and potentiates response to NE	CNS stimulation
Amphetamines	Block catecholamine reuptake, promote release of dopamine, and directly stimulate receptors	CNS stimulation plus other effects
Deprenyl	Increases concentrations of catecholamine transmitters in brain	Antidepressant, antiparkinsonism
Ephedrine	Stimulates postsynaptic NE receptors and release of NE	Bronchodilator, cardiac stimulant
Reserpine (rauwolfia)	Depletes the vesicles of their supply of bound NE by blocking active reuptake	Lowered blood pressure and reduced CNS activity levels
Phenoxybenzamine	Blocks transmitter molecules from exciting NE receptors	Blockage of the sympathetic nervous system; lower blood pressure
Chlorpromazine	Blocks transmitter molecules from exciting NE receptors	Sedates psychotics; antiemetic; relief of intractable hiccups
Lithium salts	Increase the metabolism of NE	Calm the manic patient
Curare	Binds to postsynaptic receptors, thus blocks transmitter action of acetylcholine	Paralysis of skeletal muscle
Carbamate and organophosphate insecticides	Inhibit activity of acetylcholinesterase enzyme, allowing increased concentration of acetylcholine at synapse	Intense synaptic activity; convulsions; death
Fluoxetine (Prozac)	Blocks serotonin reuptake	Antidepressant
Atropine	Blocks the effects of acetylcholine at postsynaptic nerve endings	Many effects, including hallucinations
Biological warfare (nerve) gases	Causes massive inhibition of acetylcholinesterase[b]	Intense cholinergic activity; death

[a] Numerous drugs owe their pharmacological activity to the fact that they can alter conditions at the synapse, either by increasing or decreasing availability of one of the neurotransmitters or by themselves exerting a direct action on the postsynaptic receptor cells. All of the compounds in this table are of this type.

[b] In combat zones where nerve gases are expected to be used, our military personnel are given hypodermic syringes ready-loaded with atropine (which blocks acetylcholine receptors) plus 2-PAM chloride, a powerful synthetic reactivator of acetylcholinesterase. Thus, both drugs act to relieve acetylcholine overdosage.

cause depletion of NE from storage depots in the vesicles, so there is less transmitter available for release when called for. This action results in a decrease of CNS activity and a lowering of blood pressure. Finally, some of the drugs in the table (amphetamines, ephedrine) act directly on the postsynaptic receptor, taking over the function of the natural transmitter.

The compounds in Table 5.1 are remarkable examples of the progress modern medical research has achieved in the area of modification of synaptic transmission through a knowledge of transmitters, receptors, and enzymes. Indeed, the view is now well established that synapses represent drug-modifiable control points in the complex network of nerves in the human body.

5.5 Drugs and Our Response to Stress

Many a student (and teacher) has stood before a large audience and made a presentation while literally shaking with apprehension and excitement. An American president was observed on one speaking occasion to be trembling so noticeably that he had to hide his hands behind the podium. Stressful situations such as appearing before a large audience or narrowly averting a high-speed collision elicit physiological responses that can be uncomfortable at the moment but that are in reality the essence of protective measures. The "adrenaline that flows" in an exciting situation is normal, as are the shaking knees, dry mouth, and rapid pulse that it produces.

Let's take a look at the nervous system in our body that operates when we are more or less at rest and the system that takes over under acute stress situations. Figure 5.4 will help make the situation clearer.

Autonomic nerves generally regulate body processes that are not under voluntary control, such as heart rate, dilation of bronchi, digestion of food, peristalsis, secretion of glands, and dilation of the pupil of the eye. (By contrast, nerves in the **somatic division** regulate voluntary activity such as muscle movement.) According to what they do, autonomic nerves are placed in one of two great categories—the **parasympathetic division** or the **sympathetic division**. As indicated in Figure 5.4, discharge of parasympathetic nerves is associated with conservation and restoration of body processes—for example, slowing of the heart rate, secretion of digestive juices, increased peristalsis, and relaxation of sphincter muscles of the GI tract and urinary bladder. The distribution of parasympathetic nerves throughout the body is much

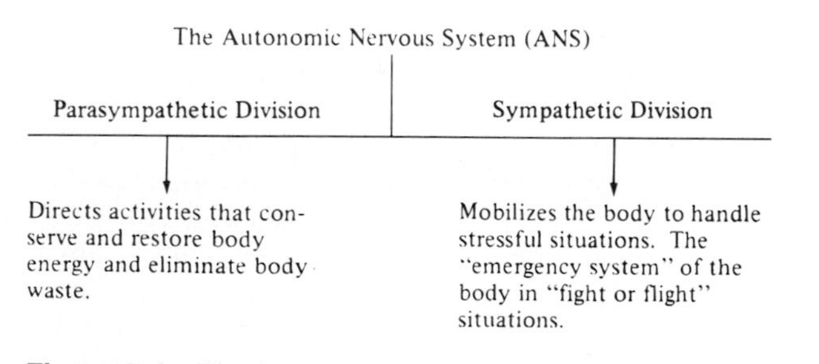

Figure 5.4 The human autonomic nervous system and its divisions.

more limited than that of sympathetic nerves. Transmission in the parasympathetic division is exclusively cholinergic.

Discharge of sympathetic nerves is intimately associated with the body's response to stress of all kinds: severe heat or cold, mental anguish, threats to physical safety, pain, loss of blood, lack of oxygen, fear, and so on. Since sympathetic nerves are extensively distributed throughout the body and have many synapses, the division tends to discharge as a whole, mobilizing the entire body to meet the stress. Table 5.2 lists some of the typical responses to sympathetic stimulation and how these responses help us cope with a temporary "fight-or-flight" stress situation.

After the stress has passed, the body must get back to normal. This happens as catecholamine release is terminated and as epinephrine and NE are broken down in the presence of MAO or are actively reabsorbed into the presynaptic membrane. In addition, the parasympathetic system comes back to life, slowing the heart, restarting digestion, constricting the pupil of the eye, and so on. Clearly, the sympathetic and parasympathetic divisions are in opposition to each other pharmacologically, the former taking over from the latter as we are faced by stress. Note that people who are continuously stressed—by their job, by drugs, by being overweight, by smoking, or by other causes—have a sympathetic nervous system that predominates most of the time, with resultant rapid heart rate, diminished GI activity, and all the other effects listed in Table 5.2. Sooner or later, health has to suffer under such conditions.

Besides nervous system activation, our bodies can respond to stress in several other ways. Stress stimulates the adrenal gland (more precisely, the adrenal cortex) to secrete cortisol and other hormones that play an important role in body strength and resistance (increased blood sugar, resistance to inflammation, wound healing, etc.). Stress generally turns off the thyroid gland; subsequently, there is diminished thyroxine secretion and increased urinary excretion of iodide.

The terms **preganglionic** and **postganglionic** describe nerve endings. A ganglion is a bunch of nerve cells and synapses all collected in one spot for easy interconnections and crossover. (Getting the "wind knocked out" results from a blow to a ganglion.) Nerve fibers leading up to the ganglion are preganglionic; those leading the impulse away from the ganglion are postganglionic. Both the sympathetic and parasympathetic divisions have many examples of pre- and postganglionic fibers.

Table 5.2 Results of Sympathetic Nerve Stimulation

Organ	*Effect*	*Results*
Eye	Dilation of pupil	Greater visual acuity
Heart	↑ Rate, ↑ output	Increased blood flow to skeletal muscle
Bronchi	Dilation	More oxygen to the lungs
Skin and peripheral blood vessels	Constriction	Minimal blood loss if injured; blood and oxygen for muscle work
Brain	Stimulation	Greater mental acuity
GI tract	↓ Peristalsis	Diversion of blood to muscle areas
Sweat glands	↑ Secretions	Facilitation of loss of heat
Salivary glands	↓ Secretion	Dry mouth
Liver	Glycogenolysis	Increased glucose in blood

5.6 Drugs That Mimic Sympathetic Discharge

As our discussion now turns to drugs, we will concentrate on those that affect the sympathetic division of the autonomic nervous system, specifically the adrenergic nerves contained therein. The catecholamines, norepinephrine and epinephrine, are the two most important naturally occurring adrenergic drugs. Under stress, our adrenal glands secrete epinephrine and NE into the bloodstream for distribution to all adrenergic receptors, thus permitting the whole-body response mentioned earlier. Dopamine, another naturally occurring catecholamine, is found distributed in the brain and other organs and is believed to be related to human motor function. In Parkinson disease, the dopamine content of the brain is found to be far below normal.

The secretion and action of these three catecholamines, epinephrine, NE, and dopamine, account for the physiological effects listed in Table 5.2. They are the keys that unlock and start our body's "emergency system."

As mentioned in Chapter 7, chemists use the structures of the naturally occurring catecholamines as the basis for their preparation of synthetic analogs that mimic the effects of natural sympathetic activity. We call these synthetic analogs **sympathomimetics**. Of course, the best-known sympathomimetics are the amphetamines, discussed in Chapter 7 (examine their pharmacology and note how similar it is to the effects listed in Table 5.2). Some additional sympathomimetics are listed in Table 5.3, along with their trade names and principal medical applications, such as relief of asthma attacks, increasing a too low blood pressure, and shrinking congested nose and sinus membranes. Some famous trade names can be recognized in the list.

Table 5.3 Sympathomimetic Drugs (Other Than Amphetamines)

Name	*Trade Name*	*Principal Use*
Isoproterenol USP and NF	Isuprel	Treatment of shock; bronchodilator
Phenylephrine USP	Neo-Synephrine	Vasoconstrictor; vasopressor; mydriatic; nasal decongestant
Phenylpropanolamine NF	Propadrine, PPA	Nasal and sinus decongestant
Methoxamine USP	Vasoxyl	To increase blood pressure (vasopressor)
Metaraminol USP	Aramine	To increase blood pressure
Pseudoephedrine	Sudafed, Novafed	Nasal decongestant; bronchodilator
Propylhexedrine NF	Benzedrex inhaler[a]	Nasal decongestant
Ephedrine USP and NF		Bronchodilator in asthma; nasal decongestant
Naphazoline USP	Privine	Nasal decongestant
Tetrahydrozoline USP	Visine	Constricts blood vessels in eye and nose
Terbutaline	Brethine, Bricanyl	Bronchodilator in asthma
Epinephrine	Adrenalin, Primatene	Bronchodilator in asthma

[a] Naturally occurring and synthetic, many trade names.

All of the Table 5.3 drugs retain their heart-stimulating capabilities; for this reason, they must be used cautiously or not at all by people with heart disease.

Some of the OTC drugs in Table 5.3 should be used with caution, since their side reactions are potentially serious. Pseudoephedrine, for example, can cause nervousness, dizziness, and sleeplessness at higher doses. It can be dangerous in persons with high blood pressure, heart disease, diabetes, or thyroid disease. Pseudoephedrine should not be used by persons already taking an MAO inhibitor drug for depression or any drug for hypertension, except on the advice of their physician. The same applies to pregnant women and nursing mothers. It is very important to read the labels on OTC drugs. The FDA has issued a public health advisory on phenylpropanolamine in drug products (see Section 7.9).

5.7 Other Synaptic Transmitters and Serotonin

We know with certainty the role of neurotransmitters such as acetylcholine, epinephrine, NE, dopamine, and perhaps 20 more, but it appears that these constitute only a small percentage of the total number of synaptic transmitters in the body. Researcher Michael Krassner of Sandoz Pharmaceuticals says that actually only 5% of the brain's neurotransmitters are known. Several hundred possible neurotransmitters are being investigated.

Gamma-aminobutyric acid[2] (**GABA**, $H_2NCH_2CH_2CH_2COOH$) plays an important role as a CNS transmitter and modifier of ganglionic transmission. There is evidence that GABA acts as the major inhibitory transmitter in the brain, accounting for about a third of all cerebral neurotransmission. Its release blocks the arousal of higher brain centers. Drugs that bind to GABA-ergic receptors can act as anticonvulsants, muscle relaxants, or antispasmodics. Barbiturate depression of the central nervous system involves GABA-ergic receptors. There is also some (but not conclusive) evidence that benzodiazepines such as Valium owe their antianxiety and muscle relaxant properties to their potentiation of GABA or GABA-ergic receptors.

Serotonin (5-hydroxytryptamine, 5 HT, Figure 3d in Appendix I) is a synaptic transmitter found in the human brain, blood and GI tract. Research on serotonin has concentrated on its actions in the brain, where it is created in the raphe nucleus, deep in the brain stem, and then sent to nerve endings in the brain and spinal cord. We know that at least 15 different serotonin receptors exist, divided into four classes and many subclasses. We now know that serotonin

1. Is found in bacteria, plants, and many animal forms.
2. Is found in humans in the intestinal mucosa, blood platelets, and CNS, where it serves as a neurotransmitter.
3. Is produced in the body from the essential amino acid tryptophan.
4. Is destroyed by MAO, antagonized by LSD, and depleted by reserpine.
5. When selectively depleted from the brains of laboratory animals, causes prolonged wakefulness in these animals.
6. Inhibits gastric secretion, stimulates smooth muscle, and is a vasoconstrictor.

[2]Not to be confused with *gamma-hydroxybutyrate* (GHB), a dangerous and illegal drug sold under many names and touted to build muscle, reduce fat, and induce sleep.

7. Occurs to the extent of several milligrams in one banana, and in avocados, plums, walnuts, pineapples, and eggplant.

Stimulation of serotonin-rich cells results in painkilling, sleepiness, and changes in appetite. The metabolism of serotonin is abnormal in migraine patients.

The study of serotonin is complicated by the fact that it acts on so many areas of the body plus the fact that its actions are often uncharacteristic and variable, differing in a human subject from one day to the next. Literally thousands of research papers have been published on serotonin, showing that it has a pharmacological effect on the heart and vascular system, on the CNS, respiration, muscle tissue, and the GI tract. Still, there are no official preparations of serotonin, and it is not used therapeutically. Serotonin poorly penetrates the blood-brain barrier.

Although serotonin has not been of great practical use, it has been an object of much theoretical discussion since it was discovered that it bears a strong chemical resemblance to such powerful hallucinogens as LSD, psilocybin (Mexican mushrooms), bufotenin, and the harmala alkaloids (see Appendix I). All of these hallucinogens possess the indolethylamine chemical moiety found in serotonin, stimulating speculation that they owe their hallucinogenic activity to a synaptic interference with the usual role played in the brain by serotonin. For a while, researchers were sure that they had found a clue to the cause of schizophrenia in humans, that is, that the schizophrenic's body was unintentionally producing an indolethylamine chemical that was interfering with serotonin's normal actions. It was felt that this reaction led to a psychosis much like that observed when LSD is taken. Unfortunately, too little evidence has been presented to support this hypothesis, and it remains only of academic interest.

Today, research continues on serotonin's role in sleep/wake patterns, regulation of body temperature, appetite, sexual activity, mood, aggression, anxiety, alcohol intake, pain response, and other responses to external stimuli. Two researchers at Princeton University, Barry Jacobs and Michael Trulson, suggested that serotonin plays a general inhibitory role in the human body, that is, it prevents us from overreacting to stimuli we encounter in daily life. Serotonin would thus modulate our behavior and maintain it within narrow limits. Support for this suggestion comes from the fact that when serotonin is blocked or reduced through diet control, the person becomes hypersensitive to virtually all environmental stimuli and becomes unusually hyperactive.

Use is made of **selective serotonin reuptake inhibitors** (SSRIs)[3] (Table 5.4) in the treatment of mental depression, social phobias, severe premenstrual distress, generalized anxiety, and almost any other condition that the smiling faces in TV commercials would have us treat. These drugs, termed **serotonergics**, act as **antidepressants** by effectively increasing the availability of serotonin in the synapse (Figure 5.2). Typically, it is 2–6 weeks before their effects are seen. Prozac (Table 5.4), the drug that changed the way we regard mental illness, and for years the best-selling antidepressant, has lost ground to newer drugs having fewer adverse effects; its patent expired in 2002 and generics are on the market. Antidepressants have been used in overeaters to reduce snacking, to treat the seriously obese, in bulimia, panic attacks, seasonal affective disorder (once known as the winter blues), and in one instance to stop a polar bear from pacing its Canadian zoo cage. The use of MAO inhibitors as antidepressants is discussed in

[3]Also termed serotonin neuronal reuptake inhibitors.

Table 5.4 Antidepressant Prescription Drugs, SSRI and Other Types

Trade Name (Manufacturer)	*Generic Name*	*Daily Dosage, mg*	*Comments*
Celexa (Forest Labs)	citalopram	20–40	For depression; do not combine with MAO drugs
Effexor (Wyeth)	venlafaxine	25–375	Inhibits norepi, serotonin, dopamine; for GAD
Lexapro (Forest Labs)	escitalopram		An isomer of Celexa, for panic disorders
Luvox (Solvay)	fluvoxamine	50	For obsessive-compulsive disorder
Paxil (GlaxoSmithKline)	paroxetine	20–60	For post-traumatic stress and other disorders
Prozac (Pfizer)	fluoxetine	20–80	Since 1988 the leading drug, but patent has expired
Wellbutrin (GlaxoSmithKline)	bupropion	150	Sold as Zyban for smoking cessation; not an SSRI, but is dopaminergic
Zoloft (Pfizer)	sertraline	50–200	Especially for severe PMS

Section 4.7. Not all the drugs in Table 5.4 are of the SSRI type, as noted under "comments." Prozac and the other drugs in Table 5.4 can cause adverse effects, including altering appetite and weight, nausea, insomnia, edginess, anxiety, and sexual problems in both achieving orgasm and maintaining erections. A meta-analysis of Prozac, reported in the *Journal of Nervous and Mental Disorders*, found that the drug is no more effective than the older antidepressants it has largely replaced. **Meta-analysis** is a newer statistical technique of combining findings from many smaller studies into one large one, rendering a more accurate overall assessment of the effectiveness of a treatment than does any single study. For a drug to show antidepressant activity, it must influence both serotoninergic and beta-adrenergic receptors.

Serotonin's supposed role in promoting and regulating sleep has earned it the nickname of *sleep juice*. Biochemist Ken Walton found evidence that serotonin is released during transcendental meditation (the increased calmness is the result of increased serotonin) and that serotonin levels fluctuate according to our individual daily (circadian) biorhythms.

Tryptophan is highly promoted by the health food store and supermarket press as an amino acid to help you sleep better (because it is converted to serotonin in the body). A degree of caution here is wise. Although most studies show that, in large enough doses, tryptophan helps induce sleep, animal studies have shown that a large intake of any amino acid may create an imbalance and interfere with the absorption or utilization of other amino acids. Just because tryptophan is a constituent of the proteins in our food does not mean it is safe in *isolation* and in large doses, says the FDA. Some evidence exists that even tryptophan can pose risks to the pancreas and to the immune system. Claims that tryptophan will alleviate your jet lag, stress, and premenstrual syndrome remain unsubstantiated.

For a discussion of aspartate and glutamate neurotransmitters and the NMDA receptor, see Section 1.6.

Nitric oxide (NO), a simple yet important molecule, is made by the human body and is now known to be a signal the body uses telling blood vessels to dilate, thus producing a drop in blood pressure. Infants with high blood pressure in the lungs obtain relief by breathing nitric oxide gas. Nitroglycerin and related artery-dilating medicines act by releasing NO. White cells use NO to defend against tumors. The anti-impotence drug Viagra was designed to function by blocking the enzyme that interferes with nitric oxide's smooth muscle-dilating effects. Paradoxically, nitric oxide is also an air pollutant found in automobile exhaust. It is not to be confused with nitrous oxide (N_2O, laughing gas).

5.8 Summary of the Role of Synaptic Transmitters in Psychedelic Drug Use

We can better understand the psychedelic action of some drugs by noting their similarity to, or action upon, a transmitter substance.

1. Norepinephrine and adrenergic neurons
   - Mescaline owes its psychotropic activity to its effect on 5-hydroxytryptamine in the midbrain.
   - Cocaine and the tricyclic antidepressants (e.g., imipramine) block active reuptake of NE at the synapse, resulting in central stimulation.
   - Amphetamines induce local release of NE in the brain. They also directly stimulate the NE receptors.
   - Amphetamines trigger the release in the brain of dopamine, a transmitter associated with the arousal and pleasure systems in the brain.
   - MDA and STP (see Table 12.2) also act by influencing NE neurons.
2. Acetylcholine and cholinergic neurons
   - Abtropine, scopolamine, and hyoscyamine all occupy the acetylcholine receptor site but do not activate it (thus, they block it).
3. Serotonin and serotonergic neurons
   - Serotonergic receptors in the brain and elsewhere are influenced by LSD, psilocybin, DMT, PCP, bufotenin, and nutmeg. These hallucinogens appear to inhibit serotonin systems and potentiate NE systems. Thus, LSD acts partly by blocking serotonin receptors and by antagonizing or modifying the actions of serotonin centrally.

Web Sites You Can Browse on Related Topics

Synapse - Neurotransmitters
http://dog.net.uk/neurotransmitters.html
Also search for "neurotransmitters"

Selective Serotonin Reuptake Inhibitors
http://panicdisorder.miningco.com/msub5c.htm
Also search for "Prozac" and "Zoloft"

Receptor Site Theory
http://www.csuchico/psy/BioPsych/neurotransmission.html
See diagram of:
http://www.sdsc.edu/10tw/week19.96/10tw.html
Also search for "receptor site"

Study Questions

1. This chapter focuses on the existence of the synapse.
   a. Where do synapses occur in the body?
   b. What constitutes a synapse? That is, what are its anatomical features and size?
   c. Is the transmission of a nerve impulse across a synapse always restricted to one direction?
   d. What is meant by a *presynaptic fiber* and a *postsynaptic fiber*?
2. Describe in general terms how a nerve impulse crosses a synapse.
3. **a.** What are the three principal neurotransmitters at synaptic junctions?
   b. What happens to them after they have done their job?
4. Let us assume that dopamine is a neurotransmitter substance in the brain that functions to maintain a sufficiently high body blood pressure, and that an enzyme is present in brain tissue that catalyzes the destruction of dopamine after it has done its job.
   a. Predict the effect of adding a drug that deactivates the enzyme.
   b. Predict the effect of adding a drug that blocks dopamine from its dopaminergic synaptic receptor.
   c. Predict the effect of adding a drug that retards active reuptake of dopamine into the presynaptic fiber.
   d. Predict the effect of adding a drug that depletes storage sites of dopamine in the presynaptic fiber.
5. Define (**a**) axon, (**b**) synapse, (**c**) active reuptake, (**d**) adrenergic, (**e**) preganglionic, (**f**) MAO, (**g**) sympathetic nerve, (**h**) sympathomimetic drug.
6. Identify the division of the autonomic nervous system, parasympathetic or sympathetic, that would be dominant in the control of the following body activities: digestion, taking a final exam, running a mile, urinating, participating in a debate, healing a broken arm, playing bridge, jogging, sleeping, and driving a car.
7. In a fight-or-flight situation, circulation of blood in the skin is reduced, the pupil of the eye is dilated, and digestion of food stops. How would these physiological changes help prepare the individual to meet the stressful situation?
8. What do the catecholamines and the sympathomimetic drugs have in common?
9. Methoxamine USP (Vasoxyl) is a vasopressor drug. What does this mean? In what kind of situation would you expect it to be used?
10. Phenylephrine USP, one of the sympathomimetics, is used as a mydriatic drug. Without looking up the definition, can you state what a mydriatic is? (You should know that it deals with the size of the pupil of the eye.)
11. A patient is having an asthma attack, and on his doctor's orders he takes one of the bronchodilators listed in Table 5.3 (e.g., ephedrine). Predict the *side effects* the patient is likely to experience.
12. True or false:
    a. Transmission across a synapse can proceed in only one direction.
    b. A transmitter substance can be released from an axon but cannot be reabsorbed.
    c. Acetylcholine is a typical adrenergic drug.
    d. Amphetamines are typical adrenergic drugs.
    e. In a fight-or-flight situation, heart rate, blood sugar, and saliva production are all increased.

f. Ephedrine, a typical sympathomimetic drug, typically dilates the bronchi.
g. The body makes serotonin from the dietary amino acid phenylalanine.
h. Serotonin is involved in the establishment of sleep/wake patterns.

13. If dopamine is indeed involved as a cause of schizophrenia, would too much or too little of it be responsible for the result?

14. A researcher wants to study the possible effect of jogging on the release of brain chemicals. He has a subject jog and then takes a blood sample. Why is it unlikely that he will find in the blood any chemical released in the brain? In what fluid would he be likely to find the chemical?

15. After you have read the information on Prozac, prepare to offer your opinions on the following questions:

a. Is it wise and proper for Prozac to be prescribed for teenagers passing through a difficult stage of life?
b. Is it wise and proper for Prozac to be prescribed for a person who wants to change his or her personality?

16. *Advanced study question*: With the help of a modern pharmacology textbook, prepare a summary of the role of acetylcholine in the human body.

17. *Advanced study question*: Contrast the synaptic roles of GABA and serotonin. What similar actions do they show? How do the results of stimulation of GABA receptors compare with the results of inhibition of serotonin?

18. St. John's wort is a wild-growing plant that contains substances believed to enhance serotonin, norepinephrine, and dopamine. Explain the pharmacologic rationale for ads that claim St. John's wort is a mood enhancer and antidepressant.

Narcotic Analgesics: Opiates and Opioids

Key Words in This Chapter

- Narcotic analgesic
- Fentanyl
- Opium poppy
- Heroin
- Codeine
- Endorphins
- Opiate addiction
- Methadone maintenance
- Agonists and antagonists
- Therapeutic community

Learning Objectives

After you complete your study of this chapter, you should be able to:

- Define narcotic analgesic.
- Tell where opium, morphine, and heroin originate.
- Explain the pharmacology of the opiates in pain relief.
- Tell the effects of opiates on respiration and the GI tract.
- Explain the origin and actions of endorphins.
- Show how Narcan acts as a narcotic antagonist.
- Describe the nature of opiate addiction, tolerance, and withdrawal.
- Describe how methadone maintenance works.

6.1 Introduction

Chronic pain, afflicting nearly one-third of the U.S. population, is the single most common reason for seeing a doctor. It is also one of the primary reasons people take medication. Relief of pain is big business in America. While minor pain can be relieved with OTC aids such as aspirin (see Chapter 14, which describes nonnarcotic OTC pain relievers), relief of the deep-seated pain of fractures, kidney stones, heart attack, gallstones, or toothache usually requires the administration of one of the **narcotic analgesics**. These powerful agents can eliminate pain without causing loss of consciousness or loss of sensory perception, although in larger doses they can cause stupor or insensibility.

Narcotic analgesic drugs and preparations in common use today are listed in Table 6.1. Additional narcotics of less importance are two semisynthetic derivatives of the baine, etorphine and diprenorphine. Etorphine, used by veterinarians

Table 6.1 Narcotic Analgesic Drugs and Preparations Available by Prescription or Over the Counter

Generic Name	*Other Name*	*Comments*
Alphaprodine	Nisentil	Injectable; given intravenously, 0.5 mg/kg.
Buprenorphine	Buprenex	Parenteral analgesic 30 times as potent as morphine.
Butorphanol	Stadol	Injectable. Dose: 2 mg.
Codeine	—	Usually comes in combination with acetaminophen or aspirin. Dose for pain relief: 30–120 mg; for cough suppression: 10–30 mg. See Section 6.6.
Concentrated opium alkaloids	Pantopon	Injectable; morphine makes up half of total alkaloid content. Dose: 5–20 mg.
Dihydrocodeine bitartrate	Compal, Synalgos-DC	Semisynthetic; related to codeine. May be dispensed in combination with either aspirin or acetaminophen.
Terpin hydrate and codeine elixir	T.H. and C.E.	In some states sold OTC. For cough suppression. Dose: 1 teaspoonful. Not for children under 12 years.
Fentanyl	Sublimaze	Very widely used in surgery. See Section 6.10.
Hydrocodone	Hycodan, Vicodin	Hycodan used as a cough suppressant, Vicodin for pain relief. Dose: 5 mg (1 tablet).
Hydromorphone	Dilaudid	Use in pregnancy only with doctor's permission. Five times as potent as morphine. Oral dose: 2 mg.
Levorphanol	Levo-Dromoral	Average adult dose: 2 mg (1 tablet).
Meperidine	Demerol	Typical dose: 80–100 mg; lasts 2–4 hours. See Section 6.10.
Methadone	Dolophine	Key drug in heroin detoxification and methadone maintenance. See Section 6.10.
Morphine	M.S.	Dose: 5–20 mg. Most abundant of the opium alkaloids. See Sections 6.6, 6.7.
Nalbuphine	Nubain	Injectable for relief of moderate to severe pain. Dose: 10 mg.
Oxycodone	Percodan, Percocet, Oxycontin	Aspirin-allergic persons should know that Percodan also contains a standard dose of aspirin. Percocet contains acetaminophen rather than aspirin. Dose: 4.5 mg (1 tablet). Oxycontin is oxycodone in a controlled-release oral tablet.

Table 6.1 continued

Generic Name	*Other Name*	*Comments*
Oxymorphone	Numorphan	Injectable. Semisynthetic substitute for morphine. Dose: 1 mg by injection.
Paregoric	Camphorated tincture of opium	In some states sold OTC. For relief of diarrhea. See Section 6.2.
Pentazocine	Talwin	Often used as a preoperative medication. Dose 30 mg by injection. See Section 6.10.
Propoxyphene	Darvon	Available in combination with aspirin and acetaminophen. Sold generically. Usual dose 65 mg. See Section 6.10.
Remifentanil	Ultiva	Selective mu opioid receptor agonist; given IV for analgesia during induction of general anesthesia.

to immobilize large wild animals, is 1,000 times as potent as morphine in its sedative, analgesic, and respiratory depressant effects.

The historical account of narcotic analgesics must include abuse as well as use, for the two are inseparably intertwined. A historical summary follows:

- Opium was used by the ancients.
- Opium and its alkaloids are introduced to America in the 18th and 19th centuries.
- Morphine is isolated from crude opium in 1803.
- The Bayer Company in Germany introduces heroin in 1898.
- Congress passes the Harrison Narcotic Act in 1914.
- Heroin and opiates go underground; stricter laws are enforced.
- Heroin is declared illegal in America in 1924; heroin clinics are closed.
- U.S. heroin and opiate addiction rates are at a low level in the 1940s and 1950s.
- There is runaway drug use in the 1960s and 1970s, with a 10-fold increase in heroin abuse between 1960 and 1969.
- Society recognizes addiction as a disease and attempts to deal with it.
- Methadone maintenance programs are established in the mid-1960s.
- The 1970 Controlled Substances Act is passed.
- Discovery and clinical use is made of narcotic antagonists.
- The brain's "own opiates" are discovered in 1975.
- The mixed agonist/antagonist type of narcotic analgesic is synthesized and used.
- The Drug-Free America Act of 1986 is passed.
- The Anti-Drug Abuse Act of 1988 is passed.

The role of the opium poppy in this story is central. All of the morphine in legitimate use today, as well as all of the heroin in illicit use, originates in this plant.

6.2 The Opium Poppy

The source of opium is the opium poppy plant, *Papaver somniferum* (see Figure 6.1). Until recently, almost all of the world's opium was grown in its native habitat—a narrow belt of mountains that stretches along the southern rim of the great Asian land mass. This opium-producing belt extends from Turkey's Anatolian plateau, through the northern reaches of the Indian subcontinent, and to the remote mountains of Burma, Thailand, and northern Laos. Besides Southwest Asia, the Middle East (Afghanistan, Turkey, Pakistan, and Lebanon), Mexico, and South America are sources of U.S. heroin. Drug lords in Colombia, who flooded the United States with cocaine in the 1980s, have added a new product line, heroin, and are shipping it to New York and other cities. Not native to Colombia, opium poppies are now found on Andean mountainsides.

Crude opium is obtained in the following way. Workers enter the fields when the seed capsules of the poppy plant have dropped their flowers but are still unripe, and they incise the capsules (see Figure 6.2), making slits with a sharp knife (horizontally in Turkey, vertically in Southeast Asia). The milky white sap that exudes from the slits turns brown upon exposure to air. It is scraped off and is formed into

Figure 6.1 A field of opium poppy plants, showing flowers and seed capsules. (Photo courtesy DEA, U.S. Department of Justice.)

Figure 6.2 Incised opium poppy seed capsule (note the corona, or crown). The exudate contains many alkaloids, including morphine, codeine, papaverine, and thebaine.

a ball (see Figure 6.3). This brown mass, the crude opium of commerce, can enter legitimate channels to yield morphine sulfate USP, or it can enter illicit channels and become heroin.

Hundreds of tons of opium are legally imported into the United States annually. A small part of it is used to make antidiarrheal preparations such as paregoric (camphorated tincture of opium); the remainder is processed by American pharmaceutical firms for the isolation of active ingredients.

Crude opium contains dozens of chemical compounds, including some 20 alkaloids. For medicine, the two most important are **morphine** (after the Greek god of dreams, Morpheus) and **codeine**, the former comprising about 10% of the weight of the crude dry opium, and the latter about 0.5%. In its chemical structure, codeine is very similar to morphine (for the chemical structure of morphine, see Figure 7a in Appendix I). The annual licit production and use of codeine in the United States is more than 121,000 pounds.

Opiates is the name given to the active narcotic analgesics in opium and to the derivatives that can be synthesized from them. (The term opioids is usually reserved for totally synthetic morphine-like drugs; examples of opioids are Demerol, Talwin, fentanyl, and methadone.)

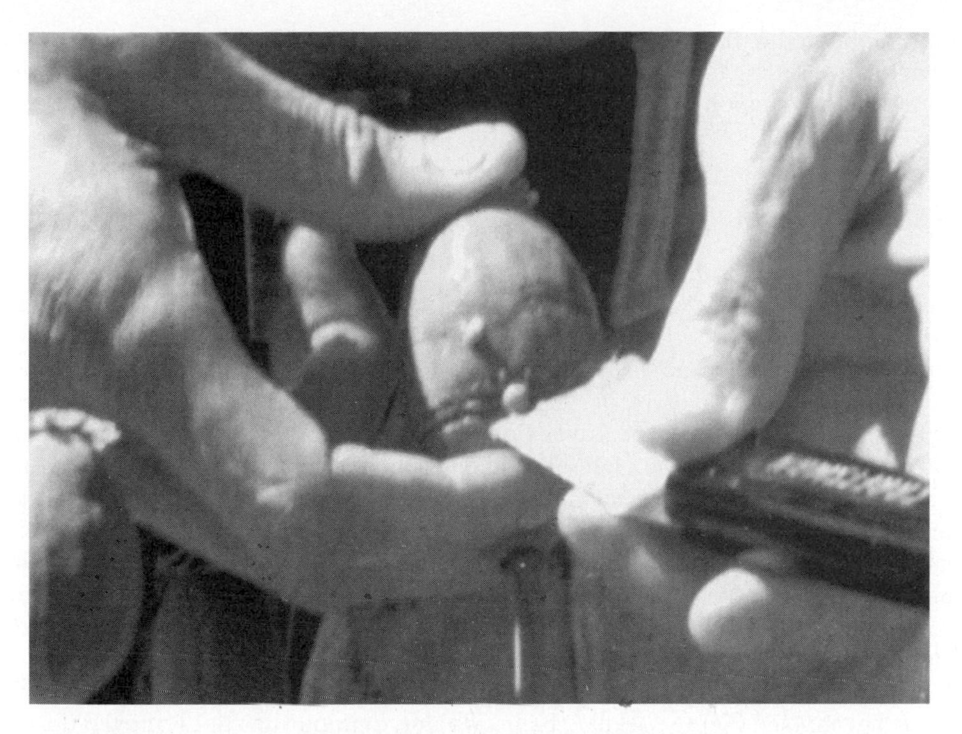

Figure 6.3 Crude opium is obtained by scraping the latex from the seed capsule. (Photo courtesy DEA, U.S. Department of Justice.)

Evidence exists that the effects of opium were known to the ancient Sumerians (4000 B.C.). The first undisputed reference to poppy juice was recorded by Theophrastus in the third century B.C. Interestingly, a terra-cotta head found at ancient Knossos in Crete wears a corona of incised poppy capsules. Arabian physicians knew of the effects of opium. Arabian traders introduced it to China, where it was used for the control of dysenteries. Spread of the opium habit throughout China was facilitated when the British East India Company found that it was necessary to ship Indian-grown opium to China in return for tea. When the Chinese government attempted to control the influx of opium, the British government went to war (First Opium War, 1841). Fifteen years later, Britain and France combined to wage a second war on China to force the Chinese to receive their diplomats and deal with their traders.

Until 1800, all of the preparations of opium were crude—that is, they were dried preparations or extracts that contained dozens of chemicals. Morphine, the first active ingredient ever to be isolated from a plant, was obtained in pure form in 1803 by the German scientist Freidrich Sertuerner (who almost killed himself by self-administration of the drug). Codeine was isolated from crude opium in 1832. Thus the opium alkaloids (alkaline, nitrogen-containing compounds from plant sources) became renowned, in pure form, as narcotic analgesics indispensable for the relief of pain. Their extensive medical use in America was engendered by the invention of the hypodermic needle in 1856 and by the great number of painfully wounded in the Civil War. Abuse of opium in this country can be traced in part to Chinese laborers who immigrated here to work on the great railroad and canal projects of the late 19th century. That opium smoking was commonplace is evident from these quotes, which appeared in the *Mammoth City (California) Herald* in 1879: "Mammoth City town lots are selling for $1,500 for 25 ft. frontage," and, "Another opium den is in the course of construction right in the center of town on main street."

6.3 Heroin

Heroin (diacetylmorphine—see Figure 7b in Appendix I for structure) does not occur as such in nature. It was first prepared in Germany in 1874 by acetylation of the two hydroxyl groups in morphine (heroin comes from the German word *heroisch*, meaning "heroic"). Heroin thus exemplifies the partially synthetic or modified naturally occurring type of drug. Heroin was introduced into medicine in 1898 by Dr. Heinrich Dreser of the Bayer Company as a cough suppressant (antitussive) and pain reliever (analgesic). Dreser touted heroin as a safe, nonaddictive substitute for morphine. For years, the medical profession remained ignorant of its potential for inducing physical dependence. In fact, it took 12 years and millions of addicted patients before doctors realized what we know today—that heroin is one of the most dangerous of the addicting drugs. A 29-year-old Spring Valley, California, heroin addict says heroin is "the purest form of evil on earth, because it just seduces you with how good it feels."

Pure heroin base (free base) is a white solid with a bitter taste; it melts at 173° C. Street heroin (see Figure 6.4) may be brown due to impurities not removed during its manufacture or the presence of a diluent (a diluting agent used to "cut" it), such

Figure 6.4 Heroin samples, including Mexican brown and Southeast Asian. Colors can be due to impurities not removed or to chemicals used to cut (dilute) the sample. For smoking, larger sample particles are used. (Source: DEA, U.S. Department of Justice.)

as cocoa or brown sugar. Southeast Asian heroin may appear as a light tan powder. Eastern U.S. street heroin is often cut with quinine, talc, flour or cornstarch, and there have been numerous cases of addicts with talc and cornstarch **emboli** (clots) in their lungs and eyes. This embolism is the result of direct injection of the heroin and its insoluble diluents into the bloodstream. Black tar heroin, formerly sold on the street with a concentration of 3%, is now 20 times that and has killed youths in Texas who mainlined it and overdosed. Plano, Texas, physician Larry Alexander tells teens that people who overdose on heroin usually die because they drown in their own vomit. (Opiates paralyze the gag reflex.) The caveat is clear. Heroin users be warned: You have no real idea of the strength of your purchase. There are no labels. Heroin has many street names, including "horse," "junk," "mud," "skag," "Big H," and "smack."

Today in the United States, possession, sale, distribution, and prescribing of heroin are totally banned. The only heroin available is the illicit stuff on the streets. In Canada, heroin is again legally available on prescription, after a 30-year hiatus. In England, sale and use of heroin is still legal, although very little of it is actually dispensed. Most of the heroin in England is illicit, originating from the Indian subcontinent and southeast Asia. The Institute for the Study of Drug Dependence in London notes that the heroin is adulterated typically with lactose, glucose, mannitol, chalk dust, caffeine, quinine, vitamin C, or talcum powder. Street heroin has recently averaged 37% purity (a range of 15–63%) and sells for £60–100 ($100–160) per gram.

6.4 The Illicit Traffic in Opiates

As a chemical substance, heroin begins in a clandestine laboratory, where the crude opium of illicit traffic is extracted and its morphine content is isolated. An acetylating agent such as acetic anhydride is used to convert the morphine to diacetylmorphine (heroin).

Today, three major supply complexes of illicit opiate traffic are recognized: (1) Southwest Asia, (2) Southeast Asia, and (3) Mexico and South America. According to the latest government figures, 58% of the heroin smuggled into the United States comes from Southeast Asia, with the remainder equally split between Mexico and Southwest Asia (see Figure 6.5).

Southwest Asian heroin is made from Southwest Asian opium and is converted into heroin primarily in Afghanistan, Pakistan, and Iran. Southeast Asian heroin is usually made from morphine derived from Southeast Asian opium and is processed in Burma, Laos, and Thailand (the Golden Triangle). Mexican heroin refers to heroin usually made from morphine derived from Mexican opium and processed in that source country. Most of the estimated 730,000 acres of opium poppy cultivated in Mexico come from the tri-state area of Sinaloa, Chihuahua, and Durango. Mexican heroin is usually trafficked directly over the U.S.-Mexican border in small amounts by couriers using either motor vehicles or body carry. It is trafficked primarily in the West but is also available in Chicago and Detroit. Mexican heroin has become more prominent in recent years, partly because of the availability of black tar heroin, a crudely processed, highly pure form of heroin that had spread to virtually all western cities by the end of 1985.

The DEA estimates that only 2–5% of smuggled heroin is intercepted. Cut street samples may contain anywhere from 1 to 67% heroin, with the average being

Figure 6.5 Pressed bricks of heroin for bulk trafficking. (Photo courtesy DEA, U.S. Department of Justice.)

about 36%; the average price has been around $2 per milligram or $2,000 per gram (calculated as pure heroin). Wholesale prices of Mexican heroin ranged from $50,000 to $250,000 per kilogram (often shortened to "kilo"), which is significantly lower than Southeast Asian, South American, or Southwest Asian heroin.

Heroin smugglers are exceptionally creative. Cyprus police discovered 4 kilos of heroin woven between the threads of three elaborate rugs. Elsewhere, 120 pounds of heroin was found in false duck eggs. The biggest heroin bust so far reported occurred in Bangkok, where 2,822 pounds of 92% pure No. 4 Thailand heroin was discovered on a ship headed for New York. The heroin was hidden in bales of rubber sheets (where dogs could not detect it) and was found on a tip. The DEA states that the approximately 600,000 U.S. addicts consume 4–6 tons of heroin a year. About half of the addicts live in New York City.

6.5 The Harrison Narcotic Act

In 1911, an international conference was called at The Hague to deal with the worldwide uncontrolled abuse of opium and its derivates. President Theodore Roosevelt pledged America's participation in attempts to correct the opium situation. In 1914, Congress passed the Harrison Narcotic Act to honor U.S. pledges given at The Hague. The act originally was not directed at the eradication of narcotic drugs but was a record-keeping and tax law for those who manufactured or distributed opium, morphine, codeine, heroin, or coca leaf products. No mention was made of addicts.

Soon, however, the Harrison Act was interpreted to deny addicts all access to legal drugs. The consequence was something never before seen in America—the establishment of an underground traffic in narcotic drugs. There were serious misgivings about the wisdom of the act. An editorial in the *Illinois Medical Journal* for June 1926 stated, "The Harrison Narcotic Act should never have been placed on the statute books of the United States." Heroin, which previously had been legally available, was banned by federal law in all forms and sources in 1924.

The Harrison Act was for 46 years the basis for federal prosecution and imprisonment of addicts, smugglers, and peddlers. The act made an immense impression on the American scene and continued in force until superseded by the 1970 Controlled Substances Act (see Chapter 3).

6.6 The Pharmacology of Morphine, Codeine, and Heroin

The opium alkaloids are **narcotic analgesics**. *Narcosis* is defined as depression of the central nervous system (brain and spinal cord) leading to analgesia, drowsiness, changes in mood, mental clouding, apathy, lethargy, and ultimately unconsciousness. It is interesting that the Harrison Narcotic Act of 1914 defined morphine, codeine, and heroin as narcotics but made the gross pharmacological error of including cocaine—a drug that is a CNS stimulant. Indeed, one version of the Brompton Cocktail, administered today to relieve pain in terminally ill cancer patients, contains morphine and cocaine, the latter to *counteract* the narcotic effects of the former. The Comprehensive Drug Abuse Prevention and Control Act of 1970 again included cocaine and coca leaves as examples of narcotic drugs. To the lawmaker, the legal definition of a narcotic apparently takes precedence over the pharmacological definition.

Morphine's ability to relieve severe pain without causing loss of consciousness makes it an extremely useful drug. Given orally, intramuscularly, or subcutaneously in 5–20-mg doses every 4 hours, morphine can ease the pain of fractures, angina pectoris, surgery, kidney stones and gallstones, and terminal illness while the patient remains awake, although often drowsy. Typical doses of morphine will not interfere with sensory perception (touch, hearing, and vision) or speech (in contrast to ethyl alcohol). Morphine is especially effective in relieving dull pain, although in larger doses it also relieves sharp pain. Serious side effects of morphine are nausea and vomiting. They occur when morphine stimulates the reflex center in the brain controlling emesis. **Some advanced information:** Exactly how do the narcotic analgesics act to produce their pharmacological effects? Research has shown that it is possible to administer a radioactively tagged opiate molecule, wait for it to become attached to its receptor sites, and then identify the receptor site areas by the use of scanning techniques that pick up the radioactivity. This technique has shown that receptors for morphine-like drugs are widely distributed in the brain and spinal cord in areas that parallel nerve pathways known to be involved in mood and in the perception of pain. Opioid receptors are also found in areas involved in the regulation of respiration, body temperature, blood pressure, hormone secretion, the control of reflexes such as vomiting and coughing, and in the GI tract and certain reproductive organs. The four types of receptors most studied are named *mu, delta, kappa*, and the orphan receptor ORL_1. Certain subtypes have also been discovered. After an opiate enters the bloodstream, it acts as an agonist at cellular opiate receptors. This action results in a reduction in membrane calcium conductance and an increase in potassium conductance with an inhibition of cAMP synthesis. Nerve cells become less excitable owing to hyperpolarization. In other words, opiates and opioids inhibit nerve functions. Specifically, this is the result of inhibited neurotransmitter release at the level of the thalamus, periaqueductal gray, and substantia gelatinosa. It is the *mu* receptor that is most commonly involved. Analgesia results from inhibition of substance P release.

Virtually all opioid and opiate painkillers act through one of the four known opioid receptors mentioned earlier. The idea that multiple opioid receptors exist arose to explain the dual actions of nalorphine (Section 6.9), which antagonizes morphine's analgesic effect but also acts as an analgesic itself. Nalorphine's analgesia is mediated through the kappa-receptor, whereas morphine's is not. The naturally occurring enkephalins (Section 6.8) appear to act via the delta-receptor. For an excellent review of opiate receptors and subtypes see **http://opioids.com/receptors/index.html**.

Codeine is only about one-twelfth (some say one-fiftieth) as effective an analgesic as morphine, but it is a better cough suppressant. Codeine is best used in combination with acetaminophen or aspirin, because the two non-narcotic analgesics definitely add to the pain-killing effect of the codeine. Combinations of codeine are effective for moderate pain; increasing their dosage above 120 mg usually will not eliminate severe pain. Side effects of codeine include dizziness, nausea, and vomiting. Intravenous use of codeine is not recommended. Orally, codeine is much better absorbed than morphine. Recent research suggests that codeine needs to be transformed in the body to morphine to have a pharmacological effect. Some people do not get relief from codeine, because they were born without the enzyme that converts it to morphine.

Percodan and methadone have about the same analgesic potency as morphine. Demerol is about one-ninth as effective.

Heroin, although highly addictive and outlawed, is an excellent pain reliever long considered by some to be more effective than morphine. It is available in 30 countries for use in pain relief and research (but not in the United States). A bill proposing the legalization of heroin for treatment of pain from cancer was introduced in Congress by proponents who claimed that heroin is more effective than morphine and can be given in smaller doses. However, in a study reported in the *New England Journal of Medicine*, Dr. C. E. Inturrisi of the Memorial Sloan-Kettering Cancer Center said, "The rationale in believing heroin itself does something morphine does not do becomes less and less likely when you consider [that] heroin must be converted into morphine first." Inturrisi found that before heroin can act in the body, it must be chemically broken down into morphine and another chemical called 6-monoacetylmorphine (6-MAM). In a double-blind test, heroin addicts could not distinguish between heroin and morphine. This study seems to show that heroin is not a more effective analgesic than morphine. Heroin does, however, pass the blood-brain barrier faster than morphine, thus giving a more acutely pleasurable rush than morphine. Morphine itself is metabolized mainly in the liver to give the pharmacologically less active demethylated compound known as normorphine.

6.7 Additional Pharmacology of Morphine

In addition to analgesia, morphine and the opioids produce the following effects in the human body:

Depression of respiration. Morphine and other opiates and opioids can interfere with respiration, even in ordinary doses. At high doses they can depress respiration acutely, even fatally—as many drug overdose deaths can attest. As a narcotic morphine can have a powerful effect on the brain generally. Breathing slows and becomes shallow, and oxygenation of the brain falls. In acute overdosage, breathing stops. (See how Narcan can reverse this condition, in the next section.)

Suppression of cough reflex. Morphine is not used for this antitussive effect, but codeine is. Depression of the brain can deactivate the reflex arc governing coughing in persons suffering from a persistent, unproductive cough. Codeine's usefulness as an antitussive is limited by the nausea it produces in many people. Of course, non-narcotic antitussives such as dextromethorphan are available.

Constriction of the pupil of the eye (miosis—even in total darkness). This symptom is a useful sign in identifying the addict. The triad of pinpoint pupils, coma, and respiratory depression strongly suggests opiate intoxication.

Reduction of peristalsis and the antidiarrheal effect. The opiates and opioids are antidiarrheal and potentially constipating. Morphine, for example, diminishes peristalsis (the rhythmic contractions of the gut that propel contents), delaying passage of material and relieving diarrhea. Paregoric (see Table 6.1) is used for this purpose. Constipation is common in long-term opiate users, as is abdominal cramping and diarrhea in the heroin withdrawal syndrome.

Reduction of secretions. Gastric, pancreatic, and bilial secretions are diminished by morphine.

Morphine's pharmacology can be summarized thusly:

Actions of morphine and opiates (generally)	
Pain relief	potent
Addiction potential	high
Respiration	↓
Cough reflex	↓
Peristalsis	↓
Constipation	↑
GI secretions	↓
Pupil of eye	constricted
Euphoria	↑

The heart and circulatory system remain relatively unaffected by morphine. A fall in blood pressure that is sometimes seen may be the result of peripheral vasodilation. Morphine's effects on the urinary system are variable. The pharmacology of codeine is similar to that of morphine. Morphine, Talwin, and the narcotic analgesics generally are contraindicated in head injuries accompanied by increased intracranial pressure, because these drugs all depress respiration, increase carbon dioxide partial pressure, and further increase intracranial pressure.

The illicit substance China White is not heroin but possibly alpha-methylfentanyl or 3-methylfentanyl. The latter is reported to be 1,000 times as powerful a narcotic as morphine. Unsubstituted fentanyl is available on prescription as the analgesic Sublimaze. For more on China White, see Section 12.8 on "designer drugs."

Case History *Heroin Death of College Seniors*

Two college men, only a month from graduation, died of heroin overdoses. Both had experimented with drugs before, including heroin, but had given assurances they were off drugs, were serious about their studies, and had plans for careers. On the way home one night they stopped at a street corner and bought a $20 balloon of heroin. Police found them two days later sprawled in their living room, not far from the spoon "cooker" and hypodermic syringe. Los Angeles dealers are now selling "black tar" heroin from Mexico with a purity level that averages 25 percent but can be higher than 50 percent. Spurred by cheaper, purer forms, opiate use in California is soaring.

6.8 Endorphins: The "Brain's Own Opiates"

In this discussion, it is important to mention a remarkable discovery. The existence of brain receptors specific for morphine suggested to scientists that perhaps the human body makes its *own* narcotic analgesics—natural, morphine-like compounds that help us deal with pain and emotional crises. Indeed, in 1975, Drs. John Hughes and Hans Kosterlitz of the University of Aberdeen, Scotland, found just such naturally occurring morphine-like substances in the brains of pigs (they are now known to occur in all vertebrates). Hughes and Kosterlitz identified these endogenous painkillers as pentapeptides (chains of five amino acids) and named them **enkephalins** (we now identify them as leucine-enkephalin and methionine-enkephalin) from the Greek word meaning "in the head." Subsequently, the sequence of amino acids in the enkephalins was shown to be identical to the sequence of amino acids 61–65 of beta-lipotropin, a large protein hormone from the pituitary gland. The enkephalins are found in high concentrations in brain areas very similar to those in which opiate receptors are found. Other fragments of beta-lipotropin were also shown to have powerful analgesic effects, especially a peptide fragment named **beta-endorphin** (from the words *endogenous morphine*).

Today we recognize four groups of endorphins (the modern scientific name is **endogenous opioid peptides**). Alpha-, beta-, gamma- and sigma-endorphins contain 16–31 amino acids and are produced naturally in the human. Dynorphin, another endogenous peptide isolated from brain tissue in 1979, is 200 times as potent in relieving pain as morphine, and 700 times as potent as enkephalin. Beta-endorphin is found primarily in the pituitary gland, whereas enkephalins and dynorphin are distributed throughout the nervous system.

Endogenous opioid peptides interact with opiate receptor neurons to reduce the intensity of pain. Altogether, there are some 18 different molecules endogenous to the brain that have the capacity to act like morphine. People with chronic pain have unusually high endorphin levels. Many pain-killing drugs, such as morphine and codeine, act like endorphins and activate the same opiate receptors.

The discovery of these naturally occurring painkillers, termed the "brain's own opiates" by the media, is of great consequence. It appears that we now understand the body's own mechanism for handling pain. We understand how some accident victims or wounded soldiers are, initially, able to disregard great pain. Animals, too, have endorphin-enkephalin receptor sites. The beta-endorphin obtained from camels is at least 30 times as powerful as human endorphin, a fact that may explain the camel's extraordinary indifference to pain. Endorphin activity builds throughout human pregnancy and reaches a peak just before and during labor. It has been speculated that the tenfold drop from peak endorphin levels within 24 hours of delivery may contribute to postpartum depression.

The fact that enkephalins also occur in the small intestine may be connected with the well-known tendency of opiates to cause constipation. It appears, too, that enkephalins modulate emotional responses, producing feelings of pleasure much as the opiates produce euphoria. Enkephalins (or endorphins) are also somehow involved in the initiation of eating, feeding duration, memory, stress, seizures, mental illness, and the use of alcoholic beverages. Some have suggested that "runner's high" is a phenomenon based on endorphin release, but there is no evidence for this statement. Others speculate that endorphins are involved in the phenomenon known as "near death experience," where they could simulate a massive narcotic-like effect on the brain. Endorphins are found in rodents and in simple animal forms.

It is possible that pain relief from both acupuncture and the placebo effect is mediated through the release of endorphins. Evidence for this theory comes from the use of naloxone (see the information later in this chapter), a drug that blocks the pain-relieving action of opiates and endorphins by preferentially occupying the receptor site. In experiments on cats and humans, acupuncture worked in the absence of naloxone but failed in its presence. When pain is significantly reduced by a placebo, it may be that the mere suggestion that pain will be relieved causes endorphins to be released, producing relief. One researcher found that a single shot of placebo could equal the relief of 6–8 mg of morphine—a low but fairly typical dose.

Endorphin researchers have located a pain control system in the periaqueductal gray, a system that runs through the central part of the brain; interestingly, it produces opioid peptides. Other investigations continue in the anticipation of finding new, potent painkillers that are nonaddicting and in the hope of explaining the role, if any, of endorphins in opiate addiction. Researchers also want to know if beta-endorphin is the primary agent and the enkephalins merely its breakdown products, or if endorphins only serve as the precursors to a neuroregulatory role played by the enkephalins. There are abnormal endorphins in the spinal fluid of schizophrenics, but giving Narcan to catatonics does not help them.

The discovery of neuronal receptor sites for opiates had led to the discovery of binding sites in the brain for other drugs such as antipsychotics, anxiolytics, antidepressants, cocaine, and cannabinoids. For more on the discovery of endorphins, use your browser to locate this URL: **`http://www.methadone.org/discover.html`**.

6.9 Narcotic Antagonists

Picture a single molecule of heroin (one of billions) that has entered an addict's bloodstream by way of a hypodermic needle and passes through the circulatory system until it reaches a particular anatomic location in the addict's brain. Here, it finds a specialized tissue called a *receptor site*, on which it just fits, in both a three-dimensional and a chemical association sense. Because the narcotic molecule fits so perfectly, the receptor site recognizes it as an opiate, and a sequence of events is triggered that ultimately results in the classic pharmacology of the opium alkaloids. The drug, termed an ***agonist***, is the biochemical key that unlocks cellular response.

Imagine that a chemist has synthesized a substance A, so closely related (three-dimensionally and chemically) to heroin or morphine that it will also fit the receptor site, perhaps even better than heroin or morphine does. But one profound difference exists. Substance A does not trigger the customary pharmacological response of a narcotic analgesic. It occupies the receptor site but has no drug effect. Substance A, called an ***antagonist***, is a key that fits the lock but fails to unlock the cellular response.

Going a step further, we can expect that if an addict is given a sufficient dose of the antagonist, his or her brain receptor sites will be occupied (saturated) with antagonist, and when the real narcotic is taken it will be unable to attach and trigger its normal response. Indeed, if we have designed a superior antagonist (with especially high affinity for the receptor site), we might even expect it to displace the narcotic drug from its prior attachment to the receptor site.

Such narcotic antagonist drugs actually exist. A few, such as **naloxone** (Narcan) and **naltrexone** (Trexan, Revex), are "pure" antagonists, fitting the site perfectly but having no narcotic effect of their own. Others, such as **nalorphine** (Nalline), **cyclazocine, butorphanol** (Stadol), and **levallorphan** (Lorfan), fit the site well but exhibit some narcotic effect.

Narcotic antagonists are in use today; they have saved the lives of addicts who overdosed and were in danger of respiratory failure. A dose of only 0.4 mg of naloxone administered to an adult heroin addict is often sufficient to counteract and reverse severe symptoms within minutes. This result can be explained by postulating such a high affinity of naloxone for the opiate receptors that it can competitively replace heroin from its bound sites. Figure 6.6 shows the competition between a drug and its antagonist for the receptor sites. Given to a cooperative addict, these antagonists can prevent the euphoric rush of subsequently injected heroin.

Theoretically, antagonists are chemical tools that could help the addict overcome the craving for heroin. If the addict is off the drug but is tempted to return, antagonists could be used as a crutch, because euphoria, physical dependence, and tolerance cannot develop after antagonist use. That is, if the opiate receptors in the addict's CNS are already saturated with high-affinity antagonist molecules, then heroin molecules cannot attach. Given to newborns of addict mothers, narcotic antagonists can reverse dangerous respiratory depression.

How applicable is the antagonist idea to the treatment of the 600,000 U.S. addicts? First, the addict must *want* to kick the addiction; antagonist use cannot be forced. Second, an addict must prefer a treatment that requires him to be opiate-free; however, heavy heroin users might find that their "heroin hunger" is too overwhelming. For these reasons, experts in the field of narcotic addiction treatment speculate that at most 15% of the addict population would select the drug-free, antagonist route of therapy.

Research continues in this fertile field. Scientists are looking for long-acting, pure antagonists that have no untoward side effects. Naltrexone, available for prescription use as Trexan, is a narcotic antagonist that can block the euphoria and other effects of heroin for 2–3 days, depending on the dose given. A New Jersey doctor has detoxified over one thousand heroin addicts by administering naltrexone while the addict is under a deep, four-hour anesthetized sleep. The addict is oblivious to the

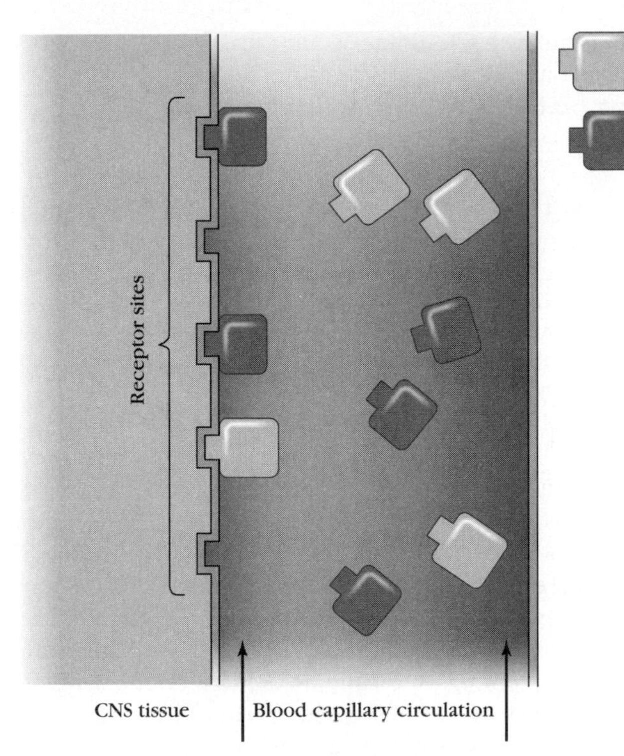

Figure 6.6 Diagrammatic representation of competition between a narcotic agonist and an antagonist for binding to the receptor sites. Which one will succeed (and therefore which pharmacological action will ensue) is determined by affinity for the receptor site plus the numbers of each kind of molecule. For example, a 1-mg dose of naloxone (the antagonist) completely blocks the effects of 25 mg of heroin (the agonist), showing the much greater affinity of naloxone for the receptor sites.

nausea, stomach cramps and irritability that would be felt without anesthesia. Patients can then receive a naltrexone slow-release implant that antagonizes heroin for about a month. Refills are possible. Patients are also strongly advised to participate in Narcotic Anonymous' twelve-step recovery program.

Trexan is intended for addicts who have been withdrawn from opiates, who are in rehabilitation programs, and who are attempting to remain narcotic-free. Incidentally, naltrexone is now approved by the FDA for use in the treatment of alcoholism.

Clonidine (Catapres), while not a typical narcotic antagonist, deserves to be mentioned here. Clonidine has been widely used to treat high blood pressure, but recently it has been employed to help addicts get through narcotic withdrawal. Given every 2–3 hours before withdrawal begins, it fools the brain into thinking it is not in withdrawal. Clonidine acts on alpha receptors in the brain to "mellow out" all of the withdrawal effects and permit the addict to stop the use of drugs cold turkey and become drug-free in 10–14 days. Unlike methadone, clonidine is not a substitute for heroin, is nonaddicting and noneuphoric, and can be used in a skin patch (Catapres TTS).

BritLofex (lofexidine), similar to Clonidine, is another alpha-2-adrenergic agonist used in Britain in methadone withdrawal. For more, see **http://opioids.co.uk/withdrawal/lofexidine.html**.

6.10 Synthetic Analgesics That Have Morphine-like Activity

In this section, we discuss drugs that do not occur in nature; they are products synthesized by chemists. Some are termed ***opioids*** because their pharmacology is so similar to that of the opium alkaloids.

Meperidine USP (Demerol) is a widely used, effective narcotic analgesic that was accidentally discovered in 1939 by researchers who thought they had prepared an atropine-like drug. Demerol, which is chemically unlike morphine, produces almost all of the important effects of morphine: analgesia, sedation, euphoria, and respiratory depression (it does not cause pinpoint pupils). Meperidine is an addicting drug to which tolerance develops and which, upon abrupt removal, produces withdrawal symptoms. As an analgesic, Demerol is only about 15% as potent as morphine. It is given in 50–100-mg doses. Its ready availability in hospitals has resulted in some physicians and nurses becoming physically dependent on the drug.

Methadone USP (Dolophine) will be discussed in the following sections. Its pharmacology is similar in all respects to that of morphine. It is addicting and produces tolerance and withdrawal symptoms. Methadone's threshold toxic dose (100–200 mg) is very close to its therapeutic dose (40–100 mg). Some addicts maintained on methadone complain of nausea, constipation, loss of sex drive, and nodding. A court decision in December 1976 stipulated that methadone could be sold legally through pharmacies, subject only to restrictions as a schedule II drug.

Propoxyphene USP (Darvon) is an analgesic drug structurally related to methadone, although much less potent in its pain-relieving properties. Darvon's pharmacology is generally similar to that of the narcotics as a group, but its abuse potential is less than that of codeine. While Darvon can produce psychological dependence, administration by the IV route can increase its potential for producing physical dependence and tolerance. Propoxyphene only partially suppresses the withdrawal syndrome in people addicted to the opiates.

Darvon is marketed either alone or in combination with aspirin or acetaminophen. The drug is available generically. Propoxyphene napsylate (Darvon-N) has been used in place of methadone in heroin maintenance (see the next section for more information). The lower water solubility of the napsylate salt gives it unique absorption characteristics.

The FDA in 1979 estimated that 1,000–2,000 deaths a year are associated with propoxyphene, either alone or in combination with other drugs (the majority appear to be suicides). Respiratory depression is the usual cause of death in propoxyphene overdosage. In 1979, the FDA ordered a new warning about the risk of death on labels for all propoxyphene products (Darvon, Darvon Compound, Darvon-N, Darvocet-N, and others). The FDA is trying to discourage physicians and consumers from unnecessary medical use of the drug and to warn them to exercise extreme care if the drug is prescribed. Suicidal and alcoholic patients should not be given propoxyphene. Patients taking tranquilizers or antidepressant drugs also should not be given this drug. The FDA has warned against the use of propoxyphene during pregnancy, since neonates can experience propoxyphene withdrawal if their mothers take the drug during pregnancy. Withdrawal signs include tremors, irritability, high-pitched crying, diarrhea, and weight loss with a ravenous appetite. Darvon is now classified as a schedule II narcotic. Propoxyphene is in the pregnancy category C (see Section 3.5).

Pentazocine (Talwin) is a schedule IV synthetic analgesic used for the relief of moderate to severe pain. A typical dose for pain relief is 30 mg, which is about equivalent to a 10-mg dose of morphine. The CNS pharmacology of Talwin is similar to that of the opioids; its use can result in respiratory depression. Talwin, designed to be of low abuse potential, does not have an untarnished image. When administered by injection, it has been shown to be capable of inducing psychological and physical dependence. The withdrawal syndrome has been seen upon abrupt cessation following lengthy parenteral use. Reports from Chicago tell of Talwin's use, mixed with the antihistamine pyribenzamine, as a cheap substitute for heroin.

Talwin was one of the first **mixed agonist/antagonist** opioids, that is, it demonstrates both agonist and antagonist properties. Talwin is an agonist because it attaches to brain receptors, producing potent analgesia. It is an antagonist because it can block the effects of morphine and partially reverse respiratory depression caused by that drug. The explanation for this dual, apparently contradictory action lies in the compound's special chemical structure that controls its binding to receptors. Talwin binds to some receptors and has a pharmacological action of its own, but its presence on other receptors keeps other drugs such as morphine from binding and working. Talwin is like a key that will fit all the door locks in a dorm but will open only a few. Other examples of mixed agonist/antagonist opioids are butorphanol (Stadol), nalbuphine (Nobain), and buprenorphine.

The last named drug, **buprenorphine** (Buprenex, Temgesic, Subutex), is being hailed as a substitute for methadone in heroin addict detoxification and maintenance, especially for non–hard-core addicts and for those under 21 years of age. Although it is an opioid, it gives only a mild high and appears to have low addiction potential. Much more data will be required, however, before buprenorphine can be judged as a valuable drug in the fight against heroin. For more on buprenorphine, see **`http://www.biopsychiatry.com/buprenorph.html`**.

Fentanyl (Sublimaze, Innovar) is a narcotic analgesic in the same category as morphine and meperidine, but it is some 100 times as powerful as morphine and about 750 times as potent as meperidine in pain relief. Marketed as Sublimaze, it is now used in an estimated 70% of all surgeries in the United States. Fentanyl also is marketed as a raspberry-flavored narcotic lozenge, on a stick, resembling a lollipop (Actiq, Anesta Corp.). Fentanyl has gained a reputation as a safe and effective drug, although it can depress respiration dangerously and slow the heart. It appears to cause less nausea and vomiting than morphine and codeine. Its duration of action is quite brief: 30–60 minutes after a single injection of 0.1 mg. Fentanyl's abuse potential is high. It is widely used in hospitals across the country, and some physicians and health professionals have become dependent on it. Fentanyl, itself a controlled substance, has become the chemical model for the creation of analogs, or "designer drugs," by which unscrupulous chemists hope to evade the Controlled Substances Act and make a lot of money. The first analog of fentanyl, alpha-methylfentanyl, was uncovered in 1979 in Orange City, California, and was found to be 200 times more potent than morphine. Since then, almost a dozen other analogs have appeared on the streets, including *para*-fluorofentanyl and the infamous 3-methylfentanyl. For more on designer drugs, see Section 12.8. As a typical narcotic analgesic, fentanyl depresses respiration. Big enough doses can stop breathing altogether. Fentanyl can cause a large drop in blood pressure, nausea, sweating, dizziness, and blurred vision.

Oxycodone and **hydrocodone** are useful pain relievers chemically derived from morphine; termed opioids, their pharmacology is basically the same as that of morphine. Oxycodone is marketed as Oxycontin (controlled-release tabs), Percodan, Percocet, and Tylox. Hydrocodone is distributed as Hycodan, Hycomine, Lorcet, Lortab, Norco, and Vicodin. Oxycodone is a pure opioid agonist used primarily as a pain reliever. In its Oxycontin form, it has been widely abused, sometimes with fatal consequences. Vicodin, a combination of hydrocodone and acetaminophen, is also a valuable analgesic, but its reputation is tarnished by the extent of its abuse and by the incidence of hearing loss apparently associated with its use.

6.11 Opiate Addiction

Although it is true that some who experiment with opiates become only casual users, if at all, it is almost always true that once opiate dependence is established, it becomes chronic—it is a career. This career is characterized by intense use interrupted by remissions and usually spans 10–20 years—in some cases, an entire lifetime.

Tolerance to and physical dependence on heroin, morphine, codeine, and the synthetic opioids can develop upon continued administration of these drugs. The typical heroin addict starts around age 18 and becomes addicted within 1 or 2 years. Two-thirds of American addicts are between 15 and 25 years of age. The serious heroin addict learns to enjoy and become dependent on the exhilarating effects of injected heroin. This condition is called *heroin hunger*. He or she may inject a combination of heroin and cocaine (or amphetamine) to experience an enhancement of the effects of both drugs. Such a combination is termed a speedball. It was a morphine/cocaine speedball that killed a well-known rock group keyboardist, according to the coroner of Martinez, California.

Whether or not a user will become addicted depends on many factors:

1. *The individual*. Research has shown that the addiction-prone personality probably does not exist—that no group of traits has been identified that is applicable to all opiate users. Yet, people differ in their ability to respond to stress, peer pressure, and drugs. This variation is a factor in addiction susceptibility.
2. *The specific opiate used*. The addiction potential of drugs differs. Heroin, producing a special euphoric "rush," is highly addictive. Codeine is far less so. Methadone maintenance is based, in part, on the assumption that methadone is less addicting than heroin.
3. *The size of the dose*. Hospital doses (15 mg three times daily) of morphine during a typical hospital stay are not likely to result in addiction. Actually, fewer than 1 percent of hospitalized patients given large doses of narcotics

Table 6.2 Aspects of the Heroin Withdrawal Syndrome

Hours Since Last Heroin Dose	*Signs and Symptoms*[a]
0–8	No withdrawal signs or symptoms
8–12	Yawning, runny nose, tearing of the eyes, sweating
12–16	Perhaps a restless sleep ("yen") lasting for several hours; addict awakes unrefreshed and miserable
16–48	Intensification of misery; loss of appetite, nausea, vomiting, diarrhea, irritability, gooseflesh
48–72	Peak syndrome; all of the preceding signs plus violent yawning, tremors ("kicking the habit"), severe sneezing, weakness, depression, intestinal spasm,back pains, alternate chills and flushing, ejaculation in men, orgasm in women, abnormal white cell counts and acid-base balance

[a]At any point during withdrawal, administration of any opiate will dramatically suppress signs and symptoms.

for 10 days or longer develop true addiction. Large doses of narcotics given to terminal cancer patients are likely to result in dependence.

4. *The frequency of administration*. The more often an opiate is administered, the greater the chance of addiction to it. Some weekend-only users of heroin (*chippers*) claim that they will never become addicted. Others claim, "It's so good; don't even try it once."
5. *The route of administration*. "Mainlining" (IV injection) of morphine or heroin is highly conducive to the development of addiction. Oral ingestion is definitely less so.

To become addicted to heroin, morphine, or codeine means that one has become physically dependent on these drugs. The body has changed in a physical and biochemical way so that it is now dependent upon the continued presence of the drug for the new physiological and mental status. Dr. Vincent Dole of Rockefeller University believes that heroin addiction is a metabolic disease analogous to diabetes mellitus. Just as the diabetic must take insulin to remain functional, the addict must take the drug.

We do not know exactly what changes take place in the body or where, but they are profound enough that if the addicting drug is abruptly withdrawn, a crisis called the **withdrawal syndrome** (see Table 6.2) ensues. How serious this crisis is depends on the individual drug, the depth of the addiction, and the addict. With some drugs, withdrawal is no worse than jittery nerves or irritability. With others, withdrawal can be life-threatening. Many of today's heroin addicts are, in fact, taking relatively little heroin, and the distress of their withdrawal has been compared to a mild case of the flu. In contrast, if an addict has been mainlining hundreds of milligrams of heroin daily, abrupt withdrawal can be a painful crisis lasting many days.

Case History *Opiate Addiction*

K., a nurse, was given a prescription for Tylenol with Codeine for a migraine headache. The prescription was for 100 tablets with unlimited refills, and she took the drugs for 3 years. When she moved to a new town, she found that she "missed" the Tylenol with Codeine. K. saw a physician, who prescribed Vicodin (hydrocodone bitartrate) for her migraine, and she began using the Vicodin to get to sleep. Not having enough pills, she began doctor-shopping. At this point, she did not consider herself an addict. Working the p.m. shift in a hospital ER, K. found a prescription pad and began writing her own. She was now taking as many as 16 Vicodin a day. After a drugstore pharmacist became suspicious and reported K. to the authorities, she entered her state's *Board of Registered Nurses* (BRN) diversion program. She was permitted to keep her nursing license but was fired when she confessed to her employer that she was an addict. Eventually, she found an employer who trusted the BRN program. K. continues her recovery, is employed, and is drug-free.

Without treatment or resumption of the drug, initial opiate withdrawal is complete in 6–10 days. However, from the following case history, it can be seen that total recovery from heroin addiction can require up to a year, including a hospital stay.

Offspring of drug or alcohol abusers are at seven times the risk of heroin addiction, according to a NIH report. You can check this subject out online at **http://www.health.org/**.

Case History *A Recovering Heroin Addict*

J., a woman in her early forties, had become deeply involved with heroin. She soon learned that heroin by mouth or by snorting was not nearly as pleasurable as the rush by injection, and she sought constantly to find a vein that would take the needle. Toward the end of her deepest addiction, she would inject 20–30 times trying to hit a vein to "get off." Finally, she went to her feet to find good veins.

J. sometimes relied on Demerol but found that very large doses (up to 1000 mg/day) were required; with less, withdrawal might be observed. For J., withdrawal was akin to a bad case of the flu. She feared heroin withdrawal all out of proportion to reality. Withdrawal was actually not life-threatening. Her pain threshold as an addict was low; after withdrawal it was dramatically acute. This situation included emotional as well as physical pain.

At the addiction treatment center, J. was given Vistaril to treat anxiety and tension, plus small doses of clonidine (see Section 6.9) for 2–3 weeks. She feels that recovery depends not so much on what kind of treatment is given but on the length of time for which one is treated. A hospital stay of 4–6 weeks is typical, and the assistance of Narcotics Anonymous is very important. She required a year free of heroin before her self-perception was again normal.

In J.'s opinion, 5-year success rates for recovery from heroin addiction are not good—10–15% at best. Heroin is difficult to kick long-term. She also has never met anyone who used heroin as a first drug.

According to one theory, withdrawal occurs because natural endorphin production by the body is strongly suppressed by the opiates taken. Thus, abrupt discontinuance of the opiate leaves the body's receptor sites with no agonists of any kind.

Studies of ex-addicts who remain completely off the drug have shown characteristic, postwithdrawal behavior. They tend to be incapable of withstanding stress. They have a poor self-image and a low pain threshold and tend to be overly concerned about personal discomfort.

Secondary withdrawals from opiates can occur. The heroin-free addict can experience all of the signs and symptoms of withdrawal weeks, months, even years after the original detoxification. The crisis, however, is not as severe. Most often, a secondary withdrawal occurs 14 days after the original withdrawal. It is believed that secondary withdrawals follow circadian rhythms.

Little evidence exists that addiction to heroin or opioids actually results in organic damage to the human body. We know that heroin typically has to be injected three to five times daily and that immediately after a fix addicts may nod off and thus be recognizable as addicts, but afterward they can act as normal as anyone. They can hold a job, perform manipulative tasks, and, as physicians, even perform surgery. As long as addicts obtain their drug and do not overdose, they can function normally in society. The director of the Public Health Services Addiction Research Center in Lexington, Kentucky, has been quoted as saying, "It is not possible to maintain that addiction causes marked physical deterioration." Cigarette smoking and alcoholism are unquestionably more damaging to one's physical health than heroin use. If this conclusion does not correlate with your mental picture of the heroin addict as dirty, sickly, infected, and living in squalor, then the difference could be environmental: dirty needles, no job, malnutrition, society's condemnation. But don't blame heroin, say some experts in justice administration; heroin as such doesn't cause any of the deleterious aspects of addiction.

Physical effects of long-term heroin use include constipation and menstrual irregularity. But even with large doses, women generally remain fertile—and pregnancy is possible. Diarrhea during withdrawal may make the contraceptive Pill ineffective.

6.12 Tolerance

Repeated administration of an opioid over a period of a few weeks will reduce the respiratory, depressant, analgesic, and euphorigenic effects initially produced by the drug. Both duration of effect and intensity of response are diminished. Larger and larger doses will then become necessary to achieve the same effects elicited by the original dose. This development of **tolerance** will continue as long as the drug is continued and will gradually disappear after complete withdrawal. Development of tolerance to an opioid does not mean that overdosage cannot occur. It can; and hundreds of people die of heroin overdosage annually. Remarkable tolerance to opiates can be achieved. In one verified case, an addict consumed 5 g (5000 mg) of morphine daily—500 times the single therapeutic dose of 10 mg.

6.13 Cross Tolerance

When tolerance to one drug automatically results in reduced susceptibility to others, we say that **cross tolerance** has developed. This situation can be explained by assuming that the liver enzymes induced in response to the first drug also have an effect upon the other drugs.

A person who has become tolerant to one opiate, say, morphine, has automatically become tolerant to all of the other opium-like drugs: heroin, codeine, Percodan, and even the synthetics such as methadone and meperidine.

Cross tolerance is a general phenomenon. It is observed with the barbiturates and between the barbiturates and alcohol. Cross tolerance develops between the meprobamates (Miltown or Equanil), alcohol, and the barbiturates. It has been shown to occur between LSD and other hallucinogens such as mescaline and psilocybin. There is no cross tolerance, however, between narcotic analgesics and barbiturates.

6.14 Detoxification

The process by which a person ends dependence on a drug is termed **detoxification**. This process may occur "cold turkey" in a hotel room, or it may take place in a jail or hospital setting with medical supervision. Drugs may be used in the detoxification process—for example, methadone for a heroin addict or tranquilizer for an alcoholic. Implied in the concept of detoxification is a limited time schedule of days or weeks, not months.[1] Detoxification does not mean that the person will stay off the drug permanently. An addict may be detoxified numerous times.

6.15 The Number of Narcotic Addicts and Their Cost to Society

While no one knows exactly how many narcotic addicts there are in this country, the DEA has estimated the national addict population to be 600,000. More than 586,000 Americans use heroin at least weekly, the Office of National Drug Control Policy reports. SAMHSA reports there were 146,000 new heroin users in 2000, and that the rate of heroin initiation is on the increase. Most addicts are found in the seven states: New York, California, Michigan, Illinois, Florida, New Jersey, and Pennsylvania. New York City has the highest number of addicts.

[1]In fact, federal regulations now specify a maximum of 3 weeks for detoxification from heroin; beyond that time, it is considered maintenance.

The American market for all illicit drugs produces annual revenues of more than $50 billion at retail. Of this amount, approximately $12 billion is spent on heroin retail sales. Where do addicts get the money to pay $25–$100 a day for their heroin? Some obtain this money by selling narcotics to others, but most must steal readily "fenceable" property, such as radios, jewelry, and small appliances. The California State Office of Narcotic Enforcement estimates that 50–75 percent of all property crimes are related to narcotics. If $12 billion a year is spent by addicts to buy heroin, and if the addict must steal $3–$5 in fenceable goods to net $1 in cash, the yearly cost to our society is $36–$60 billion.

There are also secondary costs of narcotic addiction that must be met by the taxpayer:

1. Costs to rehabilitate the addict. The U.S. Bureau of Prisons maintains drug treatment centers in certain of its correctional institutions. Centers in Kentucky and Texas treat heroin, cocaine, and amphetamine addicts of both sexes. In recent years, the federal budget has included about $160 million for the treatment of heroin addicts alone.
2. Local hospital costs (emergency room treatment of overdosage, etc.)
3. Methadone maintenance programs (discussed in the following section)
4. DEA. The DEA employs some 2,800 agents in the United States and abroad. Traffic in narcotics is a prime area of concentration.
5. State and local costs. New York and California are two of the states that have established expensive narcotic treatment and rehabilitation programs and centers. The cost per year to rehabilitate just one addict is estimated at $1,000–$3,000.
6. Alarm systems for private and public buildings

6.16 Society's Response: Heroin Clinics and Methadone Maintenance

We next discuss two social experiments through which society has attempted to deal with the problem of drug addiction:

1. Free heroin clinics
2. Methadone maintenance clinics

The rationale for free heroin is obvious: Since addicts exist, let society supply them with their drug, for free or at little cost to them. We thus eliminate their need to steal and to prostitute themselves, and we save society billions of dollars by reducing theft, law enforcement, and jail sentences. This approach, it is said, would eliminate illicit drug traffic and cut connections to organized crime.

Many Americans may be surprised to know that clinics for obtaining heroin or opiates existed in many cities in this country between 1912 and 1924 (44 clinics serving 12,000 addicts were established in such cities as New York, Chicago, and Shreveport). Some clinics actually dispensed heroin or morphine to the addict; others supplied prescriptions for opioids. In 1924, federal law made heroin illegal in all of its aspects, and the clinics were closed. Political machinations helped undermine the New York City program; in others, there were charges that clinic-supplied heroin was being resold on the street. Citizens living in proximity to the clinics complained of the influx of "undesirables." How successful these clinics were remains somewhat of a mystery.

Methadone maintenance was pioneered by the husband-wife team of Vincent Dole and Marie Nyswander Dole at Rockefeller University in 1964. Methadone, synthesized in Germany during World War II, was known to be similar to the opium alkaloids in its pharmacological actions and its addiction potential. It is totally synthetic, and while its chemical structure is not closely related to that of morphine, there are similarities. The Doles discovered that methadone could substitute pharmacologically for heroin or morphine but that the addict does not obtain the usual pleasurable effects if heroin or morphine is injected subsequent to a dose of methadone. There was evidence that daily methadone maintenance discourages further use of heroin. Methadone alone was found not to produce the exhilarating rush elicited by heroin. There is reason to believe that the receptor sites in the CNS for methadone are the same as those for the naturally occurring opiates. The Doles' early reports were highly encouraging: 80 percent of patients stopped their criminal activities and reduced or eliminated illegal drug use.

In due time, methadone maintenance clinics were established in the United States in such cities as New York, Chicago, Philadelphia, San Francisco, and San Diego, mainly with federal money. In 1977, approximately 135,000 narcotic addicts were receiving methadone in about 700 clinics (35,000 in New York City alone). This is approximately one addict in five.

Ideally, the methadone maintenance program works like this. An addict, carefully screened to make sure he or she is an addict who has tried to quit before, is evaluated by the clinic team of physician, psychologist, and clinic director. Assuming that the addict sincerely wishes to get off heroin, he or she is started on a daily dose of methadone, say 40 mg. This dose is adjusted upward or downward in order to substitute adequately for the heroin no longer being mainlined, thus preventing a withdrawal syndrome. The goal in detoxification is to get the addict totally free of drug in 21 days or less. If the process requires longer than 3 weeks, the program is considered to be a maintenance, and different federal rules apply. Some heroin addicts have been maintained on methadone for years. Personal and family counseling and job placement are offered to the recovering addict, and it is anticipated that he or she can successfully reenter society.

In actuality, the methadone-maintained addict has substituted an opioid for an opiate. He or she is now addicted to methadone. Opinions differ, but some feel that methadone addiction is more difficult to kick than heroin addiction. Furthermore, close to 30% of the methadone-maintained clinic patients may continue to use opiates, even after 6 months of treatment. The costs of operating local clinics, plus politics, can serve to defeat the concept. In addition to reducing sexual drive and inhibiting orgasm, methadone can cause some degree of constipation, sweating, and insomnia.

Ten years after their original report, Drs. Dole and Nyswander Dole admitted that their early predictions were overly optimistic.[2] Noting that some methadone maintenance clinics had lost their ability to attract heroin addicts to treatment, Dr. Dole concluded that there were more addicts on the street taking methadone

---

[2]P. V. Dole and M. Nyswander, "Methadone Maintenance Treatment. A Ten-Year Perspective," *Journal of the American Medical Association*, Vol. 235, No. 19, 1976, pp. 2117–2120.

illicitly than there were patients in methadone maintenance clinics. Illicit methadone gets on the street primarily by diversion of doses sent home with addicts for use over the weekend. In some areas, this diversion has become so significant that methadone maintenance clinics are admitting methadone addicts for first-time treatment. Thus methadone itself has become an addictive.

For all of these reasons, the early enthusiasm for this novel social experiment waned. After many years of clinical experience, it was learned that success did not lie in the methadone itself but in the scale of clinical operation, the heterogeneity of the patients and their expectations, and the varying standards of care, including compassion and skill of methadone use.

Despite its reputation, however, methadone maintenance continues to impress. The National Academy of Sciences' Institute of Medicine found that "methadone maintenance has been the most rigorously studied modality and has yielded the most incontrovertibly positive results." At proper doses methadone permits addicts to function normally without experiencing a "high" and can be taken for many years with few adverse effects. The NAS concluded that methadone maintenance can reduce addicts' use of other drugs, drug-related crime, and serum hepatitis and HIV spread. At present, fewer than 180,000 of the 1 million heroin addicts in the United States are in methadone maintenance treatment. In early 2003, federal and state officials concluded that methadone is being increasingly abused by recreational drug users, with an alarming increase in overdoses and deaths. Clinic operators realize that attempts to detoxify a methadone-maintained addict too soon can be totally counterproductive. Research indicates that about 80 percent of addicts detoxified after only 6 months are likely to return to heroin use.

The pharmacological battle against heroin continues to be fought using two drugs: methadone and **LAAM** (levoalpha-acetylmethadol, levomethadyl acetate hydrochloride oral solution, Orlaam). The great advantage of LAAM is that it can suppress opiate withdrawal symptoms for 48–72 hours (compared to 24 hours for methadone), and is effective orally. Like methadone, LAAM is an oral narcotic which itself can induce dependence. But when taken as part of a maintenance program, LAAM can block the "highs" of other opioid narcotics such as heroin. Nationwide, federal funds support 791 maintenance programs and 282 hospital programs, serving some 115,000 patients in all 50 states. LAAM is approved for use in 46 programs. Under FDA rules, programs may admit only current addicts with at least a 1-year history of addiction, plus certain pregnant women at high risk. Methadone maintenance, but not LAAM, is approved for patients under 18 who have twice failed detoxification or other drug-free treatment. Enrollment is voluntary. Medical services, HIV and other counseling, vocational rehabilitation, and pregnancy tests must be provided by the program, as must individualization of dosage, regular physician review, and observed dosing. LAAM is not approved for daily treatment—every 3 days is the goal. Frequent random urine tests are performed. Under federal law, LAAM is an "orphan" drug, that is, one which cannot be patented and which is therefore not economically attractive to a manufacturer. To offset this, the FDA gives to LAAM and certain other orphan drugs fast-track review status and tax credits and 7 years of exclusive marketing to the manufacturer.

Problem 6.1 *Methadone maintenance therapy is recognized as effective in the prevention of HIV/AIDS. Explain how this comes about.*

6.17 Therapeutic Community Approaches

Not all therapists believe that drug abuse should be treated with *another drug*. In therapeutic communities, patients live in a highly structured drug-free environment that encourages them to help themselves; live-in social dynamics are used to change social behavior. The prototype of the drug-free therapeutic community is **Synanon**, established in 1958 and dedicated to the belief that an addict can be rehabilitated by a process of socialization that involves family, systems of authority, and attack therapy groups.

The **Daytop Village** therapeutic community, begun in New York in 1963, relies on public funding and has as its goal the return of the addict to society. Through behavioral change, development of insight, job training, and general education, Daytop hopes to convert the addict to complete freedom from drugs and crime, honesty, responsibility, and social productivity. Clients remain in the therapeutic community for 3–6 months of intensive encounter therapy and then return to society.

Phoenix House, begun in 1967 in a flophouse in Manhattan, has become the nation's largest private, nonprofit program for treating drug abuse. In its 14 centers on the East and West coasts and in programs in some of the nation's largest cities, it takes almost complete control of the lives of young drug abusers, who come to live at the center for up to 20 months, or longer if needed. Phoenix House can lose 50% of new intakes in the first 30–60 days, but those youths who stay have ordered lives, a strict daily regimen, and peer policing. They cook, clean, answer the phone, and organize supplies. Parents visit regularly. The Phoenix Academy in San Diego has a waiting list; however, it does not accept youths who have a history of violence, arson, or serious psychiatric disorder.

Other therapeutic community approaches have been **Gateway House** in Chicago, Family Awareness House in California and Arizona, Marathon House in Rhode Island, and the Odyssey houses in Manhattan.

Leaders of therapeutic communities believe that, in general, drug addiction is symptomatic of a character disorder, that addicts suffer from inadequacy when dealing with stress because they are immature or lack self-esteem. Therapy, then, is based upon providing a role model with whom the addict can identify and from whom he or she can draw strength and confidence.

It has been estimated that approximately 14% of addicts who have graduated from therapeutic communities remain drug-free for 1–2 years—a significant rate compared to that of other treatment modalities. It is recognized, however, that the dropout rates can be high.

Web Sites You Can Browse on Related Topics

Methadone Maintenance

Search for "methadone maintenance"

Heroin and Opiates
http://www.drugfreeamerica.org/heroin.html

Teen Challenge Drug Program
http://www.teenchallenge.com/main/drugs

General Information; Endorphins
http://www.methadone.org/discover.html

Study Questions

1. Define narcotic analgesic.
2. Examine Table 6.1 and identify all of the narcotic analgesics that are found as such in nature (i.e., that are neither synthetic nor semisynthetic). Include dosage forms or preparations in which the drugs are found.
3. Under federal and most state laws, cocaine is defined as a narcotic. Pharmacologically, is cocaine in the same category as morphine? Explain.
4. Describe the major aspects of the pharmacology of morphine with respect to the CNS and the GI tract.
5. What differences, if any, exist between morphine and heroin in terms of chemical makeup, pharmacological effects, legal availability, and abuse?
6. What is the average percent strength of U.S. heroin street samples? Predict what would happen to an addict who unknowingly mainlined a 65% strength street sample.
7. Contrast codeine to morphine with respect to source, potency as a pain reliever, use in cough suppression, possible combination with other pain relievers, and schedule in the Controlled Substances Act.
8. Define (**a**) narcosis, (**b**) agonist, (**c**) antagonist, (**d**) opioid, (**e**) addiction, (**f**) emesis, (**g**) tolerance, (**h**) detoxification, (**i**) withdrawal syndrome.
9. Trace the route of licit morphine from the opium poppy to your neighborhood pharmacy.
10. **a.** What are endorphins?

    b. Speculate on the role endorphins play in everyday life.
11. From your own experience plus what you have read in this chapter, which of the following four drugs—tobacco, heroin, ethyl alcohol, or marijuana—produces the most physical damage to the human body? The least physical damage?
12. True or false:

    a. Meperidine is the generic name for Demerol.
    b. *Papaver somniferum* is the opium poppy.
    c. Heroin is a narcotic analgesic and antitussive.
    d. Tolerance to codeine does not develop.
    e. The triad of pinpoint pupils, coma, and respiratory depression is characteristic of aspirin poisoning.
    f. *Mainlining* means swallowing drugs rapidly.
    g. LAAM's effects are similar to those of methadone, except that LAAM's last two to three times as long.
    h. A person tolerant to Demerol is automatically tolerant to Dolophine.
    i. Methadone is not an addicting drug.
    j. Synanon is a synthetic agonist/antagonist drug.
    k. Paregoric is an OTC preparation containing morphine.
    l. In the human, morphine is one of the metabolism products of heroin.
    m. Death from heroin overdosage is far more common than death from cocaine OD.
13. Cross tolerance exists between morphine and methadone. What can we conclude about their receptor sites?
14. Name three federal acts (with dates) that have attempted to control the use or abuse of narcotic analgesics in the United States.

15. Naloxone given to a heroin addict can reverse respiratory depression and precipitate a withdrawal crisis. Explain how this result can happen.
16. In this chapter you have read the estimate of $36–$60 billion as the yearly cost to our society due to thefts and other illegal activity by heroin addicts. If one heroin addict in five enters methadone maintenance, and if the cure rate is about 47%, calculate the saving to society from reduced theft in 1 year.
17. Ideally, what is the goal of a methadone maintenance program, and how is this goal reached? Practically speaking, to what extent is methadone maintenance a solution to the problem of heroin addiction?
18. Explain (a) where and how a narcotic antagonist works and (b) how a pure antagonist differs from a nonpure one.
19. List three ways in which antagonists are used therapeutically today.
20. How does addiction differ from habituation? If addiction is a disease, is habituation also a disease?
21. Name three synthetic, licit analgesics. Which can produce physical dependence?
22. What is fentanyl, how is it used in medicine, and why is it frequently mentioned in connection with "designer" drugs?
23. List five factors that can help determine whether the use of a narcotic analgesic will actually result in the development of tolerance and physical dependence.
24. On a philosophical and practical basis, what differences, if any, exist between the British approach to the treatment of heroin addiction and the American approach?
25. Opiates produce *analgesia*, and local anesthetics produce *anesthesia*. Explain the difference between the two italicized terms.
26. Why does the human body produce endorphins during exercise?
27. Divide your class and debate each of the following propositions, pro and con: (1) The American taxpayer should simply *give* money to foreign countries to stop growing opium, which would dry up the source and also provide the poor foreign farmer with an income. (2) To reduce the availability of heroin, we should *buy up* all of a country's opium production and then destroy that opium.
28. *Advanced study question:* Talwin is a painkiller that, in 30-mg doses, produces the same analgesic effects as 10 mg of morphine. Talwin is, however, a mild narcotic antagonist. Some methadone patients given Talwin experience withdrawal symptoms. Explain how Talwin can be both a painkiller and a narcotic antagonist.
29. *Advanced study question:* The vas deferens is the tube that carries sperm from the testis to the ejaculatory mechanism. Researcher John Hughes discovered that mouse vas deferens contain receptor sites for morphine. Presumably, the same situation exists in the human male. Speculate on the reason receptor sites for opiates occur in the tissue associated with sexual activity.
30. Harm-reduction strategies for treating infections and overdoses in addicts include free distribution or exchange of fresh hypodermic syringes and even free distribution of syringes full of Narcan (Section 6.9). Are you in favor or opposed to this policy? Why?

CHAPTER 7

Cocaine, Amphetamines, Caffeine, Nicotine, and Other Stimulants

Key Words in This Chapter

- CNS stimulant
- Cocaine
- Illicit traffic
- Amphetamines
- Caffeine
- Nicotine
- Skin patch

Learning Objectives

After you complete your study of this chapter, you should be able to do the following tasks:

- Describe the pharmacologic effects of cocaine on the CNS.
- Understand the extent of illicit cocaine trafficking.
- Cite the dangers in cocaine abuse.
- List the names of important amphetamines.
- Discuss the pharmacology of the amphetamines.
- Discuss caffeine's effects on the CNS.
- Identify the pharmacological effects of nicotine on the heart and vascular system.
- Cite the evidence that nicotine is an addicting drug.

7.1 Introduction

On the street, they are called *uppers*. Heavy users of one variety are called *speed freaks*. Would-be weight losers rely heavily on them, and some long-distance truck drivers keep awake with them. To the pharmacologist, they are **central nervous system (CNS) stimulants**, and while cocaine and the amphetamines are the most

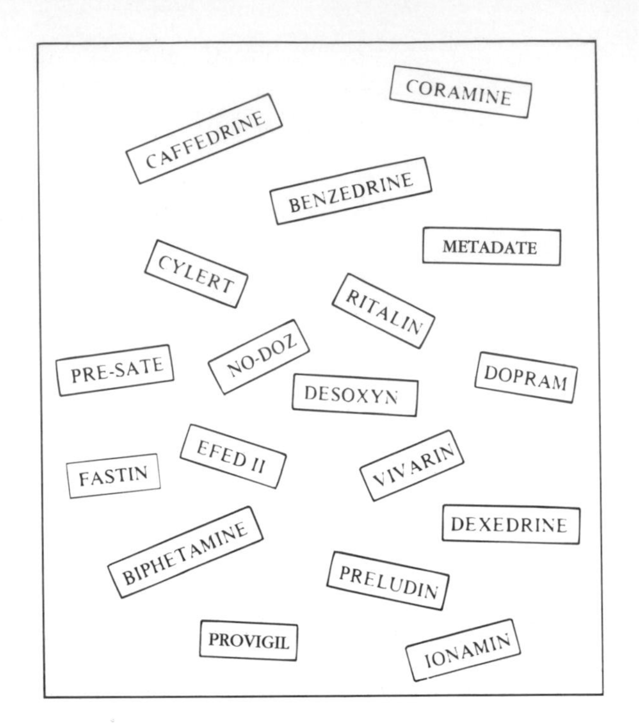

Figure 7.1 Examples of the many CNS stimulant drugs ("uppers") available by prescription or OTC.

significant examples, there are literally dozens of forms available on prescription and OTC to the American consumer (see Figure 7.1).

Observers of the American drug scene today recognize the inherent usefulness of the CNS stimulants, but they are also acutely aware that this use is overshadowed by extensive abuse. Federal agencies estimate that 22.6 million Americans have tried cocaine and that 608,000 use it once a week or more. Each day, some 1,800 Americans try cocaine for the first time. The unbelievably vast profits from cocaine trafficking corrupt everyone along its distribution path.

There has been so much abuse of amphetamines in weight control procedures that the FDA no longer recommends their use for this purpose. Even the use of caffeine, the great American wake-up drug, has been questioned by consumer advocate groups.

In this chapter, we examine the pharmacology, uses, and abuses of the most significant CNS stimulant drugs. Note that the plant source of cocaine is discussed in Chapter 2, and OTC and prescription-only diet aids are discussed in Section 15.12.

7.2 The History and Pharmacology of Cocaine

Cocaine is a New World drug, and the first reports of coca[1] chewing did not reach Europe until the 16th century. German chemists isolated pure cocaine from the leaf

[1]Currently, a distinction is drawn between *coca* (the plant leaves of *Erythroxylon coca* with less than 1% concentration of active ingredient), and *cocaine* (the alkaloid extracted from the plant and found on the street in sometimes near-pure form). Of course, coca is not to be confused with cocoa, the roasted and ground seeds of the totally unrelated chocolate tree (*Theobroma cacao*).

in 1855 (see Figure 7.2). In 1884, Sigmund Freud, after hearing exciting reports about cocaine, investigated the actions of the drug on himself and on his psychiatric patients. Largely through Freud, the drug became popular in Europe. Freud was so enthusiastic about cocaine that he wrote a "song of praise" about it, but he later discontinued all use of it personally and professionally as he came to learn more about its effects on the mind and body.[2] A contemporary, Albrecht Erlenmeyer, accused Freud of having unleashed the "third scourge of humanity" (after alcohol and the opiates).

Cocaine was an ingredient of numerous patent medicines in the late 19th century. In 1914, in enacting the Harrison Narcotic Act, the U.S. government made the serious error of designating cocaine a narcotic. It isn't. There is no way cocaine can properly be grouped with heroin or the opiates, because they are CNS depressants and cocaine is a stimulant. But the description stuck, and courts have held that it is valid to classify cocaine as a narcotic for purposes of punishment. Cocaine is a schedule II drug, along with morphine, Dexedrine, PCP, and the short-acting barbiturates. Cocaine and cocaine hydrochloride are official in the USP.

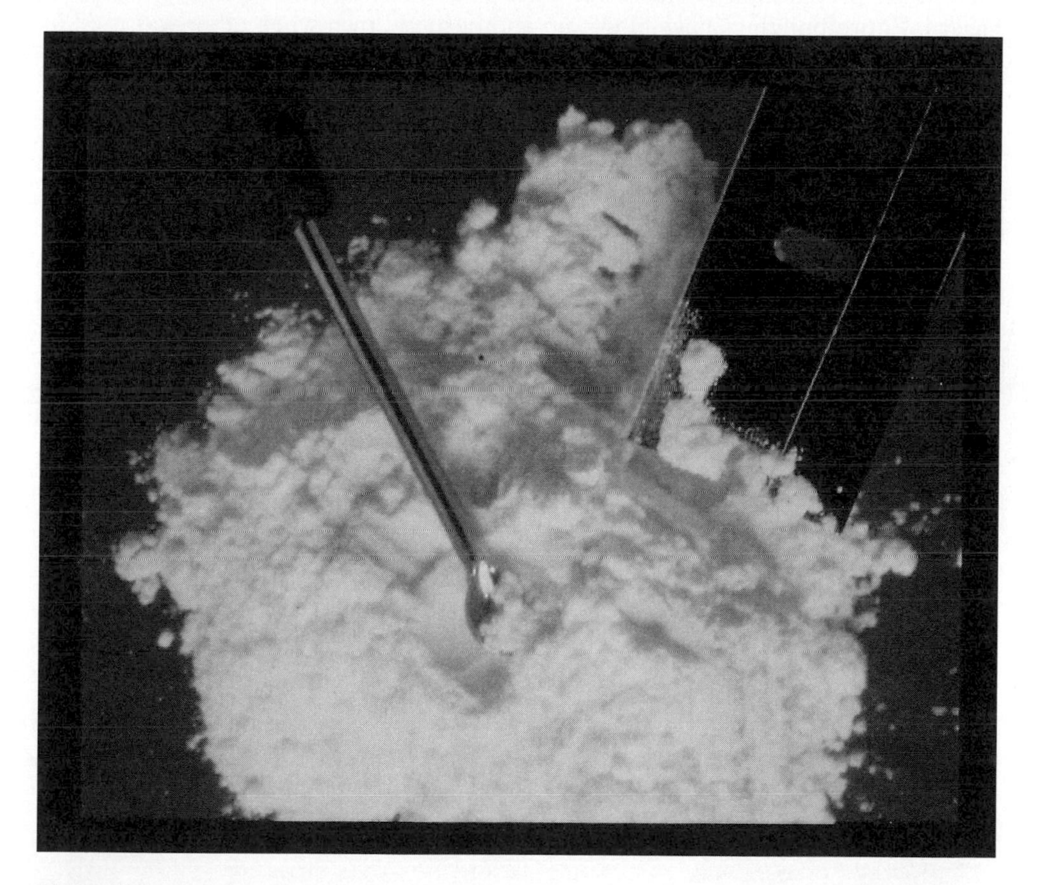

Figure 7.2 Powdered cocaine.

[2]Before this episode in his life ended, one of Freud's patients and a close friend died because of the cocaine Freud prescribed for them.

Most large hospital pharmacies have a supply of legal cocaine on hand. The drug is used in the surgery of highly vascular areas (the eye, ear, nose, and throat) and sometimes in intubations; 1 ounce is enough to perform 28 nose operations.

Cocaine has long appeared in illicit traffic as the water-soluble hydrochloride salt and in this form is sniffed into the nose or injected intravenously. In the mid-1980s, a new form of this CNS stimulant became widely available in Houston, Los Angeles, Miami, San Diego, and other cities. This form, termed **crack**, is the free-base form of cocaine (see Section 1.11). Crack is typically prepared by adding baking soda (sodium bicarbonate) to cocaine hydrochloride dissolved in water. The alkaline soda converts the water-soluble form into the water-insoluble free base, which floats to the surface and is skimmed off and dried. The dry product is rocklike and can be broken into pieces (see Figure 7.3). Crack looks like small lumps or shavings of soap but has the texture of porcelain. In some parts of the country, the lumps are called "rock" or "readyrock." In other areas, the drug is sold in 1- to 3-inch-long sticks with ridges called "french fries" or "teeth" that may be colored pink or chocolate. When heated, the crack mixture can make a crackling sound. Ether could be used in the preparation of crack, but ether constitutes an acute fire and explosion hazard. Entire buildings have blown up and burned down when ether was used in the process of making free-base cocaine. Crack is a mixture of cocaine, baking soda, salt, and other substances, but its purity has been high, averaging more than 80%. (Adulterants, however, are now increasingly being added.)

Crack can be placed into a plastic pipe, warmed to a vapor, and inhaled. This process is termed smoking, but actually nothing is combusted. Free-base cocaine is easier to heat to a vapor than cocaine hydrochloride, hence the popularity of

Figure 7.3 Wholesale quantity of crack cocaine. (Source: DEA, U.S. Department of Justice.)

free-basing. Also, smoking free-base cocaine gets a big dose of the drug to the brain much more quickly than does snorting or smoking cocaine hydrochloride. Besides, burning cocaine in a pipe or cigarettes destroys a good part of the drug.

Crack distribution and use seem to have reached the saturation point in large cities, and prices in these areas have fallen. In Detroit, a vial of crack costs $3; in Philadelphia, $2.50; in New York City, $2, and a puff on a crack pipe, 75 cents. This cheapness has transformed cocaine from a $100-a-gram indulgence into a pusher's dream and has made cocaine available to almost everyone. Crack is being mass-marketed on the streets in small vials, folding papers, or foil packets usually containing one to four pellets.

Cocaine has two main pharmacological actions: It is a powerful CNS stimulant, and it is an effective local anesthetic with its own vasoconstrictive action. We will examine the central actions first.

In doses of 25–100 mg (of pure cocaine) taken intranasally, cocaine produces its maximum central effects in 15–30 minutes. These effects were vividly described by Freud 100 years ago: exhilaration and lasting euphoria, an increase of self-control, a greater capability for work, long-lasting mental or physical work performed without fatigue, and complete elimination of the need for food and sleep.

Modern users describe cocaine's central effects as the "perfect illusion." They think they are more competent, smarter, sexier, vigilant, masterful, confident, productive, and full of energy. Thus, they say, cocaine is ego food.

Cocaine stimulates the brain by stimulating the release of and blocking the brain's reuptake of dopamine and norepinephrine (see Chapter 5), two chemicals called *catecholamines* that are naturally present in the CNS and that account for mental alertness and excitation. With more catecholamines present, the brain is stimulated eventually to the point of euphoria. In the brain, dopamine transmits impulses to the "pleasure center." Hence, cocaine, by increasing dopamine levels, induces pleasurable sensations. Figure 7.4 explains that the drug-induced euphoria occurs when cocaine blocks the normal recycling of dopamine at the synapse. For more information on this drug on the Internet, visit the URL **http://www.pslgroup.com/dg/45ebb.htm**. The *chronic* use of cocaine results in a significant depletion of dopamine in certain areas of the brain. It is believed that this depletion triggers the intense craving for more cocaine.

Cocaine is rapidly metabolized by plasma esterases and liver enzymes, and its effects wear off within an hour. Its plasma half-life is 90 minutes. Hence, a cocaine high can be experienced over and over again, many times a day. The central effects of IV cocaine are particularly short-lived. According to drug expert Sidney Cohen, in the past tolerance to and physical dependence on cocaine were believed not to occur; cocaine was thought to induce only a psychic dependence. But, says Cohen, that was because of the smaller doses usually delivered through membranes in the nose. When IV and smoking routes are used, high plasma concentrations of cocaine are achieved, and these induce the development of tolerance—and, ultimately, physical dependence (addiction). In other words, the difference lies in getting a relatively few milligrams of cocaine in one's nose or getting 3–4 g of cocaine intravenously in one day. Others confirm that heavy coke users get physically sick (suffer withdrawal) upon abrupt termination of their cocaine. Abuse of cocaine is discussed in Section 7.3. Cocaine is a short-acting drug and is flushed out of the urine within 24 hours after use (or within 2–3 days with heavier doses). See Table 16.2, "Retention Times of Drugs in Urine."

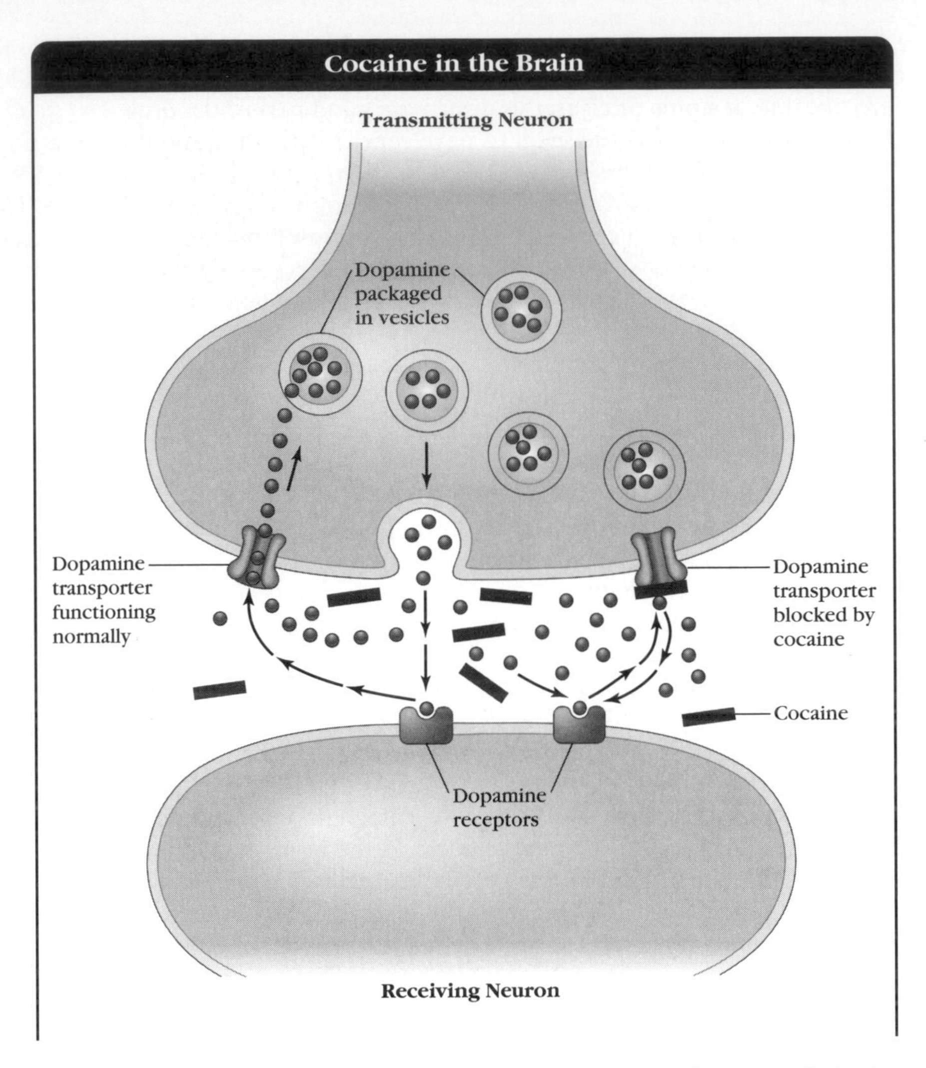

Figure 7.4 Cocaine in the Brain. The neurotransmitter dopamine transmits brain signals by flowing from one neuron into the spaces between neurons and attaching to a receptor on another neuron. Normally, dopamine is then recycled back into the transmitting neuron by a transporter molecule on the surface of the neuron. But if cocaine is present, it attaches to the transporter and blocks the normal recycling of dopamine, causing an increase of dopamine levels in the spaces between neurons, which leads to euphoria. (Source: *NIDA Notes*, National Institute on Drug Abuse, Volume 13, No. 2, June 1998.)

Cocaine has additional pharmacological actions that are mediated centrally. These include an increase in heart rate and blood pressure and hyperthermia (elevated body temperature). Cocaine also acts to dilate the pupil of the eye.

Researchers at Children's Hospital of Michigan at Detroit found that infants and children who live with crack abusers test positive for cocaine, most likely from inhaling second-hand crack smoke.

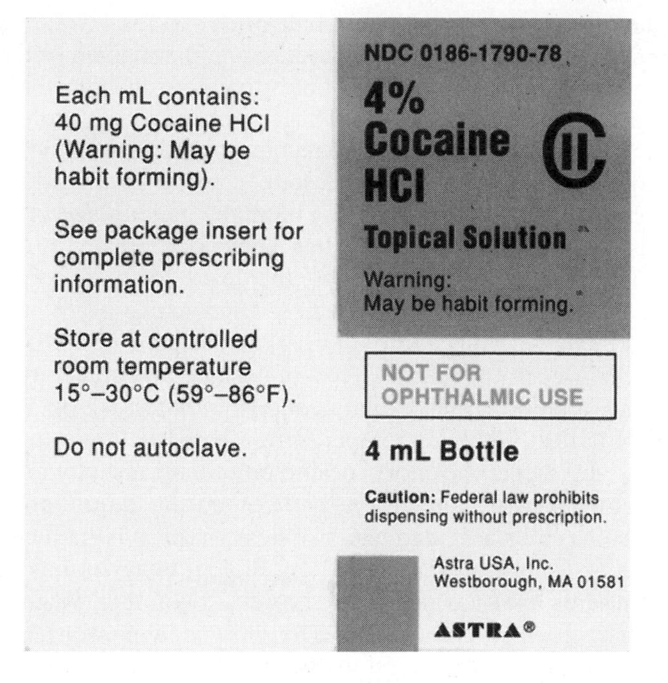

Figure 7.5 The label from a vial of cocaine for hospital use. Cocaine, as the soluble hydrochloride salt, is used as a topical solution in surgery of the ear, nose, and throat. The patient in this case underwent sinus surgery.

The second major action of cocaine is local anesthesia combined with a vasoconstrictor action. Applied to the skin or injected close to a nerve, cocaine effectively blocks the conduction of sensory impulses—an action put to good use in surgery of highly vascular areas (see Figure 7.5). By simultaneously constricting peripheral blood vessels, cocaine prolongs its anesthetic action. Cocaine is used to dull feelings when tubes are inserted down the windpipe or esophagus (i.e., intubation). For the chemical structure of cocaine, see Figure 9a in Appendix I.

7.3 Dangers in the Abuse of Cocaine

Cocaine is a seductive and intensely coercive drug, one of the most powerful pharmacological reinforcers known.[3] Once it has been experienced, the tendency to reuse it is nearly irresistible. Cocaine induces pleasurable, seldom achieved ecstatic experiences, whether by snorting (inhaling the powder into the nose through a tube or rolled-up bill), by mainlining (IV injection), or by smoking the free base. A "Speedball" used by a speed freak is a mixture of either cocaine or amphetamine and heroin.

[3]J. Spotts and F. Shontz, "Drug-Induced Ego States I. Cocaine," *International Journal of the Addictions*, Vol. 19, No. 2, 1984, p. 143. This reference is well worth reading.

Coke is the only drug laboratory animals prefer to food, water, and sex. They will self-administer the drug until they overdose and die if given free access to it. In one study, monkeys pressed a bar 12,800 times to get a single dose of cocaine. Says Ron Siegal, University of California—Los Angeles psychopharmacologist, "It doesn't matter whether he has a tail or a $100,000 income, primates like cocaine."

Spotts and Shontz point out that despite cocaine's reputation as the most sought-after drug, it is the **most unsatisfying** of all the major drugs of abuse. Cocaine produces ecstasy but never satisfaction. After a meal or after sex, our appetite or sex drive is gone. But after cocaine, there is only the desire for more cocaine.

Hence the user feels compelled to "enjoy" the cocaine again and again; in the heavy user, this means tolerance and an ever-increasing dose until finally signs of toxicity appear. However, the initial effects of cocaine and crack are vaso-constriction (a narrowing of blood vessels), dilated pupils, and increases in heart rate, blood pressure, breathing rate, and body temperature. Users lose their appetite and have trouble sleeping, and those who snort cocaine often have a runny nose. As the dose and use of cocaine increase, convulsions and respiratory failure are possible, as is cardiovascular collapse or liver damage. The heavy user may hallucinate about the presence of insects ("snow bugs") on the skin. Violent behavior has been observed. Perforated nasal septa are seen in chronic snorters. Irritability, weight loss, and poor nutrition are characteristic of cocainism. The most malignant aspect of heavy cocaine involvement is post-use depression and paranoia. The high is so euphoric, and the mental depression that follows withdrawal is so intolerable (suicidal in depth), that the user is forced to return to the drug. For some people, its use can only lead to more use, to insatiable craving, and ultimately to ruin. Four stages in cocaine's road to ruin have been described:

Euphoria → Dysphoria → Paranoia → Schizopsychosis

At high risk of heavy cocaine use are strong, ambitious, competitive, well-educated individuals who strive for perfection and are financially successful. They are intolerant of weakness or frailty in themselves or in others.

The acute toxicity of cocaine is becoming better understood, and there is now evidence that cocaine can affect the heart in a life-threatening manner. From admittedly limited studies, researchers have concluded that cocaine can cause sudden cardiac death, heart attack, irregular heartbeat, and heart muscle tissue damage. Cocaine causes the heart to beat rapidly and forcefully, but at the same time it can constrict vessels bringing blood to the heart. This can lead to fibrillation, in which the heart stops pumping effectively and quivers like Jell-O. The harmful effects of cocaine on the heart increase significantly if the cocaine user concurrently smokes tobacco. Animal tests at the NIH appear to show that the anesthetic effect of cocaine can sensitize the brain in such a way as to lead to potentially fatal seizure attacks from later small doses of the drug. In other words, using coke over a period of time can have a "kindling" effect that may mislead users into thinking they are taking a safe dose—when in fact they are gradually lowering their brain's threshold dose for seizure and sudden death. Repeated use of the drug without any problem does not guarantee freedom from seizures in the future. The next dose, taken the way you have always taken it, could produce a heart or brain seizure. In addition, some users have suffered strokes after using cocaine.

The powerful vasoconstrictor action of cocaine can result in serious problems. Upon repeated snorting of cocaine, blood vessels inside the nose are constricted to the point where no blood flows and the surrounding tissue dies. This can result in a hole clear through the septum (the part that separates the two nasal passages). Cocaine can have even graver consequences: A 34-year-old man suffered autoamputation of the penis caused by repeated insertion of cocaine into the urethra.

Cocaine or crack[4] used early in pregnancy may cause miscarriages or stillbirths; taken later in pregnancy, the drug may induce premature labor or premature delivery. Babies exposed to cocaine in the womb often do not cuddle or nurse well and may be irritable, unresponsive, and hard to take care of. Babies whose mothers took cocaine during pregnancy are 10 times as likely to die of *Sudden Infant Death Syndrome* (SIDS). Cocaine also can be passed to babies through breast milk.

Despite all of the publicity about crack babies, maternal smoking of tobacco is far more damaging to the fetus—and a greater problem for society—than mothers smoking crack, according to a study in the *Journal of Pharmacology and Experimental Therapeutics*. Each year, maternal smoking causes as many as 100,000 miscarriages and stillborns, tens of thousands of admissions to intensive care units, and extensive brain damage—a toll dwarfing that of crack use.

According to SAMHSA, the estimated number of U.S. cocaine/crack users ages 12 and older for 1998, 1997, 1996, and 1995 were:

	Lifetime Ever Used	*Used Past Year*	*Used Past Month*
1998 cocaine	23,089,000	3,811,000	1,750,000
crack	4,476,000	971,000	437,000
1997 cocaine	22,597,000	4,169,000	1,505,000
crack	4,208,000	1,375,000	604,000
1996 cocaine	22,130,000	4,033,000	1,749,000
crack	4,628,000	1,375,000	668,000
1995 cocaine	21,700,000	3,664,000	1,453,000
crack	3,895,000	1,108,000	420,000

These data indicate that cocaine/crack use has not changed significantly in the 4 years reported.

Latest available DAWN data reveal that among the most frequent annual drug episodes and mentions in *Emergency Departments* (EDs) across the conterminous United States, cocaine ranked second with a total of 144,000. Cocaine was the most frequently reported ED drug in Atlanta, Baltimore, Chicago, Miami, New Orleans, New York, Philadelphia, St. Louis, and Washington, D.C. The combination of cocaine with alcohol accelerates the heart rate far beyond the effects of either drug alone, and may be related to heart attacks and sudden death in cocaine abusers.

People seeking to overcome their dependence on cocaine can turn to *Cocaine Anonymous* (CA), now a nationwide organization that applies the same principles

[4]The reader will recall that "cocaine" implies cocaine hydrochloride; crack is cocaine free base. Obviously, it is the same drug, but the former is in the HCl salt form, and the latter is not. Actually, once in the bloodstream, the extent of protonation of the nitrogen in cocaine is fixed by the pH of the system, no matter which form of the drug is ingested.

as those used by Alcoholics Anonymous. See Section 1.9 for the telephone number and Web address. Additionally, one can contact the Cocaine Anonymous URL `http://www.ca.org/` (more than 60,000 hits last year), or contact them by e-mail at `cawso@ca.org`. Some 60,000 members attend 1,600 CA meetings across the country and in Canada. About 50% of those who regularly attend CA meetings for one year stay off the drug. This number contrasts favorably to psychotherapy or drug maintenance programs.

7.4 The Illicit Traffic in Cocaine

Based on the latest *National Narcotic Intelligence Consumers Committee* (NNICC) report, Colombian crime organizations, particularly the Cali drug mafia, presently maintain control over cocaine trafficking to the United States. The drug is shipped directly to the United States or through Central America, Mexico, and the Caribbean. It travels by land, air, and sea, and by personal courier. Peru remains the world's leading producer of coca leaf, coca paste, and cocaine base, with Bolivia second and Colombia third. Colombia, however, remains the leading processor. The federal government's NNICC report estimates that the maximum annual world production of cocaine is 760 metric tons.

> ***Problem 7.1*** *Estimate the number of 50-mg doses of cocaine that theoretically are available from 760 metric tons of cocaine. A metric ton is 2200 avoir.lb. Figure 100% purity.*

Primary importation ports in the United States are in Arizona, southern California, southern Florida, and Texas; distribution continues via Houston, Los Angeles, Miami, and New York City. Nationwide, cocaine prices range from $10,500 to $40,000 per kilogram (average purity in kilogram lots is 83%). Coca leaf intended for legitimate use in the United States is imported from Peru by Stepan Chemical Company, Maywood, New Jersey. They extract the cocaine and sell it to Mallinkrodt Company, which distributes 100% pure cocaine hydrochloride flakes for $3.50 a gram. On our streets, gram-quantity prices range from $20 to $200.

The quantities and dollar values of cocaine smuggled into the United States yearly are staggering. While no one knows how much eludes detection, we know the latest annual seizure was 115 tons. Worldwide, in one recent year, 303 tons of cocaine were seized. The United States is estimated to consume 70% of the world's production. The record U.S. bust was 21.4 tons found in a Sylmar, California, warehouse in 1989, valued at $6 billion. Another 15.7 tons were seized in Miami, hidden within concrete posts transported from Venezuela on a cargo vessel. A total of 12 metric tons of cocaine was seized off the coast of Mexico, the largest ever intercepted at sea. Cocaine has been smuggled in fake gas tanks, fake legs, coffee shipments, air-conditioning units, radiators, rental cars, backpacks, horse packs, labeled as coffee beans or crab meat, and by nice old couples driving motor homes. In attempts to defeat cocaine detection by dogs, shipments have been doused with perfumes, oils, and Dijon mustard. One suspect even sprayed his vehicle tires with female dog urine. In one of the more bizarre incidents, authorities found more than 2 kg of cocaine surgically implanted in the abdomen of an Old English Sheep Dog that arrived at New York's JFK International Airport from Bogota, Columbia. The dog survived. But a professor at a midwest state university did not survive when one of the dozen cocaine-filled balloons in his body burst on a flight from Amsterdam to Detroit. When emergency surgery was performed on the 45-year-old man, the balloons were discovered.

Besides being smuggled into this country inside every conceivable inanimate object, cocaine is "body smuggled" by "mules"—people paid $2,000 a trip to hide the drug on or inside their bodies. Mules stuff the cocaine into condoms and swallow them, intending to defecate their cargo once safely past customs. The record quantity of cocaine (found on autopsy) was 1.1 pounds stuffed into 147 condoms and swallowed, says the chief medical examiner of Dade County, Florida. Currently, mules are swallowing increased amounts. Some body packers collapse at airports, chewing their tongue and lips bloody in the throes of convulsions when the cocaine packets burst or leak inside their GI tract. Others, who get through customs, die in hotel rooms. In 1990, a Colombian airliner crashed in New York. When one critically injured survivor was operated on in a hospital, four condoms filled with a total of 1 pound of cocaine were found in his intestines. In Honduras, police captured a Colombian suspect who was hiding a "key"[5] of cocaine in his stomach, distributed among 104 paraffin and plastic capsules containing 9 or 10 g of cocaine each. Arrested on a tip, the suspect was held in a hospital where he expelled more than 60 of the capsules.

All of this shocking illegal activity is motivated by the high profits in illicit trafficking in cocaine. From 500 kg of coca leaves in Colombia, worth $500, farmer-chemists can prepare 1 kg of cocaine base worth $1,500. This base yields 1 kg of cocaine hydrochloride, worth $4,000 in South America but being sold wholesale in the United States for as much as $36,000. Each time the cocaine changes hands, it is cut with chemicals such as procaine, lidocaine, mannitol, inositol, phenylpropanolamine, or even laundry detergent, until the original sample hits the streets as cocaine hydrochloride with an average purity of 63%[6] and a value of about $100,000 per kilogram.

Coca paste is the initial product obtained when the leaves are mashed and treated with alkali to free the alkaloid. Kerosene or gasoline is used to extract the cocaine, followed by addition of sulfuric acid to precipitate a crude, semisolid product called *cocaina sulfato* (coca paste). This substance is usually sprinkled on tobacco or marijuana and smoked.

7.5 Penalties for Illegal Use of Cocaine

Cocaine is a schedule II drug under the Controlled Substances Act of 1970. The act specifies penalties for unlawful possession of cocaine and for unlawful possession with intent to distribute. The penalties are as follows:

1. Unlawful simple possession of cocaine
   First offense: Up to 1 year's imprisonment and/or fines up to a maximum of $5,000.

[5]That is, a kilogram, or 2.2 lb.

[6]Purity, of course, drops as the cocaine is cut (diluted) by middlemen on its way to local pushers.

Second offense: Twice the imprisonment and fines of the first offense. A person under 21 years of age arrested for first offense, simple possession, may have his or her record of trial and conviction cleared after satisfactory completion of probation.

2. Unlawful possession with intent to distribute
First offense: Up to 15 years' imprisonment and/or fines up to a maximum of $25,000, plus 3 years' mandatory parole.
Second offense: Maximum penalties are twice those of the first offense.

See Section 3.9 for maximum penalties for cocaine trafficking.

7.6 Amphetamines and the Stimulation of America

One of the reasons we are called a drug-taking society is our startling use of the powerful CNS stimulant drugs called the **amphetamines**. Production quotas have been reduced, but they are still set at an annual level of one-half billion dosage units (tablets, capsules, etc.), enough to supply four doses to every American 18 years of age or older. In other words, drug companies are turning out 8 tons of amphetamines each year. It is hard to believe that America needs that much chemical stimulation, but we are getting it.

7.7 The Development of Amphetamines

In the early 20th century, pharmaceutical chemists knew the chemical structure of epinephrine (adrenaline), one of the substances the body uses naturally to prepare us to meet the stress of everyday life. Epinephrine is a catecholamine (see Chapter 5), and as such has the chemical structure of a *beta-phenethylamine*. With this knowledge, researchers were able to prepare a series of chemically related compounds that had drug effects very similar to those of the natural catecholamines. The group name became *amphetamines*, based on the parent member, amphetamine (Benzedrine). These compounds were first used medicinally in 1927 and have since become the classic examples of CNS stimulant drugs. For the chemical structure of amphetamines, see Figure 11a in Appendix I. Because the nomenclature can be a little confusing, Table 7.1 is offered for clarification. (Hallucinogenic amphetamine derivatives, including ecstasy, are discussed in Section 12.6, "Other Current Drugs of Abuse.")

Dexedrine comes in 5- and 10-mg tablets and capsules and as an elixir. It is chemically identical to Benzedrine but is what chemists call the dextrorotatory isomer, whereas Benzedrine is the racemic mixture. Propylhexedrine's chemical structure is not strictly identical to the general amphetamine structure. Various dosage forms of amphetamines are encountered (see Figure 7.6).

The first use of the amphetamines was in 1932 in the Benzedrine inhaler, which was used to clear blocked nasal passages. Benzedrine shrank the nasal mucosa effectively, but when the effect wore off, some people experienced a condition worse than before. In addition, it became known that Benzedrine was a euphoriant and that one could get high by extracting the Benzedrine from the impregnated inhaler

Table 7.1 Nomenclature of Amphetamines

Generic Name	*Trade Name*	*Street Name*	*Dose, mg*
Amphetamine sulfate	Benzedrine	Bennies, uppers	5–10
Dextroamphetamine sulfate USP	Dexedrine	Dexies	5–10
Methamphetamine HCl USP	Methedrine,	Splash, crank,	5–25
	Desoxyn, etc.	crystal, speed, ice	5–25
Amphetamine + dextroamphetamine	Biphetamine	Black beauties	—
Propylhexedrine	Benzedrex inhaler		250
Hydroxyamphetamine	Paredine		1% solution

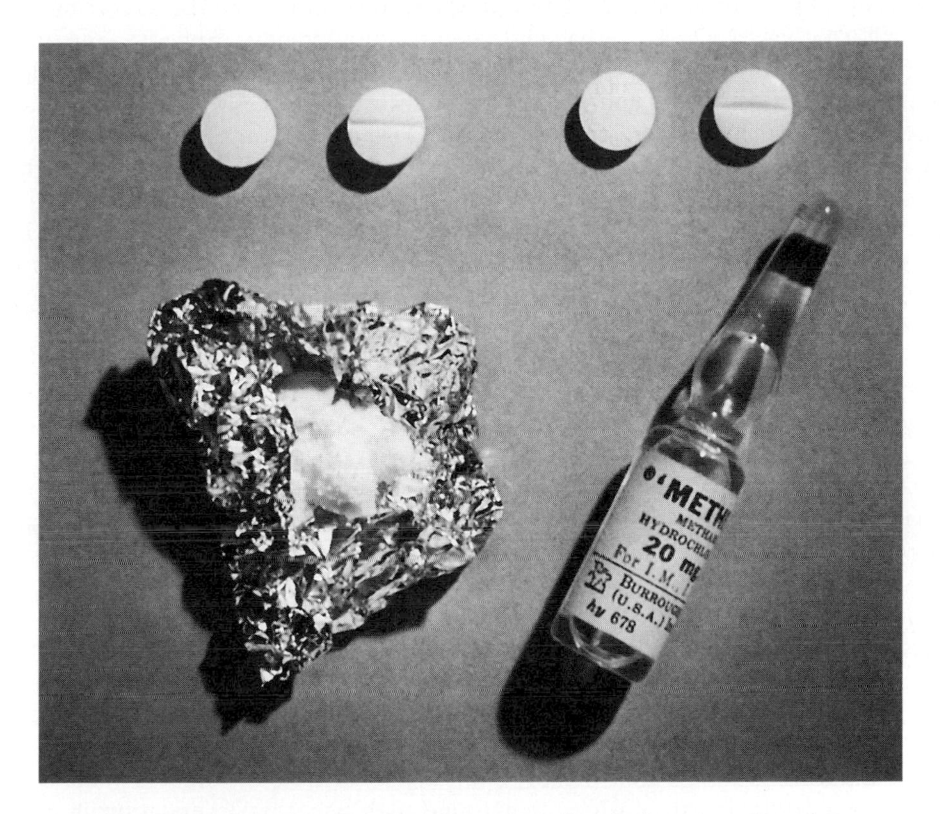

Figure 7.6 Amphetamines are found in various dosage forms, such as these tablets, injectables, and street-drug crystals.

paper using a soft drink (the carbonic acid in the soda helped to extract the basic amphetamine). The manufacturer of the inhaler soon changed to a nasal shrinker that did not have a stimulant effect on the brain, but the damage was done. America had learned about a new way to get high—the amphetamines.

An additional discovery was made in 1937 that directed attention to the amphetamines. It was found that they were useful in treating narcolepsy—an epileptic seizure disorder in which a person repeatedly falls asleep. The stimulant effect of the amphetamines helped keep the patients awake and functional. Provigil (modafinil) is marketed as a wakefulness-promoting agent having effects similar to those of sympathomimetics, such as amphetamines. The FDA approved modafinil for use in narcolepsy, but its mechanism of action is unknown.

Hyperkinetic children (so overactive that they cannot sit still or concentrate) are benefited by amphetamines such as Dexedrine. Paradoxically, these CNS stimulant drugs act to calm down a hyperactive patient, perhaps by awakening lagging control mechanisms in the brain.[7] Amphetamines or amphetamine-like drugs are still used for treating these conditions. Ritalin (methylphenidate USP) and Cylert (pemoline) are effective in the treatment of attention deficit hyperactivity disorders (ADHD) (previously known as minimal brain dysfunction) in some children. Critics of Ritalin use in hyperactive children are numerous and vocal. *Parents Against Ritalin* is an organization opposing prescription treatment of ADHD, preferring natural alternatives, such as herbal and dietary supplements. Ephedra, the most controversial supplement in the industry, has been promoted and sold as an ADHD treatment. For cautions on ephedra use see Section 2.3. Actually, Ritalin is the most prescribed stimulant drug in America, but because its structure is similar to that of the amphetamines—and because of its wider availability through legal use—it has become an important street drug of abuse. In Chicago, Ritalin is sold as "west-coast" and is injected with Talwin as a substitute for the heroin-cocaine combination known as a speedball. Ritalin highs are short-lived and soon give way to depression, apathy, and loss of concentration and memory. Ritalin can be a dangerous drug; it can cause stroke and heart attack.

In 1939, narcolepsy patients reported that their appetite was depressed after they took amphetamines. This discovery led to the application of the amphetamines as **anorectics**—appetite depressants for weight loss (more will be said about this topic later).

World War II saw heavy use of amphetamines by both sides. American, British, Japanese, and German soldiers were given amphetamines to counteract fatigue and increase endurance. In combat, U.S. soldiers received a maximum dose of 10 mg of amphetamine every 12 hours. So many amphetamines were given to Japanese home front workers that a major abuse problem was discovered after the war (550,000 addicts in a population of 88.5 million). Strict controls, strict penalties, and educational programs eliminated the Japanese problem.

Extensive use and abuse of amphetamines was recognized in the United States by the early 1950s. Because restrictions were inadequate, amphetamines were sold by the millions of dosage units to individuals at post office box numbers in the United States and Mexico, from where they found their way to workers at lunch counters, taverns, and filling stations. Truck drivers used the stimulant drugs to stay awake for many hours beyond their normal endurance—a dangerous practice that contributed to accidents.

Truckloads of amphetamines were hijacked, and the drugs were funneled into the underground distribution system.

In 1965, Congress passed the drug abuse control amendments that gave the FDA authority to regulate the manufacture and distribution of dangerous drugs

[7]Recent research in mice suggests a completely different mechanism: an increase in serotonin, with resulting inhibition of aggressive and impulsive behavior.

such as the amphetamines. Investigations, arrests, and criminal prosecutions followed. Accountability investigations revealed that some manufacturers had not kept accurate records on amphetamine distribution. (It was little wonder that a crusading reporter could set up a phone office, print phony business letterheads, and successfully order and receive a million dosage units from unscrupulous dealers.)

In the 1970s, a small amphetamine tablet known as the *minibennie* was produced in makeshift laboratories in Mexico and was widely distributed in the United States. In 2 years, 12.7 million minibennies were seized or sold, more than a third of them in California. Forensic chemists at the DEA's Special Testing and Research Laboratory in McLean, Virginia, found that many of these minibennie tablets contained no stronger stimulant than caffeine or ephedrine.

Today, with stricter accountability and distribution controls, illicit traffic in amphetamines has been greatly reduced, but it has not been eliminated. San Diego, California, because of its great number of illicit "speed" labs and the large quantities of methamphetamine manufactured (4–5 tons in a recent year), has unfortunately become known as the amphetamine capital of the world.

"Ice" (see Figure 7.7) is a new physical form of an old amphetamine. It is highly pure, recrystallized methamphetamine hydrochloride (for decades sold in powder form on the street as speed or crank), and it is smoked, not injected. Some ice comes from Hawaii, but most of it comes from Asia—Taiwan, Hong Kong, South Korea, and the Philippines. Known as *hiroppon* in South Korea and *shabu* in Taiwan, it can be synthesized readily from either ephedrine or phenylacetone and appears on the

Figure 7.7 "Ice"—so named because of its appearance—is a smokable form of methamphetamine.

street as large, individual crystalline chunks that resemble rock candy, broken glass, or pieces of ice. Although it is not a free base, "ice" can still be heated in glass pipes, melted, vaporized, and inhaled, providing a concentrated dose to the brain within seconds. The effect is intensely pleasurable: a powerful rush of euphoria, energy, or intensified sexual pleasure, followed by a high lasting 4–8 or even 12 hours (compared to the 20–30-minute high from cocaine). It is the initial rush that the ice smoker is seeking. But because tolerance builds quickly, some smokers are forced into high-dose, nonstop use (called "chasing the high"). Some ice abusers can smoke up to a gram a day.

Ice is not without its dangers. It can cause irregular heartbeat, severe anxiety, and, with chronic use, paranoid hallucinations. Hospitals in Honolulu report that some users suffer from water in the lungs and shriveled lung blood vessels. And, of course, the low that follows the ice high can be as deep and morbid as the high is euphoric. National surveys by NIDA show that use of ice is primarily concentrated in the Western region of the country. Its use in Hawaii became widespread in 1988. Ice sells for $300 to $500 a gram, or $5,000 to $7,500 per ounce.

7.8 The Pharmacology of Amphetamines

The amphetamines were designed chemically to be similar to adrenaline (epinephrine) and the other catecholamines (see Appendix I). Thus, we can expect them to have similar actions in the body, and they do. In low to moderate oral doses (1–50 mg), amphetamines generally will act to:

1. Stimulate the CNS.
   a. Elevate the mood.
   b. Improve concentration, thinking, and coordination.
   c. Produce electroencephalographic signs of arousal.
2. Delay the onset of fatigue and the need for sleep.
3. Increase blood pressure.
4. Increase respiration rate.
5. Raise blood sugar levels.
6. Dilate the bronchi.
7. Divert blood flow from internal organs to skeletal muscle.
8. Constrict nasal mucous membranes.
9. Depress the appetite.
10. Improve athletic performance.

We learned in Chapter 5 how closely these pharmacological actions mimic our body's reaction to stress. All 10 of these actions are stimulatory. They do not apply to resting states, but rather to excited states. Thus, we say that the amphetamines are CNS stimulant drugs with pressor effects on the blood system and stimulant effects that prepare one to handle stress. As discussed in Section 5.4, they may be classified as alpha$_1$- and alpha$_2$- adrenergic agonists.

High doses (100 mg or more), as, for example, in IV amphetamine abuse, can produce a pleasurable mental rush or "flash." But high chronic dosing with amphetamines is dangerous. The user becomes overactive and irritable and shows defective reasoning and judgment. Irregular heart rhythms, liver damage, paranoia, and cerebral hemorrhage can occur. More information on this topic is given in Section 7.9.

Amphetamine psychopharmacology is very similar to that of cocaine. In fact, these two psychomotor stimulants cannot be distinguished during blinded laboratory administration, by either experienced human abusers or animals trained to identify the drugs.

People who use amphetamines regularly to get going in the morning, to move ahead in their job, to compete in athletics, or to lose weight may discover that they need drugs like barbiturates to sleep at night. The downers, in turn, enhance the need for more stimulants the next morning; thus, a vicious cycle may develop.

Tolerance to amphetamines can develop upon continued use. The user realizes that the initial dose is no longer effective. But increasing the dose works only for a while; it must be increased again and again, until the user is consuming 150 mg or more daily. The amphetamine doper is highly tolerant and may ingest more than 1000 mg a day to maintain a high. Some chronic users have reportedly injected as much as 15,000 mg of amphetamine in 24 hours without observable acute illness. For neophytes, however, rapid injection of 150 mg could be fatal. There is disagreement among the experts as to whether physical dependence develops. If it does, it is much weaker than that of heroin or alcohol, and the withdrawal crisis (*crashing*) is largely emotional (severe depression, fatigue, increased appetite, high fluid intake). It is still

Case History *Methamphetamine Abuse*

C. was introduced to methamphetamine (crystal) at a friend's house when he was 19. He had previously used marijuana and alcohol. C. observed his friends separate the methamphetamine into lines and snort it through paper straws that had been cut short. But for his first use, he chose to take the drug mixed with a small glass of Pepsi. In a few minutes, a "strong sensation encompassed my body. My head began to tingle, and then a sudden burst of energy overtook me. For the next few hours, I was the most energetic that I have ever been." C. decided to snort methamphetamine. He bought a "quarter" from a friend, cut it up into lines, and snorted it. "My nose felt as if it were on fire. It burned, causing my eyes to water profusely. As the burning began to go away, the drug that had gotten stuck in the fluids in my nose began to drip down my throat. It was one of the most disgusting tastes I have ever had." Despite this, C. snorted often. Under the drug's powerful stimulant effects, C. "cleaned the whole house from top to bottom. He was able to travel nonstop from a bachelor party in one city on a Friday night to a wedding the next morning in another city many miles distant. C's use of crystal became more frequent. Within a month he was using it multiple times a day, including during work. He was now buying it from people he did not know and should not have trusted. One day C realized he was headed towards total dependence. He quit cold turkey and has not tried it in the past 2 years, although the urge occasionally returns.

a crisis, however, and so we conclude that amphetamines produce tolerance with psychological dependence. Amphetamines are metabolized more slowly than cocaine; up to 2 days are required to eliminate a single dose of Desoxyn.

Mechanism of Action. Amphetamines owe their generally excitatory action to the fact that they stimulate the release of catecholamines (dopamine, norepinephrine) into the synaptic cleft (see Chapter 5) and delay their reuptake.

7.9 The Great Amphetamine Debate

The following is a true account of an amphetamine dependency that began with a prescription for weight control. A 22-year-old California housewife and mother of three asked her doctor for advice about a minor weight problem. He prescribed Desoxyn (methamphetamine USP). As she stated later, the first tablet she took made a great impression on her. "I had never had anything like it. It was the most wonderful feeling, the most wonderful sensation. I took one of those and I think I was like a wonder woman that day. I was really hooked from that day on." She took amphetamine preparations for 6 years, gradually building up to 250 mg daily (30 tablets a day). "After a while, I just couldn't function without the pills in any way." She made the rounds of four or five doctors and found that obtaining prescriptions was "amazingly easy. I never had any problem getting prescriptions." Even after becoming dependent on amphetamines, she had no guilt feelings. "You got them from a doctor, and that made it OK, made it legal." Eventually, her mental and physical deterioration became so profound that she was forced to enter a clinic for detoxification.

Benzedrine, its potent partner **Dexedrine**, and **methamphetamine** are the amphetamines widely used today to treat obesity. Here is how they work. When a person is very excited or angry, catecholamines are secreted and are distributed to the brain and GI tract. Digestion of food is temporarily halted, the CNS is stimulated, and interest in food and eating is lost. (What a good reason for not arguing at the dinner table!) Amphetamines, being very similar to catecholamines in chemical structure, produce the same kind of GI tract suppression and CNS stimulation. Peristalsis, the rhythmic contraction of the stomach and intestines associated with digestion, is diminished. Thus, appetite is lost, and the person doesn't feel like eating. Side effects of the amphetamines include CNS excitation, wakefulness, possible palpitations, headaches, and increased blood pressure. To some people, these side effects are of no significance. To others, they are intolerable.

Research has shown that homemakers and other women who are either unemployed or retired are among the highest regular users of amphetamine diet pills. Sales workers also have high rates of use. Only about 70% of all regular users of amphetamine diet pills obtain *all* of their drugs with their own legal prescriptions.

Although amphetamine use results in some weight reduction, physicians are critical of such use for the following reasons:

1. After several week s, the user develops a tolerance to the amphetamine and may have to take ever-larger doses to get the same effect. This process has led to psychological dependence on the drug.

2. The benefits are short-term. Unless diet and exercise are also included, the weight loss will be trivial.
3. Amphetamines prescribed for weight reduction find their way into illicit drug traffic. DEA data show that in about 25% of the reported cases of amphetamine abuse, the drug was obtained through legitimate prescriptions.
4. Amphetamine use in "fat clinics" has gotten out of hand. The DEA has information on physicians who have ordered and received more than 3 million amphetamine dosage units in 1 year. A substantial amount of illegal diversion of stimulant drugs occurs through the operation of clinics devoted entirely to the treatment of obesity.
5. Amphetamines taken for weight loss cause nervousness and CNS excitation, and the user may have to take tranquilizers or other sedatives to sleep.
6. Obesity can be considered a chronic problem; it cannot be helped by amphetamines, which have only a brief effect. In fact, some doctors maintain that a would-be weight loser cannot learn how to handle hunger if it is eliminated—and that diet pills are therefore an unnecessary crutch.

At present, the FDA has not banned the use of amphetamines in weight control but has recommended a ban. Physicians still have the right to prescribe amphetamines for any condition in which they believe they could be helpful. However, as an FDA spokesman said, if a doctor prescribes amphetamines for obesity and the patient complains of serious side effects, the doctor could be in serious trouble. Wisconsin, New York, and Maryland have passed legislation that virtually bans the use of amphetamines for the treatment of obesity.

The FDA continues to approve the use of amphetamines in the treatment of narcolepsy and ADHD in children.

Table 15.6 lists some stimulant drugs that, strictly speaking, do not have the chemical structure we ascribe to the amphetamines. These prescription drugs have been used in weight control and generally have less potential for abuse, although Preludin is a schedule II drug.

OTC diet pills are sold locally and nationally, often with catchy names such as Appedrine, Dexatrim, Hungrex, and Slim One. Some contain ***phenylpropanolamine*** (PPA), a now-banned appetite suppressant drug chemically related to the amphetamines and often used as a nasal decongestant. See Sections 15.3 and 15.12 for more on phenylpropanolamine. Others contain no CNS stimulants but rather a bulking agent such as carboxymethylcellulose, which swells when in contact with water, supposedly giving the dieter a feeling of fullness and therefore of satiety. The FDA has approved the use of benzocaine, a local anesthetic, as an appetite-suppressant drug, and such an OTC preparation is now sold. Presumably, the anesthetic dulls the taste buds, and the dieter eats less because he is not enjoying his food as much. Benzocaine, of course, continues to be sold OTC in throat lozenges and in other preparations.

Problem 7.2 *America faces an epidemic of childhood obesity: as many as 14 percent of our youth are obese. Identify cogent factors on both sides of the question, Should overweight kids take diet drugs?*

Case History *Drugs in Weight Loss*

A 30-year-old wife and mother (5 feet 6 inches, 222 pounds) entered a fat clinic operated by a physician (whom she never met). She was highly motivated to lose weight. She received an injection of HCG (see Chapter 2) five times a week and a daily dose of an amphetamine sold to her by the clinic. In addition, she was placed on an approximately 800-calorie-a-day diet and was instructed to drink a gallon of water a day. She was not permitted to even touch any fatty substance in any cosmetic or food but was permitted as much black coffee as she desired (she drank up to 20 cups a day). During her visits to the clinic, she was never examined in any way except for weight. Six weeks into the program, after she had lost between 30 and 40 pounds, she fainted in the shower and was taken to a hospital, where she spent 2 weeks recovering from heart arrhythmias, electrolyte imbalance, low blood pressure, and general malnutrition. She has since regained all of the weight she lost.

7.10 Mainlining and the Speed Freak

We have described one type of amphetamine overuse, which starts with the treatment of obesity and progresses to larger and larger oral doses as tolerance develops. A second, much more serious form of abuse involves individuals who find pleasure in the euphoric rush effects of injected amphetamine. The drug of choice here is methamphetamine ("crystal"; see Table 7.1), but all of the various types have been used. Called a *speed freak*, or "tweaker," the user will mainline 20–40 mg of methamphetamine (*speed*) at first and then gradually increase the dosage to hundreds of milligrams a day as tolerance develops. The user may go on an amphetamine spree (binging), staying awake for days at a time, not wanting to come down off the euphoric high. He or she will probably eat nothing during the high and lose considerable weight. Then, exhausted and drained, the user will sleep for a day or more, awake hungry, and start the whole abusive process again. The speed freak is driven by a desire to re-experience the ecstasy of the high, which is described as being different from the high following an oral dose.

Most speed freaks and small-time dealers buy crystal in what is known as "B-balls"—one-eighth of an avoirdupois ounce (approximately 3.5 g), selling for $180. A "quarter" of crystal is 0.20 g, selling for $20. A dealer can buy a B-ball for $180 and resell it as 17 quarters (with 0.1 g left over.) That's a $160 profit.

The possibility of a psychotic breakdown accompanies the heavy use of amphetamines, either from a single large dose or from chronic moderate doses. A paranoid psychosis may develop in which the individual is convinced that people are talking about him behind his back or that he is being followed everywhere. One user climbed to the roof of a building to escape secret agents and then threw tiles down on his imaginary pursuers. Others accumulate weapons. Auditory and visual hallucinations may accompany the psychosis, giving the impression that the speed freak is a schizophrenic.

A strange behavior termed stereotypy[8] is seen in heavy amphetamine users. A simple act such as stringing and unstringing beads, pacing the floor, cleaning, or shining shoes will be carried out repeatedly for hours. Amphetamines induce stereotypy in animals, also.

Experiencing the phenomenon called **formication**, the heavy amphetamine user hallucinates about the presence of insects crawling around just under the skin. So real is this hallucination that addicts have been known to probe with a knife in attempts to eliminate the crawling bugs. A skin mole may be watched for minutes to see if it will move. Formication is also seen in alcohol-induced delirium tremens and with heavy use of cocaine.

The combination of heroin or other opiate with an amphetamine is termed a *speedball.* Taken by injection, the combination appears to offer an enhancement of the actions of both drugs.

Methamphetamine use and abuse is near epidemic in Southern California (San Diego county is known as the methamphetamine capital of the country). According to federal and local surveys,

- Methamphetamine-related admissions to emergency rooms in California jumped 49% (to 10,167) in a recent year.
- Drug tests on job applicants reveal about 5% positive for methamphetamine, according to PharmChem Corporation, Menlo Park, California.
- Thirty-five percent of admissions to California drug treatment centers are for amphetamine abuse.
- Most "meth" users are white, between 25 and 35 years old, with usage divided equally between males and females.
- Yaba, a new form of methamphetamine, is now appearing in Thai communities in northern California and Los Angeles. Whereas most meth is injected, snorted or smoked, yaba (a mixture of meth and caffeine) is taken orally as tablets small enough to fit inside a drinking straw. Users have the perception that tablets are safer (why?). The DEA says Southeast Asian meth tablets are made by large drug-trafficking organizations in Burma.

In almost every industry, more workers are abusing methamphetamine because they are deluded into believing it can help them work faster, better, and longer. Workers take the drug anywhere they don't expect to be caught—in a private office, restroom, stairwell, in a car during a break, and in airplane toilets. Typically, a worker uses speed for a while and feels superproductive. But prolonged or heavy use can make the person irritable, wired, and jittery. Arguments at work or at home may become violent and physical. Alcohol may be used at night to help induce sleep. Finally, for the addict, the drug's deleterious effects can leave the user depressed, physically exhausted, and often out of a job.

7.11 Federal Law on Amphetamine Abuse

Due to their high potential for abuse, Benzedrine, Dexedrine, Methedrine, Preludin, and Ritalin are all schedule II drugs under the 1970 Controlled Substances

[8]Called "tweaking" on the street.

Act. Prescriptions for these drugs must be written on special forms and are not refillable. They must not be telephoned in to the pharmacist but must be written in ink or typewritten and must bear the signature of the physician. It is noteworthy that federal authorities consider the amphetamines dangerous enough to be placed in the same drug schedule as morphine, cocaine, Seconal, and Quaalude.

Despite the laws enacted, "crystal" methamphetamine is back on the streets in a big way. It can be synthesized in illicit laboratories from ephedrine, pseudoephedrine (see Section 15.3), and from the now FDA-banned phenylpropanolamine (see Section 15.12). The latter two drugs, of course, were sold legally in hundreds of OTC products as decongestants and diet aids. Currently, individuals are purchasing the pills from drugstores and through mail orders by the hundreds of thousands for the purpose of chemically converting the legal drugs to street methamphetamine. Note that pseudoephedrine is a stereoisomer of ephedrine (which means it has exactly the same chemical structure, but the atoms are arranged differently in space.) PPA is closely related to ephedrine, lacking only a methyl group on the nitrogen. Some states, including California, have passed laws making ephedrine a controlled substance whose unauthorized sale is punishable. However, a "designer" crystal analog called methylmethamphetamine, or "**glass**," is now on the street, synthesized from an analog of ephedrine, *N*-methylephedrine. It is called "glass" because in its freebase form, it resembles tiny pieces of glass. Neither *N*-methylephedrine nor methylmethamphetamine were anticipated by the people who wrote the law, and are thus legal—at least until the law is revised to include them. See Section 12.8 for more on designer drugs.

Dangerous illegal amphetamines continue to be sold, and the unwary or neophyte user may encounter tragedy, as illustrated in the following news item from Pennsylvania:

> The Allegheny County Coroner's Office investigated a drug death of a female found dead in a car alongside of a convulsing male. Found in the male's pocket was a small packet of a white powder with the name SNO-SEALS[9] in light blue with the picture of a seal with a snowflake on its nose. The packet contained approximately 30 mg of a white, crystalline powder, identified as 4-bromo-2,5-dimethoxyamphetamine. Identification was based on ultraviolet and infrared data. The material was 99% pure, and was apparently taken orally, as the highest concentration in the deceased female was found in the stomach contents. The boy survived after extensive life support efforts and charcoal hemodialysis. Eight days after hospitalization, the compound was still detected in his urine.[10]

7.12 Caffeine: The Great American Stimulant

You wake up in the morning, go to the kitchen, and brew a large cup or two of coffee. Hot, full of flavor, delicious. You drink the first cup: aromatic principles, oil constituents, rich coloring agents, and—dissolved unseen in the liquid—approximately

---

[9]Waterproof packets with this brand insignia are sold in head shops and are used to package drugs other than amphetamines (e.g., cocaine).

[10]The *Toxicology Newsletter*, published by Duquesne University, Pittsburgh, Pennsylvania. Compare this drug to 2-CB (EVE, VENUS), described in Section 12.6.

150 mg of 1,3,7-trimethylxanthine, or **caffeine**. An 8-year-old is hot and thirsty. Disdaining a glass of cold water, he reaches for a 12-ounce can of a cola-type beverage. In the 360 mL he drinks, he is getting up to 50 mg of caffeine—a high dose for an 8-year-old.

Americans like you and me and the 8-year-old are consuming over 100 billion doses of caffeine *yearly*. That number is roughly 5000 tons every 12 months. Data from an NIMH study indicate that 82% of the U.S. population drink coffee, and 52% drink tea. Forty-five percent of the coffee drinkers consume three to seven (or more) cups daily. Add to these figures the consumption of caffeine in tea, cola drinks, chocolate candy, hot chocolate, and OTC preparations such as No-Doz, and one begins to perceive our tremendous consumption of this CNS stimulant. In the soft drink market, the use of caffeine has now spread to orange drinks. Table 7.2 gives the caffeine content of various beverages and preparations.

Caffeine is found in more than 60 plants and trees that have been cultivated by humans. One of these is *guarana*, a vine from the Amazon that belongs to the soapberry plant family. It packs a wallop because it is loaded with enough caffeine to resemble cocaine in its intensity of CNS stimulation.

Ripped Fuel is one of the many herbal "energy boosters" now sold in health food and convenience stores. Ripped Fuel contains strong concentrations of Ma huang (see Section 2.3) and guarana, the former supplying ephedrine and the latter, caffeine. This dangerous combination of CNS stimulants has been implicated in the death of a 15-year-old California girl as she played soccer. The energy claims made for these devious products ("Feel the energy," "The power to achieve," and "Ginseng Blast") lure youth who have no idea how dangerous they can be.

Table 7.2 Caffeine Content of Beverages or Preparations (Approximate)

Beverage or Preparation	*Caffeine Content (mg)*
Espresso, 2 fl. oz.	120–150
Coffee, 1 cup, 6 fl. oz. drip or percolated	100–150
Coffee, 1 cup, instant	50–70
Tea, 1 cup, 6 fl. oz., 5-minute steep	30–60
Coffee, decaffeinated, 1 cup	1–6[a]
Cola beverage, 12 fluid oz.	72 maximum[b]
Milk chocolate, 1 oz.	1–15
Chocolate cake, 1 slice	20–30
Hot cocoa, 1 cup	5[c]
No-Doz, 1 tablet	100
Excedrin, 1 dosage unit	65
OTC products including Anacin, Midol, and Vanquish	32
Vivarin, 1 tablet	200

[a]Some years ago, a process was developed to remove up to 97% of the caffeine from unprocessed coffee beans. FDA tests on decaffeinated coffee, both ground and instant, have confirmed the claims for the low caffeine content of such products (about 0.6 mg of caffeine per fluid ounce).

[b]Federal regulations permit a 0.02% maximum caffeine content in any soda water drink with the words "cola" or "pepper" in the name, when such beverage is obtained from kola nut extracts or other natural extracts. Manufacturers routinely add enough caffeine to bring the concentration up to a range of 33–52 mg/12 fluid ounces; they use some of the 200 tons of caffeine obtained in the process of decaffeinating coffee.

[c]One cup of cocoa may contain up to 200 mg of theobromine, a chemical relative of caffeine.

Guarana and ephedra showed up in a recent nationwide study of the incidence of side effects in dietary-supplement use. They were the supplements linked most often to reactions such as chest pains, irregular heartbeat, tightness in the chest and throat, and heart attack.

7.13 Caffeine Is a Potent Drug

To be sure, caffeine is not in the same class as amphetamines, but it is a powerful CNS stimulant capable of exciting the brain at all levels (see Figure 7.8). Caffeine's stimulation of the cortex of the brain, the center for thought processes, results in clearer thinking, greater sensitivity to stimuli, wakefulness, and better physical coordination. At doses of 150–200 mg, an electroencephalogram shows an arousal pattern. Tests have shown that typists perform more effectively after taking a dose of caffeine. Caffeine also excites the spinal cord and the centers of the brain that control breathing. This stimulation can lead not only to increased respiration but also to

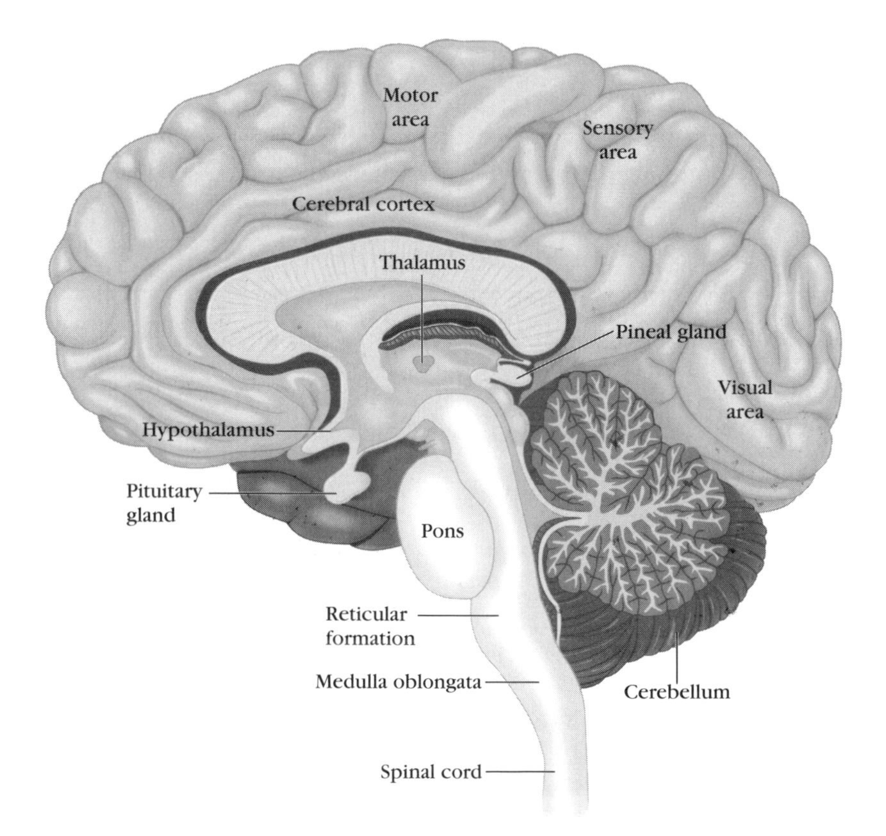

Figure 7.8 The central nervous system (CNS) consists of the brain and spinal cord. Caffeine excites the CNS at all levels. One theory of its mechanism proposes that caffeine competes with adenosine for receptors in the brain that block mood-raising cells. Caffeine can bind to these receptors but does not block mood raising the way adenosine does.

tremors or shakiness. At very high doses (several grams), caffeine stimulates the brain to the point of convulsion, and death may follow. However, in a study of 68 subjects, caffeine was found to be ineffective in "sobering up" alcohol-intoxicated persons. Caffeine's half-life in the body is 3.0–7.5 hours.

Caffeine and other **xanthines** (theophylline, theobromine) stimulate the muscles of the heart so that it functions more efficiently as a pump. However, in some individuals, this increased myocardial sensitivity shows up as an irregular heartbeat (arrhythmia). Abnormal heart rhythms are likely to be experienced by individuals who use caffeine drinks to excess. Caffeine causes constriction of blood vessels in the brain, dilation of the coronary arteries, and increased peripheral circulation. Usually, a small rise in overall blood pressure is seen.

Caffeine and the xanthines are **diuretics**. The increase in urine output caused by caffeine, however, is not striking and must be accounted for in part by the increased fluid intake.

Other actions of caffeine in the body include the following:

- *Bronchial tubes*—Dilated
- *Stomach and GI tract*—Gastric irritation; increased secretion of digestive juices (coffee is contraindicated in ulcer patients)
- *Muscles*—Increased capacity for work
- *Basal metabolic rate (BMR)*—Slightly increased
- *Embryo*—Possibly teratogenic in the first trimester of pregnancy

Researchers at Vanderbilt University conducted a careful study on the effect of caffeine on six men and three women, ages 21–30, who were not coffee drinkers. The subjects were permitted no coffee, tea, cola drinks, or drugs for 3 weeks prior to the study, and their salt intake was controlled. The placebo used was a caffeine-free rye–barley extract known in Europe as Pero. The study found that the subjects who received caffeine exhibited the following effects:

- Blood pressure was up 14%.
- Pulse rate dropped a little at first, and then rose mildly.
- Epinephrine (adrenaline) output was up 147%, and norepinephrine was up 75%. (Epinephrine and norepinephrine are catecholamines.)
- Respiration rate was up 20%.

Clearly, caffeine is a potent drug in the human. In smaller quantities in most people, its effect appears to be of little significance. In fact, the mental and physical stimulation it affords is welcomed by millions of drinkers who have become habituated to the drug. On the other hand, sensitive individuals (cardiac patients, ulcer patients, the very young and the old, light sleepers, and the debilitated) risk an increase in morbidity by ingesting this stimulant substance. The Vanderbilt University study finding of increased catecholamine output after the use of caffeine confirms the judgment of physicians who prohibit the use of caffeine beverages by their cardiac and ulcer patients. Caffeine sets the body's motor at a higher idle—a form of artificial stress—and this is hard on vulnerable systems.

As regards studies on caffeine's effects on pregnancy, the results have been conflicting and confusing. For a while, caffeine appeared exonerated from any deleterious effect, and pregnant women were counseled to consume caffeine-containing beverages if they wished. But more recently, a major study reported in *JAMA* found that pregnant women who drank from 1.5 to 3 cups of coffee a day doubled their risk of miscarriage. Three cups tripled the risk. Now, we realize that this is only one study—albeit a major one—and that cup size can vary, and that coffee contains many chemicals in addition to caffeine, and that the statistics do not *prove* anything. Nevertheless, the risk is real, and the prudent pregnant woman will probably want to limit her intake of caffeine to the minimum.

Little is known about the stimulant effect of caffeine on **children**. One study reviewed estimates of caffeine intake among American children and found that on the basis of caffeine consumed per kilogram of body weight, children ages 1–5 years had the highest intake. Children run a risk of "caffeine nerves," because one cola for a small child may have the same effect as 4 cups of coffee in an adult. The hazards of caffeine consumption in very young children are unknown.

The U.S. preoccupation with cola drinks, fostered by constant, insistent advertising, is to be deplored. A child of 8 can look forward to 60 or more years of drinking cola beverages, with their content of sugar, acid, caffeine, and empty calories. If the child continues to drink one regular 12-ounce cola every other day, he will have consumed 1.7 pounds of caffeine, a quarter ton of sugar, and enough acid to have dissolved his teeth many times over.

Some years ago, a Harvard School of Public Health study found an increased risk of pancreatic cancer among people drinking as little as 1 to 2 cups of coffee daily. The study was headline news but was later criticized for inadequate controls. Other studies subsequently failed to confirm the Harvard University findings, and caffeine fed to rodents was not found to cause cancer. Actually, the authors of the study reversed their findings in a second study carried out 5 years later. Here is an example of a situation in which it is wise not to jump to conclusions. Probably, when headline-making theories about health problems are proposed, a slower, more cautious approach to decision making is wise.

The same statement might be true for the proposed link between fibrocystic breast disease and caffeine. In this condition, noncancerous lumps appear in the breast, along with tenderness. There is some older evidence that avoiding all xanthines helps to cure or control *some* forms of the disease, but a report on 3,300 women issued by the NIH found no relationship between caffeine intake and fibrocystic disease.

7.14 Tolerance and Habituation to Caffeine

Researchers in the Vanderbilt University study concluded that tolerance to the effects of caffeine clearly develops. Most authorities would agree with this finding, although it appears that not much tolerance develops specifically to the CNS stimulant effects of caffeine. No doubt some people have become habituated to or psychologically dependent on their favorite caffeine beverage, such as the 39-year-old housewife who drank 15–18 cups of coffee a day. The term **caffeinism** is used to describe the condition of heavy use and preoccupation with this stimulant. In caffeinism, agitation, anxiety, and heart palpitations are typical. Abrupt termination of caffeine in the heavy user may lead to the following relatively mild withdrawal symptoms: restlessness, irritability, headaches, shakiness, inability to work effectively, and lethargy.

Cora Kurtz, a nutritionist at the University of Miami, takes a more critical stance against caffeine, terming it a sneaky drug to which people do not know they are addicted until they stop using it. She is critical of the caffeine-loaded drinks in which some children overindulge and that cause them to be restless, irritable, sleepless, and nervous. She states that the symptoms of caffeine withdrawal are headache, upset stomach, and cold sweats.

A Dutch study found that when heavy coffee drinkers (average 5 cups a day) gave up caffeinated coffee, more than 48% of them developed headaches starting a day or two later and lasting for up to 6 days. British researchers found that 500 mg or more of caffeine ingested daily is sufficient to cause withdrawal symptoms if caffeine intake is suddenly stopped.

Case History *Caffeine Dependence*

College student Y.B. was ingesting four No-Doz tablets and a six-pack of diet cola each day. On this regimen, she lost considerable weight. Repeated attempts to stop the ingestion of the caffeine cold turkey resulted in headache, shakiness, irritability, and nausea. A single diet cola relieved all the symptoms.

7.15 The Medical Uses of Caffeine

Chemically, caffeine is one member of a group called the **xanthines** (pronounced *ZAN-theens*), the other members being theophylline and theobromine. Medically, caffeine is important because of its CNS effects. When therapeutic effects on the heart, vascular system, or bronchi are desired, the physician must turn to theophylline or theobromine.

Caffeine, in combination with ergot in Migranal, is useful in the treatment of migraine headaches. One theory states that caffeine relieves pressure by constricting blood vessels in the brain. Caffeine's action augments that of ergot, which is also a cranial vasoconstrictor. Migraine is a disorder characterized by recurrent attacks of headache and possibly by dilation of blood vessels in the brain. An attack of migraine might be precipitated by allergy, emotional strain, fatigue, eyestrain, menstruation, or the eating of certain foods. There is a pronounced hereditary influence. Approximately 20% of adults suffer from migraine, with women being two to three times more headache-prone than men.

While caffeine by itself has no analgesic effect, in combination it can increase the painkilling action of such nonprescription analgesics as aspirin, acetaminophen, and ibuprofen. Studies have shown significant enhancement of action of such combinations compared to placebo or analgesic alone.

7.16 Strychnine

Poison. If ever there was a substance that deserves that label, it is strychnine, an alkaloid obtained from the seeds of a tree that grows in India. A human being died from a 30-mg dose of strychnine (0.5 grain). In children, as little as 15 mg may be fatal.

In one of Sir Arthur Conan Doyle's stories, Sherlock Holmes discovers a corpse with facial muscles twisted hideously into the "smile of death" and correctly diagnoses poisoning by strychnine. Holmes knew his poisons, for strychnine, a powerful CNS stimulant and convulsant, contracts all voluntary muscles including those of the face. In large enough doses, strychnine so excites the brain and spinal cord that the slightest stimulus triggers a powerful convulsion. It is during these convulsions that breathing is cut off and death follows.

Strychnine has been used as a rat and rodent poison for 400 years and is sold today in some places; products contain 0.25% strychnine. Accidental poisonings in children have occurred from the rat and rodent poison.

No rational use of strychnine in medicine exists today, but the laity persist in regarding it as some sort of tonic to stimulate the appetite and digestion, possibly because of its bitter taste. Thirty years ago, strychnine was used as an antidote in severe barbiturate overdosage. Today, it is found in some "bitters" used in certain mixed drinks.

In a recent Olympic Games, a Chinese volleyball player was sent home after she tested positive for strychnine. Apparently, some athletes believe that the CNS stimulant effect of strychnine will improve their competitive skills (see Section 17.2).

7.17 Nicotine

Among the nearly 4,000 chemicals found in tobacco smoke, nicotine is the most notorious. We will discuss it shortly, but first a few observations on tobacco and its smoke: According to the National Cancer Institute and the American Cancer Society, lung cancer has outstripped heart disease as the leading cause of smoking-related deaths in the United States. Altogether, some 425,000 Americans die each year from nicotine and tobacco-related diseases, such as heart attack, stroke, and emphysema. The chairman of the President's Cancer Panel points out that at least one-third of all cancer deaths in the United States can be attributed to smoking—that's more than 160,000 deaths a year, or 440 a day. Cigarette and cigar smoke contain polycyclic hydrocarbons (proven carcinogens), nitrosamines, and most likely numerous other carcinogens amidst the tars that result from incomplete combustion of organic material.

An estimated 21 million women smoke. Some 13% of pregnant women and 18% of pregnant teens, ages 15–19 years, are smokers. Among women, 73% say they want to quit, but only 2.5% are able to each year. A minimum of 29.6% of the U.S. population smoked in 1997. The notion that "light" cigarettes are safer has been dispelled by research studies. Little correlation exists between claimed nicotine yields from various brands and nicotine levels found in the blood of smokers. One manufacturer wants to market a "reduced-carcinogen cigarette." Critics rush to point out that there is no proof that reducing carcinogens reduces risks to smokers.

Here is a summary of tobacco use by Americans (latest data), provided by SAMHSA:

- An estimated 64 million Americans are current smokers (30% are 12 and older).
- About 4.5 million youth ages 12–17 are current smokers (20%).
- Youth ages 12–17 who currently smoke cigarettes are about 12 times as likely to use illicit drugs and 23 times as likely to drink heavily as nonsmoking youth.
- Current smoking rates for 18–25-year-olds have increased every year and are now at 40.6%.
- While more adult men smoke than women, high-school-age females smoke more than males.
- An estimated 3.2% of our population are current users of chewing tobacco (smokeless tobacco).
- Current smokers are much more likely to be heavy drinkers and illicit drug users than nonsmokers (see Figure 7.9).
- Cigarettes cost the nation over $7 a pack in terms of medical care and lost productivity. The nation's total cost for smoking is put at $157 billion yearly.
- Smoking causes an average man to lose more than 13 years of life and an average woman to lose 14.5 years of life.

Nicotine, the most abundant of the 12 volatile alkaloids in tobacco leaf, is partly destroyed at the temperature of a burning cigarette. Still, according to Dr. Neal

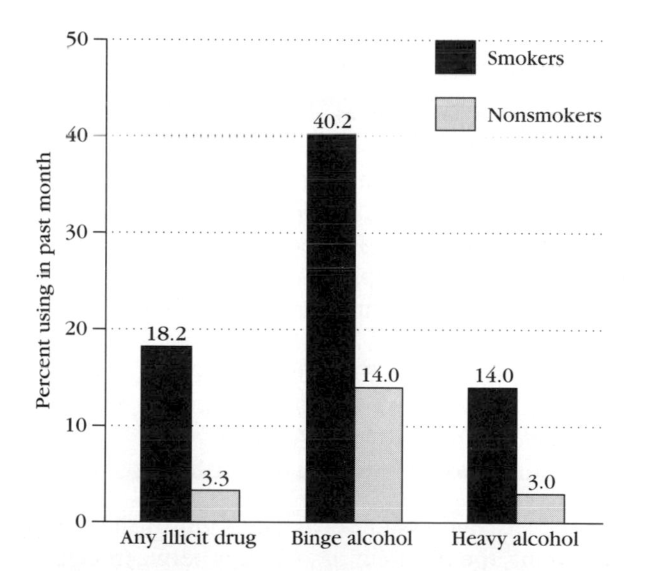

Figure 7.9 Past-Month Illicit Drug, Binge Alcohol, and Heavy Alcohol Use Among Smokers and Nonsmokers, Aged 12 or Older: 2001. (Source: Results from the 2001 National Household Survey on Drug Abuse: Vol I. Summary of National Findings, SAMHSA, 2002.)

Benowitz of the University of California—San Francisco, of the 9 to 10 mg of nicotine in each cigarette, smokers typically inhale from 1 to 3 mg. A 40- to 60-mg dose of nicotine is considered acutely toxic, and a 60-mg dose has been known to kill a human. Thus, nicotine is one of the strongest poisons Americans routinely take into their bodies. Nornicotine, another tobacco alkaloid, is now implicated in the onset of cancer, diabetes, and aging.

Case History *Winner of Cigarette Smoking Contest Loses*

A national news service reported on a contest between two young Chinese men who made a bet on who could smoke more cigarettes at one sitting, with the loser paying for all the cigarettes. After the first smoker quit at 40 cigarettes, the second continued to 100. He then collapsed and was taken to a hospital where he was pronounced dead. *For the reader*: From the information in this text, calculate how many milligrams of nicotine the deceased probably ingested in the 100 cigarettes he smoked and determine whether this amount exceeded the reported lethal dose. Incidentally, a paper in the *British Medical Journal* states that two-thirds of men in China become smokers before the age of 25—and that tobacco will kill about 100 million of the 300 million males now under 30.

With the quantities of nicotine provided by the smoking of cigarettes, the following pharmacological effects are observed:

- *CNS*—Excitement of the brain at all levels, and tremors
- *Respiration*—Markedly increased
- *Heart*—Increased rate (tachycardia)
- *Vascular system*—Increased blood pressure; constriction of peripheral blood vessels (leading to cold hands and feet)
- *Adrenal glands*—Release of epinephrine
- *GI tract*—Nausea and vomiting; increased peristalsis and activity of the intestines; diarrhea
- *Kidney*—Reduction in volume of urine flow
- *Blood*—Elevated concentration of free fatty acids (implicated in hardening of the arteries)
- *HDL*—Circulating plasma levels are diminished (see Appendix III)

The heart and the vascular system bear the brunt of the strong effects of nicotine, probably explaining the clear-cut association between cigarette smoking and heart disease. Concentrated extracts of tobacco have long been used in horticulture as insecticides and fungicides, especially in plant sprays. The extracts contain high levels of nicotine. Unfortunately, such sprays are sometimes kept in old beverage containers in the family garage, where young children mistake them for a pleasant drink—with tragic results.

7.18 Drug Interactions of Nicotine

As regards drug interactions, nicotine intake can influence the action of prescription drugs that the smoker is taking concurrently. For example, Darvon can lose its painkilling effect in a smoker, and tranquilizers such as Valium may work with diminished frequency. Smokers have been known to require one-and-a-half to two times as much theophylline, an important bronchodilator used in treating asthma, as nonsmokers to achieve the same effects. The half-life of theophylline is reduced to about 4 hours (from about 7 hours in the nonsmoker). How are these effects explained? Some researchers believe that nicotine stimulates the liver to increase its production of enzymes responsible for the metabolism of many drugs and chemicals, including pain relievers, tranquilizers, and theophylline. As a consequence, the rate of drug metabolism is accelerated, with decreased intensity and duration of action. Regardless of theories, we are sure of this statement: With certain drugs, smokers need different doses or a different frequency of administration than nonsmokers, and when smokers stop their consumption, their drug regimen may have to be adjusted (see Table 7.3). The considerable influence smoking can have on oral contraceptive users is discussed in Chapter 13.

7.19 Nicotine Is an Addicting Drug

The U.S. Surgeon General has reported irrefutable proof that cigarette smoking can lead to a **powerful addiction**. In fact, 200 research studies have shown that nicotine is addictive, and the FDA has concluded that nicotine is the chief reason people smoke cigarettes. Nicotine is a mood-altering drug, affecting the same part of the brain that

Table 7.3 Interactions Between Smoking and Drugs, Presumably Based on Alterations in Drug Metabolism

Analgesics
Propoxyphene (Darvon)—Higher rates of ineffective responses
Antipyrene—Increased hepatic metabolism
Phenacetin—Lower plasma levels
Pentazocine—Higher maintenance dose needed
Tricyclic antidepressants
Imipramine (Tofranil)—Increased clearance rate
Xanthines
Theophylline—Significantly accelerated metabolism
Tranquilizers
Diazepam (Valium)—CNS depression less frequent (probably due to increased rate of metabolism)
Chlorpromazine (Thorazine)—Reduced drowsiness

Source: *FDA Drug Bulletin*, vol. 9, no. 1, 1979.

controls other psychoactive drugs. Nicotine produces an effect that smokers are eager to repeat, and this effect places it in the same category as alcohol, heroin, and cocaine. What is more, **tolerance** to nicotine clearly develops, and one-cigarette-a-day smokers typically build up to one and sometimes three packs a day. They now ingest quantities of nicotine that would have nearly killed them when their smoking began. Second, there is a withdrawal syndrome to nicotine use, albeit a rather mild one. When many confirmed smokers (who inhale) are suddenly deprived of their nicotine, they become nervous, irritable, restless, drowsy, and depressed; they have headaches, experience difficulty concentrating, and report digestive troubles. In fact, among regular smokers, more than 75% will report four or more of such symptoms when they stop tobacco use. Nicotine administered by injection or orally can relieve these withdrawal symptoms. (Compare this situation with heroin withdrawal.) Furthermore, the extreme difficulty that the majority of heavy smokers have in quitting is strong evidence of a true addiction. Each year, 30% of smokers attempt to quit, but only 3% of these smokers are successful. Even more startling are the 50% of smokers who continue to smoke even though they have undergone major surgery for a tobacco-related disease and have been counseled to abstain because resumption would be suicidal. Further evidence of true addiction is the smokers' compulsive use of and preoccupation with obtaining a supply of their drug—and the tendency to relapse after having quit.

Based upon the number of deaths and diseases it causes, nicotine is as least as addictive as heroin, cocaine, or alcohol. If we consider the number of people who start smoking and the number of people who become addicted, nicotine is more addictive than alcohol, for about 40% of those who try cigarettes become regular users, and of the users, 90% are addicted. In contrast, only about 15% of people who drink alcoholic beverages are addicted.

Actually, then, one can make a convincing case for a clear parallel between nicotine and heroin. Just as the heroin addict can sense a subpotent batch of heroin, so can the heavy smoker perceive cigarettes deliberately manufactured to contain less than the usual level of nicotine. Both kinds of addicts will reject the subpotent samples and look for the real thing. Furthermore, there are heroin users who use it occasionally or only on weekends (chippers), and their withdrawal is no worse than a case of the flu. Similarly, some cigarette smokers are anything but dedicated to the act and are able to quit cold turkey without suffering significant withdrawal symptoms. Here's bad news for teens who smoke. John Wiencke, writing in the *Journal of the National Cancer Institute*, has evidence that starting smoking on a daily basis before age 18 causes dramatic DNA and lung damage that lasts a lifetime. He says it doesn't matter if the person is a light or heavy smoker—what matters is that they started young. He says smoking in the teen years is more addicting, a conclusion also reached in a study of seventh-graders sponsored by the National Cancer Institute. This study found that young people get hooked after only a few cigarettes and that girls become addicted more easily than boys (only about three weeks after starting only occasional smoking). For youth, here are some key signs of nicotine dependence: smoking in forbidden places (like school), irritability if smoking is denied, and difficulty in concentration. Currently, about 90 percent of smokers begin before age 19. For more on nicotine dependence, see **http://www.findhelp.com/**.

Scientists have found evidence that nicotine can reach a nonsmoking pregnant woman's fetus if she is routinely exposed to second-hand smoke. The evidence turned up in hair samples taken from newborns, reported Dr. Gideon Koren in the *JAMA*.

As in the case with alcohol, there is mounting evidence that genes influence cigarette smoking (see Figure 7.10).

Nonsmoking individuals living with heavy smokers have four times the risk of heart attack compared with those living in smoke-free environments, according to a paper in the *Journal of the American College of Cardiology*.

According to the national organization *Action on Smoking and Health* (ASH), dogs can become addicted to nicotine and have been known to retrieve cigarette butts from ashtrays to satisfy their cravings. Pet owners who smoke are urged to be careful, because nicotine is poisonous to dogs and can cause convulsions and paralysis. Passive smoke kills dogs, too. A Colorado State University researcher found that dogs whose owners smoke tobacco are at 50 percent greater risk of developing lung cancer. Cats living with smokers are more than twice as likely to develop a feline cancer as those living in smoke-free homes.

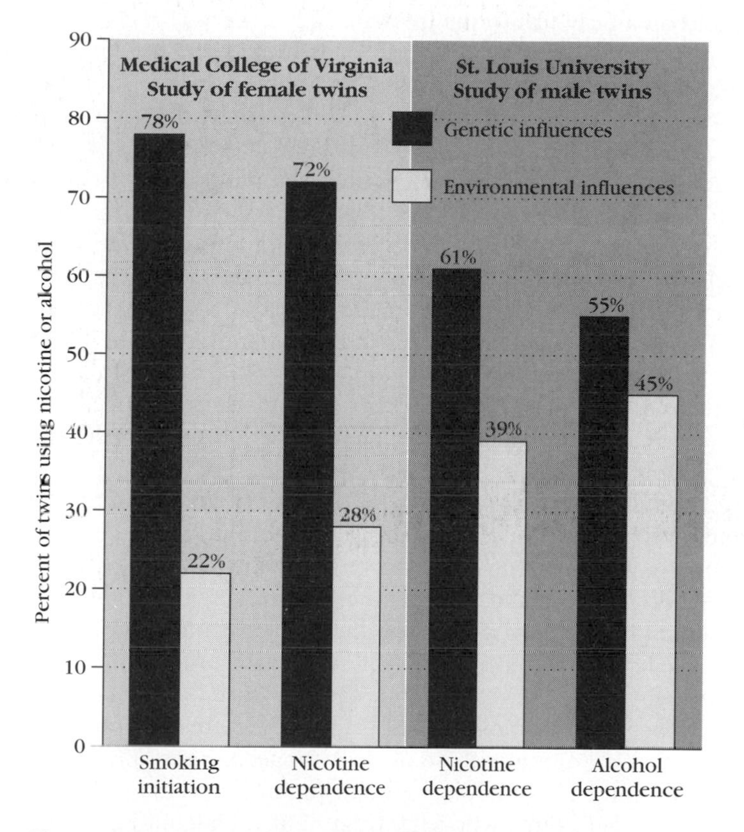

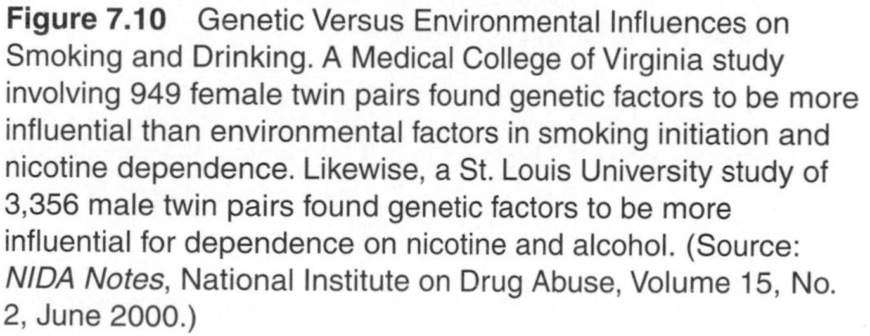
Figure 7.10 Genetic Versus Environmental Influences on Smoking and Drinking. A Medical College of Virginia study involving 949 female twin pairs found genetic factors to be more influential than environmental factors in smoking initiation and nicotine dependence. Likewise, a St. Louis University study of 3,356 male twin pairs found genetic factors to be more influential for dependence on nicotine and alcohol. (Source: *NIDA Notes*, National Institute on Drug Abuse, Volume 15, No. 2, June 2000.)

The results of rat studies conducted at the University of Chicago indicated that nicotine's mechanism of addiction involves dopamine release and prolongation. Nicotine not only turns on the brain's pleasure chemical dopamine (see Section 1.4) but it neutralizes the control mechanism that usually cuts off the dopamine high. Nicotine has "ferociously addictive power," say the researchers. NIDA's micro-magazine, *The Brain's Response to Nicotine*, aimed at children in grades 5 through 9, can be downloaded from NIDA's homepage at **`www.nida.nih.gov`**, or call NCADI at (800) 729-6686 or (301) 468-2600. Via e-mail, use **`info@health.org`**.

The National Cancer Institute issued a major report showing that **cigars** can be just as lethal as cigarettes. Citing the 4,000 chemicals—including known carcinogens—found in cigar smoke, the Institute reported that cigar smokers are 4 to 10 times more likely to die from cancer than nonsmokers. Compared to cigarettes, cigars emit relatively huge amounts of cadmium and nitroamines, both carcinogens. What's more, concentrations of carbon monoxide at two cigar social events exceeded those found on a busy California freeway.

A 2001 report from the National Cancer Institute concluded that low-tar, low-nicotine cigarettes are just as dangerous as regular cigarettes.

Illegal sales of tobacco products to minors amount to more than $1.5 billion yearly, says ASH. And the tobacco pushers spend more than $14 million a day to promote smoking. Tobacco ads and such promotional items as T-shirts and gadgets entice a significant number of teenagers to try smoking even if they were adamantly opposed to starting, according to a recent article in the *JAMA*. The study "is the first to conclusively prove that the effect of tobacco marketing happens at the very beginning and encourages teens to start the process of becoming a smoker. These were kids who said they would absolutely not start smoking even if their best friend offered them a cigarette."

Bidis are small, flavored, filterless cigarettes made in India that contain more tars and nicotine than regular cigarettes. They are wrapped in string and are becoming popular with teens.

If we can convince the teenager of today (the heavy smoker of tomorrow) that the nicotine in tobacco is really an addicting drug, that getting hooked on cigarettes is a lot like getting hooked on heroin, and that cigarette smoking is the greatest single cause of death and disability in the United States, the would-be smoker might think twice about starting in the first place. If few Americans recognize nicotine as an addicting drug, it is because we have failed to publicize that fact—especially among the young.

The National Conference on Nicotine Dependence has declared that cigarettes are a gateway for youth to more potent drugs. Cigarettes are placed in a group of four "**gateway drugs**," along with smokeless tobacco, beer, and wine coolers. The Conference says that 92% of adolescent marijuana smokers used tobacco first, and 92% of teenagers and adults who seek treatment for alcohol or drug dependence are also addicted to the nicotine in cigarettes.

7.20 Antismoking Chewing Gums, Skin Patches, Inhalers, and Nasal Sprays

Nicorette is the first FDA-approved **antismoking chewing gum**, designed as an aid in eliminating nicotine dependence. Now available OTC, the gum is chewed when the smoker feels the urge for a cigarette. Each stick of gum contains 2 mg of nicotine bound to an ion-exchange resin. Chewing hastens release of the nicotine; "parking" the gum in the buccal cavity slows release. The antismoking gum appears to work, say researchers who have studied its use in Europe and Canada. In one double-blind study, 28% of users were able to stay off cigarettes for up to a year. The gum is most advantageously used by highly motivated people who are also participating in some form of behavior modification program. Ninety-six sticks of gum cost about $20. An article in the *Los Angeles Times* reported on a five-year study of smokers who tried to quit using the gum. The study found that 10 percent of the former smokers made a habit of the gum. Note to reader: do you thinks a nicotine-gum habit would be a problem?

Nicotine skin patches (Habitrol, Nicoderm, Nicotrol, ProStep) are now approved by the FDA. The patch system is designed to wean a smoker from nicotine over a 10-week period. Three patch systems are typically offered, delivering 21, 14, or 7 mg of nicotine per day. Nicotine skin patches don't always work. Many wearers develop skin irritation at the patch site or abdominal pain, tachycardia, bizarre dreams, insomnia, or sweating. Removing a patch does not stop nicotine dosage immediately, because there is a reservoir of nicotine in the skin that continues to release the drug for several hours. And, if the would-be ex-smoker sneaks a cigarette or two while wearing the patch, overdosage on nicotine can occur. In fact, several heart attacks have been reported. Usually, the smoker will require group support in addition to the patch system. Does the patch approach work? A University of Nebraska study showed that although 61% of patch patients had quit smoking within 6 weeks, only 26% remained abstinent 6 months after treatment began. (See Section 4.2 for additional comments on skin patches.) Figure 7.11 shows how a typical skin patch is constructed. The microporous membrane is the key to the proper release of the drug.

In this connection, it should be noted that the FDA considers the supposed smoking deterrent product Bantron (containing lobeline) to be ineffective. Clonidine (see

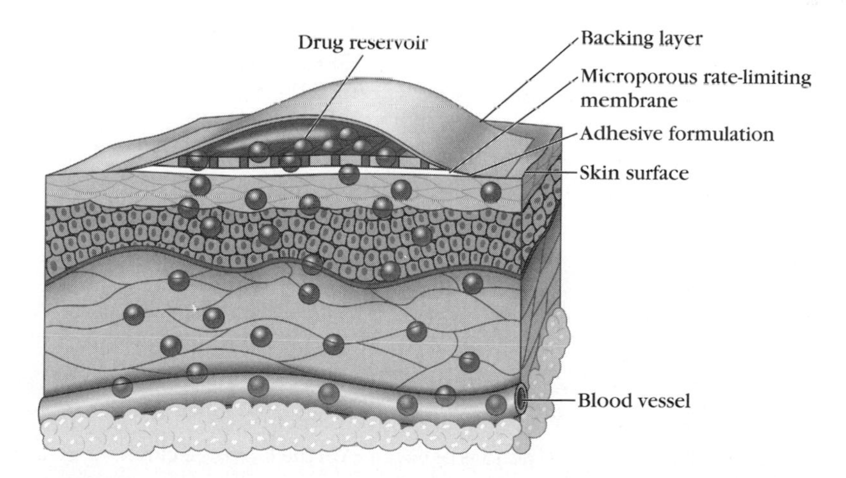

Figure 7.11 Cross section of a typical transdermal therapeutic skin patch. The microporous membrane controls the slow release of the drug (nicotine) into the skin.

Section 6.9), however, has been found to diminish withdrawal pangs when cigarettes are abruptly terminated, as has Zyban (bupropion, Wellbutrin), an older antidepressant now FDA-approved for smoking cessation. Zyban works on dopamine and norepinephrine, two brain neurotransmitters associated with cravings and withdrawal symptoms. Two cautions here: bupropion can cause adverse effects, and it should be used in conjuntion with a tobacco counseling program.

For parents: Used skin patches can retain up to 74% of nicotine after use, so dispose of used patches (and lock up the new ones) before a child can experiment. Symptoms of nicotine poisoning include vomiting, nausea, headache, dizziness, and fatigue. Remember, nicotine is a very potent substance.

Regarding smokeless tobacco products, the U.S. Surgeon General has warned that the risk of oral cancer alone is four times higher for smokeless tobacco users than nonusers, and that new nicotine delivery systems such as the "smokeless cigarette" may be "more toxic and addicting" than current tobacco products. In 2001 chewing tobacco makers agreed to pay $2.75 million to the state of California for use in anti-tobacco advertising.

Nicotrol NS, a prescription-only nasal spray that gives smokers a shot of nicotine from a bottle instead of a cigarette, has won FDA approval. The bottle holds 100 mg of nicotine that smokers can inhale to ward off cigarette cravings. The FDA warns that smokers could become as dependent on the nasal-spray nicotine as they are on cigarette nicotine, and that the spray should be used no longer than 3 months. A squirt of Nicotrol NS up each nostril provides the smoker with 1 mg of nicotine. Smokers are not to inhale the spray more than 5 times a day. A companion product, Nicotrol Inhaler, uses a plastic mouthpiece to supply nicotine to the mucous membranes of the mouth, from which it is absorbed into the body.

Web Sites You Can Browse on Related Topics

Cocaine
`http://www.drugfreeamerica.org/cocaine.html`

Also search for "crack cocaine"

Amphetamines
`http://www.amphetamines.com/`

Attention Deficit Disorder
`http://www.add.org/`

Caffeine FAQ
Search for "caffeine toxicity"

Ritalin
`http://content.health.msn.com/content/article/4046.1616`

Nicotine and Smoking
`http://www.ash.org`
`http://nicotine-anonymous.org`
`http://www.quitnet.org`
`http://www.mentalhealth.com`

Study Questions

1. What are the pharmacological actions of cocaine on the CNS, the cardiovascular system, and the eye?
2. Why has cocaine been called "ego food"?
3. By what mechanism does cocaine produce its central effects?

4. What pharmacological properties of cocaine make it useful in the surgery of highly vascularized areas such as the face?
5. Describe "crack" from the standpoint of its nature, source, cost, pharmacological action, and special hazards.
6. List and explain the factors that make cocaine a very appealing drug to many people, and compare them with the factors that make cocaine such a dangerous drug.
7. Match the name in the right-hand column with the items in the left-hand column. (A name may be used twice or not at all.)

a. Dexedrine	**1.** Benzedrine
b. Benzedrex inhaler	**2.** Speed
c. "Crystal"	**3.** Dextroamphetamine sulfate
d. Black beauties	**4.** Methamphetamine
e. Amphetamine sulfate	**5.** Ritalin
f. Desoxyn	**6.** Biphetamine
g. A general term for any amphetamine	**7.** Propylhexedrine

8. Describe the pharmacological effects of amphetamines on the CNS, blood pressure, blood sugar, bronchi, respiration, and blood flow in various parts of the body.
9. How do the amphetamines function to depress the appetite?
10. Doesn't an obese person have the right to lose weight and to use amphetamines as part of the therapy? By what right does the FDA get involved in the use of amphetamines in weight reduction?
11. Define **(a)** anorectics, **(b)** narcolepsy, **(c)** catecholamine, **(d)** stereotypy, **(e)** formication, **(f)** speed freak, **(g)** speedball.
12. True or false:
    - **a.** Tolerance to amphetamines can develop.
    - **b.** An average low dose of Dexedrine is 5 mg.
    - **c.** Amphetamines improve athletic performance.
    - **d.** Withdrawal from amphetamines has physical effects much like those of heroin withdrawal.
    - **e.** Amphetamine acts by delaying the release of catecholamines.
    - **f.** The FDA recommends the use of amphetamines in the management of obesity.
13. Describe the pharmacological effects of caffeine on the cortex of the brain, physical coordination, respiration, heart muscle, blood pressure, the kidney, bronchi, gastric secretion, and BMR.
14. In America, what is the most widely used mind-altering drug?
15. The line between habituation and true addiction has always been difficult to draw, and the situation with nicotine is no exception. However, in your opinion, what would be the differences in behavior between a cigarette smoker who is habituated and one who is truly addicted?
16. True or false:
    - **a.** A freshly brewed cup of coffee typically contains 10–15 mg of caffeine.
    - **b.** Caffeine stimulates the body's release of catecholamines.
    - **c.** Caffeine can irritate the lining of the stomach.
    - **d.** Physically speaking, the withdrawal syndrome from caffeine is relatively mild.
    - **e.** Caffeine appears in the drug Cafergot, which is used to treat high blood pressure.
    - **f.** Milligram for milligram, strychnine is a more potent poison than nicotine.
17. Considering only legitimate medical applications, which drug is most useful today—cocaine, amphetamines, caffeine, or nicotine?

18. Mark Gold has said, "Drugs are a lot stronger than people." What drugs would you say are stronger than you are?
19. My friend, J. M., a heavy smoker, complains a lot about cold hands and feet. Explain the pharmacology involved.
20. What drug interactions could occur in a smoker who was also taking (**a**) theophylline; (**b**) propoxyphene?
21. Visit a large pharmacy and check the labels of products that contain aspirin, acetaminophen, and caffeine. Which, if any, of the labels caution pregnant women and nursing mothers to avoid use? Which ingredient is responsible for this FDA-required labeling?
22. Cocaine abusers typically do not inject it but take it by snorting the hydrochloride form or inhaling the vapors of the free base. Explain the aversion to the parenteral route.
23. Tobacco advertising is totally banned in 21 countries, including Argentina, Canada, Finland, Hungary, Italy, Norway, and Poland. State reasons why you would or would not favor a similar ban in America.
24. Congress has forced manufacturers to place an addiction warning on cigarette packages. Do you expect that such a warning would deter the use of cigarettes? Explain your answer.
25. Statistics show that 320,000 Americans die each year from tobacco use, compared to 125,000 from alcohol, 4,000 from opiates, and 2,000 from cocaine. Why, then, in your opinion, does cocaine hog the headlines?
26. According to Stephen Fortmann, MD (*Journal of the American Medical Association*, Vol. 260, 1988, p. 1,575), what is the single most important behavior contributing to illness, disability, and death in the United States?
27. Sears & Roebuck's 1909 catalogue advertised their celery malt compound as a "high class preparation, . . . nerve builder, brain tonic and stimulant." It was to be used by nervous, exhausted people who could not sleep. It contained celery seed, malt, coca leaves, cinchona, senna, couch grass, Rochelle salts, iron phosphate, sugar, alcohol, and water.
    - **a.** What is/are the active ingredients?
    - **b.** Where is the false advertising?
28. *Advanced study question*: Chemists term dexedrine (see Section 7.7) a *chiral* drug. That means its molecules have handedness, much like the spatial characteristics of a right-handed glove (versus a left-handed glove.) Our text says Benzedrine is chemically identical to Dexedrine, but is the racemic mixture. Explain what a racemic mixture is.
29. A California company planned to market drinking water containing 2–4 percent nicotine as an aid to smokers trying to quit. Give reasons why the FDA should have approved this product or why it should have not.

CHAPTER 8

The Barbiturates: Battered but Still Here

Key Words in This Chapter

- Barbiturate
- CNS depressant
- Synergism
- *Rapid Eye Movement* (REM) sleep
- Abuse potential
- Tolerance
- Physical dependence
- Chloral hydrate
- Sedation

Learning Objectives

After you complete your study of this chapter, you should be able to do the following tasks:

- Classify barbiturates as short-, intermediate-, or long-acting.
- Explain the pharmacological actions of barbiturates on the CNS.
- Define synergism, especially as it relates to barbiturates and alcohol.
- Explain the mechanism by which tolerance to barbiturates develops.
- Cite alternatives to the use of barbiturates.

8.1 Introduction

The sedative-hypnotic barbiturates long occupied a leading position in the physician's armamentarium. But today, we realize that the barbiturates offer too high a danger of overdosage and potential for abuse for them to continue to be used routinely as sleep aids. The barbiturates are the classic example of a great drug discovery with intense use in medicine—but ultimate development into a social hazard.

The number of prescriptions for barbiturates has declined greatly in recent years, and physicians and the public are more aware of their dangers. DAWN data (SAMHSA) show that barbiturate-related visits to EDs fell in the 1990s, but among drugs mentioned most frequently in emergency-room episodes, phenobarbital still ranks 31st (out of 38).

8.2 Discovery

The great German organic chemist Adolph von Baeyer is credited with the discovery of the chemical structure from which all of the barbiturates (pronounced *bar-bih-CHUR-ates*) are derived. According to one story, the name *barbituric acid* was coined by von Baeyer in 1864 when, to celebrate the discovery of his new compound, he stopped at his town's tavern. By coincidence, the town's artillery garrison was there celebrating St. Barbara's Day (their patron saint). Von Baeyer's new compound was a uric acid derivative, so he combined Barbara with uric acid to get *barbituric acid* (derivatives of which we call *barbiturates*). In 1903, the barbiturate Veronal (named after peaceful Verona, Italy) was introduced into medicine, and in 1912, phenobarbital became available to the medical world.

The chemical structure of the parent barbituric acid is easy to modify, and in the following years thousands of different barbiturates were synthesized and described in the literature. However, relatively few of these compounds have survived in the marketplace. Table 8.1 lists some of the more common barbiturates available today.

As shown in Table 8.2, all of the barbiturates can be classified on the basis of how quickly and how long they act in the human body. Phenobarbital is the classic example of a long-acting barbiturate, and pentobarbital and secobarbital are the classical short-acting ones. Thiopental sodium USP (Pentothal Sodium) is a widely used IV barbiturate anesthetic often employed in dental extractions. Its onset is rapid (30 seconds), and its duration is brief (10–30 minutes). Thiopental has a sulfur atom in place of one of the oxygen atoms in the typical barbiturate nucleus (see Appendix I).

Remember that a barbiturate taken at night to induce sleep will still be in the bloodstream the next day. Seconal, for example, has a half-life of more than 20 hours. Phenobarbital's half-life is 36 hours, and its blood level remains measurable for days. Enough barbiturate can remain in the user's body the following day to

Table 8.1 Some Barbiturates Available Today

Manufacturer	*Trade Name*	*Generic Name*	*Description*	*Typical Adult Sedation Dose*
Abbott	Nembutal	Pentobarbital sodium USP	Yellow and white capsule	50 mg
Sanofi Winthrop	Mebaral	Mephobarbital	Tablet	50 mg[a]
Lilly	Seconal	Secobarbital sodium USP	Red capsule	50 mg
McNeil	Butisol	Butabarbital sodium NF	Various color tablets	50 mg
Roche	Alurate Elixir	Aprobarbital	Red liquid	40 mg
Many	Many (e.g., Luminal)	Phenobarbital USP	Tablets, elixir, injectable	65 mg

[a]In epilepsy, average adult dose: 400–600 mg.

Table 8.2 Duration of Action of Barbiturates

Category	*Time Needed to Take Effect (Hours)*	*Duration of Action (Hours)*	*Examples (Generic Name)*
Short-acting	0.25	Less than 3	Pentobarbital sodium USP, Secobarbital USP
Intermediate-acting	0.5	3–6	Amobarbital USP, Butabarbital USP, Talbutal USP
Long-acting	0.5–1.0	More than 6	Mephobarbital USP Phenobarbital USP,

make him or her feel groggy and to impair significantly his or her motor coordination (driving, working with machinery, sports, etc.). For Example, a 200-mg dose of secobarbital can interfere with driving or piloting skills for up to 22 hours. In this regard, barbiturates can act much like alcohol, producing a "hung-over" feeling the next day. The release of inhibitions observed with alcohol has its counterpart in barbiturate use. And, as with all disinhibiting drugs, there is always the possibility of violent behavior, accidents, or unplanned overdosage.

8.3 The Pharmacology of the Barbiturates

The barbiturates are CNS depressants—*downers*. In smaller doses, they sedate or quiet the person; in larger doses they are hypnotic (sleep-inducing).

Sedation with barbiturates typically requires a dose of 15 mg (0.25 grain) and is useful in cases of anxiety or to combat the excitatory effects of amphetamines. It is believed that the barbiturates act to depress nerve transmission at the point where nerves meet each other; thus, they produce a nonselective, general type of CNS depression. Inhibition of nerve transmission by barbiturates involves—at least in part—GABA-ergic receptors (see Section 5.7). Barbiturates also interfere with the oxygen-using and energy-producing systems in various tissues and cells such as the mitochondria; somehow these effects are related to CNS depression, induction of sleep, and ultimately to coma and death. We do know that CNS neurons are very sensitive to loss of oxygen.

Insomnia is one of the most common mental symptoms in our society today. Complaints of insomnia are often the result of slow sleep onset and frequent awakenings, rather than lost sleep time. More than twice as many women as men complain of insomnia; women also use considerably more sleeping pills.

Problem 8.1 *Give the possible reasons that women insomniacs outnumber men two to one.*

Geriatric patients typically require less sleep than younger adults—perhaps only 5–7 hours a day. Hypnosis (sleep induction) typically requires a dose of 0.1–0.2 g of a barbiturate. When barbiturates are used as hypnotics, it is important to recognize that they decrease the amount of time the user spends in *rapid eye movement* (**REM**) sleep. For many users, this reduction corresponds to a night of dreamless sleep. Now, to sleep without dreaming is considered a physiological abnormality. Volunteer subjects have exhibited anxiety, irritability, and difficulty in concentrating during the day when repeatedly deprived of time to dream at night. Furthermore, if after weeks of barbiturate-induced sleep the drug is abruptly discontinued, there is a large increase in REM sleep with concomitant nightmares, restlessness, and difficulty in falling asleep. The conclusion is obvious: Occasional use of a barbiturate to induce sleep may be acceptable, but chronic use can lead to trouble.

All excitable nervous tissue is depressed by barbiturates, but the activity of the brain and spinal cord is especially lowered. Because the center that regulates breathing is located in the brain, respiratory depression is a possibility in barbiturate overdosage. Indeed, the cause of death in accidental or deliberate barbiturate poisoning is commonly respiratory depression. There is no blood-brain barrier to the distribution of barbiturates in the human body.

Overdosage with short-acting barbiturates is more dangerous than overdosage with long-acting compounds such as phenobarbital, because the short-acting drugs produce a quick, high (albeit brief) blood level that can acutely affect respiration. Years ago, physicians treated severe barbiturate overdosage by administering CNS stimulants such as strychnine or ephedrine. A kind of pharmacological battle was conducted between drugs having pharmacologically opposite effects—the patient's body being the battleground. Some remarkable recoveries were achieved. In one verified case, a woman recovered from a deliberate overdosage of 4,750 mg of phenobarbital. She was given more than 170 mg of strychnine to counteract the depressant effects of the phenobarbital. It took her 5 days to recover. Strychnine, however, is a dangerous drug. Today, effective treatment of barbiturate poisoning is accomplished by the technique of hemodialysis (much like using a kidney machine for the patient with renal failure). Hemodialysis removes the barbiturate from the bloodstream and works well if treatment is begun in time.

Synergism of barbiturates with alcohol is potentially so dangerous that the combination should be avoided at all costs. Both drugs are CNS depressants. They potentiate each other to cause superdepression of the breathing center. In a typical case, a person who has been drinking returns home and takes the usual dose of barbiturate to induce sleep. He or she is literally knocked out by the combination of CNS depressants, falls into a deep sleep, and stops breathing when the respiratory reflex is eventually inhibited. Barbiturates and alcohol have accounted for the death of famous film and television personalities. Another dangerous combination is barbiturates and chloral hydrate. A famous Hollywood star died with a blood level of phenobarbital 10 times higher than normal and a simultaneous chloral hydrate level 20 times the recommended. Barbiturates are not pain relievers (analgesics), but they are useful in the control of convulsions induced by brain damage, poisonous chemicals, tetanus, epilepsy, or toxemia of pregnancy. Phenobarbital is the most useful barbiturate anticonvulsant, being inexpensive, relatively nontoxic, and effective. Mebaral (see Table 8.1) is used to control grand mal and petit mal epilepsy. (Of course, there are nonbarbiturate drugs such as Dilantin, which are also used to control epilepsy.)

8.4 Tolerance and Addiction to Barbiturates

Individuals who take daily or near-daily doses of barbiturates over a period of weeks or months will develop a **tolerance** to all barbiturates. They will have to take ever-larger doses to get the sedative or hypnotic effect desired. Sleep laboratory research on most hypnotics has found that they lose their effectiveness within 3–14 days of continuous use.

We have already discussed (in Chapter 4) the mechanism by which this tolerance develops—namely, induction of liver enzymes. Phenobarbital is a drug that hastens its own inactivation by inducing the liver to increase the number of enzymes that catalyze the metabolism of this drug (see Figure 8.1). Furthermore, chronic administration of phenobarbital to patients decreases the effects of many *other* drugs by hastening their inactivation also. These same phenobarbital-induced enzymes reduce the blood levels of phenylbutazone (an antiarthritic), phenothiazines (tranquilizers), tetracyclines (antibiotics), antipyrine (an analgesic), and coumarin anticoagulants. Conversely, habitual alcohol consumption results in induction of liver enzymes that cross over to hasten the metabolism of barbiturates and other sedatives. This action helps to explain why heavy drinkers are often tolerant to sedatives like phenobarbital when they are sober. (But note: Large doses of alcohol and barbiturates taken concurrently put a great strain on the enzyme system, and overdosage can occur.)

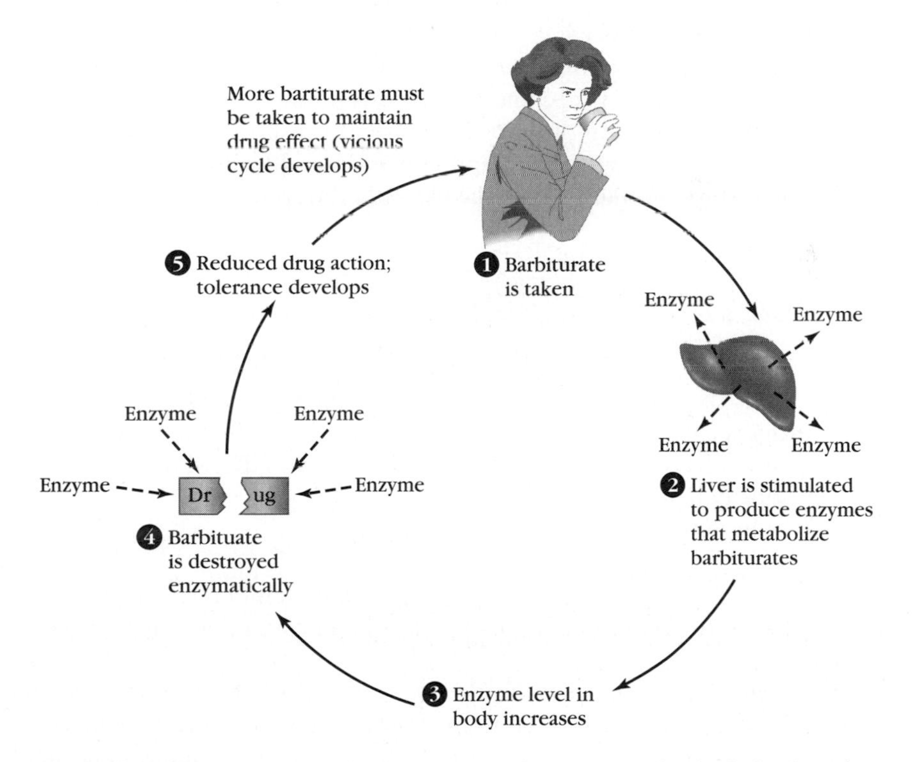

Figure 8.1 The development of tolerance to barbiturates by the induction of liver enzymes.

When the dose of barbiturate is increased to 400 mg a day for several weeks, **physical dependence** begins to develop. Note that high doses and extended periods of time are required for this process to occur. Nonetheless, addiction to barbiturates is well known, and the withdrawal syndrome has been well characterized: convulsions, insomnia, hallucinations, tremors, and possibly death. When we compare barbiturate addiction to opiate addiction, we find major differences: with opiates, tolerance and physical dependence develop with much smaller doses. But withdrawal from opiates is not as life-threatening as withdrawal from barbiturates (primarily because serious convulsions can occur in the latter).

Interestingly, the tolerance to hypnotics does not extend to the brain's respiratory center, which remains sensitive to the depressant action of the barbiturates. Hence, a person's lethal dose of barbiturate remains relatively unaffected. Also, there is no evidence of tolerance or the need for increased dosage when barbiturates are used to treat epilepsy.

Barbiturate detoxification (kicking the habit) is a difficult procedure that is best accomplished in a hospital or under a doctor's supervision. This procedure is at least as bad as kicking heroin addiction. The barbiturate withdrawal syndrome begins with anxiety, weakness, loss of appetite, tremors, and sleepiness. As the withdrawal proceeds, the symptoms become more intense and include vomiting, low blood pressure, fever, shaking, and grand mal convulsions. The most serious signs are convulsions, delirium, and lowered body temperature, which can endanger life. Convulsions are more likely to occur if larger doses of barbiturates have been taken.

Rebound hyperexcitability is a condition observed when the use of a CNS depressant drug is abruptly discontinued. The physiological functions that the drug was supposed to suppress now become greatly exaggerated.

Consequently, termination of barbiturate use can result in excessive nightmarish dreaming, restlessness, irritability, and convulsions. To eliminate these effects, there is a strong temptation to resume the use of barbiturates.

8.5 The Abuse of Barbiturates

While true addiction to barbiturates may not be common, habituation and compulsive use are. So is their use by the thrill seeker, the escapist, and the addict whose supply of heroin is cut off. There is no doubt that we are dealing here with drugs of high abuse potential. DEA officials consider the control of illegal traffic in barbiturates a high-priority item.

Patterns of barbiturate abuse are as follows:

1. *Suicide.* DAWN data show that a significant number of the barbiturate emergency-room drug mentions are associated with suicide attempts or gestures.
2. *Accidental deaths and near-deaths due to the alcohol-barbiturate combination.* These deaths are well known to coroners and medical examiners.
3. *Street use.* Barbiturates taken from home medicine cabinets, plus supplies obtained from illicit traffic, appear on the street scene as the euphoric downers. The names *bluebirds* (Amytal), *redbirds* (Seconal), and *blues and*

reds (Tuinal) attest to their popularity among those who wish to escape reality. Users and abusers run the gamut from grammar school students to housewives to elderly persons. Barbiturates are frequently used by thrill seekers in combination with opiates (a hazardous mixture, because both depress respiration) or as a substitute for heroin when the supply of the latter is cut off.

4. *Barbiturate withdrawal syndrome.* When a barbiturate is abruptly terminated, a physically dependent user encounters a serious medical emergency, believed by some experts to be more life-threatening than that of all other drugs.
5. *Drug interactions.* Barbiturate use induces the liver to synthesize enzymes that accelerate the metabolism not only of the ingested barbiturate, but of many other substances as well (tranquilizers, antidepressants, anticoagulants, heart drugs, and alcohol). While this problem usually is only minor, in some cases it can be critical.

8.6 Some Alternatives

Alternatives to the use of barbiturates as sleep aids in cases of insomnia include (1) discovering and rectifying the *cause* of the insomnia and (2) selecting a nonbarbiturate drug for sleep induction. Drugs in the **benzodiazepine** category of minor tranquilizers (Dalmane, Librium, Valium, Serax—see Chapter 10, "The Major and Minor Tranquilizers") have largely replaced barbiturates as drugs for sleep induction. Chemically, benzodiazepines are distinct from barbiturates. Clinically, they have a much wider margin of safety, induce less tolerance, have less effect on REM sleep, stimulate minimal liver enzyme induction, and show no effect on the respiratory center in hypnotic doses. They are virtually suicide-proof. (This statement does not apply if alcohol is consumed concurrently.) Few deaths have been attributed to benzodiazepines. Unfortunately, however, some benzodiazepines have suffered from too much success. Drugs such as Valium and Serax have become heavily overused (not just for insomnia), with serious consequences. Therapeutic doses of benzodiazepines for 20 weeks can induce dependence. See Chapter 10 for more information on benzodiazepines.

There are alternatives to the alternatives. **Chloral hydrate**, useful for inducing sleep, has had a long and colorful history. First synthesized in 1862, it is considered an effective sedative and hypnotic that is unlikely to induce tolerance (although the effective dose may be as much as 2 g). Chloral hydrate disturbs REM sleep less and depresses respiration less than the barbiturates but does show some drug interactions. Chloral hydrate, in combination with ethyl alcohol, constituted the famous "**Mickey Finn**" knockout drops. While the potency of the preparation was probably overrated, it is true that chloral hydrate and alcohol offer a synergistic combination of CNS depressants that makes sleep hard to resist. The modern version of the Mickey Finn appears to be alcohol spiked with minor tranquilizers such as Ativan or diazepam, combinations that can knock out the victim for up to 18 hours. Los Angeles police also report the use of scopolamine in the modern "hooker's knockout scam."

A word of caution: California state scientists say that evidence from more than 20 animal and laboratory tests indicates that chloral hydrate causes cancer in animals and is mutagenic. They suggest that epidemiological tests be done on children who have received the drug. Chloral hydrate is widely used by radiologists to keep children still during lengthy examinations and by neonatologists, who routinely give it to newborns on ventilators. Chloral hydrate use by dentists has led to a few serious overdoses.

Methyprylon (Noludar), ethychlorvynol (Placidyl), and ethinamate (Valmid) are not barbiturates but are similar in structure. They are hypnotic drugs that have been used in place of barbiturates. So, too, was methaqualone introduced as a safe hypnotic with low abuse potential. Methaqualone was a fiasco. It turned out to be one of our greatest drug problems (see Chapter 12 "Hallucinogens, Street Drugs, Designer Drugs, and Some Observations").

The risk with many sleep-inducing agents lies in the great temptation to use, reuse, and finally over-rely. It is just too easy to get to depend on a chemical for sleep. With nonbarbiturates such as Noludar, Placidyl, and Valmid, there also is the real risk of developing a dependence, either physical or psychological, with a significant withdrawal syndrome when the drug is finally discontinued. If you must use a drug to get to sleep, use it only one or two nights in succession and only occasionally.

Other nonbarbiturate drugs that can induce sleep are the antihistamines (if you have ever taken one for allergies, you know how groggy they can make you feel). OTC sleep aids are discussed in Section 15.2. Scopolamine is another nonbarbiturate sold without a prescription. This drug can interfere with REM sleep and cause dry mouth and dry throat.

8.7 Barbiturates and the Law

Because of their potential for abuse, many of the short-to-intermediate-acting barbiturates have been placed in schedule II under the 1970 Controlled Substances Act (see Chapter 3). Amobarbital, secobarbital, and pentobarbital are examples of schedule II barbiturates. Aprobarbital and butabarbital are schedule III drugs, while phenobarbital is a schedule IV drug (reflecting its lesser potential for abuse).

The maximum penalty under federal law for a first-time offense involving a schedule II barbiturate is 5 years imprisonment, a $15,000 fine, or both. These penalties are doubled for a second offense.

Federal law stipulates that prescriptions for depressant drugs such as the schedule II barbiturates (Amytal, Nembutal, and Seconal) must be typewritten or written in ink by the physician personally, must show his or her signature, and may not be refilled. Prescriptions for schedule III or IV barbiturates (aprobarbital, phenobarbital) may not be filled or refilled more than 6 months after the date originally written.

Web Sites You Can Browse on Related Topics

Barbiturates
Search for "barbiturates"

Benzodiazepines
http://www.biopsychiatry.com/benzabuse.html
Also: search for "benzodiazepines"

Clinical Pharmacology Online
http://www.cponline.gsm.com/

Study Questions

1. What is the difference between a sedative drug and a hypnotic drug?
2. A dose of hexobarbital (adjusted for body weight) that puts mice to sleep for an average of 12 minutes puts rabbits to sleep for 49 minutes, rats for 90 minutes, and dogs for 315 minutes. Considering the kinds of animals involved here and the complexity of their brains, what conclusions can be drawn?
3. Give the generic name of one ultra-short-acting, one short-acting, one intermediate-acting, and one long-acting barbiturate.
4. True or false:
   - **a.** Barbiturates generally are useful pain relievers.
   - **b.** Phenobarbital is a useful anticonvulsant.
   - **c.** Barbiturates generally have long half-lives, on the order of 20–30 hours.
   - **d.** Barbiturate pharmacology is similar to alcohol pharmacology.
   - **e.** Tolerance to barbiturates does not develop.
   - **f.** Physical dependence on barbiturates is unknown.
   - **g.** Chemically, chloral hydrate is a nonbarbiturate.
   - **h.** The reason that barbiturates and alcohol form such a deadly combination is because of their synergistic action to depress respiration.
   - **i.** In the body, phenobarbital hastens its own destruction by inducing the liver to make more of the enzymes that catalyze the destruction of barbiturates.
5. It is true that overdosage with short-acting barbiturates is more dangerous to life than overdosage with long-acting barbiturates. Why is this so?
6. What is meant by *REM sleep*, and why is it important in a discussion of barbiturates?
7. Some authorities like to term barbiturates "solid alcohol"; they call alcohol "liquid barbiturate." List the similarities in drug action for these two kinds of CNS depressants that evoke those kinds of terms.
8. According to the U.S. Department of Health and Human Services, some of the reasons people take drugs are to get high, to get rid of pain, to calm nerves, to be "cool," to experiment, because of peer pressure, to get "down," to escape, because they like it, to work better, to be alert, because they're bored, as medication, to hurt someone else, to get in a mood, for a dare, from fear of stopping, or to commit suicide. Which of these reasons are applicable or probably applicable to taking barbiturates?
9. The benzodiazepine type of tranquilizer has been promoted as an alternative to barbiturates for sedation and hypnosis. What are five advantages that benzodiazepines offer?
10. It has been said that stiffer laws, harsher penalties, and more expensive control measures have done practically nothing to restrict the abuse of drugs such as the barbiturates. In fact, it is claimed that all of the attention paid to the barbiturates in newspapers and magazines has popularized them and contributed to the spread of their abuse. Do you agree? What course of action would you suggest?
11. What is there about the pharmacology of alcohol that makes its combination with barbiturates and minor tranquilizers so potentially dangerous?
12. What risk does the average person run when he or she begins to use chemicals to get to sleep?
13. DAWN data show that the highest use of barbiturate sedatives and antidepressants is in the 30–39 age group but that the highest use of hallucinogens and amphetamines is in the 20–29 age group. Offer an explanation for the difference.

CHAPTER 9

Alcohol and 100 Million Americans

Key Words in This Chapter

- Ethyl alcohol (ethanol)
- CNS depressant
- Blood Alcohol Concentration (BAC)
- Dependence
- Fetal Alcohol Syndrome (FAS)
- Alcoholism
- Treatment
- Blackouts

Learning Objectives

After you complete your study of this chapter, you should be able to do the following tasks:

- Understand the chemical nature of ethyl alcohol and how it is denatured.
- Describe the effects of ethanol on the CNS.
- Explain the use of percentages to measure blood alcohol concentrations.
- Tell how the cardiovascular system and GI tract respond to alcohol.
- Describe the actions of alcohol as a diuretic and as a disinhibitor.
- List the steps in the body's metabolism of alcohol.
- Explain the use of Antabuse in alcoholism.
- Define FAS and cite its features.
- Discuss some theories of alcoholism including the role of genetic factors.
- Express your opinions about chemical and sociological treatments of alcoholism.

Drug Abuse Update: It is estimated that nearly one-fourth of all persons admitted to general hospitals have alcohol problems or are undiagnosed alcoholics being treated for the consequences of their drinking. Moderate to heavy consumption of alcohol is the fourth-leading cause of death in the United States. Alcohol is typically found in the offender, victim, or both in about half of all homicides and serious assaults, as well as in a high percentage of sex-related crimes and robberies. The annual health, social, and economic costs of alcohol abuse and alcoholism in the United States include 100,000 estimated deaths and about $86 billion.

9.1 Introduction

You are reading what could be the most important chapter in this book. No other drug or substance is as critical in the lives of so many people as ethyl alcohol. In spite of the decline in its use in the past decade, alcohol remains America's most widely used drug. Department of Health and Human Services statistics show that 10% of our adult population is dependent on alcohol; 3% of all deaths are attributable to it. Alcohol, it can be said, is associated with the best of times and the worst of times. Most of us can recall convivial times with friends made more enjoyable by the presence of an alcoholic beverage. But one of my students has a far different recollection: coming home to find her mother naked and drunk on the bathroom floor.

Alcohol is by far our biggest drug problem. More people are killed or disabled by it, become dependent on it, and become psychotic by abusing alcohol than all of the other drugs put together (excluding tobacco). To define the alcohol problem more clearly, consider the following facts:

1. Alcohol accounts for 11 million accidental injuries each year. Alcohol plays a major role in 39% of the highway fatalities in the United States, and in a recent year cost the lives of 16,250 persons (more than 6,000 of whom were teenagers).
2. Forty percent of the admissions to mental hospitals and more than 50 percent of the arrests each year are due to alcohol (i.e., about 2 million arrests yearly).
3. Alcoholics have a suicide rate 6–15 times greater than that of the general population, and alcoholic depression is the number one cause of suicide in the United States.
4. Alcoholism is the third-leading health problem in the United States, exceeded only by cancer and heart disease.
5. As a teratogen, alcohol is responsible for an estimated 4,000–5,000 babies being born defective each year in this country.
6. Nearly 14% of Air Force personnel have been identified as alcoholics or problem drinkers.
7. Studies have shown that 70% of all adolescents drink; the average age of initiation is 12.9 years; 62% of seventh graders and 80% of twelfth graders drink.
8. Women alone now spend $20 billion a year on alcoholic beverages.
9. There are more than 2 million alcoholic women in the United States.
10. Alcohol is a significant factor in the battered child syndrome.
11. Alcohol is a severe problem in certain parts of the American Indian population, having risen to an epidemic level. On some reservations, the alcoholism rate is as high as 25–50%.
12. Between 50% and 68% of drowning victims had been drinking.
13. Skid row alcoholics comprise only 3–5% of the U.S. alcoholic population; the remainder come from all populations, especially workers and homemakers.
14. The life expectancy of serious drinkers is estimated to be shorter by 10–12 years than that of the general public; the mortality rate for alcoholics is 2.5 times greater than expected.
15. Considering the ill effects of drinking problems on just the families of the drinkers, some 36 million Americans can be regarded as caught in alcohol's web: unhappy marriages, broken homes, desertion, divorce, impoverished families, and displaced children.

16. Students attending college get drunk more often than do their counterparts who do not attend college—41% versus 34%.
17. Nearly 7% of college freshmen who drop out do so as a result of drinking.
18. More than 105,000 Americans (nearly 300 per day) die of injuries or diseases linked to alcohol, according to the Centers for Disease Control and Prevention, Atlanta.
19. Each year college students spend $5.5 billion on alcohol, more than they spend on soft drinks, tea, milk, juice, coffee, and books combined. In the past 18 years, the percentage of college women who drink to get drunk has tripled to 35%.
20. The estimated economic cost of alcohol abuse was $184 billion for 1998 alone (or $630 for every single American).

9.2 Who Drinks the Most?

The following are highlights from the 2001 National Household survey on Drug Abuse, as reported in 2002 by the Office of Applied Studies, SAMHSA.

Current use—At least one drink in the past 30 days (includes binge and heavy use).
Binge use—Five or more drinks on the same occasion at least once in the 30 days prior to survey (includes heavy use).
Heavy use—Five or more drinks on the same occasion on at least 5 different days in the past 30 days.

A summary of the findings from the 2001 NHSDA alcohol questions is given below:

- Almost half of Americans aged 12 or older reported being current drinkers of alcohol in the 2001 survey (48.3%). This translates to an estimated 109 million people. Both the rate of alcohol use and the number of drinkers increased from 2000, when 104 million, or 46.6%, of people aged 12 or older reported drinking in the past 30 days.
- Approximately one fifth (20.5%) of persons aged 12 or older participated in binge drinking at least once in the 30 days prior to the survey. Although the number of current drinkers increased between 2000 and 2001, the number of those reporting binge drinking did not change significantly.
- Heavy drinking was reported by 5.7% of the population aged 12 or older, or 12.9 million people. These 2001 estimates are similar to the 2000 estimates.

Age

- The prevalence of current alcohol use in 2001 increased with increasing age for youths, from 2.6% at age 12 to a peak of 67.5% for persons 21 years old. Unlike prevalence patterns observed for cigarettes and illicit drugs, current alcohol use remained steady among older age groups. For people aged 21 to 25 and those aged 26 to 34, the rates of current alcohol use in 2001 were 64.3% and 59.9%, respectively. The prevalence of alcohol use was slightly lower for persons in their 40s. Past-month drinking was reported by 45.6% of respondents aged 60 to 64, and 33.0% of persons 65 or older (see Figure 9.1).

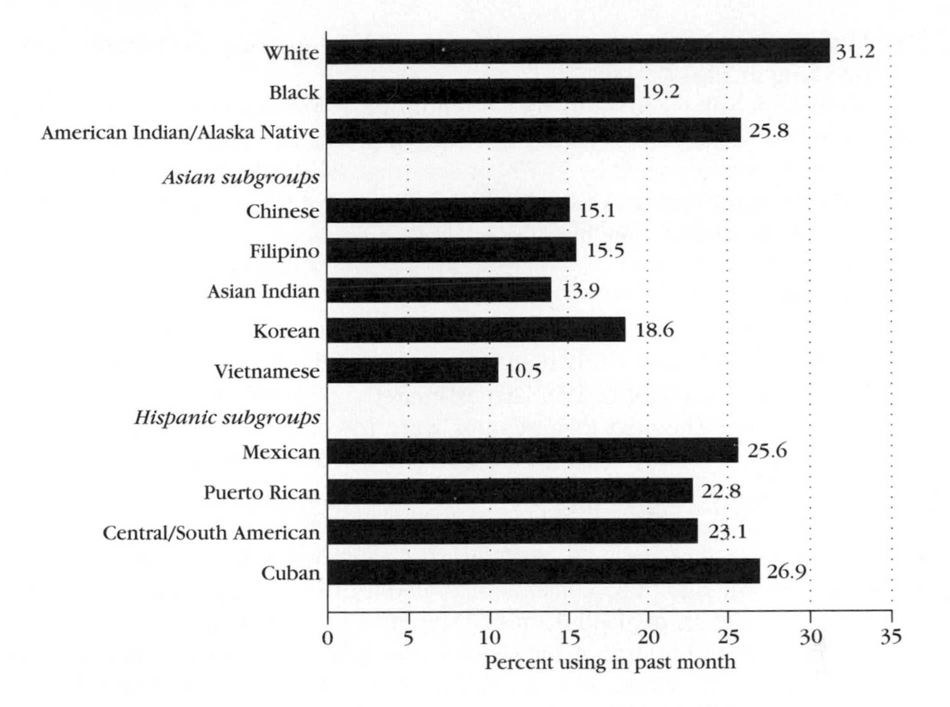

Figure 9.1 Past-Month Alcohol Use Among Youths Aged 12 to 20, by Race and Ethnicity: 2000–2001 Annual Averages. (Source: Results from 2001 National Household Survey on Drug Abuse, Vol. I. Summary of National Findings, SAMHSA, 2002.)

- The highest prevalence of both binge and heavy drinking in 2001 was for young adults aged 18 to 25, with the peak rate occurring at age 21. The rate of binge drinking was 38.7% for young adults and 48.2% at age 21. Heavy alcohol use was reported by 13.6% of persons aged 18 to 25, and by 17.8% of persons aged 21. Binge and heavy alcohol use rates decreased faster with increasing age than did rates of past-month alcohol use. While 55.2% of the population aged 45 to 49 in 2001 were current drinkers, 19.1% of persons within this age range binge drank and 5.4% drank heavily (Figure 9.1). Binge and heavy drinking were relatively rare among people aged 65 or older, with reported rates of 5.8% and 1.4%, respectively.
- Among youths aged 12 to 17, an estimated 17.3% used alcohol in the month prior to the survey interview. This rate was higher than the rate of youth alcohol use reported in 2000 (16.4%). Of all youths, 10.6% were binge drinkers, and 2.5% were heavy drinkers. These are roughly the same percentages as those reported in 2000 (10.4% and 2.6%, respectively).

Underage Alcohol Use

- About 10.1 million persons aged 12 to 20 reported drinking alcohol in the month prior to the survey interview in 2001 (28.5% of this age group). Of these, nearly 6.8 million (19.0%) were binge drinkers and 2.1 million (6.0%) were heavy drinkers. All of these 2001 rates are similar to rates observed in 2000.

- Males aged 12 to 20 were more likely than their female peers to report binge drinking in 2001 (22.0 vs. 15.9%).
- Among people aged 12 to 20, past-month alcohol use rates in 2001 ranged from 19.7% for Asians and 19.8% among blacks to 31.6% for whites. Binge drinking was reported by 21.7% of underage whites and 18.5% of underage American Indians or Alaska Natives, but only by 10.7% of underage Asians and 10.5% of underage blacks.

Race/Ethnicity

- Whites were more likely than any other racial/ethnic group to report current use of alcohol in 2001. An estimated 52.7% of whites reported past-month use. The next highest rates were for persons reporting more than one race (43.2%). The lowest current drinking rate was observed for Asians (31.9%). The rate was 35.1% for blacks and 35.0% for American Indians/Alaska Natives. Figure 9.1 is a summary.

People over the age of 65 consume less alcohol than younger adults, and they have a lower prevalence of alcohol abuse. Decreases in individual consumption may be due to the lowered tolerance for alcohol that is associated with aging. Elderly long-term alcohol abusers are at high risk for adverse health effects. Major late-life stresses may affect levels of alcohol abuse in the elderly.

Although surveys have found no major changes in women's drinking patterns during the past decade, they have identified several demographic subgroups with higher rates of drinking problems. These include the unemployed who were looking for work and those who were employed part-time outside the home. Also at higher risk are women who are divorced or separated, those who have never married, and those who are unmarried but living with a partner. Women in the last category have the highest rates of heavy drinking, drinking problems, and alcohol dependence symptoms of all the groups. Female drinkers in the 21–34 age group are least likely to report alcohol-related problems if they are married and have a stable work role.

Blacks, especially males, are at extremely high risk for acute and chronic alcohol-related diseases such as cirrhosis, alcoholic fatty liver, hepatitis, heart disease, and cancers of the mouth, larynx, tongue, esophagus, and lungs, as well as for unintentional injuries and homicide.

More than 70% of Hispanic women drink either less than once a month or not at all; in contrast, almost the same percentage of men were drinkers. As with black men, rates of heavier drinking increase sharply among Hispanic men in their thirties but decline thereafter.

Asian-Americans as a whole have the lowest level of alcohol consumption and alcohol-related problems of all the major racial and ethnic groups in the United States. This low rate of consumption and abuse may be attributed partly to cultural factors and partly to the **flushing response**. Occurring in a high proportion of Asian people, this physiological reaction is characterized by facial flushing, often accompanied by headaches, dizziness, tachycardia, itching, or other discomfort. A California study found that virtually no Chinese or Korean women were heavy drinkers.

Because of the great diversity in drinking practices among American Indian tribal groups, one cannot generalize about their drinking patterns. Some tribes are mostly abstinent, whereas others have high levels of alcohol use and abuse. A study of 11 tribal groups in Oklahoma found a wide range in percentages of deaths that were alcohol-related, from less than 1% to 24%. An estimated 75% of all traumatic deaths and suicides among Indians and Alaskan natives are alcohol-related.

Binge drinking is characteristic of many Indian tribal groups. This heavy, sporadic alcohol consumption may be the reason for the high rate of accidental deaths and homicides among American Indians.

The French are drinking one-third less than in 1955, but they are still the world's heaviest drinkers. They also show a high incidence of alcohol-related problems, with twice the rate of death from liver cirrhosis as in the United States.

In the face of vast, nearly uncontrolled American guzzling, researchers are fascinated by the 30% of American women and men who abstain from alcoholic beverages. Who are they, and why do they abstain? The 30% figure, fairly stable over the years, includes lifelong abstainers (2% to almost 20% of the general population). Some surveys show that more than 60% of people living in the eastern South Central states (Kentucky, Tennessee, Alabama, and Mississippi) are abstainers. Between one-third and one-half of the people in the West Central, Mountain, and Mid-Atlantic states are teetotalers, but rates are estimated to fall to 20% along the Pacific Coast and in New England. Usually, the more rural the area, the greater the abstention. Almost all studies show 50% higher rates of abstention among women than men. Generally, as we age, we abstain more. Religious affiliation is a factor. For example, while Jews have one of the lowest alcoholism rates in the United States, they also have one of the lowest rates of abstention (less than 7%). One study showed that almost 45% of Baptists abstain, 32% of Methodists, and 10–15% of Catholics, Lutherans, and Episcopalians. People say they refrain from drinking for religious, moral, and health reasons. Others cite the obnoxious taste of alcoholic beverages and the influence of non-drinking friends.

9.3 The Oldest Synthetic Chemical Known

Before we examine the topics of pharmacology, addiction, and treatment, we must take a look at this incredible substance, alcohol, its chemical composition, and where it comes from.

Archaeological records of the oldest civilizations show the use of beer and wine prepared by fermentation of carbohydrates. The first human brewery dates to about 3700 B.C. in Egypt. Alcohol can thus be considered the oldest substance synthesized by humans. The early beverages contained a maximum of 14% (by volume) alcohol. When distillation procedures were developed in Europe 500 years ago, the maximum alcohol content of beverages was increased to about 50%. (Since the "proof" strength of a beverage is twice the percent strength by volume, 50% is equal to 100 proof.)

Chemists term alcohol **ethyl alcohol, ethanol**, or **grain alcohol**. Its chemical structure is:

$$C_2H_5OH \quad \text{or} \quad \begin{array}{c} \quad H \;\; H \\ \quad | \quad | \\ H-C-C-OH \\ \quad | \quad | \\ \quad H \;\; H \end{array} \quad \text{or} \quad CH_3CH_2OH$$

ethyl alcohol

Ethyl alcohol is a clear, colorless liquid with a bitter taste. Figure 9.2 shows a bottle of pure, 200-proof ethyl alcohol. (The chemist terms this alcohol *absolute alcohol*; it is never used as such as a beverage.)

It is important to distinguish carefully between ethyl alcohol (grain alcohol) and its poisonous chemical cousin, methyl alcohol (wood alcohol). The latter is sometimes used in antifreeze solutions and is toxic to the human optic nerve (causing blindness) and to the body in general.[1] Isopropyl alcohol NF [$(CH_3)_2CHOH$] must also be distinguished from ethyl alcohol. When the word *alcohol* is used without further clarification, ethyl alcohol is meant. Rubbing alcohol is typically isopropyl alcohol in 70% concentration.

Most of the ethyl alcohol obtained today by fermentation goes into the making of alcoholic beverages. A far greater quantity of ethyl alcohol is made industrially by the hydration of ethylene. Industrial alcohol is extremely important to our technology, being used as a solvent, a chemical intermediate, and an ingredient in medicinal preparations.

Under federal law, ethyl alcohol to be used in beverages is subject to a tax of $13.50 per proof gallon. To prevent some of the great quantities of industrial alcohol

Figure 9.2 Absolute alcohol is 100% pure (200 proof) ethyl alcohol. If ingested, it will absorb water from mucous membranes, dehydrating them and causing pain.

[1]In the past few years, at least 192 people in China, India, Kenya, and Italy died after drinking wine or cheap beverages to which methyl alcohol had been recklessly and illegally added by profiteering traders.

from getting into beverages and escaping the tax, chemicals are added that make it unfit for human consumption. The result is **denatured alcohol**. Acetone, wood alcohol, phenol, iodine, menthol, and hydrocarbons are examples of denaturants. One should never attempt to drink denatured alcohol. There is no way for the layperson to remove the denaturant.

A variety of alcoholic beverages are marketed, depending on which carbohydrate is fermented. Table 9.1 lists a few of them. Note that "nonalcoholic" beer may contain as much as 0.5% alcohol. That is enough to cause readdiction in recovering alcoholics, according to experts on alcoholism. Also note that fortified wines are widely sold. Typically, they are strong, cheap wines, and they are often sold in lower-income, largely minority neighborhoods. San Diego city law defines fortified wine as having an alcohol concentration of greater than 15% by volume. One brand, Thunderbird, comes in screw-cap pint bottles selling for about a dollar. It is fortified to 18% alcohol. Other cheap, fortified wines are Nighttrain Express, Cool Breeze, and Wild Irish Rose. For a Web site on alcohol concentrations of various products, see **`http://sane.arf.org/infoline/line.htm`**.

Ethyl alcohol has many applications in medicine. It is a skin disinfectant; sponging with alcohol cools the skin and body; it is a solvent for drug preparations; it has been used by injection to relieve a certain kind of facial pain; and it can be used as an appetite stimulant and hypnotic. There is evidence that disinfecting health workers' hands using dollops of alcohol-based gels instead of scrubbing reduces lethal infections in hospitals.

In addition to their alcohol content, whiskey and other beverages contain certain chemicals that impart odor, flavor, and color. These chemicals are termed

Table 9.1 Alcoholic Beverages and Their Alcohol and Caloric Content

Beverage	*Alcohol, by Volume, Approx. (%)*	*Calories, Per Serving, Approx.*
Beer, regular	4.5[a]	140–200 per 12 fl oz
Beer, light	2.6–4	70–134
Beer, extra light	1.1–2.5	—
Beer, nonalcoholic	0.5 maximum	50–95 per 12 fl oz
Cisco (fortified wine)	20	—
Wine, light beverage	10–14	90 cal per 4 fl oz
Wine coolers	4–10	—
Sherry and other fortified wines	17–21	140 cal per 3.5 fl oz
Champagne	11–12	71 cal per 3 fl oz
Sake	14–16	39 cal per 1 fl oz
Tequila	40	—
Gin	40	120 cal per 1.5 fl oz
Brandy	35–40	60 cal per 4 fl oz
Vodka	40	95 cal per 1.5 fl oz
Rum	40	135 cal per 1.5 fl oz
Whiskey	40–53	130 cal per 1.5 fl oz
Hard lemonade (Alcopops)	5	—

[a] *Drug Abuse Update*, Summer 1994 (p. 6) reports that Budweiser regular contains 5.0%, Miller High Life 4.7%, Natural Light 4.7%, Coors Light 4.15%, and Coors Arctic Ice 5.5%.

congeners; they are responsible for part of the gastritis associated with drinking. Examples of congeners are methyl, butyl, and amyl alcohol, ethyl acetate, and ethyl formate. Vodka contains fewer congeners than whiskey. The "hangover" some people suffer the morning after a night of drinking alcoholic beverages can be a combination of upset stomach, headache, thirst, dizziness, fatigue, and depression. There are many old wives' remedies for preventing or curing a hangover. Most of them are worthless. For example, taking aspirin before drinking won't prevent a hangover, and dosing with coffee or stimulants won't speed up recovery. However, drinking water will counter alcohol-induced dehydration, and getting some food into your stomach *before* drinking will reduce the severity of alcohol's effects. Remember, not everyone will suffer from hangover, and some beverages, such as red wine, brandy, and whiskey, are more likely to cause hangovers.

9.4 The Pharmacology of Alcohol

9.4.1 Central Nervous System

Ethyl alcohol is a CNS depressant. The erroneous idea that it is a stimulant is based on the release of inhibition that accompanies alcohol use. The drinker may *think* he or she is a better speaker, driver, partygoer, or lover, but this is not true. Carefully performed experiments have shown that in general, alcohol increases neither physical nor mental abilities.

The brain is exquisitely sensitive to the depressant effects of ethyl alcohol. The highest centers are depressed first: speech, thought, cognition, restraint, and judgment, followed by lower brain function, respiration, and spinal cord reflexes, as the blood alcohol level rises. Depression of the respiratory reflex center by high blood alcohol levels can lead to death. In the preceding chapter, we discussed the dangerous synergistic combination of alcohol and barbiturates. We note that this dangerous combination applies to many other drugs which themselves are central nervous system depressants, including minor tranquilizers, antihistamines, opiates, and other sleeping pills. There is evidence that alcohol does not act at the synapse between nerve cells or on any receptor, but simply enters neurons by diffusion through nerve cell membranes.

A remarkable correlation can be made between the blood level of alcohol, the extent of CNS depression, and behavior. **Blood Alcohol Concentrations (BAC)** are measured in percentages, with 0.1% indicating 100 mg of C_2H_5OH in each 100 mL of blood (hence, 0.01% means 10 mg of alcohol per 100 mL of blood). A typical correlation is given in Table 9.2.

Individuals who are tolerant to alcohol can function surprisingly well at BACs as high as 0.20% and above. A study of 1,715 suspected drunk drivers found an average BAC of 0.22%. A few were actually driving their cars on public highways with BACs that would result in coma for most other people. For example, one researcher discovered that among 213 persons suspected of intoxication, 16 had BACs of between 0.40 and 0.44% but had only minor clinical signs of intoxication. On the other hand, in a study of 10,000 drivers of automobiles and bicycles, 20% of chronic drinkers arrested for signs of intoxication had a BAC of 0.05% or less. A 14-year-old Atlanta girl, who died after a night of heavy drinking, had a BAC of 0.58%, according

Table 9.2 Effect of Alcohol on Behavior

Blood Alcohol Concentration (%)		*Physical and Mental Behavior*
0.01		Clearing of the head. Slight tingling of mucous membranes.
0.02		Mild throbbing at the back of the head. A touch of dizziness. Personal appearance of no concern. Willing to talk.
0.03		Feeling of euphoria and superiority. ("Sure am glad I came to your party." "We will always be friends.")
0.04		Talking and laughing loudly. Movements a bit clumsy. Flippant remarks. ("You don't think I'm drunk, do you?")
0.05		Normal inhibitions almost eliminated. Many liberties taken. Talkative. Some loss of motor coordination.
0.07		Feeling of remoteness. Rapid pulse. Gross clumsiness.
0.10	*Legally drunk in most states*	Staggering, loud singing. Drowsiness. Rapid breathing.
0.20		Blackout level. Inability to recall events later. Easily angered. Shouting, groaning, weeping.
0.30 and above		Stupor. Breathing reflex threatened. Deep anesthesia. Death is due to paralysis of the respiratory center and is generally preceded by 5–10 hours of stupor and coma.

to the medical examiner's office. We must conclude that there is tremendous individual variation in response to this mind-altering drug, and that the potential for tolerance is great.

Some heavy drinkers experience alcohol-induced **blackouts**, or alcoholic trances—periods of time that cannot be recalled later when the person is sober. When hours or even days are blacked out of the drinker's memory, it is a very significant sign that the person has become a true alcoholic. To an observer, the person experiencing the blackout appears to be behaving normally.

Alcohol has many effects on sleep, but few of them are beneficial. While alcohol can decrease the time it takes to fall asleep, it can cause disordered sleep patterns with subsequent impairment in feeling rested. Alcohol can cause changes in REM sleep, vivid dreams, and nighttime wakefulness.

9.4.2 Cardiovascular System

At low doses, only minor effects of alcohol on the heart and vascular system are observed. Dilation of peripheral blood vessels (in the skin) produces a feeling of warmth and a flush that leads the layperson to use alcohol to "warm up" in cold weather. An hour later, however, that same person is colder than before because he or she has lost body heat through dilated blood vessels. It is better not to consume any alcohol if you want to conserve heat.

One of the physical signs of alcoholism is broken blood vessels in the upper cheeks adjacent to the nose. These result from chronic vasodilation produced by repeated alcohol ingestion. Severe alcohol intoxication results in cardiovascular depression.

There is growing evidence that drinking moderate amounts of alcohol may actually benefit the cardiovascular system. A Harvard School of Public Health study of 44,000 men found that those who drank light to moderate amounts of alcohol had a 25–40% lower chance of developing heart disease. Apparently, ethanol can raise the body's level of HDL cholesterol, the "good" type of cholesterol that protects against hardening of the arteries. Note, though, that *heavy* drinking can *cause* heart disease. Moderate drinking, say the experts, can also help prevent strokes, dementia, and the need to amputate limbs.

9.4.3 Gastrointestinal Tract

Alcohol has an irritant effect on the GI tract. This reaction is due in part to the direct toxic effect of C_2H_5OH on the stomach lining but is also the result of alcohol's ability to stimulate acid and pepsin secretion. For this reason, peptic ulcer patients must avoid alcohol. The presence of food in the GI tract tends to modify alcohol's irritant effects. Alcohol is a nauseant. A major factor in hangovers is the gastritis produced by alcohol or by congeners in the beverage. Alcohol stimulates the release of acid, and acid often injures the mucous membranes of the stomach. Thus, drinking can produce gastric hemorrhage. While alcohol alone does not cause gastric bleeding in normal subjects, it does increase blood loss in persons with aspirin-derived mucous membrane injury. Thus, the aspirin commonly taken by alcohol abusers to alleviate discomfort may actually further aggravate stomach injury.

9.4.4 Kidneys

Alcohol is a diuretic; it stimulates the production of urine. But regardless of the alcohol, the large volume of fluids often ingested as part of the alcoholic beverage probably accounts for the majority of urine output. The notoriety of this effect has given rise to the joke about the man observed pouring a freshly opened can of beer directly down the toilet. When asked what he was doing, he replied, "I'm tired of being the middleman."

9.4.5 Sexual Stimulation

Ethyl alcohol is a disinhibitor. From the behavior of some persons with a BAC of 0.05%, one might conclude that alcohol must be a potent **aphrodisiac**. However, all authorities agree that it isn't the alcohol as such that stimulates sexual function but the release of inhibition that it produces. Indeed, too much alcohol abolishes sexual function, a fact noted by William Shakespeare in *Macbeth*, Act II, Scene 3: "[drink] provokes the desire, but it takes away the performance." This statement is in keeping with our knowledge that alcohol is a general CNS depressant.

If alcohol is not an aphrodisiac, what substances are? The answer is that few, if any, true aphrodisiacs exist. The much-discussed "Spanish fly" (dried insects of the species *Cantharis*, from Spain or Russia), for example, is actually a dangerous blistering agent that is an irritant stimulant to the reproductive and urinary organs. A condition of swollen, painful genitals and ruptured membranes hardly constitutes sexual stimulation. Spanish fly is not an aphrodisiac. (See Sections 4.8 and 12.6 for discussions of drugs and sex.)

9.4.6 Liver

The liver is the organ actively involved in the metabolism of alcohol. It is drenched with alcohol shortly after the first drink is taken (the entire blood volume circulates through the liver every 4 minutes) and must bear the brunt of chronic imbibing. No wonder cells in the liver are destroyed, and scarring with fat deposition (cirrhosis) occurs in about 10% of chronic alcoholic patients. Cirrhosis and scarring occur when alcohol preempts the metabolism of fats, which then accumulate in liver cells. The fatty cells enlarge, rupture, and replace normal liver cells. The major pathway for alcohol metabolism involves the enzyme alcohol dehydrogenase. Long-term alcohol abuse reduces the digestive functions of the liver, including its ability to secrete bile. As a result, the alcoholic can become weak, lose appetite, and suffer weight loss, chronic indigestion, and constipation. Compounding the problem of morbidity is the poor nutritional status of chronic drinkers. Not only may they fail to ingest needed vitamins and protein, but their excessive alcohol and fluid intake increases the excretion of B complex vitamins, and their poor fat metabolism compounds deficiencies of vitamins A and D. Additionally, a study reported in the *JAMA* found that moderate amounts of alcohol taken by women who were also ingesting an estrogen supplement tripled blood levels of the estrogen, thus inadvertantly increasing the risk of breast cancer.

9.4.7 Epilepsy

Epileptics should avoid alcohol. For while alcohol is an anticonvulsant, when its use is stopped a hyperexcitable rebound condition can arise with an increased tendency to convulse.

9.4.8 Tolerance and Dependence

Consumption of large amounts of alcohol over long periods of time alters the sensitivity of the CNS to the effects of alcohol, and progressively larger amounts are required to produce the same effect. The inexperienced drinker achieves a much greater response to a given amount of alcohol than the experienced one. Part of this effect is due to learning what to expect; the remainder is due to the development of tolerance. Some alcoholics can accomplish difficult tasks even when their BACs are above 0.20%. Cross tolerance develops to barbiturates, other sedative-hypnotics, and general anesthetics. After many weeks or months, physical dependence on alcohol is established, and the person has become an alcohol addict.

The withdrawal syndrome from alcohol is similar to that for barbiturates. Abrupt, total cessation of drinking results in the following signs and symptoms, which commonly appear within 12–72 hours: shaking or tremors, profuse sweating, nausea, anxiety, diarrhea, hallucinations, and disorientation. In severe cases, the potentially lethal condition known as **delirium tremens** (DTs, rum fits) may occur, with possible seizures and cardiovascular collapse. It appears that delirium tremens is not simply a direct toxic effect of alcohol on the brain; in fact, it usually occurs *after* the withdrawal.[2] It can threaten life in elderly or seriously ill persons. The usual cause of death is a high fever (hyperthermia) associated with peripheral vascular collapse. After 5–7 days, withdrawal is complete. Drugs such as minor tranquilizers have been used in alcohol detoxification to help ease the withdrawal symptoms. Great care must be taken, however, to avoid creating a new dependence on the tranquilizer so used.

9.4.9 Absorption, Metabolism, and Excretion

Many people have experienced the effects of rapid absorption of alcohol taken on an empty stomach. This rapid action is explained by the fact that the stomach itself absorbs one-fourth of a dose, with the remainder rapidly absorbed in the small intestine. After one drink on an empty stomach, maximum BACs can be reached in 20–30 minutes. Food in the stomach delays absorption.

The typical drink, equivalent to two-thirds of an ounce of pure C_2H_5OH, is provided by each of the following:

- A shot of spirits (1.5 fluid ounces of 86-proof whiskey)
- 1.5 fluid ounces of gin or vodka
- A large glass of table wine or champagne (5 fluid ounces of 12% alcohol)
- 12 fluid ounces (one can) of beer (4.5% alcohol)

A 150-pound (70-kg) person who drinks at the rate of one typical drink per hour will maintain a low BAC, because the individual can metabolize about two-thirds of an ounce of straight whiskey or 8 ounces of beer per hour. As a person drinks faster than the C_2H_5OH can be burned, the drug accumulates in the blood and brain. For example, a person weighing 100 pounds will reach a BAC of about 0.22% after consuming five drinks in an hour. A 200-pound drinker will have a BAC of approximately 0.11% under the same circumstances. Actually, there are so many variables that can affect BAC—body mass, food in the stomach, tolerance—that predictions of BAC are often guesswork. Nonetheless, the state of California has published charts relating BAC to number of drinks ingested (see the inside back cover of this book).

We do not know exactly the lethal dose of alcohol for humans. Unfortunately, we have too many indications of this amount from the following tragic cases. A just-turned-21-year-old drank 10 shots of a 175-proof alcoholic beverage in quick succession to celebrate his "coming of age." He died. A 15-year-old Maryland youth died after drinking 26 shots of vodka in 90 minutes at an all-you-can-drink party. A 16-year-old Florida girl died after playing a drinking game called "Pass Out," and a 14-year-old Massachusetts girl nearly died after drinking 1 quart of Cisco, a fortified wine (see Table 9.2). A college freshman died during a Greek Week celebration; his

[2]The part of the syndrome known as delirium tremens may actually be due to the severe hypoglycemia that alcohol can induce. Alcoholic hypoglycemia must always be considered as a possible cause of unconsciousness in a person smelling of alcoholic beverage.

BAC measured 0.41%, the result of consuming the equivalent of 16 shots in one hour. The cause of death in all these cases was probably either CNS depression of respiration—or the victim choking on his or her own vomit.

Alcohol is metabolized (chemically changed) in the liver by two enzymes that act sequentially as follows:

> Alcohol quickly enters the liver and is converted to acetaldehyde, and the acetaldehyde is converted into acetic acid (occurs as the acetate). Our body's enzyme *alcohol dehydrogenase* (ADH) catalyzes the first half of alcohol metabolism, and the liver's *aldehyde dehydrogenase* (ALDH) catalyzes the second half.
>
> $$\text{Ethyl alcohol} \underset{\text{ADH}}{\rightleftharpoons} \text{Acetaldehyde} \xrightarrow[\text{ALDH}]{} \text{Acetate} \longrightarrow \underset{\text{(calories)}}{\text{Heat \& Energy}}$$

Alcohol can also be metabolized in the liver by cytochrome P450 oxidase, which may be increased after chronic imbibing.

Although alcohol is metabolized as a carbohydrate, it exhibits none of the properties expected of a food. It does not remain in the stomach long; it is not selectively secreted, nor can it be stored by the body. It is burned to the *exclusion* of fat and other substances, making it more of a fuel than a nutritional foodstuff. Each gram of ethanol yields 7 calories, called "empty" calories. It does not become a protein or carbohydrate and does not contribute positively to any type of tissue.

Excretion of alcohol occurs mainly in urine, but only about 8% of an ingested dose is actually excreted. Most of the alcohol we consume is metabolized. Alcohol and other volatile components of alcoholic beverages appear on the breath shortly after reaching the bloodstream. These facts constitute the basis for medical-legal tests for sobriety (see the discussion later in this chapter).

Alcohol and Women

Recent research suggests that women may become more intoxicated than men after ingesting the same amount of alcohol—even when body weights are the same. One explanation for this finding is that a woman generally has less body fluid and more body fat than a man of the same weight. Since alcohol does not diffuse as rapidly into body fat, its concentration in a woman's blood will be higher even if she drinks the same amount as a man. Also, many women's reactions to alcohol vary throughout the menstrual cycle. Apparently, women are often more affected by alcohol just before menstruation begins. Furthermore, researcher Frezza[3] and colleagues found that, compared to men, women have a diminished capacity to burn alcohol in the body. Both sexes have the needed ADH in the liver and in the stomach (gastric) lining, but gastric ADH activity is lower in women. Consequently, more alcohol can enter the bloodstream in a woman even when equivalent doses are consumed. Realizing this fact, women may decide to adjust their drinking patterns to avoid negative reactions. Women should also know that medications containing estrogen, such as birth control pills or hormone drugs, can affect their reaction to alcohol. With a dose of estrogen already in the body, a woman might recover more slowly from the

---

[3]M. Frezza, C. diPadova, G. Pozzato, M. Terpin. E. Barona, and C. Lieber, *New England Journal of Medicine*, Vol. 322, 1990, pp. 95–99.

effects of drinking. Women have a higher risk than men of dying in an auto crash at the same BAC. See also the discussion of fetal alcohol syndrome in the next section.

In other health consequences of alcohol use by women, research suggests that women develop alcoholic hepatitis from smaller daily amounts of alcohol than do men. Women are also more susceptible than men to degenerative diseases of the heart muscle. Alcoholic women perform worse on tests of immediate recall and psychomotor speed, and their mortality rates are higher than in men who drink heavily. The most frequent causes of death among alcoholic women are liver disease, pancreatitis, accidents or violence, suicide, cancer, and cardiovascular disease.

9.4.10 Alcohol and Cancer

Evidence for a strong link between long-term alcohol consumption and cancers of the mouth, pharynx, larynx, and esophagus has existed for many years. In addition, recent studies suggest that alcohol may play a role in cancers of the liver, pancreas, stomach, large intestine, rectum, and breast. There is no proof from animal studies that alcohol alone is carcinogenic. However, alcohol is typically ingested with other toxic substances (nitrosamines, tobacco, asbestos fibers, and polycyclic hydrocarbons) that are suspected of acting as cocarcinogens. The U.S. Department of Health and Human Services estimates that alcohol, alone or in combination with cocarcinogens, accounts for 3% of the cancer deaths in the United States yearly.

The interaction between alcohol and tobacco is particularly striking in terms of cancers of the mouth, pharynx, and larynx. Alcohol and tobacco appear to act as **cocarcinogens**, synergistically increasing the risk of oral cancer 15-fold compared with that of people who neither drink nor smoke. If you are a smoker, you are well advised not to drink any alcoholic beverage.

Approximately 4,000 chemicals are generated in the intense heat of a burning cigarette. The tar so produced is carried into the lungs on inhaled smoke and is distributed by the blood to all parts of the body. Microsomal enzymes, found mainly in the liver, convert some ingredients of tar into chemicals that are known to cause cancer. Alcohol can activate some microsomal enzymes, thus contributing to smoking-related cancers. The esophagus may be particularly susceptible because it can't easily get rid of cocarcinogens produced by microsomal enzymes. Actually, alcohol has been shown in laboratory experiments to promote throat tumors in animals exposed simultaneously to it and to tar chemicals. Alcoholics frequently are deficient in vitamin A and zinc—substances that confer a degree of protection against cancer.

9.5 Fetal Alcohol Syndrome (FAS)

Use of ethyl alcohol by women during pregnancy can result in a characteristic pattern of severe mental and physical baby defects termed **fetal alcohol syndrome (FAS)**.[4] Although FAS was "discovered" in 1972 and was named in 1973 by Dr. David W. Smith of the University of Washington, Seattle, society had long been aware of alcohol's potential for damaging the developing child. We read in *Judges*

[4]The disorder fetal alcohol effects (FAE) is less severe than FAS, consisting of low birth weight, spontaneous abortion, and some partial aspects of FAS.

13:7, "Behold thou shalt conceive and bear a son, and now drink no wine or strong drink." And in 1759, the British College of Physicians called upon Parliament to control the gin distillation business, since gin "causes weak, feeble, distemperate children who are born meagre and sickly" and unable to pass through the first stages of life.

Alcohol (or possibly its metabolite, acetaldehyde) is now recognized as a potent teratogen. It freely crosses the placental barrier. (If an expectant mother gets drunk, her fetus does too.) Fetal alcohol blood levels remain high because of a lack of liver enzymes in the early stages of development. For example, if it takes 6 hours for the mother's blood to become alcohol-free, it will take 12 hours for fetal blood to become alcohol-free.

Alcohol probably affects all organ systems in the fetus, but the tissue most severely affected is the brain. An embryo or a fetus exposed to alcohol can develop into a baby born with an abnormally small head, a joint or limb abnormality, facial irregularities, heart and genital defects, and hemangiomas (abnormal growth of blood vessels). Mentally, FAS can be manifested as mental retardation, hyperactivity, or brief attention spans and IQs 35–40 points below normal. Perhaps the most striking manifestations of FAS are seen in the face of the neonate or child (see Figure 9.3). Typically, the philtrum (the vertical groove in the middle of the upper lip) is indistinct, the upper lip is thin and red, and the nose is short and flattened. In Caucasians, the eyelids develop with an epicanthic fold, like the Mongolian race eyelid.

Apparently, FAS is much more widespread than was once believed. Dr. Ken Lyon Jones, an expert on the subject, says that FAS affects 1–2 of every 1,000 births in the United States. After Down syndrome and spina bifida (a defective closure of the bony encasement of the spinal cord), it is the third most common birth defect in the United States causing mental retardation. Findings from the National Birth Defects Monitoring Program show that from 1979 to 1993, the number of reported cases of FAS increased more than 600%. According to *U.S. News and World Report*, South Africa has the highest incidence of FAS in the world. FAS is completely preventable. Note that FAS is not always easy to diagnose, even in severe cases. A trained eye is needed.

The important question in regard to FAS is: How much alcohol is enough to damage the fetus? The answer is that no one knows for certain, but we have many indicators. Heavy drinkers are at high risk. A Boston City Hospital study found that of infants born to heavy-drinking mothers, 29% were considered to be normal, whereas 64% of those born to rare drinkers were normal. Heavy drinkers were defined as consuming up to 5.8 drinks per day on the average. Another study showed that 32% of infants born to heavy drinkers had congenital abnormalities, compared with 9% when the mothers were abstinent, and 14% when they were moderate drinkers. An NIH study showed that one or two drinks a day do not increase *overall* risk of fetal abnormality but do increase the risk of genitourinary defects and of miscarriage. Binge drinking was found to greatly increase the risk.

The mental deficiencies mentioned earlier are, of course, not apparent on morphological examination of the newborn; nonetheless, they can be real and have lifelong consequences. Light to moderate drinking by the mother prenatally can result in borderline intelligence, behavioral problems, and learning impairment as an adult.

In summary, the damages caused by the teratogen alcohol are **dose-related**. Heavy-drinking mothers are highly likely to produce a damaged child. Light-drinking mothers (two drinks a day) are at risk of producing a child with more

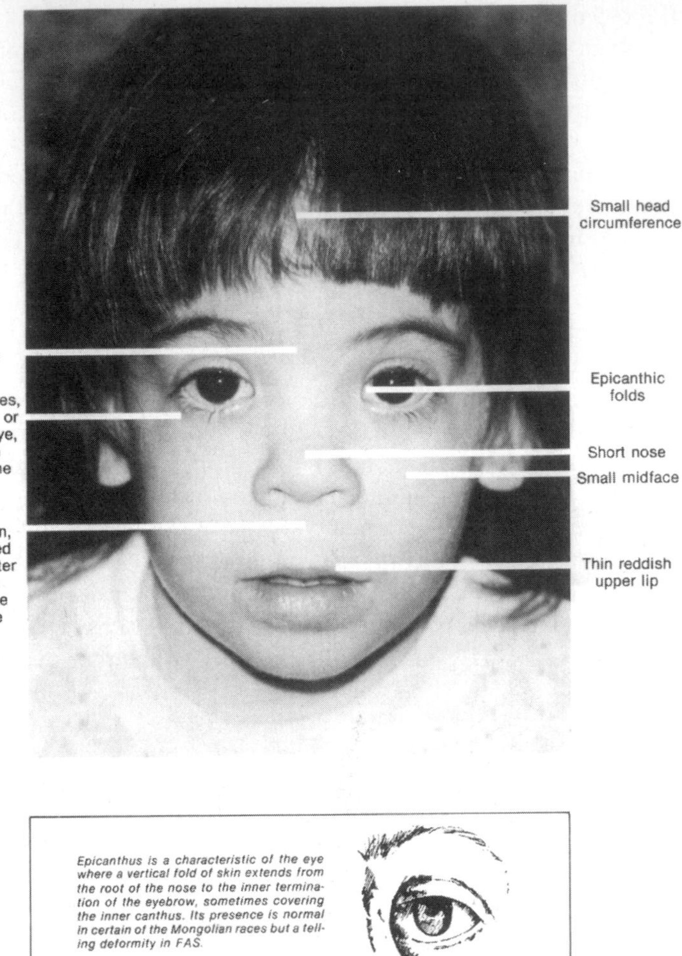

Figure 9.3 Characteristic Anatomical Defects That Are Signs of Fetal Alcohol Syndrome. Alcohol consumed by a pregnant woman can damage the developing fetus, causing fetal alcohol syndrome (FAS). (Photo courtesy of S. K. Clarren, M.D., The Children's Orthopedic Hospital and Medical Center, Seattle, Washington; used with permission.)

subtle defects, such as learning impairment later in life. Because the peak BAC is the most critical factor, pregnant women are strongly advised to avoid binge drinking. Of course, the best advice is to consume no alcohol at all during pregnancy. Alcoholic women should not conceive. In a fetus, hangovers last a lifetime. In regard to males, no relationship has been observed between an alcoholic father and an FAS child, confirming that it is the mother's BAC that is critical. However, there is now preliminary evidence that links a father's use of alcohol in the month before conception to low birth weight.

For more information on FAS, contact the Seattle Foundation, Washington Federal Building, Suite 510, 425 Pike Street, Seattle, WA 98101. For The National Organization on FAS, browse this URL: **`http://www.nofas.org/`**. Also see: **`http://www.fascenter.SAMHSA.gov/`**.

Case History *Fetal Alcohol Syndrome*

E., an acknowledged alcoholic, was drinking a bottle of vodka a day. Her mind was so jumbled by alcohol she did not even realize she was pregnant until the 16th week. She quit drinking then and there, but it was too late. Her baby was born undersized, with damaged kidneys, a poorly functioning stomach, a thin upper lip, and eyelids that dropped. Surgery restored partial vision in only one eye. The child is not severely mentally retarded but requires comprehensive care. E. has not taken a drink in the 3 1/2 years since that 16th week.

Despite widespread warnings, surveys indicate that in one recent year, 140,000 pregnant women nationwide were frequent drinkers.

9.6 Alcohol–Drug Interactions

The National Clearinghouse for Alcohol Information, a division of the *National Institute on Alcohol Abuse and Alcoholism* (NIAAA), has reported that of the 100 most frequently prescribed drugs, more than half contain at least one ingredient known to interact adversely with alcohol. Most bad effects due to alcohol-drug interactions are accidental, but the medical toll is high: an estimated 2,500 deaths and 47,000 emergency room admissions yearly.

What this means is that doctors should be asking their patients about a drinking history, if any, and both doctor and patient should be alert to possible alcohol-drug interactions before a prescription is written. Also, you should read the warnings on the labels of OTC products.

The list of drugs that interact unfavorably with alcohol is too long to include here, but the following are some of the categories of drugs that are potentially dangerous when ingested by a heavy drinker, chronic alcoholic, or recovering alcoholic. If you have doubts about the drugs you are taking, consult with your physician or pharmacist.

- *Barbiturates.* Synergism with alcohol can be lethal.
- *Analgesics.* Aspirin, for example, can combine with alcohol to cause severe stomach irritation.
- *Anesthetics.* Alcohol can potentiate the effect of an anesthetic.
- *Anti-high blood pressure drugs.* When combined with alcohol, they can lower the blood pressure too much.
- *Anticonvulsants.* Alcohol speeds up the metabolism of Dilantin, for example.
- *Antihistamines.* Patients can become overly drowsy and have an accident.
- *Minor tranquilizers.* The bad effects on performance, skills, and alertness are increased.
- *Major tranquilizers.* Phenothiazines combined with alcohol can produce possibly fatal respiratory depression.
- *Narcotics.* Alcohol intensifies the CNS depressant action.

- *Anticoagulants*. Alcohol can increase the ability of drugs such as Dicumarol and Panwarfin to stop blood clotting, leading to possible hemorrhage.
- *Antidiabetics/hypoglycemics*. Alcohol reacts unpredictably and sometimes severely with drugs such as insulin, Diabinese, and Orinase.
- *Diuretics*. Combining alcohol with drugs such as Diuril, Lasix, or Hydromox can cause a significant reduction in blood pressure.
- *Vitamins*. Continuous drinking can prevent vitamins from entering the bloodstream.

Alcohol in combination with any drug that has a depressant effect on the CNS represents a special hazard to health and safety—sometimes to life itself. The drug adds to the normal depressant effect of alcohol, further depressing the nervous system that regulates vital body functions. There is a second type of drug interaction with alcohol. If alcohol is metabolized by the same liver enzymes that metabolize another drug, then alcohol is, in effect, competing with the other drug for those enzymes. The result is that both alcohol and the other drug are metabolized more slowly. Thus, the effect of the alcohol and/or the other drug is exaggerated, because both remain in the blood for an extended period of time.

9.7 Alcoholism and the Law

Should it be a crime to be a public alcoholic, or should alcoholism be considered a disease, like diabetes or hypertension? Should a pregnant alcoholic female be jailed? These questions are of great importance to the estimated 10–12 million American alcoholics and problem drinkers and to the hundreds of thousands who are arrested or prosecuted for alcohol-related offenses.

Historically, the drunk has been treated as a criminal: jail sentences, fines, and other punitive measures. By and large this treatment has done nothing to change the course of an alcoholic's life, and he or she is repeatedly recycled through the courts and jails. It has been said, "Jails are revolving doors for public inebriates."

In 1970, a "Bill of Rights" was passed by the U.S. Congress. This landmark legislation, the **Comprehensive Alcohol Abuse and Alcoholism Prevention, Treatment, and Rehabilitation Act**, acknowledged that alcoholism is a disease and that treatment is required through health rehabilitation services. It created the NIAAA to administer all alcoholism programs and to coordinate all federal activities in this field. The federal government is now prohibited from firing or refusing to hire (for nonsecurity jobs) a person because of former drinking. The act prohibits general hospitals from discriminating against alcoholic persons in their admission policies.

The NIAAA publishes newsletters, books, and pamphlets in an ongoing educational program. Its goal is to combat alcoholism by promoting a better understanding of the problem, by destroying myths, and by encouraging the establishment of employee alcohol abuse counseling programs in private industry. The NIAAA conducts programs in collaboration with the Bureau of Indian Affairs (BIA). Its address is NIAAA, 5600 Fishers Lane, Rockville, MD 20857, and on the Web, **http://www.niaaa.nih.gov**.

Traditionally, individual state laws have made public intoxication a criminal offense, with varying degrees of punishment. However, in the past 20 years one state after another has repealed its punitive laws and has adopted statutes providing detoxification and rehabilitation treatment procedures. A model law, the **Uniform Alcoholism and Intoxication Treatment Act**, was proposed by the National Conference of Commissioners on Uniform State Laws in 1971. Today, most states, the District of Columbia, and Puerto Rico have adopted the Uniform Act or its basic provisions, thus decriminalizing public drunkenness and recognizing alcoholism as an illness. The uniform laws prescribe that intoxicated persons should not be subjected to criminal prosecution solely because of their alcohol consumption or intoxicated appearance. A person who appears to be incapacitated by alcohol may be taken into protective custody by the police, but under these laws this is not considered an arrest.

Recognition of alcoholism as a disease by sheriffs and judges has gained support, but the problem of dealing with the drunk driver remains as pressing as ever. It is unlawful in every state for a person under the influence of intoxicating beverages to operate a motor vehicle on public highways, but the definition of *under the influence* varies. Although most states currently stipulate a BAC of 0.10% for legally drunk, the U.S. Congress, in October 2000, passed a national 0.08% BAC law requiring all states to implement the 0.08 BAC limit by 2004 or lose highway funding. A poll shows that over 70% of Americans support the 0.08 BAC law. California and Utah already apply the 0.08 BAC limit. In France, the Netherlands, Belgium, Finland and Norway, it is 0.05%; in Sweden, 0.02%. The World Health Organization recommends a 0.05% limit. Investigators have found that driving skills begin to be impaired at 0.04%.

Some European countries have a national implied consent law under which drivers' licenses can be taken away when drivers are stopped by police, tested, and found to have a BAC of 0.08% or higher. **Implied consent laws** (now enacted in most U.S. states) specify that any person who operates a motor vehicle on a public highway *shall be deemed to have given his consent* to a chemical test where there are reasonable grounds to consider that person intoxicated. In Japan, if a person is found to have *any* quantity of alcohol in the blood while driving a car, he or she is automatically considered to be legally under the influence.

BAC test laws apply in all but a few states. The tests are performed on samples of breath, blood, saliva, or urine and are used to provide evidence of intoxication. More than 300 analytical methods for alcohol have been published. Popular test methods include the following.

1. *Collecting and examining exhaled breath.* Gas chromatographs are used to determine the percentage of alcohol in a sample of breath. The breath sample can be collected at the moment of testing or at a remote location and delivered to the testing laboratory. Accuracy is said to be within 5%. Criminologists have used gas chromatography for thousands of alcohol analyses, often as a source of evidence in court appearances. A hand-held alcohol breath analyzer is shown in Figure 9.4a; a digital readout gives the blood alcohol level directly.

   Note: Some defendants in driving-while-intoxicated cases claim that they were dieting at the time of the breath test and that the acetone produced from their dieting distorted test results. This claim is specious; research has shown that the level of acetone is far too low to be a factor. Also,

Figure 9.4a The Alco-Sensor IV is a portable, hand-held device for measuring breath alcohol. It can be computer interfaced and produces results of evidential quality. (Photo courtesy of Intoximeters, Inc., St. Louis, MO 63103.)

using a mouthwash will not fool a breath test. Actually, many mouthwashes are high in alcohol (Section 15.11) and may boost breath alcohol for up to 15 minutes, according to a study published in the *Journal of the American Medical Association*.

2. *Collecting and examining blood.* A sample of venous blood is withdrawn from the individual and analyzed at the laboratory. The gas chromatograph apparatus shown in Figure 9.4a can also be used for blood alcohol determinations. A direct analysis of alcohol in blood is, of course, the most accurate, because in the case of breath, results must be correlated *with* blood alcohol levels. However, it is much more convenient and less painful to give a breath sample, and so breath analysis is the most popular test, with blood a close second and urine a distant third.[5]
3. *Urinalysis.* The ratio of urine alcohol concentration to BAC is generally constant at 1 to 3, and most states permit substitution of a urine sample for blood or breath. However, according to *The Toxicology Newsletter*, published by Duquesne University School of Pharmacy, Pittsburgh, urine samples should no longer be permitted as specimens for determining the degree of intoxication. Laboratory studies have shown that urine alcohol accumulation in the bladder does not necessarily accurately reflect a corresponding blood alcohol level. Blood alcohol estimations can be made only by having the subject completely empty his bladder to eliminate prior accumulations and then submit a sample 30–60 minutes later. If, however, the subject deliberately does not completely empty his bladder, a subsequent sample may be invalid. Also, the subject may state that he or she cannot urinate within the specified time after the first voiding. On the basis of this research, it was concluded that an accurate prediction of blood alcohol from urine alcohol can be made in only about 50% of the cases.
4. *Saliva.* A 2-minute saliva test for blood alcohol has been developed (Figure 9.4b). Called the Alco-Screen test, it uses a highly specific enzyme reaction

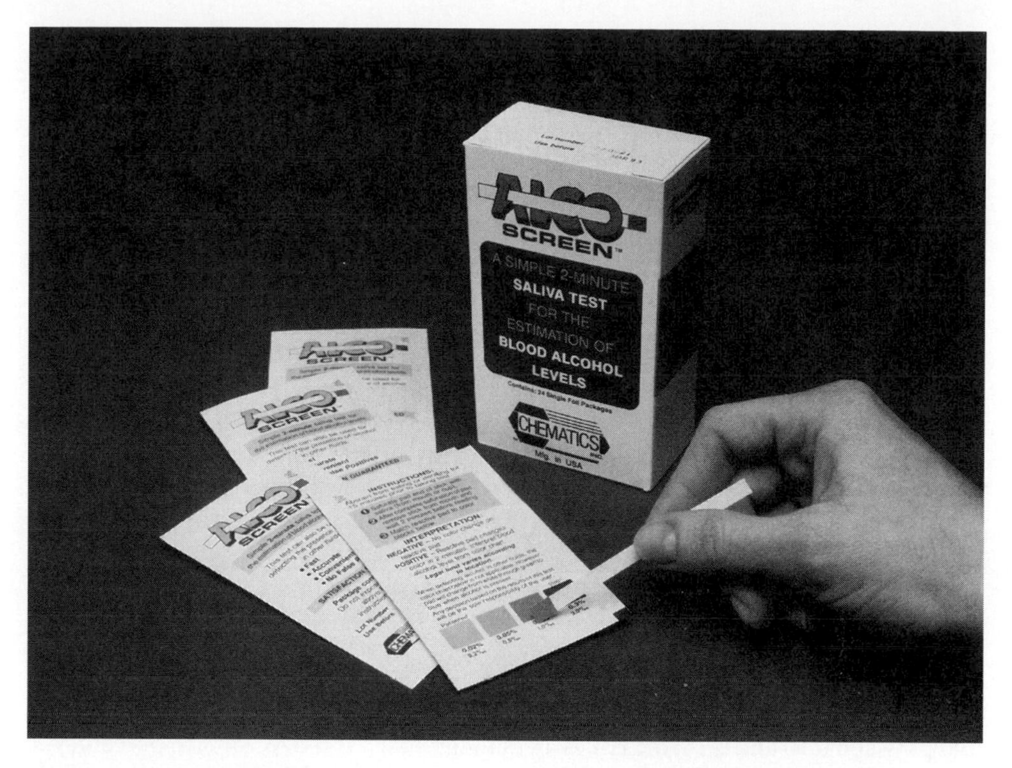

Figure 9.4b Alco-Screen is a two-minute test for blood alcohol using saliva directly from the mouth. (Photo courtesy Chematics, Inc., North Webster, IN 46555.)

coupled to a color reagent quantitatively to measure alcohol in saliva directly in the mouth. In 2 minutes, results are read as shades of green to blue using an accompanying color comparator. Since the concentration of alcohol in the saliva is comparable to that in blood, the test thus becomes a blood alcohol test. A list of interferences and limitations is available from the manufacturer.

For a discussion of the use of the GGT blood test in the detection of alcohol, see Section 16.4.

9.8 Drinking and Driving

Life expectancy in America has improved over the past 75 years for every age group except 15–24-year-olds. Their death rate is higher than it was 20 years ago. The leading cause: car crashes. During the past decade, more than 450,000 traffic accident fatalities constituted the fifth leading cause of death in the United States and the leading cause of death in the 1–35-year-old age group. Dozens of studies have shown

[5]According to Dr. Charles Winek, Chief Toxicologist, Allegheny County (PA) Department of Laboratories, U.S. apparatus for testing breath for alcohol content are typically designed on the basis of a blood-to-breath ratio (BBR) of 2100 to 1 built into the equipment. The BBR is thus the conversion factor from breath alcohol concentration to BAC. European equipment has a built-in ratio of 2300 to 1 (a better estimate, some say) and will thus overestimate BAC relative to U.S. equipment.

that at least one-third of the drivers in these fatal accidents had BACs exceeding 0.1%. It is hard to overestimate the terrible, costly impact of alcohol on the carnage on the highways.

In the early 1970s, laws were passed in about half of the U.S. states and in all 10 Canadian provinces lowering the legal drinking age to either 19 or 18 years. Previously, both New York and Louisiana had permitted 18-year-olds to drink legally for many years. Since the passage of these liberalized laws, studies have been carried out to determine what effect, if any, the lowered drinking age had on the incidence of automobile accidents. It now appears clear that there was a significant increase in auto accidents among drivers under 21 after the drinking age was lowered. For this reason, the pendulum of public opinion has swung toward a more conservative position, and now all 50 states have raised the minimum drinking age to 21 years. A subsequent study of young male drivers in Illinois showed an 8.8% reduction in single-vehicle nighttime accidents. Another study by the National Highway Traffic Safety Administration examined data from nine states that had raised their legal minimum age. The average reduction in nighttime fatal crashes among drivers in the affected age group was 28%. Additional studies in Michigan and Maine showed similar results.

Because driving is a complex physical and mental activity, your risk of a car crash is increased by any blood alcohol level, even 0.02%—the result of just one drink. Your risk of a fatal car crash is 1.4 times higher if your BAC is between 0.05 and 0.04 than if you are sober. Between 0.05 and 0.09, the risk is 11 times greater; and between 0.10 and 0.15, the risk is 48 times greater. At a BAC of 0.15 percent, you are 380 times more likely to die in such a crash than a nondrinker.

9.9 Theories of Alcoholism: What Is an Alcoholic?

If I drink more than three cans of beer each day, or if I get drunk once a month, or if I drink before noon, am I an alcoholic? Literally, hundreds of definitions of the complex illness of alcoholism have been offered, but the following would probably be accepted by most authorities:

> Alcoholism is a complex disease or behavioral disorder characterized by preoccupation with obtaining alcohol and loss of control over its consumption to the extent that use exceeds the ordinary social drinking habits of the community. It is chronic, progressive, and associated with poor health, physical disability, and impaired interpersonal relations or economic functioning.

Note that this definition does not specify the quantity of alcohol consumed as a criterion of alcoholism, because there is too much variation in how much alcoholics drink. The person who drinks only on weekends or only drinks beer may fit the definition as well as the person who starts drinking bourbon every day at 7 A.M. Our definition states that alcoholism is a progressive disease. This idea is strikingly shown in Table 9.3, which summarizes the progressively worsening behavior of an employee who is an alcoholic.

Theories about the causes of alcoholism are numerous, but no single theory has yet been proved adequate to explain this complex, addictive disease. Some of the ideas proposed to explain the origins of alcoholism are as follows:

Table 9.3 Progressively Worsening Alcoholic Employee Behavior Patterns

Employee Behavior Pattern	*Observable Signs*
Early Stage	

Drinking to relieve tension
Increase in tolerance
Memory lapses
Lying about drinking
	Absenteeism
	Tardiness (at lunchtime)
	Early departure
	General behavior:
	  Overreacts to real or imagined criticism
	  Complains of not feeling well
	Performance on job:
	  Missed deadlines
	  Lowered job efficiency

Middle Stage	

Sneaking drinks
Feeling guilty
Tremors
Loss of interest
	Absenteeism
	Frequent days off for vague ailments or implausible reasons
	General behavior:
	  Marked changes
	  Statements not dependable
	  Begins avoiding associates
	Repeated minor injuries on and off job
	Performance:
	  Spasmodic work pace
	  Lapses of attention—cannot concentrate

Late Middle Stage	

Unable to discuss problems
Efforts at self-control fail
Neglects food
Drinks alone
	Absenteeism
	Frequent time off (sometimes for several days)
	Does not return from lunch
	General behavior:
	  Domestic problems interfere with work
	Financial problems (garnishments)
	More frequent hospitalization
	Will not discuss problems
	Performance:
	  Far below what is expected

Approaching Terminal Stage	

Now thinks "my job interferes with my drinking"	Absenteeism
	Prolonged, unpredictable absences
	General behavior:
	  Completely undependable
	Repeated hospitalization
	Physical deterioration visible
	Serious financial and family problems
	Performance:
	  Uneven
	  Generally incompetent

Source: Sibyl Cline, Alcohol and Drugs At Work, Drug Abuse Council, Inc., Washington, D.C., 1975.

1. Alcoholism results from disturbances and deprivations in early infancy.
2. It is caused by identification with other people who solve their life problems through the use of alcohol.
3. A defect in genes leads to inborn errors of metabolism.
4. Alcoholism is inherited; its susceptibility runs in families.
5. It is a dysfunction of the hormone system.
6. It is a symbolic revenge against one's parents.
7. Alcoholism is a way to reduce inner fears and anxieties.
8. Some people are born with an "alcoholic personality."

The role of **genetic factors**, that is, the ability to inherit alcoholism, has been studied extensively for 80 years.[6] Most researchers are convinced that alcoholism runs in families and that genetic factors contribute substantially to a familial vulnerability to the disease. It is believed that a genetic influence is identifiable in 35–40 percent of alcoholics and alcohol abusers, and that both men and women are affected. Studies of families have revealed that first-degree relatives of alcoholics are two to seven times more likely than the general population to develop problems with alcohol sometime in their lifetimes. Twin and adoption studies confirm that our genes can play an integral role. A valuable reference to the genetics of alcoholism is the *Tenth Special Report to the U.S. Congress on Alcohol and Health*, NIAAA, June 2000, pp. 169–180.

Many people who are at risk for developing alcoholism never develop the disease. In one study, 60 percent of male alcoholics had no family history of alcohol abuse. We conclude that environmental factors and cultural norms contribute to the etiology of this disease. High levels of parental support (or lack of), close monitoring of adolescent activity by parents, the severity of the father's disease, the drinking activity of friends, parental standards, and extent of exposure to psychological stress—all can influence outcomes one way or another.

Regardless of the cause, alcoholics exhibit behavioral patterns indicative of this disease. Some typical patterns are:

1. Persistent absenteeism from work
2. Repeated accidents, particularly on the job
3. Development of a variety of excuses for drinking
4. Deterioration in the care of the home and in marital and family relationships
5. Pattern of changing from one job to another, with short stays
6. Deterioration in job performance
7. Deterioration of health
8. Repeated drunk driving arrests
9. Nervousness and irritability when not drinking

Whatever the etiology of alcoholism, the behavior of the alcoholic is characteristic. Denial is commonly observed. ("Sure I drink, but I'm not an alcoholic." "The real problem is my wife." "I can quit any time I want.") The power of an alcohol addiction is herculean, and recividism (repeated or habitual relapse into alcohol abuse) rates are among the highest of all addictions. Be sure to examine the 42-question self-test in Section 9.12, and the answers and commentaries in Appendix II.

---

[6]For example, see Schuckit, M. A. "A Long-Term Study of Sons of Alcoholics." *Alcohol Health and Research World* 19(3), 172–175, 1995, and Vaillant, G.E. *The Natural History of Alcoholism Revisited*. Cambridge, MA: Harvard University Press, 1995.

9.10 Treatment

Ten to 12 million U.S. alcoholics cannot be ignored. The pain of their alcoholism must be treated.

In the order of application, the three general steps in the difficult treatment of alcoholism are as follows:

1. Managing acute episodes of intoxication to save the alcoholic's life and to overcome the immediate effects of excess alcohol.
2. Correcting the chronic health problems associated with alcoholism.
3. Changing the long-term behavior of the alcoholic person so that destructive drinking habits are not continued.

The management of acute alcohol intoxication and the concomitant alcohol withdrawal syndrome are commonly referred to as **detoxification**. Improved detoxification procedures, using tranquilizers and anticonvulsants, have reduced the trauma of withdrawal from alcoholism. Professionals are aware, however, that great care must be taken to prevent supportive drugs from becoming substitutes for alcohol. High-potency vitamins, fluids, and care are part of the detox treatment. Some cities have detox centers to which alcoholics are taken when they are found drunk in the street.

Professionals now view the alcoholic as a physically sick person, rather than a morally corrupt one. An initial attitude of acceptance, understanding, and encouragement is essential if treatment is to be effective.

One newer procedure by which a problem-drinking family member or employee is made to recognize his or her dependence on alcohol, to admit it, and to seek help is termed **intervention**. Pioneered by Vernon E. Johnson, an Episcopalian priest who is himself a recovering alcoholic, intervention is a procedure in which close family members, clergymen, and employers together confront the alcoholic with facts about his or her illness in a receivable, undeniable manner. Citing specific events, the people present state, "We love you too much to allow your disease to progress."

Therapeutic approaches to alcoholism include **behavior therapy**, which can include assertiveness training, contingency management, deep muscle relaxation, self-control training, cognitive restructuring, and aversion therapy. Other approaches are psychotherapy, group therapy, family treatment, and drug therapy (Antabuse).

Chemical treatment of alcoholism is now well established. Naltrexone, an opiate antagonist, has been approved by the FDA to reduce craving and prevent relapse to heavy drinking. Marketed as ReVia, it is the same drug as Trexan (Section 6.9), and works by interfering with the neurotransmitter system that produces pleasurable effects. Nalmefene is another opiate antagonist, not yet FDA approved.

Since 1948, **Antabuse (disulfiram)** has been used in the treatment of alcoholism; it remains a favorite of many therapists today. Antabuse interferes with the metabolism of alcohol at step 2 in the diagram on page 233 (i.e., it blocks the metabolism of acetaldehyde). Thus, acetaldehyde will accumulate in the body if a person who is taking Antabuse drinks. Acetaldehyde is quite toxic. Its accumulation in the body causes headaches, nausea, chest pain, blurred vision, and weakness. Severe reactions can lead to unconsciousness, convulsions, and death. So if an alcoholic needs aid and support in staying away from alcohol, the physician can prescribe Antabuse. When taken before an evening out, it can be a great deterrent to drinking. Be aware, however, that alcohol can unknowingly be ingested in some OTC products (Vicks 44M is 20 proof) and that disulfiram should not be used during pregnancy. In Russia, Antabuse is surgically implanted into the buttocks, from which its slow release maintains effective blood levels. It is important that the physician determines how much Antabuse to give each particular patient so that he or she is properly, but not overly, sensitized to alcohol. For this reason, self-medication with Antabuse must be avoided.

Acamprosate is an experimental drug said to help the alcoholic recover and stay sober, possibly in 16 percent of cases. It is used in many foreign countries but is not yet FDA approved. Research continues on antidepressants, old and new, as agents to help alcoholics who also suffer from depression. The old tricyclic antidepressants imipramine and desipramine and the newer SSRIs (Section 5.7), such as sertaline and fluoxetine, show promise.

An organization that has clearly been successful in changing the long-term behavior of problem drinkers is **Alcoholics Anonymous (AA)**. Founded in 1935, this organization uses the self-help, spiritual (but not religious) approach, with regular participation in meetings and mutual aid among members (group therapy). The reader is encouraged to learn more about the philosophy of AA and about the 12 steps of treatment that form the basis of its program. Women now account for 28% of the total AA membership. Nonspiritual self-help alternative groups to AA include Organization for Sobriety, Women for Sobriety, and Rational Recovery.

Alanon is an independent organization that is much like AA but is exclusively for the support of the family of the alcoholic, especially spouses. **Alateen** has been established for the support of children of alcoholics. Other community agencies and foundations stand ready to help alcoholics and their families. Local phone directories list family alcoholism counseling and educational services, foundations, and government agencies.

Successful treatment of the alcoholic may require the combined efforts of many persons: the physician, the psychiatric counselor, community agency personnel, the employer, the family, and friends. It is recognized that in spite of considerable team effort, results can be very discouraging. Perhaps more than any other drug-dependent person, the alcoholic has a great propensity to relapse, especially if treatment is not followed by the development of close interpersonal relationships with family or friends.

"Five years without a drink" is what a study says alcoholics require in order to avoid a relapse. Harvard Medical School's George Vaillant tracked the lives of 724 alcohol abusers for 50 years and concluded that a person needs to be symptom-free for 5 years to be out of danger. Among both poor and rich, relapse was rare after 5 years, but it was common after 2 years of sobriety.

9.11 Alcohol and Advertising

Former Surgeon General C. Everett Koop had this to say about alcohol advertising: "Certain advertising and marketing practices for alcoholic beverages clearly send the wrong messages about alcohol consumption to the wrong audiences." Koop identifies the following practices:

- Directing advertising to college students and other youths who are under the legal drinking age
- Using celebrities who have strong appeal to youth

- Sponsoring and promoting events such as rock music concerts and sports competitions where the audience is largely under the legal drinking age
- Using advertising that portrays activities like race car and speedboat driving that become dangerous if alcohol has been consumed

These practices tell young people that alcohol consumption leads to athletic, social, and sexual success. They send the message that drinking is a normal and glamorous activity without negative consequences. Apparently, our teenagers believe these messages, for two-thirds of high school seniors are regular drinkers.

Beer and whiskey manufacturers heavily advertise and promote alcoholic beverages in sports contests. Actually, they have ingratiated themselves to the point where the sporting world cannot get along financially without the associated advertising revenue.

Alcoholic beverages are heavily advertised to the general public in seductive, attractive, and sexually related tones to suggest that weekends belong to a certain beer, that this brand of whiskey is what the sophisticates drink, that athletic activities and drinking go hand in hand, and that you have truly arrived if you can afford to buy case lots of a certain brand of whiskey.

Add to this the influence that beer companies have on some college campuses. Beer companies sponsor keggers and other affairs designed to encourage a young person to begin the consumption of their product—a practice that could last a lifetime. One brewery markets a new portable keg, the "Party Ball," that comes in its own foil-lined, throw-away ice bucket. Another has a TV spot that boasts that its brand "is served at 87% of the parties your parents would never attend."

Recently, the yearly budget of the National Institute on Alcohol Abuse and Alcoholism was $91 million; for one major brewery, the ad budget was $643 million. Altogether, alcohol advertisers spend more than $1 billion yearly to promote their products.

The point here is that the advertising of alcoholic products has been unbelievably successful in convincing Americans that alcohol is a part of our lives, that it is normal and healthy to drink much and often, that happiness and alcohol are intimately related, that sports and drinking go together, and that good times and alcohol are synonymous. The consequences of this culture of alcohol abuse are deep and are probably irretrievably ingrained in our society. Vernon Johnson, the originator of the intervention technique in treating alcoholics, has as one goal in his life to convince Americans that drinking, abusing alcohol, and getting drunk are not acceptable social norms in our land. Clearly, some of the greatest obstacles in achieving this goal reside in the advertising of alcoholic beverages.

9.12 A Self-Test on Alcoholism

This book has introduced the subject of alcoholism and its treatment. The following 42-question self-test on understanding alcoholism is presented to give the reader both additional perspectives and a chance to test his or her knowledge of the subject. The test was prepared by Jon R. Weinberg and Robert M. Morse and was discussed in *Alcohol Health and Research World*, Vol. I, No. 1, Fall 1976, p. 16, published by NIAAA, Washington, D.C., 20402. It is used here with permission. Answers to the questions plus commentaries will be found in Appendix II.

Understanding Alcoholism

Directions: Circle T for *True* or F for *False*.

1. **T F** Most acceptable definitions of alcoholism include reference to the approximate quantity of alcohol consumed per unit of time.
2. **T F** As well as suffering adverse consequences from his drinking, the alcoholic usually drinks according to different patterns than does the normal drinker.
3. **T F** Many health professionals consider alcoholism our No. 1 health problem.
4. **T F** Alcoholism affects approximately 1% of our adult population.
5. **T F** Alcoholism should be considered a symptom of an underlying personality or mental disorder, as opposed to a disease per se.
6. **T F** Approximately one-fourth of all alcoholics are on skid row.
7. **T F** Research has failed to establish any specific genetic, environmental, social, or personality factors as the cause of alcoholism.
8. **T F** Becoming unconscious or passing out from excessive drinking is known as an alcoholic blackout period.
9. **T F** A drinker who never consumes anything stronger than beer is probably not an alcoholic.
10. **T F** A brief drinking history should be obtained from every new patient.
11. **T F** One may be a thoroughly reliable worker on the job and still be an alcoholic.
12. **T F** The ability to confine drinking to weekends suggests that a person is probably not an alcoholic.
13. **T F** An alcoholic must hit bottom before he can begin the recovery process.
14. **T F** Alcoholics are prone to abuse any other chemical substance given them that also produces a sedative effect.
15. **T F** It is usually wise to conceal liquor when entertaining a recovering alcoholic in your home and to advise relatives of alcoholic persons to do so.
16. **T F** The suicide rate among alcoholics is markedly higher than that for the general population.
17. **T F** A person's real character emerges when under the influence of alcohol.
18. **T F** Many people who say alcoholism is an "illness" often behave toward the alcoholic as though he had a moral weakness.
19. **T F** Tranquilizing drugs, such as Librium or Valium, are often valuable in maintaining the recovering alcoholic through his first year or so of sobriety.
20. **T F** An alcoholic with more than 10 years' sobriety may safely take an occasional social drink.
21. **T F** Coming from a family background of teetotalism is relative insurance that one will not develop alcoholism.
22. **T F** The first step in psychotherapy with an alcoholic person is determining the underlying reasons for drinking.
23. **T F** Alcoholics Anonymous has been more effective than psychiatric treatment in helping alcoholics to recover.
24. **T F** The spouse of the alcoholic is often a primary cause of the alcoholism.
25. **T F** An alcoholic who has "fallen off the wagon" (relapsed into drinking) more than four times can usually be regarded as untreatable.
26. **T F** The strong resistance among alcoholics to admitting their problem is in large part due to society's attitude toward alcoholism.
27. **T F** Involuntary treatment of an unmotivated alcoholic has been shown to be effective in many cases.
28. **T F** Professionals are often wise to advise the spouse of an alcoholic to consider precipitating a crisis, often by separation from the unmotivated alcoholic, after lesser measures have failed.

29. T F Most wives of alcoholics tend to become more emotionally disturbed when their husbands are maintaining sobriety.
30. T F Treatment (versus no treatment) improves the alcoholic's chances for recovery to a greater extent than the neurotic's.
31. T F A spouse or other informant should be interviewed if possible whenever a drinking problem is suspected.
32. T F Alanon is the companion group to Alcoholics Anonymous (AA) for female alcoholics.
33. T F Alcoholics often seek help for emotional or family problems without ever mentioning a drinking problem to the interviewer.
34. T F The drug disulfiram (Antabuse) has proven to be a dangerous and unsatisfactory treatment modality for alcoholic persons.
35. T F Education about alcoholism often helps the alcoholic reduce his resistance to accepting the facts about his condition.
36. T F The alcoholic who is maintaining sobriety has no greater number of serious emotional problems than the population in general.
37. T F The physician can best help the alcoholic by adding his plea to that of the family in urging the alcoholic to quit drinking.
38. T F Alcoholism can be seen as a type of drug addiction.
39. T F A significant emotional problem or disorder generally precedes the development of alcoholism.
40. T F Physicians frequently misdiagnose psychiatric problems in alcoholic persons.
41. T F Cross-dependence (or "cross-addiction") on other sedative or tranquilizing drugs in the alcoholic may begin iatrogenically. (See Glossary for definition.)
42. T F The incidence of alcoholism and drug dependence is lower among physicians than in the general population.

Web Sites You Can Browse on Related Topics

Alcoholism—General Information
http://www.niaaa.nih.gov
http://www.health.org
http://www.iatf.org (awareness)
Also: Search "alcoholism and genetics"

Fetal Alcohol Syndrome
http://www.health.org/about/sitemap.aspx
http://www.nofas.org/

Intervention
http://www.intervention.net/

Teenage Drinking
http://www.focusas.com/Alcohol.html

Alcohol Facts for American Indian
Search for "American Indian Alcohol"

Study Questions

1. True or false:
   a. Among women drinkers, less education and a lower income are associated with greater drinking.
   b. The United States per capita consumption of alcohol is highest for beer.
   c. Grain alcohol is the same as ethyl alcohol.
   d. Wood alcohol is a synonym for ethanol.

e. One hundred percent alcohol is termed *50 proof.*
f. Denatured alcohol is poisonous and should not be drunk.
g. Regular beer contains about 4.5% alcohol by volume.
h. A typical whiskey is 86 proof.
i. A person with a BAC of 0.02% would probably be very drunk.
j. A safe alcohol intake level for a pregnant woman is three drinks a day.
k. The less people weigh, the less alcohol they can tolerate.
l. According to Figure 9.1, Vietnamese youth show the lowest past-month alcohol use.
m. The most commonly used drug in the United States is marijuana.
n. Negative risk factors for young female drinkers include being unemployed, unmarried, looking for work, divorced, or having a heavily drinking spouse or partner.

2. The highest rates of current alcohol consumption are found in the 21–25-year-old male group. Does this seem logical to you? Why?
3. What adverse effects can occur if alcohol is taken concurrently with (**a**) minor tranquilizers, (**b**) anticonvulsants, (**c**) antihistamines?
4. Define (**a**) alcoholic blackout, (**b**) cirrhosis, (**c**) aphrodisiac, (**d**) delirium tremens, (**e**) acetaldehyde, (**f**) FAS, (**g**) denaturant, (**h**) NIAAA.
5. What did the Comprehensive Alcohol Abuse and Alcoholism Prevention, Treatment, and Rehabilitation Act of 1971 do and create?
6. Why is breath analysis for alcohol preferred over urinalysis?
7. Heroin and alcohol are both addicting drugs. Both cause intense problems for our society. There are about 500,000 heroin addicts in the United States and about 20 times that many problem drinkers. Yet alcohol is a legal, socially approved substance and heroin is not. What is the reason for the difference?
8. List five uses of ethanol in medicine or technology.
9. Why can some individuals perform difficult tasks and function well with a BAC of 0.20% or more, while others have passed out with a 0.10% level?
10. What are the major pharmacological effects of alcohol on the kidney, GI tract, cardiovascular system, and brain? Which of these organs is most sensitive to alcohol?
11. Assume that individual X has a drinking problem. He is seldom absent from work, but every day on the job he sneaks one drink in the morning and several during the afternoon. He performs his job reasonably well, but he has a higher than normal accident rate. Is it proper to term this person an alcoholic? Does the evidence indicate that he is addicted to alcohol? Would he suffer withdrawal effects if all of his alcohol intake were abruptly terminated?
12. Which of the following signs and symptoms would you consider evidence of alcohol intoxication? Which might conceivably be due to a disease state or to a head injury? (**a**) Odor of the breath, (**b**) flushed appearance, (**c**) lack of muscular coordination, (**d**) speech difficulties, (**e**) disorderly or unusual conduct, (**f**) mental disturbance, (**g**) visual disorders, (**h**) sleepiness, (**i**) muscular tremors, (**j**) dizziness, (**k**) nausea.
13. Would it be fair to recommend that any person found driving with *less* than 0.05% blood alcohol be automatically *excluded* from any prosecution for drunk driving?
14. Do you believe it is fair to punish an errant driver (e.g., by taking away his license) because he refuses to submit to a breath or blood alcohol test?
15. Should the principle of implied consent be extended to the unconscious person? In other words, is it proper to withdraw a blood alcohol sample from an accident victim without his consent? Would this act constitute invasion of privacy?
16. How does Antabuse work in the treatment of alcoholism?
17. Suppose it is your job to investigate FAS in order to determine if, indeed, alcohol is causing the malformed babies. You discover that many of the women drinkers you are

studying also smoke, use caffeine (a suspected teratogen), and are exposed to environmental hazards. Describe a procedure you would follow to prove (or disprove) that alcohol, and not one of the other factors mentioned, was causing the syndrome.

18. My friend Fred classifies himself as a normal drinker. Is there such a thing as a "normal" drinker?

19. Do you feel it was wise to raise the minimum legal drinking age in all states to 21 years? Provide reasons for your opinion.

20. What role does heredity play in determining susceptibility to alcoholism?

21. If we call whiskey "hard liquor," is it correct to refer to beer and wine as "soft liquor"?

22. A person voluntarily and successfully gives up alcohol for the 40 days of Lent. This proves he is not an alcoholic. True or false?

23. A recent (4–3) Supreme Court decision permits the Veterans Administration to regard alcoholic drinking as "willful misconduct" under the rules of eligibility for certain VA benefits. Using this, the VA refused benefits to two veterans who missed the application deadline. The veterans' lawyers argued that alcoholism is a disease and that it was alcoholism that kept the two veterans from applying within the deadline. Do you agree with the court decision? Do you agree that lung cancer victims should be refused hospital care because they willfully and knowingly smoked tobacco? Should hospitals have the right to refuse to treat auto accident victims who recklessly exceeded the speed limit?

24. Monozygotic (identical) twins are produced from the same fertilized egg and share identical DNA. Dizygotic (fraternal) twins come from separate eggs and do not share identical DNA. In studies on alcoholism, the concordance rate of alcoholism in monozygotic twins was found to be higher (54%) than in dizygotic twins (30%). What conclusions do these data support?

25. Identify three groups of people who should never drink alcoholic beverages under any circumstances. Explain your answer.

26. A New Jersey woman driver had an accident and sustained injuries. The police charged her only with failure to wear a seat belt, but the doctor who treated her found incidentally that her BAC was 0.26%. He reported this fact to the police, who then charged the woman with DUI. She was convicted and sentenced. Do you believe that the doctor violated doctor-patient privileges?

27. *Advanced study question*: If we accept the fact that sugar does not *cause* diabetes and that cholesterol by itself does not *cause* hardening of the arteries, can we also state that alcohol by itself does not *cause* alcoholism? (Do most of the people who drink become alcoholics?)

28. *Advanced study*: Contact social service agencies in your community and compile a list of services available to the problem drinker or to members of his or her family. Cite organizations, hospitals, care centers, and counseling programs—public, private, and military. Report their costs and effectiveness, if such data are available.

29. *Advanced study*: With the help of a medical dictionary or pharmacology textbook, define *Korsakoff's psychosis* and list its behavioral manifestations.

30. *Advanced study question*: Federal regulators want to keep a 1937 law that forbids beer makers printing the percent alcohol content on the label of their product. They claim a "strength war" would result if beer drinkers knew which beer was the strongest. A beer maker is challenging the law, claiming freedom of speech and the consumer's right to know. What is your opinion? (In Canada, beer labeling is required by law.)

31. Pharmacologically speaking, why has alcohol been termed the "sharpest double-edged sword in medicine"?

CHAPTER 10

Anxiolytics and Hypnotics: The Minor Tranquilizers Antipsychotics: The Major Tranquilizers

Key Words in This Chapter

- Anxiolytic
- Antipsychotic
- Benzodiazepine
- Major tranquilizer
- Neurotic
- Affective disorder
- Hypnotic
- Minor tranquilizer
- Manic-depressive
- Psychotic
- Lithium

Learning Objectives

- Distinguish between the medical uses of major and minor tranquilizers.
- Recall names of the more famous examples of each type.
- Distinguish between neuroses and psychoses.
- Recognize the potential for abuse of and dependence on minor tranquilizers.
- Explain how the major and minor tranquilizers act in the body.
- Understand the use of lithium compounds in the treatment of the manic-depressive.

10.1 Introduction

In this chapter we learn about some of the most successful, useful, and sometimes controversial drugs in America. As antipsychotics, the major tranquilizers have become cornerstone drugs. As anxiolytics and hypnotics (definitions later), the **benzodiazepine-type** minor tranquilizers are now preeminent, and when introduced in the 1960s, were the physicians' answer to the dangerous barbiturates, an instant therapeutic success, heavily prescribed by doctors, and eagerly sought by patients. The benzodiazepines are CNS depressants (see Glossary) prescribed for treating anxiety, muscle spasms, and sleep disorders. They are exemplified by Valium (subsequent pages will give a full listing of generic and trade names). The benzodiazepine tranquilizers were so safe that it appeared virtually impossible to commit suicide with them. Today, in any given year, about 11% of all Americans will take a minor tranquilizer. About 80 million prescriptions for benzodiazepines are written each year in the United States.

But as we have seen before, there can be a later unfortunate phase in the pattern of new drug introduction. The safety of the minor tranquilizers, combined with their usefulness and acceptance, disarmed physicians—who began to prescribe them casually and often without supervision. In too many patients, occasional use became months-long or year-long use, and the destructive potential of these drugs became evident. Valium, in particular, seemed to be in everyone's medicine chest and pocketbook, as well as on the street. Still, the manufacturers fought to keep their tranquilizers from being scheduled under the 1970 Controlled Substances Act.

More than 15 years of use passed before society recognized that although brief use of minor tranquilizers can be beneficial, long-term use or heavy dosage can and does lead to physical addiction with a withdrawal syndrome that is worse than that from heroin—indeed, one that is life-threatening. It took that long to realize that there is little reason to use a minor tranquilizer for longer than a month, and that to do so can increase the risk of dependence. By the 1980s, self-help groups such as Pill Addicts Anonymous had been organized to help people deal with the very drugs they were relying on to cope with their original problems. It is now known that benzodiazepine addicts need as much support as alcoholics, especially since secondary symptoms of withdrawal from benzodiazepines commonly last for 18 months to 2 years.

Benzodiazepine tranquilizers continue to be popular. Recently, 5 of the top 62 most-prescribed generic drugs in America were benzodiazepines. It is clear that in the past 25 years, tranquilizers have become intrinsic to our way of life.

Studies have shown that unexpectedly high rates of minor tranquilizer use occur among housewives, retired persons, and others who are unemployed. For example, although females who are unemployed or are not in the labor force make up only 26% of the general population, they account for 46% of minor tranquilizer use. On the other hand, skilled and semiskilled workers have one of the lowest use rates.

What have we learned from the record of the minor tranquilizer? Are drugs the answer to solving emotional problems? Is relief "just a pill away"? Perhaps the story of these drugs has taught us that "safe and effective" is not necessarily good; that drugs can change cultural values; that there can be a potential for danger that is revealed only after years of drug use; that some tension and stress are normal and we should think twice about relying on a chemical crutch; that psychoactive drugs must never be prescribed casually and without supervision; and that the character needed to handle one's problems cannot be acquired by the use of any medication.

10.2 Getting the Names Straight

In this chapter, we first examine the minor tranquilizers, the type most people have heard about, along with their medical uses. Then we study the major tranquilizers and their more serious uses. But first, we must tackle the problem of nomenclature.

Some experts in the field don't like the term *tranquilizer*, preferring the semantically correct **antianxiety agents** or **ataractics** (also called *mood stabilizers* or *anxiolytics*). The major tranquilizers are often termed **neuroleptics** or **antipsychotics**. Nevertheless, the terms *minor tranquilizer* and *major tranquilizer* have become so well established and are so often used that they appear to be here to stay.

Minor tranquilizers are useful in calming anxious patients, neurotics fearful of life's situations, or those so tense that they cannot function or sleep normally. The minor tranquilizers are of little or no benefit in the treatment of psychoses. Table 10.1 lists some commonly used minor tranquilizers; for the chemical structure of Valium, a typical benzodiazepine, see Figure 8a in Appendix I.

Major tranquilizers are used in the seriously ill mental patient—the so-called psychotic. These antipsychotic drugs can calm schizophrenic patients to the extent that they become manageable or even well enough to continue their medication at home. They reduce hallucinations and periods of delusions, and if a patient is very withdrawn, they may stimulate him or her to lead a more normally active life. The major tranquilizers listed in Table 10.2 represent different chemical types: phenothiazines, thioxanthenes, butyrophenones, and others. For flunitrazapam, see Section 12.6.

Table 10.1 Some Important Minor Tranquilizers

Generic Name	*Trade Name*	*Half-life (hr)*	*Typical Oral Adult Dose (mg)*
	Benzodiazepine type		
Alprazolam	Xanax	4–20	0.25–4.0 daily[a]
Chlordiazepoxide	Librium	24–72	5–25 three to four times daily
Clonazepam	Klonopin	24–72	1.5–20 daily[a]
Clorazepate	Tranxene	24–72	15–60 daily[a]
Diazepam	Valium	24–72	2–10 two to four times daily
Estazolam	ProSom	10–24	1–2
Flurazepam	Dalmane	24–72	15–30 at bedtime
Lorazepam	Ativan	4–20	2–6 daily[a]
Midazolam	Versed	1–12	Administered parenterally
Nitrazepam	Mogadon	25	5 daily
Oxazepam	Serax	4–20	10–15 three to four times daily
Prazepam	Centrax	24–72	20–60 daily[a]
Temazepam	Restoril	4–20	15–30 before retiring
Triazolam	Halcion	4–20	0.25–0.50 before retiring
	Propanediol carbamate type		
Meprobamate	Equanil, Miltown	10	400 three times daily
Carisoprodol	Soma	ca. 10	350 four times daily

[a]In divided doses.

Table 10.2 Some Major Tranquilizers and Their Doses

Generic Name	*Trade Name*	*Usual Daily Antipsychotic Dose (mg)*
Chlorpromazine[a]	Thorazine	75–200
Clozaril	Clozapine	25 mg, monitored
Droperidol	Inapsine, Innovar	Depends on use
Fluphenazine[a]	Prolixin, Permitil	2–10
Haloperidol	Haldol	2–6
Loxapine	Loxitane	60–100
Mesoridazine[a]	Serentil	100–400
Molindone	Moban	50–75
Olanzapine	Zyprexa	5–10 daily
Perphenazine[a]	Trilafon, Etrafon	8–32
Prochlorperazine[a]	Compazine	75–100
Risperidone	Risperdal	3 max. twice daily
Thioridazine[a]	Mellaril	100–600
Thiothixene	Navane	6–30
Trifluoperazine[a]	Stelazine	4–15
Ziprasidone	Geodon	20 twice daily

[a]A drug of the phenothiazine chemical type.

Earlier discussions used the terms **psychotic** and **neurotic**; these terms need some clarification.

Briefly, a psychotic is one who has lost touch with reality to the extent that he or she cannot meet the ordinary demands of life. Possibly suffering from hallucinations, delusions, and paranoia, this person is seriously ill and usually requires hospitalization. The two most common psychoses are schizophrenia and manic-depression. In the latter, there are wild mood swings from profound depression to unrealistic euphoria.

A neurotic is a person who technically is mentally ill but who has not lost contact with reality. In the case of neuroses, the personality is not essentially changed, but the person has thoughts, wishes, and emotions that conflict with one another. Neuroses can be so severe that they border on psychoses, or they can be of the mild type that most people experience at times. In an **anxiety neurosis**, the patient complains of worry and of feelings of fear and panic ("I'll never get that job; I'm not good enough," or "I can't drive; I know I'll have an accident.") A **phobia** designates a neurosis in which the person fears and avoids certain situations (elevator rides, tunnels, plane rides, etc.).

Alternatively, the person may complain of having an imaginary physical ailment (e.g., heart disease, cancer). In an **obsessional neurosis**, the person compulsively and persistently carries out some thought or act, such as washing the hands over and over again in fear of germs.

To the neurotic, these feelings are real, and when they are accompanied by palpitations of the heart, choking sensations, dizziness, or breathlessness, treatment is essential.

Table 10.3 lists various mental illnesses and chances for recovery.

Table 10.3 Hospitalization for Mental Illness by Age, Sex, and Chance of Recovery

Condition	*Usual Age at First Admission*	*First Admissions (%)*			*Hospitalized Mental Patients (%)*	*Chances of Recovery*
		Men	*Women*	*Both*		
Schizophrenia	16–35	18.6	26.8	22.6	45.6	Poor after age 50; long hospitalization
Manic-depressive psychosis	35–40	3.1	5.3	4.2	7.6	Poor; long hospitalization
Psychosis with mental deficiency	15–44	2.7	1.8	2.3	6.0	Poor; long hospitalization
Alcoholic psychosis	25–54	21.8	5.2	13.7	3.0	Fair; short hospitalization
Involutional psychosis	48–58	3.0	8.3	5.6	3.0	Poor; die soon; short hospitalization
Senile psychosis	60+	21.1	20.5	20.8	12.2	Poor; die soon; short hospitalization
Personality disorders (nonalcoholic)	15–34	7.2	3.9	5.6	4.0	Fair; short hospitalization
Psychoneurotic	25–44	6.3	12.4	9.3	6.0	Fair; short hospitalization
Other disorders, each of low incidence	All ages	16.2	15.8	15.9	12.6	Generally poor; long hospitalization

Source: John J. Hanlon and George E. Pickett, *Public Health Administration and Practice*, 8th edition, Times Mirror/Mosby College Publishing. St. Louis, 1984. Used by permission.

10.3 The Psychopharmacology of the Minor Tranquilizers

Today, when we refer to the minor tranquilizers, we mean primarily the **benzodiazepines**, a distinct chemical category. (A second chemical type is the propanediol carbamates.) The benzodiazepines constitute the main group of drugs for the control of anxiety neuroses. They are also prescribed for tension, muscle spasm, skin rashes, alcohol withdrawal, alcoholism, agitation, hallucinations, and convulsions—and frequently for insomnia. Benzodiazepine receptors are located predominantly—if not exclusively—in the brain, which explains why benzodiazepines have no action on other tissues such as the heart and skeletal muscle. The brain receptors are in the limbic region (the part that deals with emotional regulation and control) and are quite specific; no other drugs bind to these receptors. Benzodiazepines probably act as antianxiety, anti-insomnia, and anticonvulsant agents by virtue of their ability to facilitate GABA function (see Section 5.7). They increase the affinity of GABA for its receptor by displacing an endogenous inhibitor of GABA binding. This process results in decreased excitation of many nerve systems, reduced anxiety, diminished alertness, and an increase in the threshold of convulsion. A natural anxiolytic substance has yet to be identified.

Benzodiazepine tranquilizers are generally soluble in human fat tissue and typically have a long **half-life** in the human body. Table 10.1 shows that most minor tranquilizer half-lives are 1–3 days. This means that if one takes a dose of Dalmane on Sunday at 11 P.M., only half of the dose has been excreted by Monday night at the *earliest*, and one has studied or worked all day Monday under the CNS-depressing influence of this drug.

Minor tranquilizers such as Valium are intended for the treatment of anxiety—that unpleasant emotional state characterized by unrest, panic, or fear of the future or the unknown. True anxiety is a psychiatric disorder; it can be incapacitating. Here, the minor tranquilizers are the drugs of choice. Says UCLA's Dr. Sidney Cohen, "Until better drugs come along, this is the best group for severe or incapacitating anxiety, panic, and tension states that are severe enough to be treated by medicines. They're really valuable." The minor tranquilizers are *not* intended for the tension or concern associated with the stress of everyday life. Such stress is normal and can be a useful motivator to achieve worthy goals. Being anxious about getting a good job, for example, has stimulated many a student to high achievement in college. The big question then becomes: When should I tough it out, and for how long, or when do I reach for the bottle of pills?

The benzodiazepines are much safer to use than the barbiturates, especially for patients with self-destructive inclinations. From 200 to 1,000 times the therapeutic dose has been taken deliberately, with subsequent recovery. Valium does not significantly depress respiration. This result does not apply, however, when alcohol is also in the body, because the two are synergistic—dangerously potentiating each other. Valium is not especially effective in inducing the production of liver microsomal enzymes; this reduces the chance for drug interactions.

Benzodiazepines have long been used as sleeping pills (hypnotics). They act rapidly and do not depress **REM sleep** when taken in the usual doses. Some benzodiazepines, however, have been developed especially as sleeping pills. The most important example is triazolam (Halcion), a benzodiazepine with a short half-life ($t_{\frac{1}{2}}$ = 2.3–3.6 hours) and a 0.125-mg dose that after use as a hypnotic leaves little, if any, daytime sedation. Triazolam is very popular, with over 9 million prescriptions written for it annually in the United States and sales of $265 million. However, serious questions are being raised about the safety of triazolam. Several thousand adverse reaction reports received by doctors who have prescribed this hypnotic reveal adverse effects that include amnesia, bizarre behavior, suicide, threatened suicide, agitation, paranoia, confusion, nightmares, and severe depression. Further, benzodiazepines with short half-lives sometimes produce rebound insomnia and anxiety that leave the user worse off than before treatment and encourage continued use to the point of dependence. As a sleep aid, Halcion becomes largely ineffective after 2 weeks of continued use. Halcion is in pregnancy category X (see Section 3.5), its use in pregnancy is clearly contraindicated. For all of these reasons, use of triazolam for insomnia should be made with great care and supervision if at all. Great Britain has banned the use of Halcion, and Upjohn (Halcion's manufacturer) has decided to market the drug in packages of 10 or 30 tablets and in 0.125 mg doses.

Case History *One Patient, Four Drugs*

A young woman author had difficulty sleeping, apparently because of the considerable stress of family affairs. Her physician therapist prescribed Halcion (triazolam, 0.5 mg), assuring her that this benzodiazepine had a "high margin of safety, no hangovers, and fewer problems with dependency." After a week of taking Halcion, the woman was experiencing unusual physical and emotional reactions. Her heart pounded, her mouth was dry, and she had become extremely fearful and anxious about any event or news story that was even remotely disturbing. Her therapist decided to treat the anxiety by prescribing a second benzodiazepine, Xanax (alprazolam, 0.75 mg daily). The Halcion was continued. Weeks later, the patient's condition was worse. Sleep was impossible without the Halcion; anxiety was unbearable without the Xanax. For the first time in her life, the woman considered suicide. Her therapist decided that a third drug was indicated and prescribed the major tranquilizer Mellaril (thioridazine, 10 mg) to be taken once or twice a day. The patient continued to deteriorate. She had lost 20 lb and was psychologically depressed. Her therapist prescribed a fourth drug, the antidepressant Elavil (amitriptyline, 50 mg). The patient was now taking concurrently and daily four psychoactive drugs—Halcion, Xanax, Mellaril, and Elavil—but was not getting better. Her recovery from polydrug dependence was sparked by the realization that her baby (whom she was nursing) was being drugged by way of breast milk. Without consulting her therapist, she stopped all four drugs, experienced a withdrawal that lasted 2 weeks, but came out of it feeling that she had "been given back her life."

Dalmane and **Valium** are two additional benzodiazepines that are widely prescribed for insomnia. Both have longer half-lives than Halcion and can produce daytime sedation, drowsiness, and impaired motor coordination. Approximately 5 mg of Valium causes driving impairment equivalent to a BAC of 0.07%. In one test, 12 hours after taking a 15-mg dose of Dalmane, a driver's performance was significantly impaired.

Ambien (zolpidem) is a nonbenzodiazepine hypnotic prescribed in 5–10 mg doses for short-term treatment of insomnia.

Tolerance to benzodiazepines occurs. If high doses are given over a long period, **physical dependence** can also occur. According to testimony given to the U.S. Congress in 1979, Valium withdrawal symptoms occur after taking large doses for a few days, moderate doses for a few months, and minimal doses for a few years. Withdrawal from benzodiazepines is similar to that from barbiturates; it is characterized by anxiety, restlessness, tremors, nausea, cramps, diarrhea, muscle spasms, tics (spasmodic twitches), moodiness, confusion, disorganized thinking, racing thoughts, bizarre dreams, hallucinations, paranoia, violence, depression, and possibly grand mal seizures. The period of acute withdrawal from benzodiazepines is 21–28 days (unlike the 72-hour period of withdrawal from alcohol). Many people have died of sedation drug abstinence; it can be a life-threatening situation. As mentioned earlier, secondary withdrawal from benzodiazepines commonly lasts for 1.5–2.0 years. If someone has been taking significant doses of a benzodiazepine type of minor tranquilizer for a long time and wishes to stop the drug, it would be wise to reduce the dose gradually, rather than cold turkey, in order to prevent possible withdrawal symptoms. Reduction of dosage should be 2 mg or less per week, typically, and should be done under a physician's supervision.

Flumazenil is a benzodiazepine antagonist now marketed for use in the United States. It can be used in benzodiazepine overdosage and to reverse benzodiazepine-induced sedation and respiratory depression.

Cross tolerance develops between benzodiazepines, meprobamates, barbiturates, and alcohol. Valium metabolism in the liver gives rise to three or four products that are as active as Valium itself and have half-lives of 1–8 days. This finding explains the prolonged CNS depression that can occur after taking a dose of Valium.

Benzodiazepines are often used just before surgical anesthesia for their calming effect and the amnesia that accompanies parenteral use. They are used to treat violent reactions to PCP. They are anticonvulsants useful in the treatment of petit mal (a type of epileptic seizure).

Benzodiazepines such as oxazepam (Serax) have been used to help the alcoholic get through the withdrawal from alcohol. Serax is effective, but experts in the field state that too many patients become addicted to it, taking it for months or a year and developing great tolerance to and physical dependence on it. Ironically, subsequent withdrawal from Serax is much more difficult and life-threatening than the original withdrawal from alcohol.

CAUTION: *Adverse effects of minor tranquilizers can be serious. The benzodiazepine and meprobamate types of minor tranquilizers are to be avoided during pregnancy because there is evidence that they can cause birth defects. This evidence is* *based on findings with only a few of the minor tranquilizers but appears to be applicable to all. If a patient is maintained on minor tranquilizers for longer than a month, blood and liver tests should be done to ensure that no damage to these organs is occurring. Other adverse effects to be expected in some users of minor tranquilizers are drowsiness (ask your doctor if you can cut down on the dose to reduce this), dizziness, stomach upset, blurred vision, headache, skin rash, double vision, slurred speech, mental confusion, and changes in sex drive. Of course, not every patient will experience all these adverse effects. In a very few users, the minor tranquilizers produce totally unexpected effects, including anger, insomnia, depression, hyperexcitation, and nightmares. This anomaly is just another indication of biological variation; it shows how differently people can react to the same drug.*

Drug Interactions. Oversedation can occur if a minor tranquilizer is combined with another brain depressant such as an antihistamine, a narcotic analgesic, or especially with alcohol. If a benzodiazepine type of minor tranquilizer is combined with a monoamine oxidase inhibitor drug such as Parnate or Nardil, oversedation may occur. There is evidence suggesting that combining alcohol with triazolam (Halcion) for sleep on jet-lag flights can result in amnesia. Actually, the use of benzodiazepines alone has been known to cause amnesia. Upjohn Company, the makers of Halcion, advises against combining the drug with alcohol.

Meprobamate (Miltown, Equanil) and other propanediol carbamate–type tranquilizers act much like the barbiturates, inducing sleep in large doses, suppressing REM sleep, and inducing the production of liver enzymes that hasten their own metabolism. Meprobamate can also induce tolerance, physical dependence, and psychological dependence. At doses of 3200 mg a day for a month, barbiturate-like withdrawal symptoms occur (convulsions, coma, and psychotic behavior). Meprobamates, however, have a much weaker effect in depressing the respiratory center than barbiturates, making it much more difficult to commit suicide using them.

10.4 NIDA Data on Minor Tranquilizers

Every year, the *National Institute on Drug Abuse* (NIDA) and the United States Department of *Health and Human Services* (HHS) publishes a *Statistical Series* of drug abuse information collected through the *Drug Abuse Warning Network* (DAWN). Note: DAWN is a unique, large-scale, ongoing data set that reports data on drug-related ER episodes. As with any sample, the results are estimates of the values that would be obtained if complete data were actually collected from all of the hospitals from which the sample was drawn. See Appendix III for an extensive DAWN data summary.

For 2001, DAWN reported the top emergency-department drug mentions among tranquilizers were alprazolam (25,000 mentions), clonazepam (19,100), lorazepam (11,900), and diazepam (11,400). In 2000, diazepam ranked among the top ten drugs mentioned in the 26 cities in DAWN's mortality data. Also in 2000, initiation of tranquilizer use had increased to 973,000 (in the coterminous U.S.). Regarding abuse and nonmedical use of tranquilizers, DAWN reported that males outreported females 1.6 to 1.0, whites outreported blacks 10 to 1, and the highest number of mentions occurred in the 18–25 age group (748,000).

Problem 10.1 *On the basis of these DAWN data alone, describe the typical U.S. abuser of tranquilizers.*

10.5 A Revolution in Medicine

The major tranquilizers are the antipsychotics, drugs used to treat the schizophrenic, the catatonic, the manic-depressive, the paranoiac, the acute alcoholic, or the agitated senile patient. Here, the word *tranquilizer* isn't really accurate because while the agitated schizophrenic may be calmed, the withdrawn psychotic may be brought *out of* his depression. Thus, *mood-stabilizing drugs* or *neuroleptics* would be more suitable terms.

Antipsychotic drugs work on the CNS to alleviate the panic, hostility, fear, and agitation associated with acute and chronic psychoses. They can relieve the hallucinations and delusions that plague the psychotic; they can normalize and organize the person's thinking and behavior patterns.

If ever there was a landmark in the history of chemotherapy, it was the introduction of the phenothiazines and related antipsychotics in the 1950s. For hundreds of years, the schizophrenic or other psychotic had been "treated" by being placed in isolation or knocked out with sedatives. Dr. Sidney Cohen has vividly described the pre-1950 situation:

> The major problem was violence. It was the violence of beat-up patients, beat-up staff, rooms torn apart, windows broken, toilets stuffed, clothes torn off, excrement thrown around and the all-day dehumanization of everyone. It was a time when knives and forks could not be provided at meals—curtains, wall pictures, and anything but nailed-down furniture were not possible to use. There were the "pack" rooms with their row on row of slabs and tubs for what was euphemistically called "hydrotherapy." There were the seclusion rooms furnished with nothing but a mattress and an out-of-reach light bulb where a creature, nude or in rags, paced like a caged animal, shouting back at his hallucinations. There were the insulin suite, the lobotomy ward—and always the interminable locking and unlocking of every door in the place. In such a chaotic situation, the "good" patients were the mute, posturing catatonics.[1]

Chlorpromazine (see Table 10.2), a drug developed in France, changed the entire mental illness therapy picture. Doctors discovered that psychotic patients treated with chlorpromazine became quiet, tractable, and quite disinterested in events going on around them while retaining consciousness. Severely withdrawn schizophrenics became active, and the wildly hallucinogenic ones calmed down. So intense was the impact of this drug on the field of psychiatry that the events have been called a revolution in medicine.

Inpatient populations of mental hospitals began to decline, ultimately to the point where some large hospitals closed for lack of patients. Inpatients became outpatients, sent home with a supply of their antipsychotic medicine. It was soon discovered that many of these discharged patients failed to continue taking their drug, relapsed, and had to be readmitted; nonetheless, hospitalizations were greatly reduced. As expected, there were some burned-out, chronic schizophrenics who did not benefit from chemotherapy—and even a few who got worse.

---

[1]Sidney Cohen, "The Major Tranquilizers," *Drug Abuse and Alcoholism Newsletter,* Vista Hill Foundation, Vol. 4, No. 10, 1975, p. 1.

Intense research by pharmaceutical companies wanting a piece of the therapeutic action yielded many more phenothiazine antipsychotic drugs (see Table 10.2). In addition, the chemically unrelated drug reserpine (Serpasil) was discovered and found to have a powerful antipsychotic effect. Use of reserpine, however, has declined because it has too many adverse side effects. The history of reserpine is discussed in Chapter 2.

For the chemical structure of Thorazine (chlorpromazine), see Figure 8c in Appendix I.

10.6 How Do the Major Tranquilizers Work?

Because schizophrenia is a disorder of the higher centers of the brain (the cortex), it is logical to assume that the antipsychotic agents work by acting on such higher centers. Researchers pursuing this line of reasoning have accumulated considerable evidence that the major tranquilizers act at cortical synapses where dopamine is the transmitter agent. The phenothiazines block dopamine receptors, stifling the action of dopamine or any other drug that could act on these receptors. Clozapine (Clozaril) is another known dopamine antagonist that is beneficial to the schizophrenic. In Chapter 5 we referred to a dopamine hypothesis of mental illness in which an abnormal amount of this neurotransmitter is theorized to be the cause of schizophrenia. The theory fits in beautifully with the evidence that psychotic behavior is *relieved* by the dopamine-blocking antipsychotic drugs. It is certain that dopamine will assume ever-increasing importance in normal and abnormal human physiology as the research evidence continues to accumulate.

The main action of the major tranquilizers is to reduce the schizophrenic's symptoms. But there are numerous other effects on the body, and these effects are summarized in Table 10.4.

Although the major tranquilizers generally are considered safe drugs, they occasionally produce some very irritating adverse side effects. As shown in Table 10.4, their use can result in a condition much like that of Parkinson disease, in which the person suffers from loss of muscle tone in the face, drooling saliva, a short, shuffling gait, muscular rigidity, slowness of movement, involuntary tremor, and a characteristic pill-rolling movement of the fingers and thumbs. This syndrome has been given the name **tardive dyskinesia**. Strong evidence implicates a dopamine deficiency as the cause of Parkinson disease. Because the phenothiazines block dopamine receptors (a situation that is identical to having a deficiency of endogenous dopamine), it is only natural to expect Parkinson-like side effects in phenothiazine use.

Other side effects of antipsychotic drugs are possible restlessness, postural hypotension (severe dizziness when suddenly standing up from a reclining position, due to a fall in blood pressure), dry mouth, urinary retention, and, rarely, jaundice and damage to blood cells. Long-term treatment with high doses of chlorpromazine can sensitize the patient's skin to sunlight. This photosensitization can occur with other drugs, too, and is manifested as extreme reddening and inflammation of the skin after brief exposure to the sun's rays. Treatment consists of staying out of the sun or discontinuing the use of the drug. Table 10.5 lists some additional photosensitizing drugs.

If one major tranquilizer produces disturbing side effects, the prescribing physician has a dozen or more alternative drugs from which to choose. He or she even has

Table 10.4 Drug Actions of the Major Tranquilizers

Center Where Drugs Act	*Pharmacological Effects Observed*	*Mechanism of Action*
Cortex and higher brain centers	Calmed psychotic behavior. Sedation of normal individuals. Reduced response to stimuli. Lowered spontaneous motor activity. Reduced attentiveness.	Probably by blocking dopamine receptors
Vomiting reflex center	Powerful antiemetic effect in the control of nausea and vomiting.	Dopaminergic blockage
Lower brain centers	Symptoms of Parkinsonism; trembling of the limbs, difficulty in speaking, slow movements.	Dopaminergic blockage
Autonomic nervous system	Reduced activity of the gut, possible constipation, lowered appetite, reduced sex drive.	Complex

the option of switching to a drug such as Haldol, which is in an entirely different chemical class.

It should be noted that the antipsychotic effect of the phenothiazines is slow to develop. Often, several weeks of medication are necessary before the psychotic begins to respond to therapy. Similarly, after complete termination of use, some amounts of the drug will remain in the body for a month or more.

Problem 10.2 *If phenothiazines remain in the body for 4–6 weeks, they must be stored somewhere or bound to something. Suggest where this location might be, and predict the water versus fat solubility of these substances.*

Table 10.5 A Few Common Photosensitizers

These are just a few of the more commonly used drugs that can cause photosensitivity reactions in some people:

Brand Name	*Generic Name*	*Therapeutic Class*
Motrin	ibuprofen	NSAID, antiarthritic
Crystodigin	digitoxin	antiarrhythmic
Sinequan	doxepin	antidepressant
Cordarone	amiodarone	antiarrhythmic
Bactrim	trimethoprim	antibiotic
Diabinese	chlorpromamide	antidiabetic (oral)
Feldene	piroxicam	NSAID, antiarthritic
Vibramycin	doxycycline	antibiotic
Phenergan	promethazine	antihistamine

Source: FDA Consumer, May 1996, page 29.

Fortunately, tolerance, physical dependence, addiction, and withdrawal syndrome are not important problems encountered in the use of major tranquilizers. No one becomes dependent on the major tranquilizers, and they are not street drugs of abuse.

> ***CAUTION:*** *Alcohol must never be combined with any of the major tranquilizers, because alcohol can increase their sedative action and accentuate their depressant effects on brain function and blood pressure. Conversely, these neuroleptics can increase the intoxicating effects of alcohol.*

10.7 Lithium

Chemistry students carry around with them a periodic chart of the elements that provides information on all of the elements known to exist in the universe. The first two, the simplest elements on the chart, are hydrogen and helium; the third is lithium, a simple atom with only three electrons and three protons. Lithium forms many compounds, and one of these—lithium carbonate USP—has found use in psychotherapy. Its chemical formula is Li_2CO_3.

It is natural to have mood swings most of the time; they are usually reasonable, and we can handle them. **Manic-depressives**, however, suffer from a psychosis in which these mood swings become so extreme that they cannot function normally in everyday life. The typical manic stage is characterized by euphoria, decreased sleep, excessive drinking, rapid talk, flight of ideas, irritability, rambling on about grandiose plans, impulsive spending, and an extraordinary sense of benevolence and religious enlightenment. This mania can last for 3–6 months. Manics may quit their jobs and give away all of their belongings because they feel invincible or because they feel God is protecting them. In the extreme manic phase of the condition, the person engages in a frenzy of activity, singing, shouting, making obscene proposals, tearing at clothing, or disarranging his or her room. The person is too excited to eat or sleep.

In the depression stage, which lasts for 6–9 months, there are feelings of rejection, lack of energy, and an increased risk of suicide. If manic-depressives are untreated, and if they do not kill themselves, they will emerge from one stage and slip into another.

The incidence of manic-depression[2] is greatest among the higher social and professional groups and is twice as great in women as in men. Heredity appears to play a role, as does environment. About 1.15 million Americans are manic-depressives, but only about a third are diagnosed and take lithium.

It is in the treatment of the manic phase that lithium carbonate has found application.[3] With lithium, cures are not achieved, but mood swings become stabilized, sleep becomes possible, and the patient can take time to eat. The drug must be taken prophylactically to prevent the occurrence of manic episodes; it does not act rapidly enough to terminate an episode once begun, even if one could get the patient to

[2]The newer term is bipolar affective disorder.

[3]The discovery of lithium as an effective antimania drug was serendipitous. John Cade, an Australian psychiatrist, believed that uric acid might be the key substance involved in mania. He chose to administer the uric acid as the lithium salt, along with some lithium carbonate. His patients improved dramatically. Only later was it discovered that it was the lithium—not the uric acid—that was responsible for the improvement.

take it. Lithium is effective in about 70% of the manic-depressives who use it. It can have troubling side effects: tremors, nausea, increased urination, and, more seriously, hair loss, memory loss, and kidney damage. However, many patients have been maintained on lithium carbonate for years without serious problems. Depakote (divalproex sodium), a valproic acid compound, may be helpful in the 30–40% of manic depressives who do not respond to lithium.

The mechanism of action of lithium in manic-depression is unknown. It has been suggested that lithium works by modulating serum calcium levels (an increase in serum calcium is associated with the manic stage, and drugs such as calcium lactate and vitamin D, which raise calcium levels, increase the severity and frequency of manic attacks). The makers of Eskalith (see the PDR) state that lithium alters sodium transport in nerve cells and causes a shift toward intraneuronal metabolism of catecholamines. Less norepinephrine and dopamine in the brain correlates with reduced mania.

The lithium ion is chemically and physiologically similar to the sodium ion and can act in place of the latter in some biological systems. However, patients on a salt-free diet should not be given lithium, nor should lithium chloride ever be used as a salt substitute. Lithium is secreted in human milk. Women should not nurse while taking lithium, nor should lithium be used during the first trimester of pregnancy.

Lithium has been found helpful in the treatment of cluster headache, a condition in which severe pain occurs only on one side of the head. Cluster headache is often accompanied by lacrimation, runny nose, and eye muscle problems; in the past, it was treated with drugs used to relieve migraine. Scientists at the Lithium Information Center at the University of Wisconsin Center for Health Sciences maintain a computerized library of more than 8,300 articles and studies on the use of lithium.

Web Sites You Can Browse on Related Topics

Benzodiazepines
http://www.adhl.org/benzodi.html
http://www.biopsychiatry.com/benzos.html
http://www.breggin.com/bzbkexcerpt.html

Bipolar Disorder
http://www.psycom.net/depression.central.bipolar.html
Also: Search for "bipolar disorder"

Schizophrenia
http://familydoctor.org/handouts/266.html
Also: Search "schizophrenia"

Study Questions

1. Check the one drug that is best suited to treat each condition.

Condition	Minor Tranquilizer	Major Tranquilizer	Lithium
a. Severe anxiety	_____	_____	_____
b. Schizophrenia	_____	_____	_____

Condition	Minor Tranquilizer	Major Tranquilizer	Lithium
c. Convulsions	_____	_____	_____
d. Psychoses in general	_____	_____	_____
e. Neuroses in general	_____	_____	_____
f. Manic-depression	_____	_____	_____
g. Panic claustrophobia	_____	_____	_____
h. Catatonia	_____	_____	_____
i. Senile psychosis	_____	_____	_____
j. Hallucinations	_____	_____	_____

2. True or false:

a. The minor tranquilizers are also known as the *anxiolytics*.
b. The major tranquilizers are also known as the *antipsychotics*.
c. Compazine is an example of a minor tranquilizer.
d. Librium is an example of a major tranquilizer.
e. Neurotics are usually in contact with reality, whereas psychotics are delusional.
f. The minor tranquilizers are CNS depressant drugs.
g. The minor tranquilizers are intended to treat everyday emotional stresses.
h. The typical dose of Valium for anxiety is about 5 mg.
i. Doses of major tranquilizers are generally larger than doses of minor tranquilizers.
j. Alcohol does not potentiate the CNS depressant effects of the minor tranquilizers.
k. Dalmane is a benzodiazepine used especially as a hypnotic.
l. Long-term use of minor tranquilizers is generally wise and is to be encouraged.
m. Benzodiazepines act as CNS depressants by *blocking* the action of GABA.
n. Combining alcohol with a major tranquilizer such as Trilafon is generally safe.

3. List three types of neuroses, with one example of each.

4. Miltown and Equanil are meprobamate-type minor tranquilizers. Cite three pharmacological differences between the meprobamate type and the Valium or Librium (benzodiazepine) type of minor tranquilizer. (*Hint*: Check dosage, too.)

5. List six conditions for which minor tranquilizers are prescribed.

6. As sleeping pills, why are the benzodiazepines so much safer to use than the barbiturates?

7. Check Table 10.1 for the half-life of Centrax. If a student took a 20-mg dose of Centrax before an exam at 10 A.M. on Monday, what is the least amount that would still be in his body Tuesday at 9 A.M.?

8. Write a brief paragraph describing the development of physical dependence on Valium and the nature of the withdrawal syndrome from this drug.

9. In your opinion, do the advantages of the minor tranquilizers outweigh their abuse potential?

10. Diazepam is virtually suicide-proof, and yet people use it successfully to kill themselves. What drug combination is a likely explanation for this apparent contradiction?

11. What chemotherapeutic revolution in medicine occurred in the 1950s?

12. How is dopamine related to the mechanism of action of the major tranquilizers?

13. Define (**a**) photosensitization, (**b**) Parkinson disease, (**c**) mania, (**d**) tardive dyskinesia.

14. What is the difference between taking a drug therapeutically and taking it prophytactically?

15. Table 10.3 states that manic-depressives account for 4.2% of all first admissions to mental hospitals. Which condition accounts for the greatest percentage of admissions? The second greatest? What are the chances of recovery from the first and second most common mental illnesses? Again referring to Table 10.3, if 22.6% of all first admissions are diagnosed as schizophrenics but 45.6% of hospitalized mental patients are schizophrenics, what conclusions can be drawn?

16. There are a dozen or more major tranquilizers on the market. What is the value of having so many? Why not just use the original, chlorpromazine?

17. What is the use of lithium in medicine today?

18. Prescription and OTC sleep aids are big business today. With the help of this chapter, Section 15.2, and *The Merck Manual*, make a list of at least five *types* of hypnotic drugs people use, both prescription and OTC.

19. *Advanced study question*: In discussing the pharmacology of the phenothiazines, we stated that their antipsychotic effect usually doesn't start until after the patient has taken them for several weeks. What could be happening in the body to cause this delay?

20. *Advanced study question*: Valium, Librium, etc., are effective antianxiety drugs that bind to brain cell receptors. Does this fact suggest that the human body itself manufactures *natural* antianxiety agents? If so, what would be the best way to look for them?

21. *Advanced study question*: Approximately two-thirds of the people who abuse prescription drugs are women, says Christine Walker, program director for Options. She believes that this is because doctors tend to prescribe psychoactive drugs more often for women than for men. Do you agree that women's complaints are often treated as emotional illnesses, whereas men are told to exercise or are put through more scientific diagnostic procedures? What evidence do you have for your conclusions?

CHAPTER 11

Marijuana

Key Words in This Chapter

- Marijuana
- Tetrahydrocannabinol
- Amotivational syndrome
- Decriminalization
- Marijuana as medicine
- Sinsemilla

Learning Objectives

After you complete your study of this chapter, you should be able to do the following tasks:

- Appreciate the status of America's most popular illicit drug.
- Voice an intelligent opinion on the safety of marijuana.
- Discuss marijuana's effects on brain, memory, cardiovascular and respiratory systems, and motor coordination.
- Tell how long marijuana remains in the body.
- Cite some uses of marijuana in medicine.

11.1 Introduction

> ***News Item:*** In Alabama, a man bought a pound of marijuana in a motel room for $900. The man was a Vietnam veteran, had a criminal record, but had a home, wife, young son, and had been clean for the past 13 years. He planned to keep some of the pot and sell the rest. He was caught in a sting. The seller was a freed felon working for the local county drug task force. Tried, convicted, and fined $25,000, the pot buyer was sentenced to life without parole in a maximum security prison, where he remains. The informer was paid $100 by the county for his services.

It is difficult to overestimate the tremendous consequences marijuana has had and continues to have on American life. From the 1930s and 1940s, when the U.S. Commissioner of Narcotics promoted fear of the drug, to the present research into its possible acute and chronic effects in humans, marijuana has occupied the attention of the general public, police officials, government agencies, doctors, research chemists, pharmacologists, and psychiatrists. For a long time, Americans have reacted forcefully to the "threat of the weed." Use of the "green death" has motivated parents to turn their children in to the police. Some states have adopted the harshest of laws to "control" marijuana. In at least 15 states, laws now require life sentences for certain nonviolent marijuana offenses. In Montana, a life sentence can be imposed for growing just one marijuana plant or selling one joint.

As described in Chapter 2, **tetrahydrocannabinol (THC)**—the main psychoactive ingredient in marijuana—is found in the leaves and stems of the plant, in the compressed resinous from called *hashish*, and in hash oil (see Figures 11.1 and 11.2). All three of these forms are Schedule I controlled substances under federal law.

Figure 11.1 Marijuana cigarettes ("joints") compared to a tobacco cigarette. (Photo courtesy DEA, U.S. Department of Justice.)

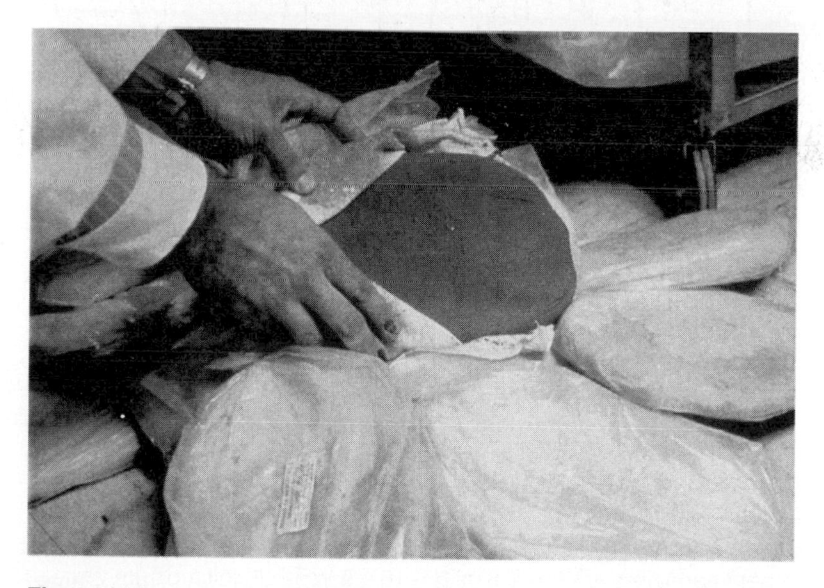

Figure 11.2 Interdicted hashish. This form of cannabis, collected from the flowering tops of the female hemp plant, is far more potent than marijuana. (Photo courtesy DEA, U.S. Department of Justice.)

SAMHSA and the DEA have provided the following summary of marijuana and hashish use:

- In 2000, an estimated 11.5 million Americans were current (during the past month) users. This number represents 5.4% of the population ages 12 and older.
- Marijuana is by far our most commonly used illicit drug. In 2001 approximately 76% of current (past month) illicit drug users were marijuana/hashish users. About 50% consumed only marijuana.
- Among youths ages 12 or 13, 4.1% were past-month pot users.
- See Figure 11.3 for annual numbers of new users of marijuana.
- Marijuana is considered a "gateway" drug to the world of illicit drug abuse.

Most of the marijuana available in the United States comes from Mexico, but two other sources have become very important. An indoor-grown, high-potency, "BC Bud" pot (15–25% THC) enters the U.S. market from Canada, where marijuana trafficking has become a $1 billion business. Domestic cannabis production, indoor and outdoor, is increasing. The five leading states for indoor growing are California, Florida, Oregon, Washington, and Wisconsin. The major outdoor-growing states for 2000 were California, Hawaii, Kentucky, and Tennessee. Altogether, domestic pot production is a $15–20-billion-a-year business, which places it, in terms of gross revenue, right up there with Exxon and General Motors.

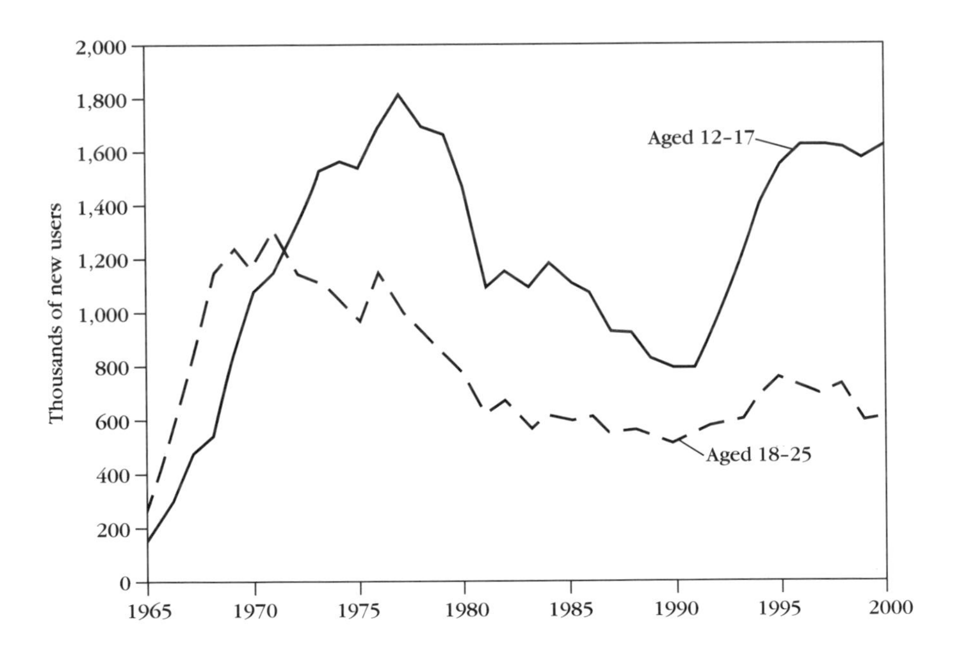

Figure 11.3 Annual Numbers of New Users of Marijuana: 1965–2000. An estimated 2.4 million Americans used marijuana for the first time in 2000. The highest use has been among 12–17-year-olds, with their numbers reaching 1.6 million per year. There are 0.5–0.8 million 18–25-year-old initiates per year, and the average age of initiation has dropped to 17.5 years. (Source: Results from the 2001 National Household Survey on Drug Abuse: Vol. I, SAMHSA, 2002.)

Figure 11.4 Hollowed-out cigars packed with marijuana are called **blunts**, and are gaining in popularity. (Photo courtesy DEA, U.S. Department of Justice.)

In America there is a resurgence in marijuana's popularity, owing to a relaxed public perception of harm, popularization by the media and by prolegalization groups, and with the trend of smoking of **blunts**—marijuana-filled cigars made by hollowing out commercial cigars and replacing the tobacco with marijuana (Figure 11.4). One blunt can contain as much pot as six regular joints. In Philadelphia, a cigar filled with a combination of marijuana and either PCP or cocaine is known as a "woolah" blunt.

Use of cannabis in the United States, and the potency (percentage of THC) and cost of various forms, are given in Tables 11.1 and 11.2. We note from Table 11.2 that marijuana and sinsemilla (Section 2.4) potency are at record highs.

11.2 The Status of the Pot Controversy

Compared to cigarettes, marijuana smoking in America is still in its infancy, for it has been widely indulged in for only about a third of a century and its effects have been seriously studied for far less time. And that means that society has yet to discover the long-term effects of THC and the 60 other psychoactive ingredients in this plant product. When your children's children are in a position to "Just Say No" to pot, they will know a lot more about it than we do, and the present controversy about its effects on the human body will be largely settled.

Evidence continues to accumulate, however, that the pro-marijuana advocate is in an untenable position. It is now impossible to consider marijuana a safe drug unless one ignores the present data. Most of these data relate to short-term effects, but we do have some good evidence of the long-term effects of pot, especially on the

Table 11.1 Marijuana and Hashish Use in the U.S. Population 12 and Over

Age	*Used in Past Year*	*Used in Past Month*	*Ever Used*
12–17	3,197,000	1,878,000	3,855,000
18–25	6,739,000	3,855,000	12,474,000
26–34	3,326,000	1,894,000	16,569,000
35 & older	5,412,000	3,390,000	39,171,000
Total	18,710,000	11,016,000	72,070,000
Male	10.8%	6.7%	38.5%
Female	6.5%	3.5%	27.9%

Source: U.S. Substance Abuse and Mental Health Services Administration (SAMHSA). Preliminary Results from the 1998 National Household Survey on Drug Abuse, July 1999. More than 25,500 persons were interviewed from three major race and ethnic groups in 13 states, and the results extrapolated to the total population.

Table 11.2 Marijuana: Annual Price and Potency Data (National Range, in Dollars)

Type	*Quantity*	*1992*	*1993*	*1994*	*1995*	*1996*	*2000*
Commercial	Pound	300–3000	300–5000	285–4000	300–4000	200–4000	400–2000
Grade[a]	Ounce	40–450	25–450	40–450	40–400	—	
Potency (THC)[a,b]		3.84%	4.18%	4.06%	3.33%	4.62%	6.07%
Sinsemilla	Pound	650–9600	1000–9500	900–9500	800–8000	700–8000	900–6000
	Ounce	125–650	75–1000	100–1000	100–900	—	
Potency (THC)		8.57%	5.45%	7.35%	7.5%	8.95%	13.2%

[a]Includes kilogram bricks, loose plant material and buds (colas).
[b]Due to the lag time in reporting potency, percentages are updated continuously for up to 2 years. The highest potency reported for the fourth quarter of calendar year 1995 was 20.3% The highest sinsemilla content ever was 29.9% (Copper Center, Alaska, 1993). Hashish from Europe has had THC contents of nearly 40%. Year 2000 potency data from University of Mississippi's MJ Potency Monitoring Project.
Source: The NNICC REPORT 1996, and *Drug Trafficking in the United States*, Sept. 2001, DEA Intelligence Division, Washington, D.C. 20537.

lung and possibly on the fetus. One must also recognize that the potency of the marijuana sold today is greater than it was even a decade ago. As noted in Table 11.2, street pot now averages 6.1% THC, and sinsemilla, 13.2%. This means that the present-day pot smoker can experience a more intense intoxicated state that lasts longer. We also recognize that all of the early research on marijuana was conducted using 1–2% THC samples, and we may be underestimating present-day marijuana effects.

In this chapter, I present an update on marijuana in the areas of extent and use, recent research findings, established effects on the body, areas of contention, regulatory laws, and the probable future of America's third most popular recreational drug. Objectivity in reporting is difficult, but no pertinent information has been consciously omitted. For a discussion of the plant source of marijuana, *Cannabis sativa*, see Chapter 2.

11.3 How Marijuana Affects the Body

11.3.1 Brain

The intake of 5–10 mg of THC into the bloodstream is sufficient to "get high." A marijuana high is a combination of sedation, tranquilization, and mild hallucination. The user experiences mood changes involving mild euphoria, a relaxed dreamy reverie, a feeling of well-being, heightened appreciation of sounds and colors, remembrance of pleasure but perhaps also imperception of time and space, a lag between thought and reaction, less-precise thinking, and a flattened effect. He or she may experience hunger, thirst, and uncontrollable laughter, or nausea, dizziness, and dryness of the mouth.

Physiologically, a few minutes after inhalation of marijuana smoke, the user's heart rate accelerates, bronchial passages relax, and blood vessels in the eye enlarge. When the euphoria passes, the user may feel sleepy or depressed.

Not all of these effects are experienced by every user. Infrequently, panic reactions are reported, with feelings of persecution, confusion, and fear. This reaction seems to occur especially in the neophyte user taking high doses. In general, high doses of THC can cause illusions, pseudohallucinations, and paranoid thinking.

Researchers at the National Institute of Mental Health have isolated and cloned a gene that gives rise to receptor molecules (termed CB_1 and CB_2) to which cannabinoids bind in the brain. This special cannabinoid receptor is bound up with proteins in cell membranes called G proteins (see Section 1.7). These proteins relay chemical signals to the cell when marijuana binds to the receptor. Thus, it is likely that there is a *natural* cannabinoid substance in the human body, but it is not one of the known neurotransmitters. Actually, an endogenous (naturally occurring) cannabinoidlike substance has been found in the brain. Termed **amandamide**, it is a fatty-acid chemical that activates the same cell-membrane receptors that are targeted by THC. Research suggests that upon release in the brain, amandamide regulates memory, appetite, mood, learning, and motor coordination. It can also block the action of brain dopamine. For more, search the Internet for "amandamide." Marijuana receptors are particularly common in the hippocampus, a region of the brain involved in memory; they are also located in complex regions of the brain that control motor skills and cognition.

11.3.2 Memory and Perception

There is now general agreement that even moderate doses of pot cause short-term memory loss. Moderate doses impair intellectual functioning in several areas, including the ability to read with comprehension, to acquire, store, and recall information, and to communicate clearly. Attention span, tracking, and perception may be impaired. These reactions are all dose-related and temporary but are important in adolescents who are especially vulnerable to any long-term learning difficulties. Impaired concentration and memory deficits in the classroom slow the development of not only cognitive skills (mathematics and reasoning), but also coping skills, such as handling stress and the pressures of social adjustment and peer relationships. This means that young people can turn to a drug as a crutch to help in coping with the difficult growing-up process.

11.3.3 Cardiovascular System

The cannabinols in pot cause dose-related tachycardia, increasing the heart rate to as high as 160 beats per minute. Long-term marijuana users develop a tolerance to this effect. There appears to be no damage to the healthy heart. Blood vessels in the eye are dilated, giving the bloodshot eye effect. Ears become red and warm. Postural hypotension (dizziness upon standing) is seen.

11.3.4 Respiratory System

In moderate doses, THC causes bronchodilation; it might be beneficial in asthma treatment. Heavy use has the opposite effect, causing slight obstruction of air passages. Heavy pot users experience inflammation of the bronchi, sore throat, and inflamed sinuses. Daily smoking of one joint reduces the lungs' vital capacity (volume of air the lungs can expel after one breath) as much as the daily smoking of 16 tobacco cigarettes. It appears that hot pot smoke is more detrimental to the lungs than hot cigarette smoke. Marijuana smoke has more tars than cigarette smoke, and the practice of smoking the joint down to the end ensures that the tars get into the smoker's lungs. Dr. Lyle Arnold of the University of California at San Diego's School of Medicine says that marijuana smoke condensate contains 50–100% more mutagens than tobacco smoke, as measured by the nationally recognized Ames test. Dr. Sidney Cohen believes that the carcinogenicity of pot smoke is equivalent to that of cigarette smoke (both vapors contain carcinogenic polycyclic hydrocarbons such as benzopyrene). Researchers at the UCLA School of Medicine report that marijuana joints release five times as much carbon monoxide into the bloodstream and three times as much tar into the lungs as tobacco cigarettes. They concluded that three to four joints a day do as much bronchial damage as 20 tobacco cigarettes. On biopsy, lungs of pot smokers show extensive microscopic abnormalities associated with the development of bronchitis, emphysema, and lung cancer. The combined use of marijuana and tobacco is worse than that of either substance alone. Rodent skins painted with marijuana tars develop tumors just as they do with tobacco tars, but direct confirmation of the generation of human lung tumors by marijuana has not yet been made. Can a person test positive for THC as a result of involuntary or passive inhalation of residual smoke from someone else's joint? PharmChem laboratories researchers in Menlo Park, California, say yes, but only if very sensitive analytical methods are used.

11.3.5 Motor Coordination[1]

Marijuana unquestionably impairs driving ability, even after ordinary social use. Evidence for this conclusion is based on results from laboratory simulators, test course performance, actual street driving, and a national study of drivers involved in fatal accidents. Some 60–80 percent of marijuana users, when questioned, say that they sometimes drive while stoned. A 1976 study of Boston-area drivers involved in fatal accidents found that pot smokers were overrepresented compared to non–pot smokers of the same age and sex. Pot causes marked deterioration in the performance of airplane pilots under flight simulator test conditions and impaired ability to handle emergency situations.

Drug interactions between marijuana and alcohol have been noted. The combination is especially effective in impairing motor coordination and steadiness on the feet. Three drinks (1.5 fluid ounces of alcohol each), together with marijuana, can cause intense nausea and vomiting.

11.3.6 Reproductive Functions

Research reports on the effect of marijuana on reproduction are incomplete and sometimes contradictory, and because of this their significance is not clear. Pot appears to produce a brief fall in sperm production, and it alters sperm shape and mobility. In animals and probably in humans, THC reduces testosterone levels, but the significance of this finding remains in doubt. Pubescent boys appear to be especially at risk from a lowered testosterone level. Long-term administration of THC to mice caused an increase in abnormal ova (5.5% above the control group). Studies of women who used marijuana regularly during pregnancy showed that stillbirths and neonatal deaths increased. Three studies have implicated marijuana in a syndrome identical to the fetal alcohol syndrome (see Chapter 9). However, polydrug use in the women studied makes it difficult to pinpoint teratogenicity. Use of pot prior to pregnancy has been associated with decreased fertility and should be avoided by both partners if pregnancy is desired.

11.3.7 Amotivational Syndrome

In the amotivational syndrome, it is suspected that regular use of marijuana dulls the mind of the smoker, blunts enthusiasm, takes away drive, and makes the person content to sit around all day, red eyed but happy and hungry. Actually, there appears to be little unequivocal evidence for this effect. Studies of college students have not generally found lowered intellectual performance as measured by grades achieved. The much-discussed research done in Jamaica, Greece, and Costa Rica on natives who had long histories of marijuana use failed to give clear support to the idea of an amotivational syndrome. In a 1980 survey of high school daily pot users, some 42% said that they experienced loss of energy, and 34% said that they felt the drug had harmed their work or school achievement. But scientifically speaking, it is conceivable that their amotivation would have occurred without the use of drugs. We cannot draw any broad conclusions about the amotivational syndrome at this point.

---

[1]This term refers to motor nerves, that is, nerves, controlling skeletal muscle activity.

11.3.8 Tolerance

Although tolerance to most of the effects of THC can develop, it does not necessarily lead to increased drug taking. For most, regular heavy use leads to a mild physical and psychological dependence. (Others strongly disagree. A 19-year-old Californian says, "People say that marijuana is not addicting, but it is extremely addicting.") The marijuana abstinence syndrome can be seen after *exceptionally* high doses and is characterized by irritability, restlessness, sleep disturbances, weight loss, GI upset, tremors, sweating, and anorexia. Symptoms typically start within 12 hours after the last dose and are most intense between days 3 and 4.

11.3.9 Other Actions

Some research has found an association between marijuana use and chromosome change and inhibition of the immune system. However, at this time, there is no convincing, unequivocal evidence that marijuana causes clinically significant chromosome damage or suppresses the immune response.

Burnout is a term first used by marijuana smokers themselves to describe the effect of long use. Young people who smoke marijuana heavily over long periods of time can become dull, slow moving, and inattentive. Burned-out users can be so unaware of their surroundings that they do not respond when friends speak to them, and they do not realize they have a problem.

The most common adverse reaction to marijuana is a state of anxiety or panic, sometimes accompanied by paranoid thoughts; these can range from general suspicion to a fear of losing control and going crazy. Acute anxiety reactions are usually experienced by novice users, especially those who have ingested one of the more potent pot samples that are so readily available. Panic symptoms usually disappear in a few hours as the drug's effects wear off; most of the time they can be handled by simple reassurance. Marijuana emergency room visits increased 40% in a recent year, however.

11.3.10 Summary of Effects in Humans

Marijuana has a unique pharmacology unlike that of any other drug. The acute toxicity of pot is low; its most significant adverse acute effects are tachycardia, impairment of short-term cognitive functioning, and impairment of motor skills. Chronic use may lead to respiratory problems as minor as sinusitis or as major as cancer. Evidence is mounting that pot use during pregnancy is hazardous to the fetus. All of the adverse effects of marijuana, including panic attacks, are made worse by the higher strength of the products now available on the street.

11.4 What Are the Risks?

In terms of what we know now, the short-term risks to physical health in normal adults from occasional social smoking of marijuana appear to be minimal. Long-term smoking, especially in heavier amounts, can cause bronchitis and general irritation to the respiratory tract. Although there is no proof, very long-term inhalation of the mutagens and carcinogens in pot smoke likely predisposes to cancer of the lung. Preteenage boys, whose need for testosterone is critical, are probably at increased risk if they smoke marijuana, as are adults with marginal fertility. People suffering from chronic heart disease, psychoses, or epilepsy should think thrice about smoking pot. Based on the evidence we have (admittedly inconclusive), pregnant and nursing women should not smoke pot, nor should people taking medicines for diabetes or epilepsy. In addition, those who drive vehicles or operate machinery while under the influence of THC or its metabolites are at increased risk of accidents.

Marijuana smokers run a slight risk of poisoning from contaminants in their pot samples. In 1982, the Centers for Disease Control reported on dozens of cases of *Salmonella* intestinal infections caused by the handling of pot that had been adulterated with animal manure. Pot decreases the production of stomach acid, which is an important defense against infectious organisms that are swallowed. Regular pot smokers are especially susceptible to infections.

11.5 How Long Does Marijuana Last in the Body?

In regard to its fate in the body, inhaled THC reaches the brain quickly and produces a euphoric high in 20–30 minutes. Its obvious effects disappear in 2–3 hours, but residual effects are demonstrable up to 10 hours after smoking a joint. Taken orally, its onset is 0.5 to 1.0 hour and lasts 2–4 hours. THC is very fat-soluble, and it and its breakdown products are stored in the brain, lungs, testes, ovaries, and body fat in general. Their slow release from these tissues permits their detection in the urine up to 10 days following the last dose (see Table 16.2). THC ingested by smoking or eating marijuana is biotransformed in the liver to 11-hydroxy-THC and then to 9-carboxy-THC. Also formed are 8-hydroxy-THC and 8,11-dihydroxy-THC. These compounds are referred to as metabolites of THC, and their detection in the body or urine is taken as definitive proof of marijuana use. The primary urinary metabolite of THC in humans is the 9-carboxy compound. Unlike THC and the hydroxylated derivatives, which are psychoactive, 9-carboxy-THC is inactive. Chemistry laboratories are now equipped to detect THC and its metabolites in nanogram (billionths of a gram) quantities in the urine of pot smokers. Poisonlab of San Diego states that, by using radioimmunoassay, they can detect THC in the urine "weeks after a single use." Studies have shown that because of its fat solubility, THC in active forms may be retained in the body for as long as 45 days after smoking. See Section 16.4, "What Drugs Are Detected and for How Long," for a discussion of marijuana urinalysis. See Appendix I for the chemical structure of THC (Figure 13) and pertinent discussion.

11.6 Marijuana and the Law: Decriminalization

In every nation where marijuana has been used, there has been at one time or another a law against it. Penalties have ranged from mild sentences to the death penalty for selling hashish.

In the United States, the first anti-marijuana law was enacted in Louisiana in 1927. Within 10 years, every state had passed a law prohibiting pot. A great part of the zealous campaign against pot was carried out by the then commissioner of the newly formed Federal Bureau of Narcotics, Harry J. Anslinger. His campaign against marijuana, using such terms as *killer weed, murder*, and *insanity*, obviously influenced the nation's opinion of pot. So did a 1928 book titled *Dope*, which said, "You can grow enough marijuana in a window box to drive the whole population of the United States stark, staring, raving mad . . . but when you have once chosen marijuana, you have selected murder and torture and hideous cruelty to be your bosom friends."

This campaign led to the passage of unbelievably harsh federal and state laws. The federal Marijuana Tax Act, passed by Congress in 1937, regulated marijuana use until it was repealed by the more lenient Comprehensive Drug Abuse Prevention and Control Act of 1970.

In the early 1970s, anti-pot sentiment began changing as the antiestablishment use of pot gave way to recreational use by that "nice kid next door." In 1972, the National Commission on Marijuana and Drug Abuse proposed that the possession and sale of *small quantities* of marijuana should be legalized. A wave of pro-pot sentiment appeared, stimulated in part by an organization called the *National Organization for the Reform of Marijuana Laws* (NORML). This organization, with a large staff and budget, remains very active. Between 1973 and 1978, 11 states decriminalized marijuana. Several additional states passed laws permitting local options for **decriminalization**. Seattle, Washington, and Ann Arbor, Michigan, are among the cities with such ordinances. Alaska permits the personal possession of up to 4 ounces of marijuana.

Decriminalization is not to be confused with legalization (see Section 12.14, "Drugs in Our Society: Some Observations"). In most states, alcohol is a legal substance that adults can purchase whenever they want to. Heroin (its purchase, sale, or possession) is illegal in America. In essence, to decriminalize a drug means to ease the legal penalties for a *still illegal* substance. Easing has two general effects: (1) possession of small amounts is usually considered a *misdemeanor* and not a *felony*; and (2) penalties are reduced to simple fines that do not exceed $100 or $200 (and sometimes as little as $5). Furthermore, the misdemeanor often does not become a part of the person's permanent record. The "small amount" referred to is usually any amount less than 1 ounce (28.35 g).

A decriminalized substance is still illegal, and the new laws still provide stiff penalties for acts more serious than simple possession, such as:

- Possession of a large quantity of pot
- Possession of concentrated forms of marijuana, such as hashish
- Offering, transporting, selling, or giving away larger quantities

For these more serious acts, fines, jail sentences, or felony charges may still be levied. (See Section 3.9 for provisions of the 1986 Act.)

Decriminalization is moderate reform. Since the late 1970s, some persons have called for radical reform, that is, legalization of pot. They contended that as long as marijuana is illegal, there will be police spying, surveillance, illegal entries, and other threats to our society. Furthermore, they stated that attempts to enforce an unenforceable law result in loss of respect for the law in general and for the people who must enforce the law. Initiatives were placed on some state ballots to permit cultivation, transportation, and use of pot.

However, the pro-pot movement rather suddenly lost its momentum. In the mid-1970s, the American Council on Marijuana was established to oppose NORML's efforts at decriminalization. The council decried the proliferation of the drug paraphernalia industry, the increased use of pot by the very young, and the higher potency of the marijuana that was so readily available. In the 1980s, articles appeared in the public and professional press pointing out the known and suspected toxic potential of the substance, especially in the very young smoker. Since 1978, no additional state has decriminalized marijuana, and it seems likely that no others will.

Antipot parent groups have become an effective force in the past 10 years. Approximately 4,000 parent groups exist nationwide, and they have called for the passage of laws to control head shops and the sale of "frisbee pot pipes," "practice pot," and other forms of drug paraphernalia that they claim lure children into the drug culture. Their lobbying has been effective; about half of the states now have laws severely restricting or banning the sale of drug paraphernalia. These parent groups are in direct opposition to NORML and to anyone else who believes that marijuana laws should be relaxed.

In the fall 2002 elections, Nevada voters refused to allow their state to legalize possession of up to 3 ounces of marijuana. A failed Arizona proposal would have downgraded small-scale possession of pot to the equivalent of a traffic violation. A failed Ohio measure would have forced judges to order treatment instead of jail for many drug offenders.

A record number of U.S. college students want marijuana legalized, according to UCLA researchers. One of the nation's most comprehensive assessments of student attitudes (240,000 students in 473 colleges) found recently that 33.8% of college freshmen support legalization of marijuana. (In contrast, only 28.5% said it is important to keep up with politics.)

Case History *Pot Smoker*

R., a 34-year-old female student, speaks of a good childhood, good parents, but much shyness in her younger years. She did not try pot until she was about 21 or 22 years old. A "clean kid" in high school, she used no alcohol until her senior year. R. was always interested in people who were different or were underdogs; she considers herself an independent thinker, even a maverick. R.'s first joint was smoked with a roommate at a college party. She had no fear in trying it, was not concerned that it was illegal, did not get sick, but did get high. She enjoyed the first experience and continued to smoke with friends for 2 or 3 years, but only occasionally and only on weekends. When her girlfriends left, R. increased her use of pot. She made new friends, including a man to whom she later became engaged. He was a pothead. R. became less close to her family and closer to her boyfriend and his group of about 30 who were all potheads. She began to grow the MJ plant in her house and yard using artificial light and learned how to identify male and female plants. R. never tried LSD, although it was there. Their main dealers in pot were fellow students at State

U. MJ cost $22–$25 per quarter bag (a plastic baggie). Pot was never pressed on her but was offered, typically, in a setting that was seductive. There was a sexual seduction ritual in which pot released inhibitions; everyone would get stoned. Tolerance developed in some people in her group. One male student could study quite effectively while stoned; he graduated with honors and was hired immediately upon graduation by a large company. Others could not study while stoned. R. was heavily into pot for some 2.5–3 years, smoking in the morning before school and getting stoned every afternoon and evening. She and her fiancé smoked up to three joints a day. Her friends admired her ability to roll her own joints, but much of her pot was smoked using a bong. They needed to sleep a lot while stoned. They kept their stash (one "big bag") in the refrigerator. R. says she smoked pot because it made her feel relaxed and accepted by her social group. If she stopped using pot she would not experience withdrawal, but she was clearly dependent on it; she would miss the ritual of getting out the stash, setting up the bong, the "feel" of the tight bag of pot, and the preparation of the leaf. During this period, R. and her fiancé had no social life outside the drug-using circle. Although she drove a car while stoned and had one near accident, she did not believe that MJ was a bad drug; she felt no guilt in using it while she was in love with her intended. However, she came to realize that culturally she was missing much, that she could not study while stoned, and that she was now "living for the weed." Her termination of MJ use coincided with her breakup with her fiancé. She moved in with her grandmother and stopped MJ use cold turkey. There was no physical withdrawal, but she craved the drug, at times looking for old joints in her jewelry chest, pockets, and purses. The first two months were the most difficult. There was no weight gain, no substituting of alcohol or other drugs. In the 6 years since she broke off her engagement, she returned to MJ once for about a year. She would get stoned, but there was more lethargy and a spaced out feeling. She stopped the second time because of her job. Her coworkers were motivated, and she admired them. She cold turkeyed again (somewhat easier this time) and feels she will not smoke pot again. R. feels that pot is a stepping stone to harder drugs. Much depends on the "gang" you are going with.

11.7 Marijuana as Medicine

Considerable medico-legal activity continues to classify marijuana as a valuable medicine that should be legally available for the treatment of human illness—especially the nausea and vomiting in AIDS and cancer patients undergoing chemotherapy—and in glaucoma. The push to entitle marijuana is being made by patients, physicians, NORML, the interested public, and their financial backers, successfully using ballot propositions in states such as Arizona and California. Pro-pot Web sites abound (search under "medicinal.marijuana").

Legally, the heart of the matter is the conflict between federal law governing interstate transportation, sale, and use and local and state law that permit it. Both sides appear adamant.

Federal law makes marijuana a Schedule I drug—that is, having a high potential for abuse and no currently accepted medical use in the United States. The U.S. Supreme Court has ruled that medicinal use is not a valid exception to the federal law that classifies marijuana as an illegal substance. However, 27 states and the District of Columbia have approved ballot initiatives or passed laws permitting some medical use of MJ. Seven states—Arkansas, California, Colorado, Hawaii, Maine, Oregon, and Washington—permit home cultivation. San Diego has passed an ordinance allowing medical patients to possess 1 pound of marijuana and to have up to 24 indoor plants. Thus conflicting jurisdictions have created confusion and contention at local levels.

The case against marijuana health claims, made by the DEA (**`http://www.usdoj.gov/dea/pubs/sayit/myths.htm`**) appears compelling:

- Major medical and health organizations and the vast majority of recognized medical doctors and researchers have concluded that smoking MJ is not a safe and effective therapy. These organizations include the American Medical Association, the American Cancer Society, the National Sclerosis Association, the American Glaucoma Association, the American Academy of Ophthalmology, the National Eye Institute, the National Cancer Institute, the Food and Drug Administration, the Drug Enforcement Administration, and the U.S. Public Health Service.
- There are now more than 10,000 scientific studies that prove that marijuana is a harmful drug. Conversely, there is not one reliable study that demonstrates that MJ has any medical value whatsoever.
- In 1996 the American Glaucoma Association stated that no form of MJ is suitable for glaucoma treatment.
- Rather than helping, MJ worsens AIDS, cancer, and glaucoma.
- Scientific studies indicate that smoking MJ damages the immune system and can cause cancer.
- The well-funded medical-marijuana movement has helped contribute to the changing attitude among our youth that MJ is harmless, and to a softening antidrug attitude generally.

The prescription-only oral drug Marinol (dronabinol, Roxane) provides pure, synthetic THC, the main psychoactive ingredient in marijuana. Pro-pot advocates claim that Marinol doesn't work because it lacks the complete drug found in MJ smoke. The makers of Marinol, however, warn that it can have adverse reactions, such as a disturbing psychotomimetic response and possible physical and psychological dependence.

11.8 Summary

The cannabinols in marijuana, which have their own special brain receptors, produce a pharmacology that is unique and distinctive. Marijuana continues to be a popular drug with American teenagers, its use rivaling that of the ubiquitous ethyl alcohol. Ever higher THC strengths are being seen, especially since the introduction of sinsemilla. Pot smoke's dangerous effects on the lungs are now unquestioned; cigarette for cigarette, it's worse than tobacco. Marijuana's CNS depressant action can impair driving performance and short-term memory. Other purported effects on the body remain controversial or unproven.

Web Sites You Can Browse on Related Topics

Marijuana
http://www.marijuana-anonymous.org/
http://www.marijuananews.com/medical_cannabis.htm

Marinol
http://www.marinol.com/patient/pat03.htm
http://www.algis.com/news/pr/1997/pr970809/html

Medical marijuana
http://www.grandpaspotbook.org

Decriminalization
http://www.drugwatch.org
http://www.drugfreeamerica.org/
http://www.hightimes.com/htsite/home/index.php

SAMHSA (for publications)
www.health.org/

Study Questions

1. It has been estimated that approximately 13,000 metric tons of marijuana are used illicitly in this country each year. There are 2,200 avoirdupois pounds in 1 metric ton and 16 ounces to a pound, so if the street price is $40 an ounce, calculate the retail value of the pot sold tax-free every year.
2. **a.** Sixty-four percent of young adults say that they have used marijuana, but only about one in six of these say that they have used it only once or twice. Do you believe that this statement supports the conclusion that few people can try pot only a few times and then give it up?

   b. Examine Figure 11.3. What factors could explain the peak of the late 1970s and the current peak of use?
3. What is the main psychoactive ingredient in marijuana? Are there many chemical ingredients in marijuana? (You may wish to check Chapter 2 for data.)
4. Pharmacologically, what does it mean to get "high" on marijuana?
5. Explain the effects of moderate doses of marijuana on (**a**) short-term memory, (**b**) reading comprehension, and (**c**) attention span and perception.
6. What are the acute effects of marijuana on motor coordination?
7. Describe the acute effects of marijuana on the cardiovascular system.
8. True or false:

   a. America's most popular illicit drug is marijuana.

   b. Most of our imported marijuana comes from Mexico.

c. In 1998, over 18 million Americans (all ages) smoked pot at least once.
d. THC can cause either bronchodilation or bronchoconstriction, depending on the amount used and the length of use.
e. Pot, like LSD, is a typical hallucinogen.
f. Pot smoke contains mutagens and carcinogens.
g. There is unequivocal evidence that pot causes the amotivational syndrome.
h. Tolerance to the effects of marijuana can develop.

9. What is the connection between the high fat solubility of THC and its fate in the body? How do we analyze for cannabinoids in the body?
10. Define (**a**) immune response, (**b**) motor coordination, (**c**) misdemeanor.
11. Identify persons who are at especially increased risk if they smoke marijuana.
12. **a.** How does decriminalization differ from legalization?
    b. How many states had decriminalized marijuana by 1978?
    c. How many states have decriminalized pot since 1979?
13. Marijuana is America's third most widely used recreational drug. What are the first two?
14. In certain parts of Michigan, the penalty for possession of small amounts of pot is a $5 fine. Is this penalty effective in controlling pot use? Give reasons to support your answer.
15. There are some who forcibly argue: Remove the legal sanctions against marijuana and you will demythologize it; you will take away its glory and attraction for youth, and the use of marijuana will naturally decline. Do you agree? Why or why not?
16. From all that you have read and heard, do you believe that marijuana has any real medical value?
17. The best estimate for the half-life of THC is 18–24 hours. This means that half (50%) of the THC will be gone from the body after 1 day. Calculate how much (total) will be gone after 2 days, 3 days, and 4 days.
18. A Gallup poll found support for decriminalization of marijuana in all major population groups except Southerners, persons 50 years of age or older, and persons whose education ended at the grade-school level. Explain why these exceptions might have been predicted.
19. The *California Research Advisory Panel* (CRAP) uses public tax monies to provide marijuana capsules and cigarettes to thousands of cancer chemotherapy patients to control nausea and vomiting. Each dose costs the taxpayer between $3.75 and $6.50, but patients find it more practical to obtain their marijuana through illicit channels. What is your opinion on the wisdom of this tax-supported program?
20. In the *Berkeley Wellness Letter*, we read that habitual users of marijuana may *never* be rid of THC in their body fat stores. This statement cannot be made about cocaine or nicotine. Explain the difference.
21. *Advanced study question*: From historical, medical, and social perspectives, what analogies, if any, do you perceive between tobacco smoking and marijuana smoking? What major differences?
22. *Advanced study question*: What signs do you see in society that suggest that Americans are becoming less tolerant of drug use and abuse? (*Hint*: What are MADD and SADD? What is happening to smoking in public places and to the trend toward decriminalization?)

CHAPTER 12

Hallucinogens, Street Drugs, Designer Drugs, Club Drugs, Predatory Drugs, and Some Observations

Key Words in This Chapter

- LSD
- Flashback
- PCP
- Hallucinogenic amphetamine derivative
- Huffing
- Look-alikes
- Smart pills
- Hallucinogen
- Street drug
- Methaqualone
- Ecstasy
- Designer drugs
- Drugs and society
- Legalization of drugs

Learning Objectives

After you complete your study of this chapter, you should be able to do the following tasks:

- Define hallucinogen.
- Give an overview of the events leading to the discovery of LSD.
- Discuss the physiological and psychological responses to a dose of LSD.
- Explain the evolution of PCP as a street drug.
- Explain how methaqualone became a social hazard.
- Understand the abuse potential of hallucinogenic amphetamine derivatives such as MDA, MDM, MDMA, and XTC.
- Identify hazards in the use of solvents and inhalants.
- Explain what designer drugs are.
- Explain what look-alikes contain.
- Cite some approaches used in the treatment and prevention of drug abuse.

12.1 Introduction

In 1954, A. Hoffer and H. Osmond introduced a new word into the English language. This word, **hallucinogen**, described drugs that in small doses could alter perception, thought, and mood, creating illusions in the mind of the user. Today, we can define hallucinations as perception without any sensory stimuli.

Although only mescaline, LSD, and adrenochrome (an epinephrine derivative) were originally described as hallucinogens, humans had been aware of a hallucinogenic effect in dozens of natural drugs for thousands of years. A few of these ancient mind-altering drugs are the following:

1. **Ololiuqui**, the seeds from the morning glory *Rivea corymbosa*. This drug is the ancient Aztec decoction used by native priests to commune with their gods and receive messages from them. Thousands of visions and satanic hallucinations appeared after the use of this vine-like plant. Hofmann, the discoverer of LSD, proved the presence of a lysergic acid derivative (the simple amide) in ololiuqui, probably accounting for some of its hallucinogenic effects.
2. **Cohoba**, the snuff of ancient Haiti. Haitians were inhaling this pulverized seed, using special bifurcated tubes, when Columbus discovered the New World. They became intoxicated, had visions, and were inclined to prophesy under the influence of cohoba. It was also used to induce bravery before battle. Fish and Horning showed in 1956 that bufotenine was a major alkaloidal constituent of cohoba. Another constituent of cohoba, *dimethyltryptamine* (DMT), is today made synthetically and appears on the streets as a hallucinogenic drug.
3. **Harmala alkaloids**, found in plants used by the tribes of Peru, Ecuador, Colombia, and Brazil as a hallucinogen. One of the many psychic uses of harmine, a harmala alkaloid, was for guidance in choosing a new spouse. The chemistry of the indole system of the harmala alkaloids is very similar to that of LSD. These same harmala alkaloids are found in a plant that grows throughout the Mediterranean area and that was known to the noted first-century botanist Dioscorides. During World War II, the Nazis used harmine as a truth serum.
4. **Scopolamine**, found in the famous plant *Datura stramonium* (jimsonweed), in henbane, and in the arboreal beauty *Methysticodendron amnesium*. Hyoscyamine and atropine are often found in combination with scopolamine in plant sources. Datura plants have been used for thousands of years because of their effects on the mind. Scopolamine is today official in the USP and is the active ingredient in some OTC sleep aids. High doses of scopolamine, instead of sedating, can cause hallucinations. Scopolamine combined with morphine can be used for inducing *twilight sleep*, a form of amnesia in obstetrics. Birth pains are felt but are not remembered afterward.
5. **Peyote** (mescaline) and the psilocybe sacred mushrooms (psilocybin and psilocin). For a discussion of the history and current status of these plants and their hallucinogenic constituents, see Chapter 2.
6. **Yagé**, a vine ingested by natives of the Upper Amazon; said to confer telepathic powers.

The only chemically induced hallucinations experienced in the United States in the nineteenth and early twentieth centuries were those of the priests in the Native American Church, who used peyote in their sacraments. On college campuses, the only mind-altering drug of significance was alcohol. All of this was changed and the course of history forever altered by an accidental discovery made in a chemistry laboratory near Basel, Switzerland, in 1943.

12.2 The Discovery of LSD

Dr. Albert Hofmann worked for the Sandoz Pharmaceutical Company. An expert in the field of ergot alkaloid chemistry, Hofmann knew how to obtain lysergic acid from the ergot fungus, which grows as a parasite on the rye plant (see Chapter 2). Ciba, a competing pharmaceutical company, had marketed Coramine, a diethylamide derivative of nicotinic acid. This prompted Hofmann to synthesize the diethylamide derivative of lysergic acid, using the acid azide. The year was 1938, and the new compound, along with others, was examined pharmacologically and placed "on the shelf."

Five years later, on the afternoon of April 16, 1943, Hofmann again had an opportunity to handle *lysergic acid diethylamide* (we shall now refer to it as LSD[1]). Ordinary manipulation of the compound was sufficient to cause accidental ingestion (probably through the skin), and the world's first LSD trip became history. Although not fully understanding what was happening, scientist Hofmann took careful note of his mental and physical reactions. His account was published by his colleague, W. A. Stoll (although not until 1947), in the *Swiss Archives of Neurology and Psychiatry*, from which I have translated the following.[2]

> Last Friday, the 16th of April, I was forced to interrupt my work in the laboratory in the middle of the afternoon, and go home to seek care, since I was overcome by a remarkable uneasiness combined with a slight dizziness. At home I lay down and fell into a not unpleasant, intoxicated-like state which was characterized by an extremely exciting fantasy. In a twilight condition with closed eyes (I found the daylight to be annoyingly bright), there crowded before me without interruption, fantastic pictures of extraordinary plasticity, with an intensive, kaleidoscopic play of colors. After about two hours this condition disappeared.

Suspecting that the LSD compound was the cause of this unique experience, Hofmann several days later intentionally took 250 μg, a heavy dose. The subsequent effects were even more intense than before. After 40 minutes he noted slight dizziness, unrest, difficulty in concentrating, visual disturbances, and an inclination to laugh.

> Here stop the laboratory notes. The last words can be written only with great effort. I asked my lab helper to accompany me home, since I believed the process would take the same course as the disturbance on the Friday before. However, already on the bicycle ride home it was clear that all symptoms were more intense than the first time. I already had great trouble speaking clearly and my field of vision wavered and was distorted as a picture in a curved mirror. Also I had the feeling of not leaving the spot, whereas my lab helper later told me we had traveled at a brisk pace.

---

[1]Sometimes termed LSD-25 because it was the 25th derivative of lysergic acid synthesized.

[2]W. A. Stoll, "Lysergsäure-diäthylamide, ein Phantastikum aus der Mutterkorngruppe," *Swiss Archives of Neurology and Psychiatry*, Vol. 60, 1947, p. 279.

> So far as I can remember, the following symptoms were more pronounced during the height of the crisis and before the physician came: dizziness, visual disturbances, the faces of those present appeared to me as colored grimaces; strong motor unrest alternating with paralysis; the head, the entire body and all of the limbs appeared at times heavy, as if filled with metal; cramps in the calves, hands at times numb, cold; a metallic taste on the tongue; throat dry, constricted; a feeling of suffocation; alternately stupefied, then again clearly aware of the situation, noting as though I were a neutral observer, standing outside myself, that I shouted half crazily or chattered unintelligibly.

Six hours after he had taken the second dose of LSD, Hofmann's condition had greatly improved—although there were still some manifestations of intoxication.

> The visual disturbances were still pronounced. Everything appeared to waver, and proportions were distorted, similar to a reflection in moving water. In addition, everything was drenched in changing colors of disagreeable, predominantly poisonous green and blue hues. Colorful, very plastic and fantastic images passed before my closed eyes. It was especially noteworthy that all acoustical perceptions, perchance the noise of a passing car, were translated into optical sensations, so that through each tone and noise, a corresponding colored picture, kaleidoscopically changing in form and color, was elicited.

Later it became clear to Hofmann and those around him that he had discovered a powerful mind-altering substance that, in microgram doses, could temporarily change a normal, healthy adult into a schizophrenic-like psychotic.

The first LSD self-experiment published by a psychiatrist appeared in a Swiss journal in 1947. W. A. Stoll reported his impressions after swallowing 60 μg of LSD. In a darkened room, Stoll said that he saw an unbelievable profusion of optical hallucinations that appeared and vanished with great speed. He saw circles, vortices, sparks, showers, crosses, and spirals in a constant, racing flux. Then followed more highly organized visions of rows of arches, a sea of roofs, desert landscapes, and starry skies of great splendor. He hallucinated about a landscape of skyscrapers, the dark roofs of a Spanish city, and a garden trellis laced with falling red, yellow, and green lights. He believed that his hands were attached to some distant body. Later, his euphoria turned into depression. The bright red and yellows became blue, violet, and dark green. He saw a gloomy battlefield and sacrificial fires. Noises in the room evoked simultaneous changes in the optical hallucination (synesthesia). So did pressure on the eyeball. The experiment began at 8 A.M. At about 3–4 P.M. Stoll experienced depression and considered with interest the possibility of committing suicide. That evening he was again euphoric, since the effects of the LSD had mostly worn off. He felt that he had experienced a great epoch of his life and was tempted to repeat the experiment.

After completing its own pharmacological examination, the Sandoz Company elected to make LSD available free of charge to qualified experimental and clinical laboratories throughout the world. At that time, it must have been difficult to appreciate the magnitude of the LSD discovery, although there was one piece of information that couldn't be ignored. It took only unbelievably tiny doses for LSD to induce its model, schizophrenic-like psychoses in humans. For a 70-kg adult male, only 100–200 *micro*grams was required (at the most). That is only 0.1–0.2 mg. Put another way, that is one three-millionth of a pound, which meant that 2 pounds of

LSD would theoretically be enough to put all of the inhabitants of New York City and vicinity into a psychotic-like state for half a day. Now, the reader may wonder just how a dose of only 100 μg can be enough for distribution to all parts of the body, to the bloodstream, and to the brain. More astonishingly, an expert once calculated that with that dose, only about 9 μg actually deposits in the portion of the brain where it is believed to have its action. Well, it may help to know that 9 μg (0.000009 g), as small as it is, still contains 10 thousand trillion molecules of LSD. If LSD acts at receptor sites at specific synapses, the ability of 9 μg to affect brain function becomes a little more understandable.

On a weight basis, LSD as a hallucinogen is some 5,000 times more active than mescaline.

12.3 LSD Comes to the United States

In 1949, the first LSD was shipped to the United States, where researchers had an opportunity to confirm the parallel between LSD-induced symptoms and the delusions of schizophrenia. Work quickly confirmed all that had been established in Europe, and soon a theory emerged—that LSD was a clue to a possible biochemical etiology of mental illness. In other words, there might exist in the bloodstream of a schizophrenic a psychotomimetic compound, something like LSD, produced in tiny amounts by an inborn error of metabolism or by some other quirk of human physiology. A great deal of support for this theory arose when it was discovered that LSD, psilocybin, psilocin, harmine, DMT, and other hallucinogens all contained the same chemical grouping (an indolethylamine), as did serotonin (which was found to be inhibited by LSD, etc.) At this point, serotonin, as a naturally occurring brain amine, came to be linked with mental health.

While this theory was being investigated, other work was undertaken to learn more about the effects of LSD on animal nervous systems. LSD was fed to spiders, fighting fish, carp, and goats. A 5000-kg elephant keeled over in a motionless stupor shortly after receiving a 0.297-g dose of LSD. (Most animals die from a lethal dose of LSD due to respiratory arrest.) Because there were few restrictions in the 1950s on the use of humans as guinea pigs, LSD was given to many human subjects. Physiological responses in humans were found to be:

> Dilated pupil, flushed face, chilliness, a rise in body temperature and blood sugar, an increase in heart rate and blood pressure, goose flesh, salivation, perspiratio, rapid development of tolerance, but no addiction. Cross tolerance to mescaline, and psilocybin develops.

Most of a dose of LSD is metabolized in the liver to various transformation products. Only 1–10% is eliminated from the body as unchanged LSD.

Researchers learned that in a typical LSD trip, which lasted for an average of 8–12 hours,

> Both visual and auditory hallucinations occur; unusual patterns are seen; smells are "felt" and sounds are "seen" (termed "synesthesia"); the person can "step outside" himself and look at his body (depersonalization); he can experience two feelings at the same time (happy and sad, elated and depressed); the arms and legs feel as

> heavy as lead; minutes seem like hours; consciousness is retained and the person can think logically up to a point; there can be great understanding, a sense of rebirth, of new insights.

Some investigators got the idea that LSD might stimulate creativity among artists. However, studies of paintings, writings, and other art forms done under the influence of LSD showed that artistic creativity is not heightened and in most cases is in fact diminished.

Taking an LSD trip became common practice as psychologists and psychiatrists sought to create therapeutic applications for this powerful new tool. Although some trips were found to be blissful, with pleasant sensations and imagery, others were "bummers," with terrifying images, full of dread, horror, and panic. Clearly, the **set** of the person's mind and the **setting** in which the LSD trip is taken can strongly influence the entire experience. People mentally unstable or depressed to begin with were bad risks when it came to taking LSD, and in a few such subjects, a psychotic crisis was precipitated that required hospitalization.

12.4 The Use of LSD in Psychotherapy

One of the primary areas of psychotherapeutic research with LSD was alcoholism. It was expected that LSD would provide insights to alcoholics, allowing them to "step outside" themselves and look at their illness and thus be persuaded to change their drinking habits. A Canadian, Dr. Humphrey Osmond, and others treated nearly 1,000 hard-core alcoholics with LSD. Although they concluded that LSD therapy had been quite successful, LSD has not developed into a significant therapeutic agent in this area.

Similarly, the use of LSD in autistic children, juvenile delinquents, narcotic addicts, epileptics, depressives, and schizophrenics has been only moderately successful in some cases. There have been accounts of highly successful individual cases, such as in a particular narcotic addict, a sexually maladjusted person, or a criminal—but in general, LSD psychotherapy is limited.

LSD has been used in terminal cancer patients who experience persistent pain. These people get pain relief during the LSD therapy and afterward do not seem to mind the pain as much as before. It is just less important to them. Of all the areas of LSD use, this seems to be the most rewarding, although it is limited in numbers. LSD has never been official in the USP or NF.

Because tolerance to the effects of LSD develops quickly, it cannot be used continually. As few as three daily doses can induce tolerance, and 3 or more days of abstinence may be required to overcome the effects. Tolerance does not develop with occasional use.

Flashbacks, also termed hallucinogenic persisting perception disorder (HPPD), have been found to occur in a small percentage of people who have taken LSD. In this phenomenon, the user spontaneously reexperiences LSD-like symptoms weeks or months after he or she has stopped taking the drug. Flashbacks can occur spontaneously or they can be triggered by the use of other drugs (especially pot or hashish), stress, fatigue, or movement from a light to dark environment. These flashbacks can last from seconds to hours. In a very few cases, the inability to escape the recurrent panic of the flashback has caused the person to commit suicide.

Throughout the 1950s and 1960s, research on LSD continued, and more than 4,000 papers appeared in the scientific literature. Lysergic acid, from which LSD is synthesized, became available in large quantities through fermentation techniques, eliminating reliance on the ergot plant fungus. Chemists synthesized analogs of LSD, looking for structure–activity relationships.

Conflicting reports were published about the ability of LSD to cause birth defects or induce chromosome damage. At the present time, however, scientific opinion is that LSD does neither.

In the 1950s, LSD was the subject of a number of sensational stories published in magazines in Canada, the United States, and Germany. There was a "sensational eyewitness account" by a painter, "My Twelve Hours as a Madman" by a newspaper reporter, "One Woman's Courageous Experiment with Psychiatry's Newest Drug," and a story on how a Hollywood movie star put his life back together with LSD. These lurid reports served to make LSD a household word and encouraged some people to use it to solve their personal problems. Broad-scale self-experimentation followed. LSD was taken in the expectation of achieving deeper religious experience, and philosophers and religious scholars argued whether the religious trip was genuine, that is, comparable to spontaneous enlightenment.

In the early 1950s, CIA investigators first began to experiment with LSD. The infamous MK-ULTRA drug and mind control program was authorized in 1953 during the cold war with the Soviet Union.[3] The program soon grew into a mammoth undertaking that investigated the psychotomimetic effects not only of LSD, but also of MDA, mescaline, PCP, heroin, cocaine, procaine, and quinuclidinyl benzilate (BZ)—in fact, almost any psychotropic, hallucinogenic chemical known or about to be synthesized. At first, LSD was perceived as a fabulous new agent for unconventional warfare, a secret drug for behavior control. Numerous grants were made to universities and other organizations to conduct medical research on LSD and other hallucinogens. In effect, a network of physicians and scientists was created to gather intelligence for the CIA. LSD was given to thousands of subjects, some without their knowledge. A few of those who took LSD experienced frightening visions and developed fears of insanity, leading to mental breakdowns. One otherwise quiet, well-adjusted family man who unwittingly took LSD fell into a deep depression that lasted for several weeks and culminated in a headlong plunge through a closed window to his death 10 floors below. Not until 15 years later was the secret nature of the experiment revealed; only then did the man's family understand what had really happened. During the early 1960s, the CIA and the military phased out their LSD tests in favor of more powerful chemicals such as BZ, which became the Army's standard incapacitating agent.

Albert Hofmann, the discoverer of LSD, found that the drug is quite sensitive to air and light, breaking down by chemical oxidation to inactive products. LSD is stable only if stored in oxygen-free ampules protected from light. Street samples of

[3]The story of the infamous MK-ULTRA program is revealed in detail in *Acid Dreams: The Complete Social History of LSD: The CIA, The Sixties, and Beyond*, by Martin Lee and Bruce Shlain, Grove Weidenfeld, New York, 1985.

LSD on sugar cubes or blotting paper decompose within weeks or months, according to Hofmann.[4]

Meanwhile, researchers discovered that LSD had an effect on two neurotransmitters in the brain. It interferes with or blocks serotonin receptors, thus limiting natural serotonin activity, and it potentiates norepinephrine systems, further upsetting the mental balance that had previously existed. Despite all of the research, however, no one knows the exact mechanism of action by which LSD accomplishes its hallucinogenic effects.

In regard to the theory to which LSD gave impetus—that some schizophrenia is caused by a tiny amount of abnormal chemical accidentally manufactured in the schizophrenic's body—years of intensive research and effort have not borne fruit. Proof is still lacking, in part caused by the immense difficulty of trying to prove the existence of a few micrograms of an unknown chemical in the entire brain. Nonetheless, the work with LSD has focused attention on brain chemicals and the roles they play in mental and physical health. We know much more about serotonin, norepinephrine, dopamine, and dopa because of the fascinating series of events that began in Hofmann's laboratory in 1943.

12.5 LSD Goes Underground

In 1960, no one was being arrested for possession of LSD, no one was jumping off buildings because of it, it was not being synthesized in clandestine laboratories, and it was not an integral part of a drug subculture. Today, all of that has changed. What caused the downfall of LSD from a promising research tool to an underground drug?

As a Harvard professor, Timothy Leary, Ph.D., was an effective, respected, and uncontroversial clinical psychologist who taught classes and collaborated on several textbooks. After a trip to Mexico, where he experienced the effects of psilocybin, Leary and his fellow faculty member, Dr. Richard Alpert, began a large-scale investigation on the effects of hallucinogens (or **psychedelics**—a term coined by Osmond in 1956 for "mind-expanding" drugs), using inmates at a Massachusetts prison.

Gradually, Leary became more and more involved with LSD and psilocybin. He held sessions on and off campus that involved personal and graduate student use of the drugs. LSD's reputation for expanding consciousness spread, and later, when undergraduate students became involved, Leary was publicly challenged. Newspapers picked up the story, and Leary's harassment (and LSD's mythologization) began. After another trip to Mexico, Leary and Alpert established the *International Federation for Internal Freedom* (IFIF) and continued their promotion of LSD and psilocybin as mind-expanding, mood-altering drugs that could change, integrate, and recircuit the human nervous system.

In May 1963, after their dismissal from Harvard, Leary and Alpert moved to Mexico. Their subsequent expulsion from Mexico and the national publicity that followed, coming at a time of impending social upheaval in the United States, set the stage for LSD's big play on the national drug scene. Major magazines did feature

[4] A. Hofmann, *LSD: My Problem Child*, McGraw-Hill, New York, 1980, p. 72.

stories on Leary and LSD. His pro-LSD stand became a crusade. Publicity helped create a growing underground movement that included intellectuals, students, and hippie groups across the nation. Newspapers carried stories of LSD users going blind after staring at the sun, of others trying to fly off buildings, and of others committing murder. The bad publicity helped set the public against this "bad" drug.

Leary appeared before congressional committees and admitted to having used LSD more than 500 times. Finally, the line was drawn between the establishment on the one hand and the turned-on generation on the other.

For much more information on the story of Dr. Timothy Leary and LSD, the reader is encouraged to consult *LSD: My Problem Child* (see footnote 4). Here, Dr. Hofmann recounts his personal meetings with Leary and discusses Leary's requests to purchase 100 g of LSD and 25 kg of psilocybin from Sandoz. (The order was never filled.) See also David E. Nichols' article at **http://www.heffter.org/essay.html**.

In the mid-1960s Augustus Owsley Stanley III and his chemist friends synthesized some 4 million hits of LSD (sold at $2 each) and helped make the San Francisco district of Haight-Ashbury the world's original psychedelic supermarket, the place where acid was first sold on a mass scale. Trippers hailed the drug as an "elixir of truth, a psychic solvent that could cleanse the heart of greed and envy, and break the barriers of separateness." Also in the mid-1960s, laws were being passed everywhere to make the use, possession, or sale of LSD a punishable offense. In 1968 the federal Drug Abuse Control Amendments were modified to make possession of LSD a misdemeanor and sale a felony. Caught up in the sweep of the laws were other hallucinogens: peyote, mescaline, psilocybin, psilocin, and DMT. Believing that the original laws were not doing the job, legislators passed even stiffer laws, and today the federal penalty for first-time unlawful possession of LSD is a maximum $5,000 fine and 1 year in jail. First-time offenders under 21 years of age can have their record expunged. See Section 3.9 for penalties for repeat offenders and for those trafficking in LSD (as opposed to simple possession). LSD is a Schedule I substance under the 1970 Controlled Substances Act.

In 1966, the Sandoz Pharmaceutical Company, the last U.S. supplier of legal LSD, stopped all distribution and sent their supplies to the federal government. Today, NIDA is the only source of legal LSD and some 600 other drugs for research use.

By 1967, the possession and sale of psychedelics had been criminalized in all states. LSD was now totally underground.

Just how LSD acts to produce hallucinations and altered perception remains in doubt. Pharmacologically, researchers have shown that LSD has high binding potential for eight of the fifteen subtypes of serotonin receptors (see Section 5.7), plus high affinity for alpha-2-adrenergic and dopamine D1 and D2 receptors. See Chapter 5 for a more complete discussion of serotonin and the role it appears to play in modulating our responses to stimuli.

Successful treatment of a bad LSD trip can often be accomplished in a conservative manner by friends of the user who "talk him down" in familiar surroundings. Minor tranquilizers or barbiturates, taken by injection, have proved useful, but the use of an antipsychotic drug (major tranquilizer) should be restricted to serious episodes of prolonged psychotic behavior. Actually, there is no antidote to LSD once it is in the body.

Because a dose of LSD is far too small to be weighed directly, the chemical is dissolved and the solution is impregnated into small squares of porous (blotter-like) paper, candy, thin squares of gelatin ("windowpane acid"), aspirin, and even the backs of postage stamps. Obviously, it is very easy to transport large doses ("hits") of this hallucinogen, and detection is difficult. LSD blotter paper (see Figure 12.1) is the most commonly sold form; the paper is typically imprinted with cartoon characters, animals, maps, stars, or other forms of art and sells for \$1–\$4 a hit. *Microdots* are tiny, 3-mm-diameter tablets of LSD. The DEA says there is no truth to the stories that LSD is marketed as a kind of "tattoo" that is stuck on the skin for dermal absorption. The Do It Now Foundation, using data collected between 1980 and 1988, reported that 83% of the street samples sold as LSD were indeed LSD. Considering the rip-offs on street sales, this is a high rate of authenticity.

Today, the strength of a typical street dose of LSD ranges from 20–80 μg, contrasted with the 150–300 μg of the 1960s. Of course, taking several doses will increase the risk of a bad trip (see the following case history). Buyers of street LSD

Figure 12.1 A collage of LSD blotter paper. A solution of the drug in water is soaked into the paper and allowed to dry. An average effective oral dose is 30–50 *micrograms* per perforated square.

have no way of knowing how potent their batch will be. Taken orally by "microdot" tablet or perforated paper squares, the minimum dose of LSD required to produce hallucinations is 25 μg; a dose of as little as 10 μg can produce mild euphoria. LSD is cheap to buy. The average street price is $5 per hit; dealers pay less than $1 in wholesale lots of 1,000 or more. When compared to marijuana, which sells for $40–$450 per ounce, LSD is perceived by many drug users as a bargain, especially considering that higher doses can give effects that last up to 12 hours. LSD use in America peaked in the 1960s and 1970s, fell in the 1980s, and has now made a comeback. Students report that LSD is easier to buy today.

Figure 12.2 shows that after lower use in the 1980s and 1990s, LSD use has risen dramatically during recent years.

The starting chemical for illicit LSD synthesis is ergotamine tartrate (see Section 7.15), smuggled into the United States from Europe, Mexico, Costa Rica, and Africa. Upon hydrolysis, ergotamine yields lysergic acid, from which the *N,N*-diethylamide derivative is made. The synthesis from ergotamine is a delicate and demanding process, one that an inexperienced "garage chemist" would find very difficult if not impossible to perform. This fact explains why there are so few illicit LSD labs. Synthesis of LSD is concentrated on the West Coast, where the drug is distributed in final dosage form or in wholesale liquid or powder forms for repackaging across the country. LSD soaked onto paper rather quickly decomposes, so this dosage form must be made up close to sale. The DEA says that LSD is now sold in virtually every state in the Union.

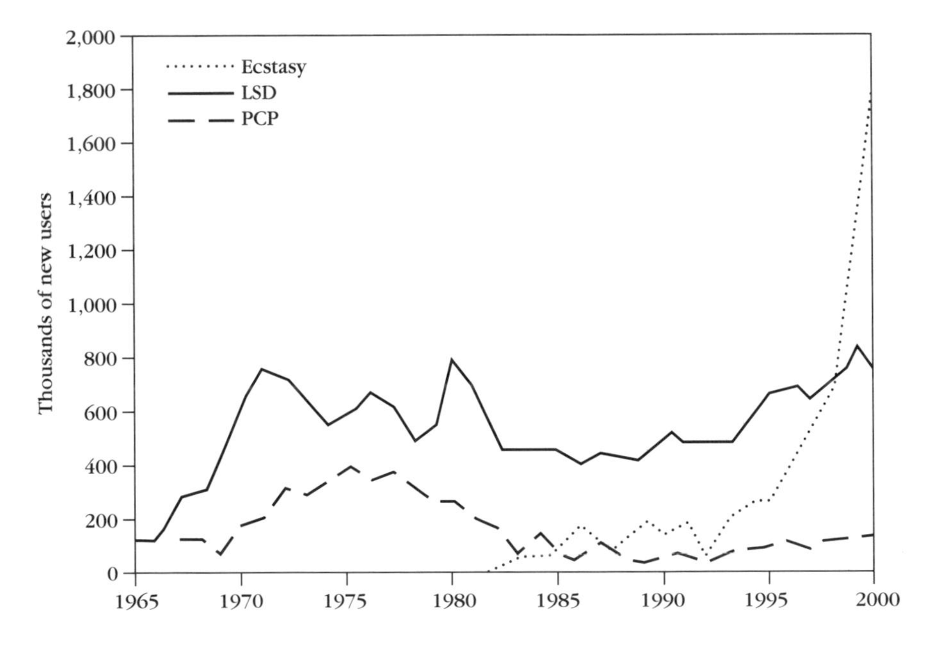

Figure 12.2 Annual Numbers of New users of Ecstasy, LSD, and PCP: 1965–2000. (Source: Results from the 2001 National Household Survey on Drug Abuse: Vol. I, SAMHSA, 2002.)

Case History *LSD Trip*

N., now a college student, took LSD twice at age 18. The first trip was brief and pleasant and prompted a second experience several months later. N. used only "some" alcohol and pot in high school but had smoked tobacco cigarettes heavily since age 13. With her mother away and the house to herself, N. bought two doses of LSD ($10 each) from a man who openly sold pot, acid, and coke to high school students. The LSD, wrapped in foil, was soaked into blotter paper imprinted with a small picture of a globe; she was told to store it in the refrigerator. The blotter paper, only 1 cm square, actually contained four hits, but N. and her girlfriend each took an entire square in their mouth and sucked on it to extract the LSD. (N. says some of her friends place the paper square under their eyelid for maximum absorption.) It was 5 P.M. By 5:45, N.'s eyes were watering as she watched TV. She felt the drug's effects coming on and could not stop grinning. Her dog entered the room. From the tip of his tail she saw purple, pink, and green laser beams shooting out, exploding as they hit the walls and ceiling. She had difficulty walking and opening a can of dog food. The dog food appeared to crawl out of the can, flopping over the sides. She could blink her eyes to make the hallucinations disappear. She was not frightened as she recognized that this was not real. An object tossed through the air trailed a series of images, much like a stroboscopic picture. A Mozart recording was unusually pleasing to the ear, but a commercial rock album "made for tripping" did nothing. Hours later her heart pounded so violently that she feared a heart attack. She experienced pain in her elbows and knees. She rolled M&Ms in her mouth; they seemed to become part of her body, and she felt that she had chocolate in her veins and arteries. "What if I become chocolate and someone wraps me up?" she wondered. She stared at her face in a mirror, her skin melted away, and she saw her skull. She hallucinated a woman, all in white, standing outside her bedroom door. Becoming scared, N. blinked her eyes, but the woman would not go away. N. felt she was traveling in space and might be sucked into a black hole. After 12 hours there was no letup, and N. considered going to a hospital for treatment. Her girlfriend was freaking out, and this added to her fright. Incapable of sleep, N. thought she would kill herself if this thing went on, especially if she were sent to jail. In all, it was a 30-hour ordeal, with hallucinations during the first 7 hours. She took two weeks to return to normal. N. says she would *never* take LSD again. The dealer who sold her the LSD was later arrested and sent to jail.

12.6 Other Current Drugs of Abuse

An indication of what drugs are being heavily used, and where, comes from data supplied by Project DAWN, a federal program of monitoring U.S. hospital emergency departments (EDs) and medical examiners (MEs) for reports of adverse drug reactions. Although it is necessary to use caution in drawing conclusions from DAWN data, it is possible to see patterns of drug abuse as well as apparent differences in these patterns from region to region. Table 12.1 ranks the top drugs mentioned most frequently by emergency departments in 2000.

Mescaline, psilocybin, and LSD opened the door to the use of psychedelics in our society. For one reason or another—low supplies, poor quality, boredom—the members of the drug subculture are always seeking different and better compounds with which to alter their consciousness. But new psychedelics are not coming from nature (that source seems to be exhausted). They are inadvertently coming from the reputable pharmaceutical firms that synthesize and introduce new drugs for possible therapeutic applications. In this section, we shall examine several synthetic compounds that are currently popular as psychedelic drugs and that have come by this route, as well as some amphetamine derivatives.

12.6.1 PCP

What a strange and powerful drug PCP is. At various times it can be a depressant, stimulant, hallucinogen, convulsant, or anesthetic. It is regarded by some experts as

Table 12.1 Drugs Mentioned Most Frequently by Emergency Departments, Year End 2000.

Drug name	*Number of Mentions*	*Percentage of Episodes*	*Rank*
Alcohol in combination	204,500	34.0	1
Cocaine	174,000	29.1	2
Heroin/Morphine	97,300	16.2	3
Marijuana/Hashish	96,400	16.0	4
Acetaminophen	33,600	5.6	5
Alprazolam	22,100	3.7	6
Hydrocodone	19,200	3.2	7
Clonazepam	18,000	3.0	8
Amphetamine	16,200	2.7	9
Aspirin	15,600	2.6	10
Methamphetamine/Speed	13,500	2.2	11
Diazepam	12,100	2.1	12
Oxycodone	10,800	1.8	13
Lorazepam	10,700	1.8	14
Carisoprodol	9,500	1.6	15
Fluoxetine	7,940	1.3	16
OTC sleep aids	6,600	1.1	17
PCP/PCP combinations	6,580	1.1	18
D-Propoxyphene	6,500	1.1	19
Amitriptyline	6,450	1.1	20
LSD	4,000	0.7	21
Acetaminophen/Codeine	3,850	0.6	22

Source: Year-End 2000 Emergency Department Data from DAWN, SAMHSA, D-18, 2001.

the most dangerous drug on the illicit market today. Fortunately, PCP use has fallen greatly from its highs in the late 1980s, concurrent with the increasing availability of crack cocaine. Unfortunately, the DEA has evidence that use is again on the increase, especially in such cities as Boston, Baltimore, Chicago, Los Angeles, New Orleans, New York, San Francisco, St. Louis, and Washington, D.C. Synthesis and trafficking of PCP is centered in Los Angeles.

PCP (**phencyclidine** or *p*henyl*c*yclohexyl*p*iperidine) was marketed by the Parke-Davis Company in 1963 as Sernyl. It was used as an anesthetic but produced such strange side effects (incoherent speech, delirium, hallucinations) that its use in humans was discontinued in 1965. In 1967 it again appeared, reintroduced by Parke-Davis as an anesthetic called Sernylan, for veterinary use only. PCP is believed to have hit the streets for the first time at the 1967 Monterey Pop Festival. It appeared in San Francisco as the *peace pill* and in 1968 in the east as *hog*. On the street, it is known as *killer weed, green, super pot, ozone, wack, embalming fluid*, and *rocket fuel*. PCP is sold admixed with marijuana or in variously shaped tablets or capsules. *Angel dust* is mint leaves, parsley, or low-grade marijuana laced with PCP. A small packet, weighing about 1.5 g and selling for $10, can be turned into 5 or 10 cigarettes, and one such cigarette is all that is needed for a dreamy state that lasts for up to 8 hours.

Because PCP is available in liquid form, it can be soaked into marijuana joints or into Sherman brand cigarettes (the products are called *Sherms*). DAWN data show that PCP ranks fifth in hospital ER mentions.

With a low to moderate dose (5–10 mg, producing in the adult a 30–100 ng/mL serum level), PCP users can show agitation, excitement, gross incoordination, a spaced-out floating euphoria, an inability to speak, sweating, and a loss of pain sensation, leading to stupor, vomiting, hypersalivation, repetitive motor movements, and an eyes-open coma. They may stumble or crawl about (**zombie walking**). Mentally, this dose of PCP produces distorted images, extreme feelings of apathy, depersonalization, and drowsiness.

Thus PCP—and its chemical cousin ketamine (Section 12.13)—are termed **dissociative drugs**. PCP and ketamine do not possess the indole chemical structure (Appendix 1) and do not produce typical LSD hallucinations. They do, however, induce amnesia.

Drowsy, intoxicated users may inexplicably turn violent and irrational, inflicting physical harm on themselves or others. One user became so disoriented that she jumped into a swimming pool and tried to exit from the bottom. Another was found head down in a garbage-filled dumpster, nude and unmindful of the cold weather, the discomfort of his upside down position, and the smell of the garbage. What is more, the strength of PCP-intoxicated persons is described as incredible. There have been cases of PCP users able to break open handcuffs locked behind their back, unmindful of the ripped-off skin or broken wrists. Another, while in restraints, attacked a gurney, and bent it. The incredible strength and violent behavior of some PCP users have led police to develop and use the Taser, an electric device that can be used at long range to shock the uncontrollable person into submission.

A mild dose of PCP is 2–5 mg. Marijuana joints have been analyzed that were laced with as much as 75 mg. Overdosage with PCP can produce days-long coma, very high blood pressure, muscular rigidity, and convulsions. Life support systems and intensive medical management are required to keep the person alive. There is no specific antidote for PCP; the physician can only treat the symptoms—for example, by giving diazepam to control convulsions.

There is some evidence that PCP acts by disrupting neurotransmitter function. Although PCP does inhibit reuptake of norepinephrine, dopamine, and serotonin, its effects are not affected by known transmitter agonists or antagonists.

Tolerance to PCP develops, and ever larger doses are required to achieve the desired effects. PCP depresses respiration and heart rate, and its combination with alcohol, another depressant, can trigger an overdose.

Case History *PCP Intoxication*

Two boys, 14 and 17 years old, were brought to Boston City Hospital by friends. The boys had ingested 6 and 15 small yellow tablets, respectively, an hour earlier. The drug was known to them as *THC*, an older name for PCP. The boys' vital signs were normal, but they had an inflamed lining of the nose, dry tongue, and tearing eyes. They produced no spontaneous speech and grunted in response to questions but were not hallucinating. They had difficulty walking and showed little response to painful stimuli. Perhaps their most unique response to PCP was **nystagmus**, bursts of irregular, shuddery jerks of the eyes when gazing in a particular direction. The boys were much improved after 24 hours, but the nystagmus was still present on the fourth hospital day. Strikingly abnormal eye movements are part of the clinical diagnosis of PCP poisoning.

After repeated PCP intoxication, users may develop a psychosis not unlike that seen in schizophrenia, and it may be 1 or 2 years before they feel normal again. Users can experience flashbacks, and for a long while they may feel depressed and spaced out. This reaction appears to be due to the storage of the fully psychoactive drug in body fat tissues. Ironically, because PCP abuse can cause schizophrenic behavior, the drug has become a useful tool for inducing schizophrenia, allowing scientists to study the disease.

Hindsight is unfair, but we now know how wise the manufacturer would have been *not* to have reintroduced PCP after the first marketing established its potential as a psychedelic drug of abuse.

PCP and its chemical analogs are now Schedule I drugs. They are regulated under the Psychotropic Substances Act of 1978, as amended by the 1986 Anti-Drug Abuse Act. See Section 3.9 for penalties for trafficking in PCP or simple possession. The act requires that the U.S. Attorney General be notified of commercial transactions in piperidine, a key chemical intermediate in the synthesis of PCP.

12.6.2 Methaqualone (Formerly Traded as Quaalude, Mequin, Sopor)

When methaqualone, a nonbarbiturate sedative-hypnotic, was introduced in 1965, it was hailed as the long-awaited replacement for the dangerous barbiturates. Methaqualone was touted as the low–abuse potential, addiction-free, safe, side-effect-free example of modern pharmacology. How tragically wrong that information was. Methaqualone ("ludes") developed into one of our nation's greatest social hazards, involving medical and law enforcement personnel from the local to the federal level. Reaching its destructive zenith in 1980–1981, when about 1 billion tablets a year were sold on the street, methaqualone earned its reputation as an "epidemic of horror." It threatens life from overdosage, serious or fatal accident, or a severe withdrawal syndrome that can include seizures.

Methaqualone (*meth-ACK-wa-lone*) is chemically distinct from the barbiturates and has a greater variety of actions. It has antispasmodic, anticonvulsant, local anesthetic, and antitussive activity. Through its general depression of the CNS, it reduces the heart rate, respiration, and muscular coordination. Five years after its introduction, however, it had become widely abused by American drug takers who called it the "love drug." Methaqualone's explosive invasion of the drug culture stemmed in part from the popular view among abusers that it was a powerful aphrodisiac. But, as with alcohol, this effect is explained by the release of inhibitions as feelings of relaxation, confidence, and euphoria set in (abusers term this a high, but they are actually undergoing CNS depression).

Tolerance of and serious addiction to methaqualone can develop after a month's regular use. Regular use of 75 mg a day can easily swell to 750 mg a day; some abusers take 2000 mg a day. Abrupt withdrawal is characterized by severe grand mal seizures and is life-threatening. Hence, detoxification is best carried out in a hospital. Because methaqualone and alcohol are metabolized by the same liver enzymes, the danger of overdosage from a combination of these two drugs is serious.

As methaqualone's bad reputation developed, licit manufacturers dropped it, one by one, until only Lemmon Pharmaceuticals remained; they changed the trade name to Mequin. Ultimately, all legal manufacture was stopped, but existing stocks on shelves were allowed to remain. The supply vacuum was filled by illicit methaqualone tablets smuggled in from Colombia. Dealers in Colombia purchased bulk methaqualone from legitimate manufacturers in West Germany, Austria, Hungary, Switzerland, and China. In the early 1980s, more than 100 tons (1.2 billion 75-mg doses) of methaqualone entered the United States illegally each year. As DAWN data showed an alarming rate of methaqualone injuries, the DEA began diplomatic negotiations with the foreign countries that were supplying the bulk methaqualone to Colombia. This effort, plus meetings with international drug control agencies, reduced the flow of methaqualone to the United States. Currently, methaqualone use is down dramatically for all age groups, but it still ranks 85th on DAWN's list of hospital emergency department mentions.

In retrospect, we see a drug, introduced to an all-too-accepting medical profession as the safe answer to barbiturates, that turned out to be extremely addicting. This situation is another example of an approved drug that became a severe social hazard, readily found in schools, at parties, and in bars, and that has cost our society hundreds of lives and billions of dollars.

12.6.3 Rohypnol (Flunitrazepam) Street Names: Rophies, Roofies, Date Rape Drug

This benzodiazepine-type minor tranquilizer—more potent than Valium—is abused in virtually every southern state from California to Florida. Rohypnol is known for its cheap ($5 a hit) high, similar to that from Quaalude. The drug is neither made nor marketed legally in the United States, but is smuggled in from Mexico and South America, where it is sold legally. Flunitrazepam is used widely in Texas, and is Florida's fastest growing drug problem. For $1.50 to $5 each, it is sold as a small, white, 2-mg tablet, giving no taste or odor when dissolved in a drink. Users say flunitrazepam gives a sleepy, relaxed, drunk feeling that lasts from 2 to 8 hours, but when taken with alcohol it can cause disinhibition and amnesia. Rohypnol has gained a reputation as a date rape drug. In Miami, a "roofie" is more likely to be clonazepam. Rohypnol, Ecstasy, and other drugs are now known collectively as **predatory drugs** (Figure 12.3).

12.6.4 Hallucinogenic Amphetamine Derivatives

Since the late 1960s, people seeking the chemical pursuit of ecstasy have taken any of a half dozen drugs that are closely related chemically to the amphetamines and

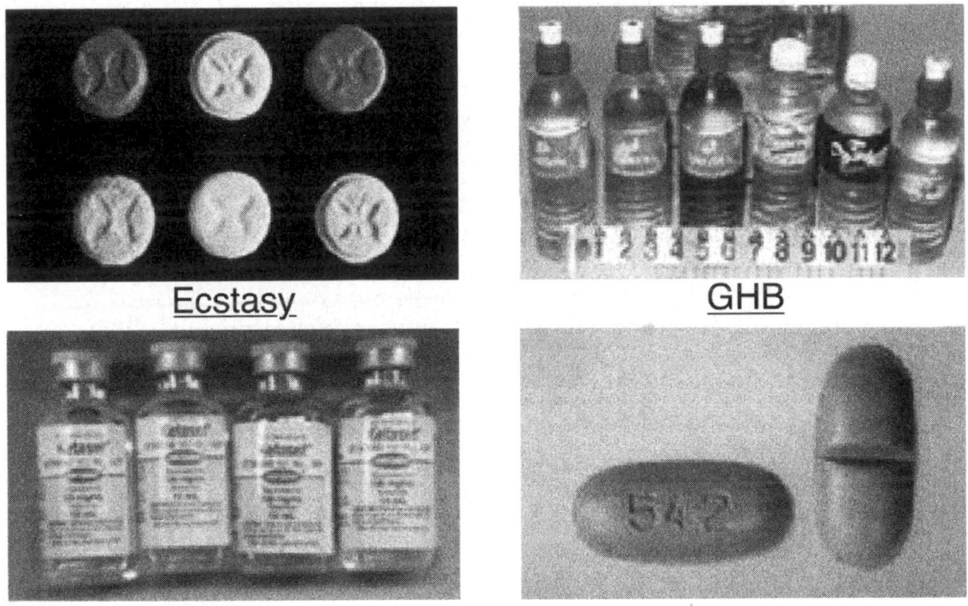

Figure 12.3 The Threat of Predatory Drugs. Predatory drugs are chemicals or drugs that can be used to facilitate sexual assault. In the late 1990s, officials noted a new, disturbing trend of rape cases involving the drugging of victims with Rohypnol, ketamine, and GHB (and its analogues GBL and BD14).

The Dangers of Predatory Drugs:

1. These drugs render the victim incapable of resisting sexual advances.
2. Dissolved in beer, liquor, juices or sodas, these drugs are indiscernable. Victims may be unaware they have been given a drug, and may be unaware of the attack until 8–12 hours after it occurred.
3. These predatory drugs are metabolized quickly, leaving no evidence.
4. Memory impairment caused by the drugs also eliminates evidence. Learn more about how to protect yourself from predatory drugs at `www.rainn.org` or `www.911rape.org`

mescaline (see Appendix I for the chemical structures). Because these compounds have a special stimulatory effect on the brain, they are termed **hallucinogenic amphetamines**. Their symbols, names, and doses are listed in Table 12.2.

MDA first appeared in San Francisco during the 1967 "Summer of Love" and gained a reputation for eliciting a sensual, easily managed euphoria. But then the "mellow drug of America" lost its appeal, and it has largely been replaced by its close chemical relative **MDMA** (also known as **Ecstasy** or Ecstacy, XTC, Adam, "E," Doctor, and MDM). Both drugs are modified amphetamines and can cause muscle tension, sweating, insomnia, tremors, a fast heart rate, and paranoia. But it is for their psychic effects that they have become popular street drugs. They induce a feeling of peace, openness, insight, delight, self-awareness, and hallucinations, all without dissociation or anxiety, say the users. Effects last 4–8 hours. According to the DEA, the new club drug Molly (1-(3-trifluoromethylphenyl)piperazine) is an extremely dangerous, more intense form of Ecstasy. It has been placed in Schedule I. Search **www.dea.gov** for more information. *Entactogens* is a term used by some psychiatrists to classify Ecstasy and related drugs. The term means "touching within." See **http://www.dancesafe.org/2cb.html**. Some users take MDMA all day for its CNS-stimulant euphoric effects or as an aphrodisiac. Hundreds of thousands of doses of MDMA are being sold on the street, typically as a white powder costing about $50 a gram or as tablets selling for $10 to $30 a dose. It can be inhaled, injected, or swallowed. SAMHSA says that in 2001 an estimated 8.1 million Americans aged 12 or older had tried MDMA at least once during their lifetime. The number of current users was estimated to be 786,000. In October 2002, three Israeli nationals, bound by ship for the United States, were arrested in possession of 1.4 million MDMA tablets worth $42 million.

Both MDA and MDMA have now been placed on Schedule I of the Controlled Substances Act, and possession of either is potentially a felony. However, before this scheduling was made, MDMA gained the respectability of therapeutic use by psychiatrists treating depressed people. Dozens of therapists reported good results in many cases.

Table 12.2 Hallucinogenic Amphetamine Derivatives

		Typical Dosage
MDA[a]	3, 4-Methylenedioxyamphetamine	50–150 mg
MDMA (or MDM, Ecstasy, Doctor, XTC, ADAM, E)	N-Methyl-3, 4-methylenedioxyamphetamine (or methylenedioxymethamphetamine)	50–160 mg
MMDA	3-Methoxy-4, 5-methylenedioxyamphetamine	—
STP (DOM)	4-Methyl-2, 5-dimethoxyamphetamine	10 mg max.
TMA	3, 4, 5-Trimethoxyamphetamine (or alpha-methylmescaline)	—

[a]In 1987, two analogs of MDA, N-hydroxy MDA and N-ethyl MDA, were temporarily placed in Schedule I of the Controlled Substances Act. MDEA is made and sold in the Netherlands.

The DEA considers both MDA and MDMA to be dangerous drugs that can cause a long-lasting reduction in the brain's supply of serotonin. An MDA dose of 7.5 mg per kilogram of body weight in humans is close to the lethal dose. One study showed that some 90% of MDA/MDMA street samples are authentic.

Isoproscoline is a synthetic analogue of mescaline in which the 4-methoxy group has been replaced with a 4-isopropoxy group.

Club drugs and **raves** are attracting the attention of NIDA as an alarming rise in the use of certain dangerous, mind-altering substances has been detected by the nation's monitoring mechanisms. Club drugs are psychotropic chemicals used by young adults (mostly 14–25 years of age) at all-night dance parties, such as "raves" and "trances," held in warehouses, dance clubs, or bars. These parties are marked by thumping electronic music and elaborate light shows. Not everyone attending a rave or trance uses drugs, but those who do typically experience increased stamina and intoxicating highs that intensify the rave or trance experience. Atlanta, Chicago, Detroit, Miami, and Newark have reported widespread drug use at rave and club scenes.

Popular club drugs include MDMA, GHB (Section 12.12), ketamine (Section 12.13), methamphetamine (Section 7.8) and Rohypnol. All of these drugs have the potential for serious adverse effects in the human body, including, variously, fast heart rate, high blood pressure, brain damage, and amnesia. In Britain, increasing numbers of deaths are reported following the use of Ecstasy; victims experienced convulsion, dilated pupils, low blood pressure, tachycardia, high temperature, coma, and death from lung failure. For more intriguing information on rave drugs see `http://www.urban75.com/Drugs/drugclubs.html`. (Note that the first "D" is capitalized, the second, lowercased.) It is of interest that the popularity of MDMA as a club drug began in Europe in the late 1980s—about five years earlier than in the United States.

Paramethoxyamphetamine (PMA, 4-methoxyamphetamine) is an illicit, synthetic hallucinogen having a pharmacology similar to MDMA. A Schedule I drug, PMA is potent and potentially lethal

STP (dimethoxyamphetamine, also known as DOM) has gained an aura of mystery, since no one really knows what the initials stand for. Achieving notoriety in the summer of 1967, STP got the reputation of being a powerful CNS stimulant capable of causing auditory and visual hallucinations, telepathic powers, and a euphoria lasting for up to 14 hours. It is less popular today, possibly because of its high incidence of bad trips. STP is a synthetic substance, active in doses of a few milligrams. On a weight basis, therefore, it is about one-tenth as active as LSD.

Knowing that STP is an amphetamine, we could have predicted many of its pharmacological effects: an increase in heart rate and blood pressure, dilation of the pupil of the eye, bronchodilation, decreased appetite, and CNS stimulation.

The abuse potential for all hallucinogenic amphetamine derivatives is high, and chemical dependence is the most serious threat in the use of these drugs. Users come to depend upon the feelings of peace, insight, openness, and self-awareness they experience from these amphetamine derivatives. They term MDMA a consciousness-effective drug. Because these drugs are sympathomimetics related to amphetamine, they should not be used by anyone with high blood pressure, heart or cardiovascular disease, diabetes, or hypoglycemia. If you have a history of seizures, stay away from these drugs. Breast-feeding mothers and women who are pregnant or think they are pregnant should not ingest any of these drugs. Users should not attempt to drive a car or use machinery even though they think they are perfectly capable of doing so. Severe, day-long panic and anxiety attacks have been attributed to DOM.

Drug Interactions: It is critical that these drugs not be combined with MAO inhibitors such as Parnate or Nardil because dangerously high blood pressure could result. Combining any of these drugs with amphetamines or any other brain stimulant could result in dangerous overstimulation.

12.6.5 2-CB (Nexus, Eve, Venus), a Beta-Phenethylamine

Closely related chemically to STP is the psychedelic **2-CB** (4-bromo-2,5-dimethoxyphenethylamine). See Figure 11a in Appendix I and accompanying discussion for the chemistry of beta-phenethylamines. In doses of 8–30 mg, 2-CB can produce powerful visual experiences with vivid colors and intricate patterning. It has been termed a "psychedelic empathogen."

12.6.6 Indolethylamine Hallucinogens

In Appendix I, Figure 3 shows the presence of the **indolethylamine moiety** in some important biochemical substances. This same indolethylamine structure occurs in the common amino acid tryptophan, which can be considered the potential biogenic precursor of *N,N*-dimethyltryptamine (DMT), *N,N*-diethyltryptamine (DET), serotonin, the alkaloid ibogaine, the harmala alkaloids, psilocybin, lysergic acid amide (and LSD), bufotenine, melatonin, and 6-hydroxytryptamine. With the exception of serotonin, all of these indolethylamine derivatives are hallucinogenic; serotonin itself is an important neurotransmitter (see Section 5.7).

DMT occurs naturally in plants growing in the West Indies, but for the American drug scene it is made synthetically in illicit labs. A uniquely rapidly acting psychedelic, DMT produces effects within seconds after it enters the blood stream; its effects last for only 15–30 minutes. Additionally, two synthetic analogs, DET and DPT, have appeared on the street.

Known on the street as "Trip" and "ET," alpha-ethyltryptamine was first sold in the 1960s as the legal drug Monase. It was used to treat depression but after 1 year was taken off the market because of its toxicity (including a death).

Bufotenine (5-hydroxy-*N,N*-dimethyltryptamine) occurs in the skin glands of certain toads of the family *Bufonidae*. (Commonly, these amphibians are incorrectly referred to as frogs.) Bufotenine is recognized as a potent hallucinogen, and we now have reports from Colorado of "frog" licking as a novel route of administration of this substance! In Sonoma, California, a married couple was convicted of possession of bufotenine, which they obtained by squeezing venom from their four Colorado River toads, drying it, and then smoking it for a psychedelic high.

The analogous 5-methoxydimethyltryptamine (5-MeODMT) is found in snuffs used by Amazonian Indians. It has only brief psychotropic action.

Case History *Toad Secretion Fatalities*

> The Centers for Disease Control and Prevention reported that four men died in one 6-month period after they used a purported aphrodisiac made from the dried secretion of a toad. The brown, rock-like substance, sold under the name "Stone" in containers without labels listing ingredients or directions for use, was applied to the genitals. At first it induced vomiting; subsequently, severely erratic heartbeats resulted in death. A 17-year-old survived, but doctors worked 12 hours to restore a normal heartbeat.

12.7 Solvents and Inhalants

Inhalants are substances whose vapors can be breathed into the lungs to produce a mind-altering effect. Included in this category are aerosol propellants, toluene, butane, nitrous oxide, benzene, carbon tetrachloride, cleaning fluid, correction fluid, lighter fluid, shoe polish, fingernail polish, gasoline, glue, paint thinner, spot remover, and varnish and the volatile solvents in rubber cement, writing markers, and hair sprays. Pharmacologically, these substances are best described as brain depressants and deliriants. Most of the sniffing of solvents and inhalants, termed "huffing," is done by teenagers (or younger children) who seem to want to put themselves into a delirious, semiconscious state of altered awareness. They call this getting high. One sniffer said that when he inhaled gasoline he felt like he was floating through the air. Actually, these persons are replacing the oxygen in their inhaled breath with a hydrocarbon or other solvent that can have a brain depressant action or a heart stimulant action. As more of the substance is inhaled, the person can become restless, excited, confused, disoriented, and finally comatose. When these chemicals are inhaled there is a real danger of dying of asphyxiation. Examples of this "sudden sniffing death" are numerous. Three young men in Los Angeles died after inhaling "laughing gas" from an 80-pound canister; a 220-pound 16-year-old died after inhaling gas from an aerosol can of air freshener; a 15-year-old boy died when he inhaled propane from a backyard gas grill (called "torch breathing"); and a 22-year-old physics major at an eastern institute of technology died after using a plastic bag to inhale nitrous oxide.

In one recent year The International Institute for Inhalant Abuse reported the death of 60 American youths from "huffing" inhalants. The Institute also reported that 1 in 5 eighth graders had used inhalants at least once. Between 1994 and 2000, the number of new inhalant users increased more than 50 percent, from 618,000 new users in 1994 to 979,000 in 2000. SAMHSA says these estimates were higher than a previous peak of 662,000 in 1978.

Nitrous oxide (laughing gas, N_2O, "Whippets") has been used in anesthesia for 100 years and continues to have wide application in hospitals and in the practice of oral surgery. In hospital surgical procedures, N_2O is typically administered by inhalation in combination with oxygen and other agents; it is not a potent anesthetic, and its effects wear off rather quickly. Nitrous oxide is a common drug of abuse. The gas is inhaled from cylinders, balloons that have been filled with it, or aerosol spray cans in which it is used as a propellant. Nitrous oxide induces pleasurable sensations, including auditory illusions, giddiness, exhilaration, or disinhibition. Users say they can get "high" on it. Although nitrous oxide obviously is not acutely poisonous, there is always the possibility that the abuser will inhale so much of it (to the exclusion of oxygen) that asphyxiation and irreversible brain damage will result. There have been literature reports that link nitrous oxide with infertility, birth defects, and spontaneous abortion. These, however, are from chronic exposure, as in the workplace. The occasional minor inhalation of nitrous oxide probably will produce no significant bad effects. As a gas, nitrous oxide is colorless, nonflammable, sweet-tasting, but essentially odorless.

Case History *Deaths from Huffing*

Sixteen-year-old Mike and his friend discovered that butane could be bought openly in stores near their home. It came in small aerosol type cans and was intended for use in refilling cigarette lighters. Labels on the cans indicated butane's toxicity and the importance of calling a doctor if the chemical was inhaled. Mike and his friend inhaled the butane to get high, Mike became ill and fell into a coma. Despite quick response by an emergency medical unit, Mike died within the hour of acute respiratory distress and cardiac arrest.

S., a 17-year-old wrestler and 4-H Club member, had previously been found huffing from an aerosol can of fabric protector, but reassured his parents that "you have to do it a long time before anything happens to you." Not long after, S. was found dead behind a 7-Eleven store, again with the aerosol fabric protector and a plastic bag. On autopsy, physicians noted that his lungs were filled with foam.

A 15-year-old girl was killed while riding as a passenger in a Mustang that crossed a highway median and crashed into an oncoming pickup. The boy driving the Mustang had blacked out after huffing Memorex Duster, a spray used to clean computer keyboards.

Hydrocarbons of the gasoline, benzene, paint thinner, and varnish types can depress the brain but also can sensitize the heart to the point where fibrillation and sudden death can occur. With smaller doses, eye irritation, headache, nausea, dizziness, and weakness can be expected. Glue sniffing in children can cause exhilaration, euphoria, excitement, slurred speech, and double vision. Stupor and unconsciousness may follow.

Propellants in common use until recently are the fluorochlorohydrocarbons (similar to the freons). Sniffing such aerosol sprays can be extremely hazardous because the heart can be sensitized to the point where fibrillation occurs. This is more likely to happen with high doses of aerosol sniffed in confined spaces. The abuse potential for these solvents and inhalants is high, and no matter what solvent or chemical is inhaled, asphyxiation is an ever-present hazard. All of the solvents (with the exception of carbon tetrachloride and certain aerosols) are highly flammable; a lighted cigarette could easily ignite them. Carbon tetrachloride and benzene are two volatile solvents that show the additional serious potential of carcinogenicity. This has been clearly demonstrated in laboratory animals and is presumed true in humans also.

Yet another category of inhalants is the **alkyl nitrites**. These include amyl nitrite, butyl nitrite, and isoamyl nitrite. They appear on the street in glass vials or ampules that can easily be broken or crushed to release the volatile liquid (hence the names "poppers" or "pearls"). Amyl nitrite, long used to relieve the pain of angina pectoris, has become popular as a means of enhancing sexual response. Amyl nitrite is a volatile liquid that, when inhaled, dilates blood vessels in the heart, brain, and other body organs. This action results in increased cardiac blood flow and relief of anginal pain. Use also appears to prolong the pleasure of orgasm in some individuals, and a new method for drug abuse has evolved based on this discovery. The drug, with a quasi-illicit status, is sold at high profit in adult bookstores and porno shops as RAM, TNT, Rush, Locker Room, or Moon Gas. Nitrites are sniffed in non-sex-oriented situations, too, for the high they produce. This high is the giddiness associated with a drop in blood pressure. Side effects of the use of amyl nitrite can include a dramatic fall in blood pressure with consequent dizziness and possible fainting. A throbbing headache results from expanded blood vessels in the brain. It is of interest that amyl nitrite is chemically related to nitroglycerin, another vasodilator used to relieve the symptoms of angina pectoris.

According to a NIDA survey, 6.9% of high school seniors have used poppers at least once. In 1990 Congress enacted a ban on alkyl nitrite products.

A potentially dangerous combination is the sexual potency drug **Viagra** (sildenafil citrate) with nitrates and nitrites such as amyl nitrite, butyl nitrite, or nitroglycerin. All of these drugs are powerful vasodilators, markedly lowering blood pressure. When combined, a dangerous, potentially fatal, drop in blood pressure may threaten. The acutely low blood pressure can manifest itself in faintness, dizziness, and mental and visual blurring.

On the Internet, teenage "huffing" is discussed at this URL. `http://www.teennewhorizons.com/huffing.htm`.

Drug Interactions: Most of the solvents and inhalants discussed here depress the brain, and the use of any other depressant such as alcohol will add to that depression. The use of epinephrine, amphetamines, or caffeine during the exposure to gasoline or the aerosol propellants can increase the chance of ventricular fibrillation.

12.8 Designer Drugs

When a legislature enacts a law controlling a certain drug—that is, controlling its manufacture, possession, or sale—the nature of the drug is specified. Its specific chemical structure is identified. Recognizing that, entrepreneurial chemists in clandestine laboratories have attempted to design and synthesize new substances that differ chemically only slightly from the original drug, yet keep all or nearly all of the pharmacological activity of the original. The idea was, if the new "designer drug" is in any way different, it cannot be included as a controlled substance under the original act. This devious rationale was effective until passage of the Anti-Drug Abuse Act of 1986, which anticipated these designer drugs and termed them **Controlled**

Substance Analogs. These analogs are not named as such under the original 1986 law, but because they are found in illicit traffic, have no legitimate medical use, and are structurally or pharmacologically similar to Schedule I or II substances, *they are treated under the Act as if they were controlled substances* in Schedule I. Of course, this anticipatory approach won't stop designer drugs from reaching the street; it just makes it possible to send the pushers to jail.

Fentanyl (see Section 6.10) is a narcotic analgesic used legitimately in surgery. Underground chemists have designed at least six analogs of it for sale to heroin addicts: alpha-methylfentanyl, *para*-fluorofentanyl, alphamethylacetylfentanyl, 3-methylfentanyl, and two others that may have been unexpected byproducts. Alpha-methylfentanyl, the first designer drug to appear on the street, was uncovered in 1979 in Orange City, California; it is 200 times more potent than morphine and very dangerous to use. The analog 3-methylfentanyl is 3,000 times more potent than morphine. These two drugs have caused more than 100 overdose deaths in California. Most of the designer drugs based on fentanyl are now Schedule I controlled substances. For chemical structures, see Figure 12 in Appendix I.

Since fentanyl analogs are so powerful and their dose so small, only tiny amounts of them will occur in body fluids, and this makes testing for them very difficult. Further, fentanyl analogs do not give positive urine tests for opiates (although a radioimmunoassay for them has been developed).

China White is an illicit street drug sold as synthetic heroin. Chemical analyses have shown that it is not actually heroin but one of several other compounds. One report identified China White as alpha-methylfentanyl (or 3-methylfentanyl). Another identified it as MPTP (methylphenyltetrahydropyridine). Obviously, there are no labeling requirements for street drugs, and it is impossible to prove the original nature of China White. Users of it or of any fentanyl analog are risking real danger because these drugs possess all of the pharmacological and toxicological actions of the classic narcotics and are far more potent. For example, they can cause severe respiratory depression.

MPTP is a designer drug disaster. An underground chemist in some illicit California laboratory apparently attempted to synthesize MPPP (methylphenylpropionoxypiperidine), an analog (designer drug) of the well-known controlled opioid Demerol. Through an error in the synthesis the unexpected compound MPTP was produced, and when it hit the streets it produced a crippling condition that closely resembles Parkinsonism. Hundreds of drug abusers—mostly heroin addicts—who unknowingly ingested MPTP became violent, hallucinatory, and eventually brain damaged and paralyzed. The condition appears to be irreversible. MPTP originally hit the streets as a tan or off-white powder but now is reported to be pure white and crystalline. It does not have the talcum consistency of fentanyl. Users have reported loss of memory, a burning sensation produced immediately after injection, drooling of saliva, sweating, and involuntary jerking of the arms and legs.

Scientists are very interested in this situation because they see possible clues to the explanation of the cause of Parkinsonism in the general population. Five legitimate chemists who worked with MPTP years before designer drugs were even dreamed of have come forward with the information that they have developed Parkinson-like symptoms to varying degrees. Here is an example of people unwittingly offering themselves as guinea pigs in drug testing, a protocol that is prohibited by law for the rest of the population.

In 1984 another Demerol analog (designer drug), 1-(2-phenylethyl)-4-phenylacetoxypiperidine (PEPAOP), was found in a sample of "synthetic" heroin.

To help you recognize the multitude of street drug names, here is a Web site that lists thousands of names:

`http://www.whitehousedrugpolicy.gov/streetterms/default.asp`.

12.9 Look-Alike Drugs

Do you know what a peashooter is? You may not, for there is a new definition for this old term. "**Peashooters**" are look-alike drugs, replicas of pharmaceutical amphetamines, cocaine, and prescription downers and uppers. These drugs mimic the size, color, shape, and effects of the controlled substance but contain only legally salable substitutes.

We have been experiencing a peashooter explosion. Adults and youngsters have been bombarded with ads for over 100 mail-order products, including Cocaine, Toot, CocoSnow, Pseudo-Speed, PseudoCaine, Milky Trails, Florida Snow, Supercaine, Pink Ladies, Speckled Pups, Co-Cokaine, and Amphetrazine.

Until recently, at least, look-alike drugs were legal, for although they look almost identical to the pharmaceutical drug, they contain only nonprescription, OTC-salable drugs. What is more, many of their ingredients are found in legal OTC appetite suppressants and nasal decongestants. Sedative look-alikes contain 25–50 mg of the antihistamine doxylamine, also found in the proprietaries Formula 44 and Nyquil.

What is so disturbing about look-alikes is the deceptive tactics employed in their promotion. Free samples are offered to youngsters. Hence, look-alike drugs often provide children with their first introduction to drug abuse and are liable to lead to abuse of other drugs. Pushers tell schoolchildren that the look-alike is "100% legal" and not as potent or dangerous as the real thing. Nonetheless, serious poisonings and deaths have resulted. The American Medical Association reported on 12 deaths from the use of counterfeit "black beauties" that looked just like the real biphetamine. There have been many other reports of toxic reactions to look-alikes. One authority said, "It's like playing Russian roulette, taking these kinds of drugs." Teenagers who take look-alike downers often ingest them with alcohol—a dangerous combination that can have fatal consequences.

Cocaine look-alikes are generally sold as "incense." They contain no cocaine but have large amounts of caffeine, phenylpropanolamine (PPA), and benzocaine. Table 5.3 notes that PPA is a sympathomimetic used as a decongestant.

Speed look-alikes typically contain caffeine (100–200 mg), ephedrine (12–15 mg), and PPA (35–50 mg). Quaalude peashooters are more like counterfeit drugs; they look much like a 300-mg methaqualone tablet and even have "Lemon 714" or "Lennon 714" stamped on them (note that the real manufacturer is the Lemmon Company). The ingredients are typically acetaminophen and aspirin.

Look-alike drugs have netted millions of dollars for the more than 350 mail-order firms that sell them. Advertisements have appeared in mainstream publications and student newspapers. Society, however, has responded. Forty-seven states have now enacted look-alike substances laws; that of Illinois specifies a 5-year prison term and a $20,000 fine for possession, manufacture, advertising, or distribution of look-alikes. These laws specify that look-alikes can no longer physically resemble the controlled substance. While there is no federal look-alike law as such, there is a federal law that limits look-alikes to contain no more than one active ingredient.

12.10 Dimethyl Sulfoxide

Dimethyl sulfoxide [$(CH_3)_2SO$, DMSO] is an industrial solvent derived from wood during paper production and used since the 1940s as a degreaser or cleaning solvent. In 1963 the University of Oregon Medical School reported that when rubbed on the skin DMSO showed remarkable penetrating power and was quickly absorbed into the body, where it relieved pain and inflammation. In a short time word spread about this "miracle" drug, and it soon became a "sure cure" for arthritis, sprains, burns, herpes infections, and high blood pressure.

Whether applied to the skin or taken internally, DMSO rapidly enters the bloodstream and soon appears on the breath with an unmistakable garlic-like odor. When used topically (e.g., in an ointment) it breaks down the skin's natural barrier to bacteria. Some people are severely allergic to DMSO.

The FDA has been looking at DMSO for many years but has steadfastly refused to approve it for uses for which it has not been shown to be safe and effective. For example, the FDA says that DMSO's powerful penetrating action could cause an insecticide on a gardener's skin to be carried accidentally into his or her bloodstream. The FDA has approved DMSO for use in certain bladder conditions and as a veterinary medicine for topical use in nonbreeding dogs and horses. Meanwhile, the 99% industrial solvent solution continues to be sold at roadside stands, at least one used vacuum cleaner store, and gas stations at exorbitant prices. It attracts a large following, and it is legal in Florida and Oregon. One DMSO advocate claims that the FDA is engaged in a "witch-hunt" against the drug. Little progress is being made on a final solution of the DMSO controversy.

12.11 "Smart Pills"

It was a 1988 FDA decision to allow the importation and private use of non-approved medications (if they have been deemed safe in other countries) that opened the doors to "smart pills"—drugs or nutrients that supposedly boost memory, intelligence, and brain power generally. Also termed *nootropics*, these substances are touted to improve problem-solving ability, spur productivity, eliminate depression, increase concentration, sharpen memory, enhance music enjoyment, and even assist sexual performance.

As can be expected, there are individuals who swear by these pills, citing "research" done in Europe as well as their own and others' personal experience. Some of the most promoted smart pills are choline, Hydergine, Deprenyl, phosphatidylserine, and piractum. Others that have been mentioned include acetylcarnitine, centrophenoxine, the herbal ginkgo biloba, and vasopressin, and there are dozens more. (You will note that some of these drugs have legitimate use in modern medicine. Deprenyl is used in treating Parkinsonism; Hydergine is available on prescription for elderly patients with declining mental capacity.)

Do smart pills really work? The Center for Science in the Public Interest, a Washington-based consumer organization, reviewed the research (or lack thereof) supporting claims for "memory pills," and concluded, "Don't waste your money." A new study, reported in the *Journal of the American Medical Association*, suggests that ginkgo supplements do nothing to quickly improve memory in healthy people.

One is moved to ask the following questions of the proponents of smart pills: Can data from rodent studies be extrapolated to the human brain? If we remember better, are we necessarily more intelligent? Is it memory or attentiveness we're testing? Is coffee a cheaper way to get the same effect? Where have you published results from double-blind, placebo-controlled studies? What are the known adverse effects of drugs such as Deprenyl, hydergine, and piractum?

DHEA (dehydroepiandrosterone) is a hormone made in the human that the body uses to make other substances. DHEA is touted by health-food stores as a veritable fountain of youth, but experts in the field recommand avoiding it.

12.12 Gamma-Hydroxybutyrate

The Centers for Disease Control and Prevention have received reports from eight states of poisonings from the drug gamma-hydroxybutyrate (GHB). An equivalent name for this substance is gamma-hydroxybutyric acid. GHB is a normal component of mammalian metabolism. This drug, once sold in health-food stores as a sedative, is now banned but continues to be sold on the Internet as a "substitute" anabolic steroid, muscle enhancer, and sleep inducer. It is not to be confused with the neurotransmitter gamma-aminobutyric acid (see Section 5.7)

Chemically, GHB is neither a steroid nor an amino acid. Toxicologically, it is a highly addictive poison capable of causing severe heart and respiratory problems, vomiting, dizziness, loss of consciousness, tremors, and low blood pressure. Despite claims of some health food stores, there are no data to show that it stimulates growth hormone production. Used in nightclubs and raves, GHB produces giddiness and euphoria, followed by possible nausea, vomiting, and sudden, instant unconsciousness. GHB is illegal except for research purposes in the United States. It has been implicated in a date-rape drug death of a 15-year-old.

Blue Nitro is an illicit product containing gamma-butyrolactone (GBL), a solvent for paint thinners. When ingested, this lactone is converted by the body to GHB. Blue Nitro combined with alcohol can be deadly.

Case Histories *GHB Intoxication*

(a) A Georgia teenager getting ready for his prom drank a concoction of water and Somatomax PM, a powder containing GHB his friend had bought at a health food store. Instead of getting the high he expected, the teenager fell into a coma 20 minutes after taking the drink. His parents soon found him, and with emergency treatment he recovered.

(b) A 15-year-old died of an overdose of GHB at an impromptu rave party in the Southern California high desert.

12.13 Other Drugs

Ketamine (trade names Ketalar, Ketaset) is a dissociative, rapid-acting, nonbarbiturate general anesthetic that is infrequently used in humans but is popular in veterinary medicine. Ketamine is chemically related to PCP, and like PCP, it can cause emergence reactions in recovery from anesthesia. A small percentage of humans anesthetized with ketamine will experience delirium, hallucinations, or simple dreamlike states as the effects of the anesthetic wear off. In a few patients the emergence reaction is accompanied by confusion or irrational behavior. These psychotropic reactions of ketamine and its use on the campus scene have won it the sobriquet of *psychedelic anesthetic*. Ketamine is sold on the streets of Washington, D.C. as "Special K" for snuffing or smoking. It is one of the drugs used in the rave culture. An excellent URL for information about it is **`http://www.erowid.org/chemicals/ketamine/`**.

Khat (pronounced "cot"), an ancient drug popular in East Africa and Southern Arabia, is a natural CNS stimulant found in the leaves of the shrub *Catha edulis*. Many Somalians chew the leaves to sharpen their thinking and allay hunger. Also known as "Qat," the drug is sold in markets all over Yemen. The active ingredient in khat is the alkaloid cathinone, a chemical relative of ephedrine. The methyl derivative of cathinone, methylcathinone, was synthesized in 1957 in Great Britain and patented as a diet aid and antidepressant, but was never marketed because of its addiction potential. After the formula for methylcathinone became known, it was illicitly made and distributed under the name *cat*. See the accompanying discussion of cat.

Cat, or **methylcathinone**, is a synthetic CNS stimulant of high potency and dependence liability. It appears to have hit the streets first in Ann Arbor, Michigan; its use is frequent in Michigan's upper peninsula. Cat is usually snorted but can also be injected or drunk with a beverage. It produces an initial euphoric burst of energy, talkativeness, and hyperactivity. Users typically progress to binging, but discover that after the binge they might experience excruciating nervousness, anxiety, paranoia, and hallucinations. Appetite loss can lead to massive weight loss. Manufacture of cat or possession with intent to distribute is a violation of federal law and is punishable by a prison term of up to 20 years and a fine of up to $1 million. You can browse for cat at the following URL: **`http://www.drugs.indiana.edu/pubs/factline/cat.html`**.

Kava (or kava-kava or "awa") is typically a water extract prepared from the root of the pepper plant (*Piper methysticum*), used for hundreds of years in rituals and in medicine throughout the South Pacific. Kavalactones, the active ingredients of popular kava tea, are said to produce CNS depression, calming, sociability, muscle relaxation, and to allay anxiety. The FDA warns that kava herbal supplement may be linked to serious liver injury. A San Francisco driver and native Tongan faced DUI charges after drinking kava tea.

12.14 Drugs in Our Society: Some Observations

From this perspective, it would be fair to say that America is awash in drugs. Prompted by incessant, insidious advertising, the American public continues to buy over-the-counter drugs at a staggering rate (see Chapter 15). From the National Household Survey on Drug Abuse we learn that in 2001 15.9 million Americans age

12 or older (7.1% of the population) used an illicit drug in the month prior to the interview. Among young adults 18–25, 18.8% are current illicit users. Almost half of all Americans age 12 or older (48%, or 109 million persons) are current drinkers. Among ages 12–20, nearly 7 million, or 19%, are binge drinkers, and 6% are heavy drinkers. In 2001, more than 1 in 10 Americans (25 million persons) reported driving under the influence of alcohol at least once in the prior twelve months. That rate is increasing. Past-year abuse or dependence is reported in Figure 12.4.

An estimated 66 million Americans 12 or older reported current use of a tobacco product in 2001. That's 29% of the population, and includes smokeless tobacco.

In 2000, more than 2 million youths aged 12–17 reported using inhalants at least once in their lifetime. Youths who reported an average grade of D or below were more than three times as likely to have used inhalants as youths with an average grade of A. See Figure 12.5 for types of inhalant chemicals.

A survey of hundreds of California physicians who were drug abusers found that the seven most-often abused drugs were (in order) alcohol, Vicodin, Demerol, fentanyl, cocaine, amphetamine, and marijuana.

Ecstasy use is on the rise. In 2000 an estimated 1.9 million people used Ecstasy (MDMA) for the first time (up from 0.7 million only two years before). Overall, *the*

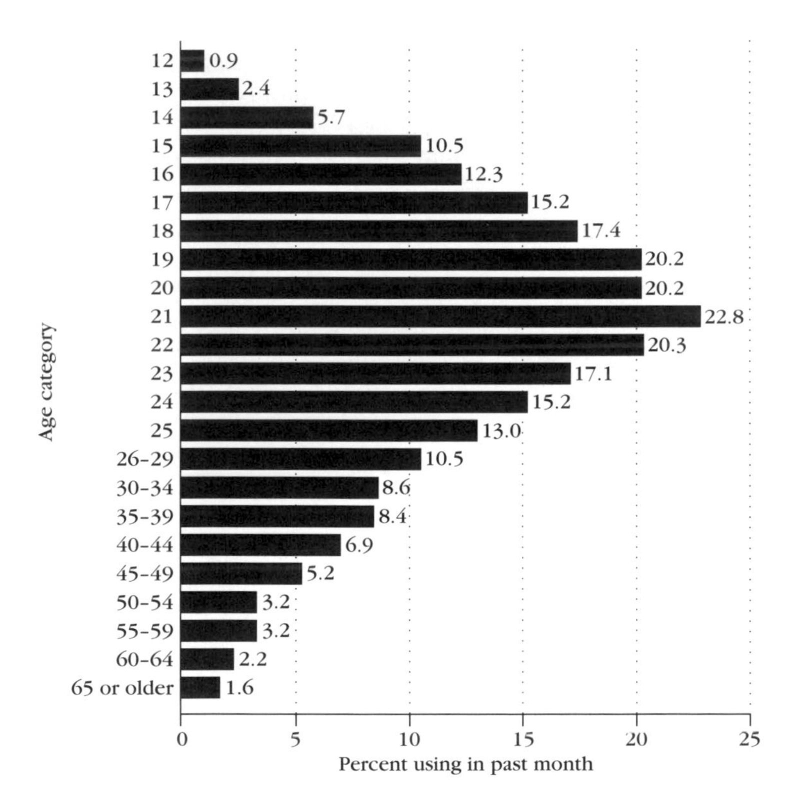

Figure 12.4 Percentages of persons aged 12 or older reporting past-year abuse of or dependence on alcohol or any illicit drug, by detailed age categories: 2001. (Source: 2001 National Household Survey on Drug Abuse, SAMHSA, October 2002.)

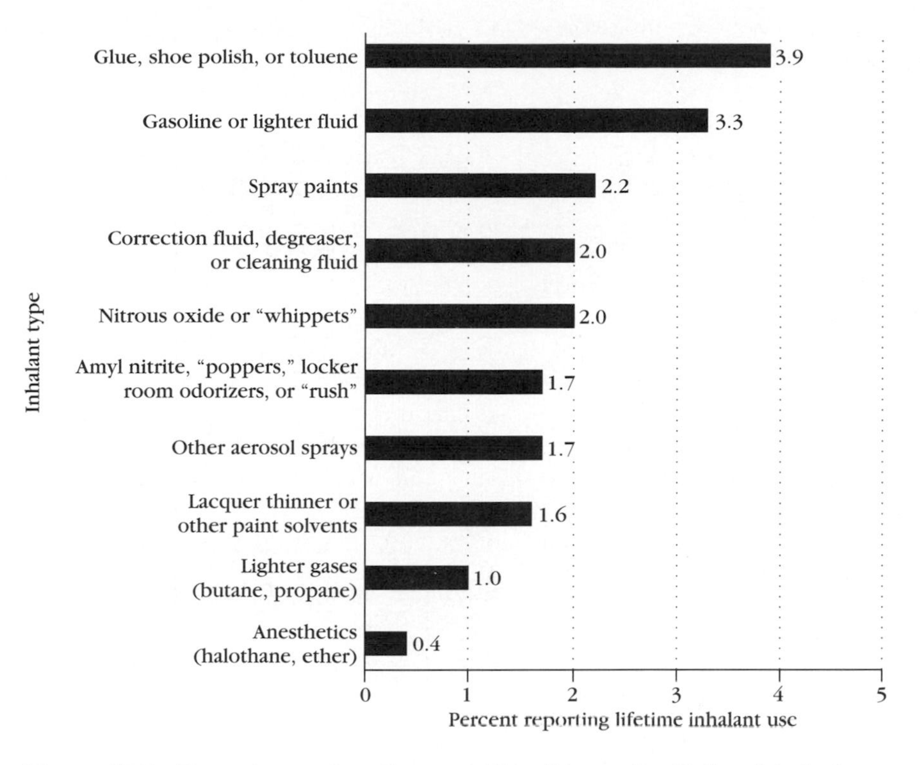

Figure 12.5 Percentages of youths aged 12 to 17 reporting lifetime inhalant use, by inhalant type: 2000. (Source: 2000 National Household Survey on Drug Abuse, SAMHSA, March 2002.)

number of persons with substance dependence or abuse increased from 14.5 million in 2000 to 16.6 million in 2001. For society, the medical and economic implications are great. See Table 12.3 for Emergency Department (ED) data on other club drugs.

Americans spent $53.7 billion in one recent year on illegal drugs; of that, $38 billion was spent on cocaine, $9.6 billion on heroin, and $7 billion on marijuana.

What factors might account for the upswing in drug use? The advertising of alcohol and tobacco products in this country is insidious and inexcusable. Young children recognized the cartoon character used in cigarette ads more readily than they recognized Mickey Mouse. Someone estimated that a typical child growing up today will have witnessed 10,000 beer commercials on TV by his or her 21st birthday. From my close contact with many hundreds of college students, I have concluded that drug use is indigenous in a significant segment of our society. Drug taking is accepted, enjoyed, and expected. Some vigorously defend their usage (at least for now). After discussions with them, I cannot imagine that their attitudes could be changed at this stage.

America's appetite for new drugs and new ways to alter consciousness is seemingly insatiable. Rohypnol, GHB, khat, methylcathinone, smart pills, TRIP, MDMA (Ecstasy), toad venom, ice, rave parties, all discussed earlier in this book, are new and eagerly embraced. Inhalants are a good case in point. Who would have expected or predicted that rubber cement, hair spray, spot remover, and fingernail polish

Table 12.3 Presence of other drugs in ED visits involving club drugs: estimates for the coterminous U.S., 2001

Drug categories and episode characteristics	*GHB*	*Ketamine*	*LSD*	*MDMA*	*Methamphetamine*
Drug concomitance					
Single-drug visits	870 (26%)	175 (26%)	671 (24%)	799 (14%)	6,715 (45%)
Multi-drug visits	2,469 (74%)	504 (74%)	2,150 (76%)	4,743 (86%)	8,208 (55%)
TOTAL CLUB DRUG VISITS	3,340	679	2,821	5,542	14,923
Selected drug combinations					
GHB	—	16 (2%)	2 (0%)	393 (7%)	55 (0%)
Ketamine	16 (0%)	—	50 (2%)	262 (5%)	65 (0%)
LSD	2 (0%)	56 (8%)	—	449 (8%)	82 (1%)
MDMA (Ecstasy)	393 (12%)	262 (39%)	449 (16%)	—	279 (2%)
Methamphetamine	55 (2%)	65 (10%)	82 (3%)	279 (5%)	—
Alcohol-in-combination	1,797 (54%)	222 (33%)	1,022 (36%)	2,666 (48%)	3,424 (23%)
Amphetemines	111 (3%)	8 (1%)	89 (3%)	234 (4%)	130 (1%)
Cocaine	174 (5%)	93 (14%)	713 (25%)	1,606 (29%)	1,968 (13%)
Heroin	16 (0%)	117 (17%)	194 (7%)	455 (8%)	537 (4%)
Marijuana	475 (14%)	85 (12%)	1,554 (55%)	1,836 (33%)	3,338 (22%)

Source: Office of Applied Studies, SAMHSA. Drug Abuse Warning Network, 2001 (03/2002 update).

remover would develop into social hazards, not to mention the gas grill out on the patio?

Local, national, and international drug trafficking has assumed gigantic proportions. According to the NNICC prepared by the DEA, the amount of cocaine seized worldwide in 1996 exceeded 200 metric tons. U.S. Federal law enforcement agencies alone seized 135 metric tons in fiscal year 2000. And that is just how much was intercepted. How much got through? Drug distribution networks in cities, states, and nations are elaborate, and the top-level dealers are nearly impossible to find and prosecute. For example, a report states that the close-knit group that synthesizes and distributes LSD from the San Francisco Bay area has eluded detection for over 10 years. Then consider the great number of nations around the world that are distributing drugs. The illicit drug in your neighborhood might have come from Burma, Thailand, China, Vietnam, North or South Korea, Malaysia, Laos, Singapore, Australia, Hong Kong, Afghanistan, Pakistan, Iran, Lebanon, Turkey, the Balkans, Belize, Mexico, Colombia, Guatemala, Ecuador, Denmark, or Nigeria, according to the NNICC. With such a fantastic supply, reservoir drug availability is assured, and with the trafficking skills of the smugglers, drug distribution is guaranteed.

Great ingenuity is applied in the smuggling of drugs across our borders, often with comparable audacity, as witness the cocaine in the dog's abdomen and the hundred drug-filled balloons and capsules swallowed by one human "mule." The DEA has lawfully seized and sold fast airplanes used in drug trafficking, only to later seize the same airplane a second—and even a third—time.

Highly intricate schemes exist throughout the world for laundering drug money. So great is the volume of small bills that drug couriers are forced to take them out of the country jam-packed in suitcases.

Gateway drugs are recognized. Marijuana, the first illicit drug most American youth experiment with, can be the gateway to more dangerous drugs. The comment made by the recovering heroin addict (see the Case History, Section 6.11) is revealing: "I've never met anyone who used heroin as a first drug."

Some interesting information comes from the national surveys. As a post-high school experience, marriage is consistently associated with declines in alcohol use in general, heavy drinking in particular, marijuana use, and use of other illicit drugs. Cocaine is a drug used more frequently among people in their twenties than among those in their late teens; this fact continues to distinguish cocaine from all other illicit drugs. Ecstasy (MDMA, Section 12.6) has only recently become important; annual use is highest among 19–22-year-olds (i.e., college age). This agrees with data I have received from England. Regarding male-female differences, males are more likely to use most illicit drugs. For example, annual marijuana use among high school seniors in 1994 was reported by 35% of males but only 26% of females. The only exceptions to this rule occur for stimulants, sedatives, and tranquilizers, where the sexes attain near parity or female use is slightly higher.

Legalization of today's illicit drugs is receiving serious attention as the answer to our intolerable worldwide drug problems. Important officials in the government and the media have called for repeal of laws regulating marijuana, cocaine, heroin, and other drugs. They say that drug laws—not drugs themselves—cause the most damage to society. Repeal the laws, the argument goes, and drug empires would collapse, black market drugs would disappear, addicts would no longer commit crimes, and our courts and prisons would no longer be overwhelmed. "Let's take the profit out of drug trafficking" is a popular theme. Prolegalization voices recall the failure of alcohol prohibition and point to the failure of present-day controls that result in surges of killings by drug dealers, overcrowded jails, money laundering, and an annual cost to the nation of about $60 billion from drug-related crimes, police actions, lost productivity, injuries, and other damages. It is argued that if Americans can legally kill themselves with alcohol and tobacco, why not also with cocaine and heroin? Besides, consider the tax revenues that could be realized. Furthermore, they argue, alcohol and tobacco are legal on the one hand, but marijuana and heroin are illegal on the other; this does not make sense since all of these drugs have high abuse potential. Another prolegalization argument says that criminalizing drugs such as marijuana and cocaine means creating addicts and putting them into the hands of unscrupulous drug barons; if these drugs were legal, this would not happen. Legalization followed by education would work, they say, citing the clear reduction in our use of tobacco products and the increased consumption of wine (at the expense of whiskey).

Not so, say the law-and-order voices. Legalizing drugs would be like legalizing murder and rape (it "solves" the problem by redefining it). To legalize our illicit drugs would result in their unlimited access, and that would create problems far worse than the present ones. Legal cocaine, marijuana, LSD, heroin, or amphetamines would mean that highly damaging substances would be cheap, pure, and far more widely available, and that would result in sharp increases in addiction, overdoses, family disintegration, hospital costs, and property damage. Heroin's pleasures, for example, are so great that if it were legalized in the United States we could expect a surge of heroin use that would make our present problem pale in comparison. Further, the counter argument goes, the fact that we are burdened with socially acceptable deadly substances such as tobacco and alcohol does not make it logical to add more. A pertinent URL is: **http://www.nationalfamilies.org/publications/catalogue.html**.

Critics from both sides of the question cite Switzerland's approach to the problem. An early five-year policy of legalizing drugs produced an increase in crime and a burden on the Swiss health-care system as users and addicts moved in from all over Europe. Recently the Swiss approved the legal sale of pot so long as it would not be used as a hallucinogen. As a result, the Swiss cannabis business has mushroomed, with 300 "hemp shops" in existence. Official figures show 600 acres of fields yielding pot for about 600,000 regular or occasional users.

If legalization of one or more illicit drugs were to occur (and frankly, the nation does not seem anywhere near this step), we would be forced to decide how to implement it. Who would import and assay the drug, and set the price and the tax rate? At what age would minors be excluded? How would we punish a person who sold to a minor? Would driving under the influence be a crime? Would sales on Election Day? Would the newly legalized drug be sold in vending machines, supermarkets, state stores? Would a manufacturer be permitted to advertise its brand of drug? Who would pay the cost of the bureaucracy required to oversee all of this? Would we have homegrown cocaine, or continue to import it? Would we open our air lanes to direct flights from South America? Would Medicare and Blue Cross pay for hospitalization from drug overdose?

Case History *Marijuana Sold Retail*

Venlo, a city of 65,000 in the southern Netherlands, has long participated in its government's liberal drug policies. Venlo has five licensed coffee shops in which customers can select their favorite brands of marijuana and hashish from among heaping plastic Tupperware-like containers. Sale of cannabis remains illegal in neighboring Germany, and of course increasing numbers of Germans are driving the short distance to Venlo to purchase 5 g of the drug, the maximum sold in the licensed shops. Because buyers usually want more than 5 g, they next visit some of the 65 illegal shops that have sprung up in Venlo. Further, street hawkers aggressively approach passersby offering all kinds of drugs. To counter this situation, Venlo town officials plan to open drive-through shops near the German border from which tourists can purchase drugs from their cars—and away from downtown Venlo. Ironically, selling small amounts of "soft" drugs at retail is tolerated by the Dutch government, but the acquisition of bulk, wholesale amounts by the same licensed shops is actually illegal and thus not to be tolerated. Nonetheless, officials so far have turned a blind eye.

Everyone on both sides of the legalization question agrees that education about the effects of drugs is essential. After all, if there were no demand for drugs in America, there would be no drug problem. Education must begin at a very early age and must never stop.

In one school district, 3,500 seventh graders were subjected to Life Skills Training. They were taught how to deal with stress, family matters, ego, and how to avoid drugs. They were followed up in the eighth and ninth grades. There was a 40% reduction in alcohol, marijuana, and tobacco use.

Partnership for a Drug-Free America, a nonprofit coalition, released a study on the Internet (see **`http://www.drugfreeamerica.org`**) that concludes the more adolescents hear from parents about the risks of drugs, the less likely they are to use drugs.

In a novel project, dozens of British schoolchildren were taken to Amsterdam's cannabis cafes one summer to "educate them about drugs" and as a reward for completing a weekend and residential course on the effects of drugs. Upon return to Britain, the children gave presentations to younger pupils on what they had learned. A national parent-teachers association criticized the project as giving too much information to naive children.

Another approach advocates the use of the media to help fight drug use. Ads on TV, radio, films, video, the Internet, CD-ROM, and multimedia are promoted as the fastest, strongest, and most effective educational tool to counter drug abuse. NIDA notes that media advertising works best when combined with other preventive programs in the community. NIDA concludes that certain types of ads achieve better results. For example, messages that encourage audiences to think about issues, as opposed to celebrities delivering slogans, tend to produce longer-lasting change. Similarly, research-based material works better than shock tactics.

Problem 12.1 *To which age group and socioeconomic profile would you direct media efforts in drug abuse prevention? Explain your answer.*

12.15 The Treatment of Drug Abuse

In the first large-scale study designed to evaluate drug-abuse treatment outcomes among adolescents in age-specific treatment programs, NIDA-supported researchers found that longer stays in these treatment programs can effectively decrease drug and alcohol use and criminal activity and improve school performance and psychological adjustment (Figure 12.6).

The study addressed peer relationships, educational concerns, and family issues such as parent-child relationships and parental substance abuse. Participation in group therapy and a twelve-step program were also included. Considering the year before treatment to the year after, the adolescents showed significant declines in the use of marijuana and alcohol, the major drugs of abuse for this age group. Adolescents also reported fewer thoughts of suicide, lower hostility, and higher self-esteem. This community-based program did not, however, significantly improve on cocaine and hallucinogen abuse.

	Before	After
DRUG USE	Percentage	
Weekly marijuana use	80.4%	43.8%
Heavy drinking	33.8	20.3
Hallucinogen use	31.0	26.8
Stimulant use	19.1	15.3
SCHOOL PERFORMANCE		
Regular attendance	62.6	74.0
Grades average or better	53.4	79.6
CRIMINAL ACTIVITIES		
Any illegal act	75.6	52.8
Any arrest	50.3	33.9

An evaluation of more than 1,100 adolescents who received substance abuse treatment in residential, short-term inpatient, or drug-free outpatient programs found improvement in rates of drug use and social behavior. Some 53 percent of those treated met or exceeded the minimum recommended stay in treatment.

Figure 12.6 Behaviors of adolescents before and one year after treatment. (Source: NIDA Notes. National Institute on Drug Abuse, NIH, USDHHS, April 2002.)

Web Sites You Can Browse on Related Topics

LSD-MKULTRA Story
Search "MKULTRA"

Inhalants
http://www.inhalants.org
Also: search "huffing"

PCP (phencyclidine)
http://www.nida.nih.gov/infofax/pcp.html
http://www.drugfreeamerica.org.pcp.html

Designer Drugs
http://www.designer-drugs.com/synth/index.html

Rohypnol
http://www.teenchallenge.com/main/drugs/rohypnol.htm

Legalization of Drugs—"Anti" Stand
http://www.usdoj.gov/dea/demand/druglegal/

Drug-Related Street Names
Search "drug street names"

Psychoactive Drugs
http://www.erowid.org/psychoactives/psychoactives.html

Club Drugs
http://www.usdoj.gov/dea/concern/clubdrugs.html

Study Questions

1. Name six ancient mind-altering substances.
2. True or false:
   a. LSD *as such* occurs in nature.
   b. A typical dose required for an LSD trip is about 100 mg.
   c. LSD produces an increase in body temperature and heart rate.
   d. LSD users rapidly develop a tolerance to the drug.
   e. LSD is a generally successful drug in treating alcoholism.
   f. LSD's mechanism of action appears to be linked to serotonin receptors in the brain.
   g. LSD is a Schedule II drug.
   h. One dosage form for LSD is known as "blotter acid."
   i. LSD is easy to synthesize.
   j. In Table 12.2, Ecstasy is another name for MDA.
3. Examine Table 12.1 and answer the following.
   a. With which drug would alcohol likely form a dangerous combination?
   b. Name the benzodiazepine that is cited most frequently in emergency rooms.
   c. Explain why OTC sleep aids rank much higher in emergency room visits than OTC diet aids.
   d. Give one explanation of why the commonly used and apparently safe antihistamine diphenhydramine (Benadryl) ranks relatively high (number 14) on the list.
4. Define (a) hallucinogen, (b) depersonalization, (c) flashback, (d) psychedelic, (e) set and setting.
5. True or false:
   a. In smaller doses, PCP produces a spaced-out, floating euphoria.
   b. Overdosage with PCP produces an eyes-open coma with very high blood pressure.

c. Nystagmus consists of jerky movement of arms and hands.
d. Quaalude and Mequin are former trade names for PCP.
e. Methaqualone is a nonbarbiturate sedative-hypnotic drug.
f. Experience has shown that tolerance and serious addiction to methaqualone can develop.
g. Seizures are typically seen in abrupt withdrawal from barbiturates and methaqualone.

6. a. In what pharmacological category do we place drugs such as MDA, MDM, and DOM?
   b. What are the pharmacological effects of MDA on the eye, heart, and brain?
7. Do PCP, methaqualone, and MDA pass the blood-brain barrier? Would we expect them to pass the placental barrier?
8. Explain what is meant by *look-alike drugs*.
9. Name three legal OTC drugs that are typically found in look-alike speed products.
10. It has been said that the teenage and young adult years are high-risk years. Explain what this means.
11. This chapter has indicated that for some drugs there is a clear-cut pattern of synthesis → development → marketing → use abuse. If high-abuse-potential drugs are to be kept off the market, at what step in the process would they best be eliminated?
12. Match the drug in the right-hand column with the correct statement in the left-hand column. A drug may be used more than once.

a. The drug used by most Americans	1. DMSO
b. The drug that is used most by all age groups	2. amyl nitrite
	3. alcohol
c. An industrial solvent used by some to treat arthritis	4. lookalike drugs
d. An indolethylamine hallucinogen	5. marijuana
e. Peashooters	6. DMT
f. An inhalant used in sexually oriented situations	7. tranquilizers
g. The drug that, after alcohol and cigarettes, is used most by Americans	8. cocaine

13. At times, PCP use results in CNS depression. At other times, it causes stimulation. How can one drug have such opposite effects? Is alcohol a similar type of drug?
14. Why would California pass a law requiring typewriter correction fluid to be water soluble and not solvent based?
15. Name six chemicals that have been abused as inhalants. What are the most serious hazards in the use of inhalants?
16. Define "designer drug." Cite two legitimate prescription drugs that have been the chemical basis for designer drugs.
17. In Section 12.11 we learned that one of the drugs promoted as a "smart pill" is Hydergine. Check a pharmacology textbook to find the current medical use of Hydergine.
18. Take a stand either for or against legalization of cocaine, marijuana, and heroin. Prepare six cogent arguments in support of your position.
19. *Advanced study*: Tolerance to the effects of almost all hallucinogens develops quickly. A period of about 5 days is required before the user will again react to the drug. Suggest a mechanism in the brain that would explain this reaction. (*Hint*: Consider what is happening at the synapse.)
20. *Advanced study*: An international expert on drug dependence believes that getting hooked on drugs provides satisfaction and punishment at the same time. In other words, the drug that gives the pleasure is also a threat that the addict uses to compensate for

guilt feelings. It is the completeness of this system that makes attempts at therapy so difficult. Express your agreement or disagreement with this viewpoint.

21. The law in the Netherlands formerly permitted unrestricted sale of most street drugs in small amounts in bars, discos, and other places frequented by 18–25-year-olds. (Trafficking in larger quantities is forbidden.) The theory was that the open sale of small amounts takes drugs and the drug user out of the "gutter" of illegality. Do you agree that the United States should adopt a similar approach? State your reasons.

22. Here are seven examples of drug prevention programs that have been described as effective. Rank them in order of effectiveness, with "1" being the program you feel would work the best. Be prepared to defend your ranking.

- **a.** Airing TV public announcements that depict drug use as immature.
- **b.** Counseling youths who are experimenting but have not yet become dependent.
- **c.** Showing films that dramatically portray possible dangers of overprescribing drugs.
- **d.** Outlawing head shops and the drug paraphernalia they sell.
- **e.** Teaching values to young people through exercises in value clarification.
- **f.** Distributing pamphlets and booklets that describe drugs.
- **g.** Publishing beverage industry ads urging moderation in alcohol use.

23. In England, government antiheroin campaigns have used three approaches: (a) advertisements in youth magazines, in the press, and on TV portraying the inevitability of dependence and deterioration after taking heroin; (b) videos and teaching packs for use in schools and youth work settings designed specially to give youngsters the social skills to refuse drug offers from their peers; and (c) traditional materials based on "shock tactics." Which of these would you favor, and why?

24. If marijuana became legal in this country, would you be in favor of unrestricted advertising by the distributors?

25. "Brain Power" is NIDA's program to provide drug-abuse education materials for students and teachers. Using NIDA's Web site (**`www.drugabuse.gov`**), evaluate the program, and report to your class.

26. An article in a major U.S. newspaper states, "The drug trade begins with experimentation in youth." Do you agree or disagree? Reasons?

CHAPTER 13 The Pill: An Update

Key Words in This Chapter

- Oral contraceptive
- Progesterone
- Progestin
- Morning-after pill
- Adverse effects
- MiniPill
- Pills for men

Learning Objectives

After you complete your study of this chapter, you should be able to do the following tasks:

- Explain how progestins are related to progesterone.
- Explain how the Pill works to prevent conception.
- Discuss the risk to the cardiovascular system from using the Pill.
- State the effect of age and smoking on the risk of Pill use.
- Explain why the estrogen content of the Pill has been reduced over the years.
- Describe the nature of the MiniPill.
- Give the name and source of one male contraceptive Pill.

13.1 Introduction

A woman's childbearing years are usually reckoned as ranging from 16 to 45.[1] Today in the United States, there are roughly 60 million women in this age bracket, and of these, an estimated 16.8 million take the Pill—twice as many as a decade ago. That's a hefty 28% and represents a tremendous impact on our society in terms of population control, age distribution, zero population growth, what people purchase, possible health effects, and drug dollars. According to one survey, women now remain on the Pill an average of 4.8 years. About 500,000 women between the ages of 45 and 50 take the Pill. Worldwide, an estimated 100 million women use the Pill.

The **Pill** (or *anovulatory agent*, or *oral contraceptive*) is big business because it is the most effective means of birth control available short of sterilization. In fact, used under *controlled* conditions, it is virtually 100% effective. To put this issue in perspective, consider 100 sexually active women who are studied for 1 year. If these

[1]In the United States, 48 years is the average age of onset of menopause.

women used no contraception at all, about 70–80 would become pregnant. If all of them faithfully took the Pill, their pregnancy *rate* would be less than 1 per 100 woman-years. By comparison, women using an *intrauterine device* (IUD) have a pregnancy rate of 2–3 per 100 woman-years. For diaphragms in combination with a gel or cream, the rate can be as high as 20 pregnancies per 100 woman-years, depending on the self-discipline of the user.

Another reason the Pill has received universal acceptance is its modest cost: about $4–$5 for a month's supply. But all is not rosy with the Pill. Within 3 years of its introduction, its tendency to cause blood clots was known, and research today continues to uncover additional reasons for caution in its use, especially among certain high-risk women.

In this chapter we bring you up to date in the areas of product changes, new information on side effects, and the latest recommendations on who should not take the Pill.

13.2 How the Pill Works

Long before the first oral contraceptive, norethynodrel (Enovid), was approved for use in 1960, researchers had known that certain hormones from the pituitary gland regulated ovarian function, and that ovulation and the menstrual cycle could be manipulated by means of drugs. They knew, for example, that a woman does not ovulate while she is pregnant because her ovaries are secreting **progesterone**, a hormone of pregnancy. Progesterone, circulating in the bloodstream for 9 months, affects the hypothalamus, which in turn affects the pituitary gland, stopping release of the hormones (see gonadotropins, Section 2.5) that induce ovulation. Figure 13.1 helps to explain this process.

When progesterone was injected into nonpregnant women, it worked well in preventing ovulation. But when given orally, it was destroyed by digestive juices and showed little or no activity. For this reason, researchers synthesized progesterone-like

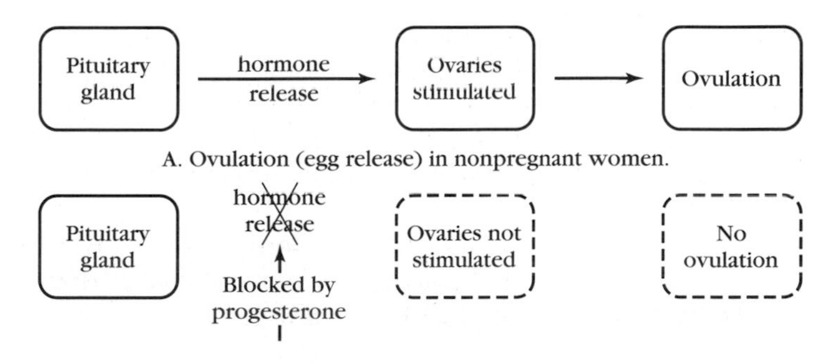

Figure 13.1 The interrelationships among the pituitary gland, the ovaries, and progesterone. Obviously, no ovulation means no egg to fertilize, which means that pregnancy is impossible no matter how many sperm are present.

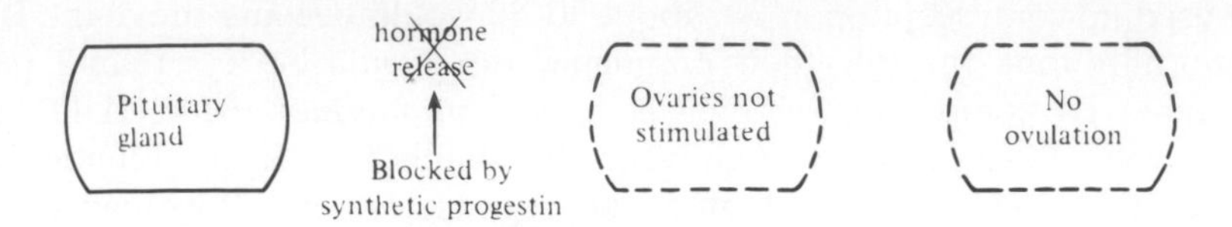

Figure 13.2 The Pill works by supplying synthetic progestins in place of nature's progesterone.

drugs (now called **progestins**, or progestogens) that could be taken orally and that acted just like natural progesterone to inhibit ovulation. Enovid was such a progestin. This means that in Figure 13.1, part B, we can substitute *synthetic progestins* for *progesterone*, thus explaining how the Pill works (see Figure 13.2). Besides inhibiting ovulation, oral contraceptives are believed to have a direct effect on the uterus, thinning the lining and making it less receptive to a fertilized egg. Some of these drugs also act to alter the cervical mucosa.

Because the use of progestins alone resulted in spotting or "break-through" bleeding, it was decided to add an estrogen to the Pill, resulting in what we now call the *combination* Pill (for a discussion of estrogens, see Chapter 2). The estrogen commonly used in today's Pill is an estradiol derivative; it provides a product that gives excellent control over ovulation with minimal spotting.

13.2.1 Summary

Use of the Pill makes pregnancy impossible because ovulation is inhibited. Today's combination Pill contains a synthetic progestin and smaller amounts of estrogen. Taken once a day, these two chemicals act to suppress the release of the hormones that are needed to start the process of ovulation.

> ***CAUTION:*** *Although the Pill can prevent pregnancy very effectively, it cannot protect against any of the sexually transmitted diseases—herpes, AIDS, chlamydia, syphilis, gonorrhea, or venereal warts.*

Another type of contraceptive pill was introduced in France in 1989. Called the "abortion pill," RU-486 (mifepristone) is an antiprogesterone drug, a kind of antihormone that works to block the development of the female hormone progesterone (see Section 13.6). It prevents implantation of any fertilized egg into the wall of the uterus. Used in combination with prostaglandin E, the drug is claimed to be 95% effective.

In about one-half of 1% of the abortions performed so far, RU-486 patients have had bleeding more severe than that of a heavy menstrual period. About 1 in 500 women need a blood transfusion after taking RU-486.

13.3 Products Available

After the G. D. Searle Company introduced the first oral contraceptive in 1960, many other companies introduced competing products. Today, dozens of preparations are available to physicians and their patients (see Table 13.1).

Table 13.1 Representative Oral Contraceptives

Trade name	*Manufacturer*	*Progestin Used*	*Estrogen Used*
Brevicon 21 Day	Watson	Norethindrone, 0.5 mg	Ethinyl estradiol, 0.035 mg
Levite	Berlex	Levonorgestrel, 0.100 mg	Ethinyl estradiol, 0.020 mg
Micronor	Ortho-McNeil	Norethindrone, 0.35 mg	None
Nordette-21	Monarch	Levonorgestrel, 0.15 mg	Ethinyl estradiol, 0.030 mg
Ovrette	Wyeth-Ayerst	Norgestrel, 0.075 mg	None
Norinyl 1 + 50	Watson	Norethindrone, 1.00 mg	Mestranol, 0.050 mg

Dozens of preparations are available. They differ in the kinds of drugs used, their strengths, and the possible inclusion of a 7-day supply of inert ingredient tablets in the dispenser.

Among the many oral contraceptive preparations marketed, there is variation in the progestin used, the strength of the progestin and estrogen, and the dosage schedules recommended. For example, Parke-Davis's Norlestrin is offered in five different preparations, containing from 1.0 to 2.5 mg of progestin and 0.050 mg of estrogen, with or without iron. The iron is included to facilitate drug administration via a 28-day regimen and is not intended to serve any therapeutic purpose.

13.3.1 Dosage Regimens

Basically, the combination Pill is taken once a day for 21 days to suppress ovulation. Then for 7 days, either no tablet or a placebo is taken. Menstruation occurs during the 7-day time period, usually within 1–3 days after taking the last tablet. Some women experience spotting or breakthrough bleeding during the 21-day time period, but this usually stops within two to three courses of tablet taking. Dosage schedules differ slightly, depending on whether the tablets are being taken on an initial or subsequent cycle. To assist the user in proper administration, manufacturers offer specially designed tablet dispensers—numbered, day-coded, or color-coded. If one tablet is missed, there is little chance that ovulation will occur, but as more tablets are missed, the possibility of ovulation increases. So does the likelihood of spotting or breakthrough bleeding. If two or more tablets are missed in a row, breakthrough bleeding should be expected. Women should remember that there is no proper use of estrogen during pregnancy. *If you are pregnant, do not take the Pill.* Estrogen taken early in pregnancy or just before conception can cause serious birth defects in the developing embryo.

The newest approach to oral contraception is to mimic the varying hormonal levels during the menstrual cycle. For most of their product lifetime, oral contraceptives have contained a fixed dose of active ingredients. Now products are appearing that attempt to simulate the normal human sex hormone cycle, with its ups and downs of endogenous estrogen and progesterone. These products, called **triphasic** (or **multiphasic**) **oral contraceptives**, have varying dosage regimens that change once or twice in every 21-day cycle. For example, Ortho-Novum 10/11 is a two-dose contraceptive that provides two different doses of norethindrone during the 21 days the pills are taken. Syntex's Trinorinyl is a three-dose product. So is Ortho-Novum 7-7-7, a product designed to deliver three graduated estrogen-to-progestin ratios during the 21-day dosage period. The three dosage phases of Triphasil-21 are diagrammed in Figure 13.3. Note that progestin levels are increased to mimic the natural condition.

While these new multiphasic oral contraceptives more closely conform to the hormonal cycle, there unfortunately is some preliminary evidence suggesting that triphasics may be linked to the development of noncancerous ovarian cysts. The FDA has ordered drug companies to conduct additional studies to gather more data.

Oral contraceptives are sometimes taken immediately after the birth of a child, assuming the mother does *not* want to nurse her child. Since suppression of lactation is possible when the Pill is taken immediately postpartum, mothers wishing to nurse should not take the Pill until lactation has been well established, usually 4 weeks postpartum.

The FDA has approved Lunelle, a once-a-month injection that combines a microcrystalline suspension of medroxyprogesterone (the progestin) and estradiol (the estrogen). Costing $25–$35 a month, the intramuscular injections are given by a health-care provider every 28–30 days, and no longer than 33 days apart. Rules are in place to prevent Lunelle from being given to a pregnant woman.

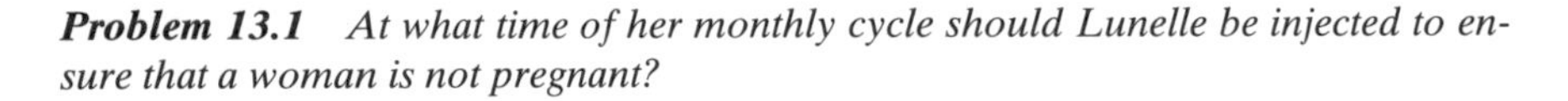

Problem 13.1 *At what time of her monthly cycle should Lunelle be injected to ensure that a woman is not pregnant?*

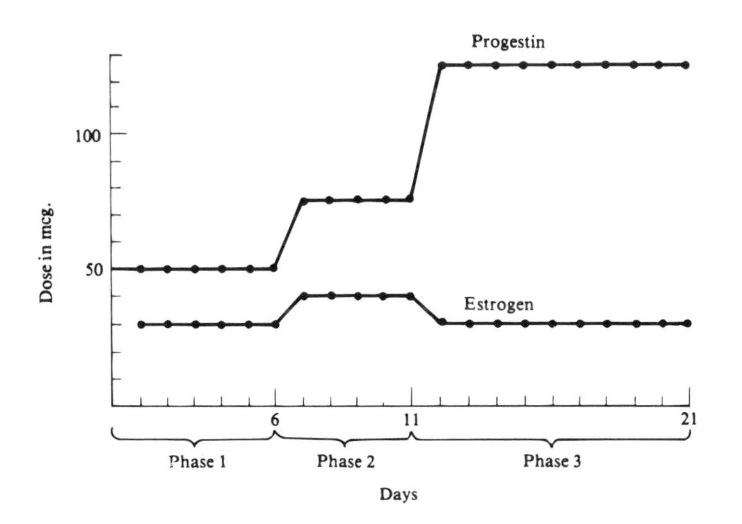

Figure 13.3 The three dosage phases of the oral contraceptive Triphasil-21. The upper line is the dose level of progestin (levonorgestrol), and the lower line is the estrogen (ethinyl estradiol) dose level.

Mirena IUD is a newly approved barrier device designed for five years of use, but that could last seven. This IUD is coated with progesterone-levonorgestrol; it acts by thickening cervical mucous.

After 25 years of testing and controversy, Depo-Provera (medroxyprogesterone acetate), an oral contraceptive given every 3 months by injection, has finally won FDA approval for marketing in the United States. Depo-Provera also has approval for use in the treatment of cancer of the kidney or the lining of the uterus. Researchers at Johns Hopkins University are investigating the use of Depo-Provera to reduce testosterone levels in men who are repeat-offender rapists.

It is of interest to note how much the dose of estrogen in the combination Pill has decreased since the Pill was first introduced. From the 0.080–0.100 mg first employed in the 1960s, the estrogen content has been reduced to 0.030 mg or less in some of the newest products (see Table 13.1). It will become clear later in this chapter that high estrogen levels seem to be correlated with higher risks.

13.4 Possible Adverse Effects of the Pill

Millions of women have used the Pill in the past 30 years, and a great deal of information has been collected about its side effects. About 40% of Pill users have adverse effects of some kind. A poll commissioned by The Association of Reproductive Health Professionals found that about 1 in 4 women who use the Pill say they stopped taking it because of adverse effects. Adverse reactions to the Pill can be grouped into two categories: minor side effects and major side effects.

Minor side effects are well known to many women. Possible minor side effects include weight gain, breast tenderness, nausea, allergic rash, vomiting, dizziness, headache, change in the color of the pigment in facial skin, intolerance to contact lenses, aggravation of preexisting varicose veins, and mental depression. Vomiting can occur in up to 10% of users and mental depression in about 5%. Irregular menstrual flow may occur but clears up after several cycles. Minor side effects are often of little concern, considering the advantages gained.

Major side effects are a different matter. The two most often mentioned are blood clots (thromboembolism) and cancer. These are discussed in this section. Other potentially serious side effects are hypertension, increase in blood cholesterol, impairment of glucose tolerance, vitamin B_6 deficiency, and bleeding liver tumors (rare).

13.4.1 Blood Clots and the Pill

Within 3 years of their introduction, oral contraceptives were linked statistically to an increased incidence of blood clotting, with resultant strokes, phlebitis, and blood clots in the lung(s). Many retrospective studies have now been carried out in Great Britain and the United States, and an increased risk of clotting from the Pill is well-established. Furthermore, analysis of mortality trends in 21 countries shows a relationship between increased mortality among women ages 15–44 and the introduction and use of oral contraceptives in those countries. For women aged 16–40, it is estimated that about 1 in 2,000 using oral contraceptives will be hospitalized each year because of abnormal clotting in veins or lungs. This number represents a ten-fold increased risk compared to that of nonusers. It has been estimated that Pill users are twice as likely as nonusers to have a stroke due to a bursting blood vessel in the brain. Note that smoking increases the risk of circulatory disorders.

Advanced age and smoking greatly increase the risk of fatal heart attack in Pill users, as noted in Table 13.2.

Most manufacturers of oral contraceptives have published a long list of possible side effects, dangers, and risks inherent in Pill use. You can read about these side effects in the *Physicians' Desk Reference*. Table 13.3 shows the annual risk of death associated with Pill use. Note the greatly increased risk to smokers, *especially* smokers older than 35 years of age.

Pill manufacturers state that you should *not* use oral contraceptives if you have a history of blood clots, chest pains on exertion, known or suspected cancer, unusual vaginal bleeding that has not been diagnosed by your doctor, known or suspected pregnancy, or history of heart attack, or if you have scanty or irregular periods or are a young woman without a regular cycle. Also, risk generally is increased if you are obese or have diabetes, hypertension, or hypercholesterolemia.

There is one piece of good news for ex-users of the Pill. According to the Harvard Nurses' Health Study, women who have stopped taking birth control drugs, even if they took them for a decade or more, are at no more risk of heart disease

Table 13.2 Yearly Risk of Fatal Heart-Attack Risk Among Oral Contraceptive Users

	Women Ages 30–39	*Women Ages 40–44*
Oral contraceptive users who smoke	1 in 10,000	1 in 1,600
Oral contraceptive users who do not smoke	1 in 55,000	1 in 9,000
Nonsmokers, nonusers	1 in 80,000	1 in 14,000

Table 13.3 Birth-Related or Method-Related Deaths Associated with Fertility Control, According to Age

	Age (in Years)					
	15–19	*20–24*	*25–29*	*30–34*	*35–39*	*40–44*
No fertility control	7.0	7.4	9.1	14.8	25.7	28.2
Oral contraceptives						
Nonsmoker	0.3	0.5	0.9	1.9	13.8	31.6
Smoker	2.2	3.4	6.6	13.5	51.1	117.2
IUD	0.8	0.8	1.0	1.0	1.4	1.4
Condom	1.1	1.6	0.7	0.2	0.3	0.4

Annual estimates per 100,000 nonsterile women.

than women who never took the Pill at all. This conclusion was based on a study of 120,000 nurses ages 30–55.

Supporters of the use of oral contraceptives point out that the risk of death from becoming pregnant is far greater than that from being nonpregnant and using the Pill. Note, however, from Table 13.3 that this statement is not true after age 40.

In the combination Pill used today, estrogen is the villain. It is responsible for most of the side effects we have been discussing. For this reason, manufacturers have reduced the estrogen content of oral contraceptives through the years until now, it is often less than 0.1 mg per tablet.

As we learned in Chapter 2, estrogen in menopausal replacement therapy is implicated in cancer of the female reproductive tract. Animal studies show this clearly, but the relationship in the human is more difficult to prove. At this time, no statistical association has been reported suggesting an increased risk of uterine cancer in users of combination-type or progesterone-only types of oral contraceptives, although individual cases have been reported. (See Section 2.5 for contrasting data on the use of estrogen during menopause.)

Several research studies have shown no risk of breast cancer to women who are users of oral contraceptives. A recent NIH study of 9,200 women ages 35–46 years who used oral contraceptives showed no increased risk, compared to women who haven't been on the Pill. This disputes a 1996 analysis of 54 other studies that suggested a doubling of the risk of breast cancer. A companion editorial to the NIH study noted that oral contraceptives do carry other health risks including lung embolism, stroke, and liver and cervical cancer.

Some studies actually show an apparent protective effect of oral contraceptives against cancer of the ovaries and the endometrium (the lining of the uterus). This finding applies only to the estrogen-progesterone combination type of Pill.

Other studies appear to show a statistical relationship between oral contraceptive use and cancer of the cervix. If this sounds like a contradiction to the earlier statement, it is helpful to draw a distinction between the endometrium of the uterus and the cervix (the opening in the neck of the uterus). A study of 600 women in Los Angeles disclosed that among those who had abnormal Pap smears to begin with, use of the Pill was six times more likely to result in conversion to cancer. But if the Pap smear was normal initially, Pill use was inconsequential.

Skin cancer and the Pill were linked in a 10-year study of nearly 18,000 patients of a prepaid health plan in California. Dr. Savitri Ramcharan, research director of the contraceptive drug study, said that women who use birth control pills for more than 4 years expose themselves to almost twice the risk of developing the often fatal skin cancer malignant melanoma. Dr. Ramcharan, however, cautioned that her data, the first to link the Pill with skin cancer, were "suggestive but far from definitive."

As before, reports can be cited that exonerate oral contraceptives from any role in cancer induction in the uterus or breast. Thus again, the doctor and the patient are faced with a decision based on suspected risks and known benefits. A woman who can ill afford to become pregnant may easily conclude that the Pill's advantages outweigh its risks. On the other hand, a healthy 21-year-old woman may want to consider other contraceptive techniques before deciding on long-term progestin-estrogen use.

In this connection, we should mention the **MiniPill**. This "new" approach is actually the old initial concept of progestin alone without any estrogen. It works in part by preventing release of an egg from the ovary but also by keeping sperm from reaching the egg and by making the uterus less receptive to any fertilized egg that reaches it. In the MiniPill, we have come full circle to a product that eliminates the ingredient that is suspected of causing most of the adverse side effects—estrogen. Micronor and Ovrette, listed in Table 13.1, are commercial products that fit our description of the MiniPill, as is Nor-Q.D.

MiniPills are taken every day of the year without interruption, a major difference from the combination products. That their effectiveness is lower than that of the combination Pill is shown by these figures:

- Use of the *combination Pill* results in fewer than 1 pregnancy per 100 woman-years.
- Use of the *MiniPill* results in fewer than 3 pregnancies per 100 woman-years.

Further, more spotting and breakthrough bleeding can be expected with the MiniPill. Why is it used if it's less effective in general? The answer: Because it eliminates estrogen. It seems obvious that both the manufacturers and the FDA are becoming increasingly suspicious of the (detrimental) role played by estrogen in our society. This concern is reflected in the generally lower quantities of estrogen found in combination products marketed in recent years. It should not be assumed that the progestin-only pill is free of excess risk (norethindrone can produce some masculinizing effects). The MiniPill, often suggested for use after a patient has experienced one of the estrogen-induced side effects, may become the oral contraceptive of choice for women older than 35 and for women with a history of headache, hypertension, or serious varicose veins.

At this point, let us stress that the very serious side effects of oral contraceptives are uncommon events. The great majority of users will never experience them. Just the same, women should be alert to what *can* happen and should be urged to keep in touch with their doctors throughout their use of the Pill. Users of oral contraceptives should learn all they can about these potent drugs. For example, they should know that the drug tends to cause a deficiency of vitamin B_6 in as many as 15% of users. This result may be linked to the mental depression that is observed in 5% of cases. Women on the Pill may wish to seek nutrition counseling.

Oral contraceptive users who get drunk remain intoxicated longer than nonusers. University of Oklahoma investigators discovered that the Pill user metabolizes ethyl alcohol more slowly, permitting blood alcohol levels to remain higher for a longer period of time.

For a free reprint of the article "The Pill: 30 Years of Safety Concerns," write the FDA, HFE-88, 5600 Fishers Lane, Rockville, MD 20857. Ask for publication number FDA92-3193.

13.5 Recommendations

Oral contraceptives are not 100% safe, but no drug is. Most women are content to take the small risk of use because they want the real advantages; this is the essence of the matter. High-risk women must be identified before they begin oral contraceptive use and must be switched to other means of contraception if possible. The physician is placed in the key role of identifying factors that could predispose the user to increased risk. This fact makes it necessary to take a careful history, conduct a thorough physical examination, and obtain necessary laboratory data.

Probably 85% of sexually active women in the United States are good, healthy candidates for oral contraception. For the others, the following recommendations may help identify their risk status:

1. Oral contraceptives are probably the best choice for healthy young women who are sexually active. The high risk of pregnancy in young teenagers makes them good candidates for Pill use.
2. Women older than 40 should use some other method of contraception because of the higher risk of heart attacks and/or fatal blood clots associated with the use of the Pill in this age group.
3. While there is no confirmed evidence that oral contraceptive use causes cancer, women with preexisting abnormal uterine conditions or a family history of breast cancer subject themselves to increased risk if they take the Pill.
4. The Pill should never be taken by pregnant women because it is known to have a potential teratogenic effect. This fact is so important that women who even suspect they are pregnant should consult their doctor at once before even one additional oral contraceptive tablet is taken. See also point 5.
5. Women who discontinue taking the Pill in anticipation of becoming pregnant should then use some *other* means of contraception for 3 months because of the possibility that some teratogenic chemicals from the Pill will remain in their body for that length of time and could damage the embryo.
6. First-time Pill users should select a brand with 0.05 mg or less of estrogen. Physicians should prescribe products containing the least amount of estrogen that will accomplish the task. This recommendation reflects our growing suspicions of the deleterious role of estrogen.
7. Birth control drugs should be discontinued at least 4 weeks before types of surgery that involve an increased chance of blood clotting or prolonged immobilization. The risk of thromboembolism is increased four- to six-fold in such situations.
8. Women who experienced an onset of jaundice or diabetes mellitus in previous pregnancies are poor candidates for the Pill, because studies have shown that these tendencies may be aggravated into full-blown illnesses under Pill use.
9. Women who are presently nursing their babies are not candidates for the Pill, because its use can result in decreased quantities of milk as well as diminished protein and fat content.
10. Add the fact that a woman is a smoker to all of the other considerations, and you increase by a factor of as much as 10 her risk for coronary artery disease while on the Pill.
11. Women with gallbladder disease are not candidates for the Pill.
12. Women with a history of psychic depression should be watched carefully while on the Pill.
13. Oral contraceptives should be taken only under a doctor's supervision.

13.6 Ovral as a Morning-After Pill

Since the 1970s doctors have prescribed Ovral (norgestrel) as a morning-after pill. College student health services across the nation have dispensed it for use within 72 hours of intercourse to prevent implantation of a possibly fertilized egg. One gynecologist calls Ovral use *interception*; that is, acting after fertilization but before implantation. Now the FDA says taking 6 additional birth control pills is safe and effective as a morning-after drug, if given in the proper dose. The FDA has approved Preven for use within 72 hours of unprotected sex, followed by 2 more doses 12 hours later. It has also approved Plan B (trade name; contains levonorgestrel, 0.75 mg) tablets for use within 72 hours.

> ***CAUTION:*** *With these chemicals, it is wise not to plan your own medical treatment. Safe and effective doses depend on the brand, and even some physicians are not clear on the regimen. Remember that all of these drugs are oral contraceptives first and morning-after pills secondarily. Typically they contain estrogen and progestin; about half of the women who use them suffer nausea, and about 20% vomit because of the stronger doses of active ingredients. Browse the Internet site* **`http://ec.princeton.edu/`** *for more on emergency contraception.*

Opponents of the use of oral contraceptives as morning-after pills claim that the latter are abortifacients—in the same category as methotrexate and the controversial French drug RU-486. Actually, RU-486 (mefipristone), in a dose of 600 mg, is now approved by the FDA for termination of early stage pregnancy. This blocks the action of progesterone, a hormone required to maintain a pregnancy. As the uterine lining disintegrates, the fetus starves and dies from lack of nutrients. RU-486 mimics a miscarriage, prompting bleeding over the course of 9–16 days. A total of three visits to the physician is required, over a period of about 14 days.

The use of prostaglandins and the former use of DES as morning-after drugs are discussed in Section 2.5.

13.7 Oral Contraceptives for Men

It is easier to develop a Pill for women than for men because women ovulate only once a month, whereas men manufacture sperm continually all year long. Just the same, much research money has been spent to develop a Pill for men that, taken orally, would render them infertile—but only as long as they continued to take the drug. Some years ago, a large U.S. pharmaceutical manufacturer thought it had developed the perfect drug for men, because it fulfilled all of the requirements in laboratory animals. Alas, clinical testing in humans disclosed that if alcohol was consumed concurrently with the drug, the men became ill. Another research idea was discarded.

The best bet so far for fertility control in males is **gossypol**, a compound that occurs naturally in the seed, stem, and root of the common cotton plant. It was first identified as an antifertility agent in males by the Chinese in the 1950s.

Research in the United States has begun to elucidate gossypol's mechanism of action. It inhibits the enzyme lactate dehydrogenase X, which plays a critical role in the functioning of sperm and sperm-generating cells. Men treated with gossypol show low sperm counts in seminal fluid. Fortunately, gossypol has no effect on sex hormone levels or on libido in males. On the other hand, one research report states that gossypol can cause irreversible sterility in some men.

In Chinese clinical tests, gossypol given daily made 99.89% of the 4,000 males tested infertile but also produced minor side effects such as nausea, loss of appetite, and inability to perspire. Gossypol should be inexpensive to produce, and its supply should be limitless. It would, of course, have to win FDA approval before it could be marketed in this country.

For more on male fertility control, see Immucon Corporation, at **`http://www.immucon.com/index.html`**. Another informative site is **`http://www.malecontraceptives.org/`**.

13.8 Product Disclosure (Patient Package Insert)

Federal regulations now require that the manufacturer of an oral contraceptive make available to all users a summary of the clinical pharmacology, contraindications to use, and precautions in the use of their product. Users will perhaps find the lengthy summaries printed in fine type difficult to read, but just the same they are valuable documents for the establishment of risk/benefit considerations.

They also represent a major breakthrough in product disclosure for the patient. Now the prospective drug user can read for herself the risks involved in drug taking, which can help make the decision.

In a subtle way, these product disclosure inserts are shifting responsibility for decision-making from the FDA to the patient and doctor. The FDA appears to be saying, "Research has discovered that the Pill can be a dangerous drug in certain women. Here are the facts as we know them. It is up to you, individually, to consult with your doctor, establish your personal risks and benefits, and then decide if you want to use this form of contraception."

13.9 Summary

Oral contraceptives retain their preeminence as the most effective means of reversible birth control. Physiologically, about 85% of sexually active females are candidates for their use. Research has made it clear, however, that the risk of blood clotting (thromboembolic disease) is increased 4–10 times in users of the combination Pill. That the Pill increases the incidence of uterine cancer is not proved, but it is suspected. Women at high risk from oral contraceptive use include those older than 40, smokers, those who are pregnant, and those with a history of cardiovascular, reproductive system, or liver disease. Because disclosure of product information is now mandatory, the patient herself can help make the ultimate decision as to whether she should or should not use oral contraceptives.

It is the estrogen content of the combination Pill that appears to cause most of the adverse side effects. For this reason, the estrogen content of oral contraceptives has been reduced to the lowest levels ever and has even been eliminated altogether in the new MiniPill.

Web Sites You Can Browse on Related Topics

Oral Contraceptives
http://www.fda.gov/womens/default.htm
Also: search "oral contraceptives"

Morning-after Pill
Search "morning after pill"

Male Contraceptives
http://www.gumption.org/mcip/paper.html
Also: search "male contraceptives"

Study Questions

1. **a.** Explain how the Pill works as a contraceptive.
   b. Why is the Pill called an *anovulatory* agent?
2. What is the difference between a retrospective study of drug effects and a prospective study? Which do you believe is more likely to uncover the truth? Why?
3. A doctor-spokesman for the FDA has said that oral contraceptives are "safe enough" for continued use by a large number of American women. Is that a callous viewpoint? Shouldn't the FDA insist on 100% safety before approving a drug for general use?
4. Explain the following statistic: Users of a certain Pill have a pregnancy rate of 0.3 per 100 woman-years.
5. If the natural hormone progesterone works so well as a contraceptive and can be synthesized in the laboratory, why isn't it used in today's oral contraceptives?
6. What appears to be the trend in regard to the *amount* of estrogen used in the combination-type Pill? Explain this trend.
7. Suppose that a woman forgets to take her Pill for 2 or 3 days in a row. What effect will this have on her body?
8. **a.** What are the common minor side effects from the use of oral contraceptives?
   b. What are the serious side effects of Pill use, and what evidence do we have for them?
9. If 5,000 women were studied for 4 years and were reported to have a total of 100 pregnancies, what would be the pregnancy rate per 100 woman-years?
10. Some years ago, medical researchers told Congress that about 400 deaths per year can be linked to Pill use in the United States. What percentage is this of the estimated 10 million women now using oral contraceptives? In your opinion, does this death rate represent an unacceptable risk to the American woman?
11. Studies show that by using the rhythm method, women can be expected to have 20 pregnancies per 100 woman-years. One large study of women who used oral contraceptives revealed a pregnancy rate of 0.13 per 100 woman-years. Apply these rates to the 50 million women of childbearing age in America and calculate the theoretical number of pregnancies that would result if all of the women used the rhythm method for 1 year versus the number of pregnancies resulting from oral contraceptive use for 1 year. Is it reasonable to expect that all 50 million women could tolerate the Pill?
12. Approximately 5.3% of all U.S. females are ages 40–44 (a total of 5.65 million women). Using the data in Table 13.2, calculate the total number of fatal heart attacks expected in 1 year if 20% of these women used the Pill (and 80% did not). Make the same calculation if none of these women used the Pill.

13. Most package inserts for oral contraceptives contain a graph showing that the annual death rate for 35–39-year-old nonsterile women who use *no* contraceptive methods whatever is 19.5 per 100,000. For women of the same age who use abortion only as a birth control, the death rate is 9.5 per 100,000. And for those who use oral contraceptives, the rate is 7.5 per 100,000. Explain this order of death rates. (*Hint:* Consider the effects of pregnancy taken to full term or near-full term.)
14. How does the MiniPill differ from the combination Pill? What are the advantages and disadvantages of the MiniPill?
15. The text lists 13 recommendations about risk in the use of the Pill. What category of women would appear to be at highest risk? Explain your selection.
16. Women take oral contraceptives not because they have a disease to be cured or a condition that requires therapy, but to prevent the occurrence of a condition. What other drugs or preparations are taken for a similar reason?
17. For each term in the left-hand column, indicate the correct definition from the right-hand column.

    a. Pap smear
    b. Combination Pill
    c. Gossypol
    d. Postpartum
    e. Progestin
    f. Estrogen
    g. Thromboembolism

    1. A male oral contraceptive made from cotton
    2. Synthetic or natural agents having progesterone-like activity
    3. An oral contraceptive containing both progestin and an estrogen
    4. A blood clot obstructing a blood vessel
    5. Occurring after childbirth
    6. A procedure for the detection and diagnosis of various conditions, such as cancer of the female reproductive tract
    7. Female sex hormone

18. *Advanced study question*: With the Pill, a trend might have started in America, namely, the patient helping to decide which drug to use or whether to use a drug at all. In other words, a patient may now contribute directly to making a medical decision formerly made by the physician alone. What would a woman need to know about herself and about a particular brand of anovulatory agent to equip herself to help make the decision to use it or not?

Nonsteroidal, Anti-inflammatory Drugs (NSAIDs): Aspirin, Acetaminophen, Ibuprofen, Naproxen Sodium, Ketoprofen, Fenoprofen, and Celebrex and Vioxx

Key Words in This Chapter

- Analgesic
- Acetaminophen
- Aspirin
- Ibuprofen
- Salicylate
- Naproxen sodium
- Side effects
- NSAID
- Ketoprofen
- Fenoprofen
- COX-2 inhibitors

Learning Objectives

After you complete your study of this chapter, you should be able to do the following tasks:

- Identify the effects aspirin has on pain, fever, and inflammation.
- Give the reason aspirin should not be used after a tonsillectomy, in ulcer patients, or in pregnant women.
- Discuss aspirin's mechanism of action in pain and fever reduction.
- Identify the possible link between aspirin and Reye syndrome.
- Draw pharmacological distinctions between acetaminophen and the salicylates.
- Give 2 medical uses of acetaminophen.
- Describe the danger in taking excessive doses of acetaminophen.
- Describe the distinction between ibuprofen and the salicylates.
- Give 2 medical uses of ibuprofen.

14.1 Introduction

To relieve the pain of their headache, backache, dysmenorrhea, and other low-to-moderate intensity painful conditions, Americans can now choose from among at least six OTC nonnarcotic analgesics: aspirin, acetaminophen, ibuprofen, naproxen sodium, ketoprofen, and fenoprofen. These drugs do not relieve intense pain, as, for example, the pain of fractures, kidney stones, or heart attack. They are but six of at least a dozen agents tagged **nonsteroidal anti-inflammatory drugs** (NSAIDs). These NSAIDs are thus distinguished from anti-inflammatory steroids such as Aristocort, Decadron, and Hydrocortone. In addition to these six, more than two dozen prescription-only NSAIDs are listed in the PDR, some of which are simply stronger doses of the OTC chemical. Arava (leflunomide) is a new drug competing with methotrexate for use in severe, active rheumatoid arthritis patients who do not respond to other medications.

OTC analgesics are heavily advertised, and Americans consume them in great quantities. The FDA estimates that we swallow 80 billion aspirin tablets a year. McNeil Consumer Products Co., the makers of Tylenol brand of acetaminophen, claims that Americans take 8 to 9 billion tablets of the drug yearly. More than 200 other products also contain acetaminophen.

Each pain reliever is discussed on the following pages, but first some general comments. The OTC analgesics under discussion represent three distinct chemical categories of drugs. Aspirin is a salicylate; acetaminophen is an aniline derivative; and ibuprofen, naproxen sodium, ketoprofen, and fenoprofen are propionic acid derivatives. The last four listed were formerly available only on prescription but now are sold OTC. With one exception, all of these drugs are effective analgesics (relieve pain), antipyretics (lower fever), and anti-inflammatories. The exception is acetaminophen, which is too weak an anti-inflammatory to be useful as such clinically. With the exception of acetaminophen, all of these drugs have a tendency to irritate the lining of the stomach—with aspirin being the worst offender. Persons who are allergic to aspirin should not take any of these drugs without first consulting a doctor. Of these pain relievers, only acetaminophen has no effect on bleeding time. (In lay terms, it does not "thin" the blood.)

As you will conclude from the following discussions, all of these pain relievers can pose some health risk for the user. This is especially true for the heavy or chronic user, and for the user who concurrently ingests alcohol. "Don't take over-the-counter pain relievers like candy," warn the experts. "I think people are taking over-the-counter drugs way too casually," says Peter Woo, Director of the Pain Consultation and Management Service at the University of California, San Francisco. Other experts decry taking analgesics for routine athletic aches.[1]

Heavy or long-term use of aspirin, acetaminophen, ibuprofen, and naproxen sodium can cause kidney failure, the National Kidney Foundation has concluded. Of special concern, says the foundation, are products that use a combination of analgesic chemicals, such as putting aspirin and acetaminophen together. What is more, as many as 8% to 10% of new cases of chronic kidney failure may be related to the misuse of pain relievers, says the foundation. The foundation stresses, however, that these analgesics are generally safe when taken in small doses for a short period of time by people with no underlying kidney disease. A maker of a combination product points out that the FDA has found combination products to be safe and effective when used as directed.

Although some of these are old drugs, new discoveries about them continue to be made. These findings include how aspirin and acetaminophen work, aspirin's link to Reye syndrome, and the possible use of aspirin in the prevention of heart attacks.

14.2 The Discovery of Aspirin

Aspirin is a **salicylate**. This simply means that it, along with oil of wintergreen (methyl salicylate), is a derivative of salicylic acid, a chemical first prepared in 1838. It is important to emphasize the salicylate structure, for then we can distinguish a compound such as acetaminophen, which is not a salicylate and does not have all the side reactions of the salicylates. Aspirin, or *acetylsalicylic acid* (ASA), was first synthesized in 1853 by a French chemist, Charles Gerhardt, but was not introduced into therapy until 1899. The word *aspirin* was coined by Heinrich Dreser (the same man who introduced heroin to the medical profession in 1898). Dreser took the *a* from *acetyl* and added the suffix *spirin* from the spirea plant, a natural source of salicylic acid. Other natural sources of salicylic acid are willow and poplar trees.

The Bayer Chemical Company introduced aspirin to Europe around 1900. The Bayer trademark was awarded to the United States as part of Germany's World War I reparations, and the American firm of Sterling Drugs, Inc., bought the trademark at auction. Mass-produced and distributed by Sterling, aspirin soon became a household remedy. The Bayer trade name is carefully and continuously fostered by extensive advertising. If you would like to start your own business making and selling aspirin, go ahead. The formula is in the public domain (the patents expired in 1917).

All aspirin sold in the United States must conform to USP standards of purity and effectiveness. Whether you pay premium prices or buy the cheapest generic brand, you are getting the same drug.

---

[1]Some ads border on the ridiculous: "Did you exercise today? Do your muscles ache a little? Quick, take one of our pain relievers."

The discovery was made early that aspirin could reduce the body temperature of a febrile (feverish) person, although it had no effect on normal body temperature. This is termed the *antipyretic* action of aspirin. The discovery was also made that aspirin could eliminate pain, especially the minor type, as in headaches. This is termed the *analgesic* action of aspirin. When aspirin acts to reduce swollen and painful joints, it is said to exert an *anti-inflammatory* action.

Buffered aspirin is a relatively recent product variation. To make a buffered aspirin product, the manufacturer adds alkaline chemicals (such as aluminum glycinate or aluminum hydroxide) to neutralize ("buffer") the acidity of the aspirin when it contacts the stomach lining. Any aspirin product can be buffered; buffering is not limited to the product Bufferin. There is evidence that strong buffering reduces the gastric irritant side effects of aspirin but does not eliminate them. Apparently most brands of aspirin do not contain enough buffering agent to have much effect.

Few drugs have enjoyed the popularity of aspirin. On any given day, on the average, one person in four will take one aspirin tablet, amounting to a daily consumption of 20 tons of acetylsalicylic acid in the United States alone.

14.3 The Medical Uses of Aspirin

When you buy an aspirin-containing preparation in America, you have more than 400 products from which to choose. To listen to the advertisements, each one is better, faster, more effective, or more extra-strength than the next. In an effort to discover the facts, the FDA empaneled a group of experts to look at the safety and effectiveness of internal analgesics. Some of the conclusions reached by this panel are the following:

1. Aspirin is safe and effective as a painkiller and fever reducer and is "acceptable" for use in reducing inflammation when such use is medically supervised.
2. Advertisers for aspirin products should be permitted to claim only that they are "for the temporary relief of minor aches, pains, and headache." Ads for fever-reducing products should be limited to the claim, "for the reduction of fever."
3. Use of buffered aspirin is not objectionable. However, the combination of aspirin with an antacid for the treatment of heartburn, sour stomach, or acid indigestion is irrational.
4. Aspirin interferes with blood clotting. If taken in the last 3 months of pregnancy, it can prolong pregnancy and labor and cause bleeding before and after delivery.
5. Labels on chewable aspirin and aspirin-containing gum should advise against use for 7 days after tonsillectomy or oral surgery (because of item 4).
6. Labels on aspirin-containing products should warn against use by people who are allergic to aspirin, have asthma, or are in the last 3 months of pregnancy.[2]

[2]We might well add that you should *never* give aspirin to children suffering from chicken pox, the flu, or flu-like conditions. See Reye syndrome, Section 14.4.

7. Labels on aspirin products should warn against use by people who have stomach distress, ulcers, or bleeding problems or who are currently taking anti-coagulant drugs.
8. The United States should adopt a standard aspirin tablet containing 325 mg (i.e., 5 grains) of the drug. The oral dosage schedule for aspirin should be 325–650 mg every 4 hours, not to exceed 4000 mg in 24 hours. Table 14.1 provides further dosage schedules for nonstandard aspirin tablets.

It is clear, then, that aspirin is an incontrovertibly effective analgesic, antipyretic, and anti-inflammatory drug. The kinds of pain that are relieved by aspirin are of the minor, or dull, low-intensity type: headache, muscle, or joint ache. The sharp pains of kidney stones, colic, and so on are not relieved much by aspirin.

For a long time, practically nothing was known about how aspirin worked to relieve pain. Now, there is evidence to indicate that aspirin interferes with the synthesis of prostaglandins, natural body chemicals that are necessary for the perception of pain. It accomplishes this by inhibiting cyclooxygenase, the enzyme believed to be the key to triggering pain and inflammation. Cyclooxygenase, in its COX-1 and COX-2 forms, catalyzes the conversion of arachidonic acid to prostaglandin H_2. Aspirin, buprofen, naproxen, and some other NSAIDs inhibit normal activity of both COX-1 and COX-2. There may also be a direct effect of the drug on the CNS, helping to produce analgesia. As noted in point 8 in the preceding list, use of aspirin as a pain reliever should be limited to a maximum of 4000 mg in any 24-hour period, for reasons discussed later. No painkiller should be taken for longer than 10 days by adults or 5 days by children. Prostaglandin inhibition also appears to be the mechanism by which aspirin manifests its anti-inflammatory action.

Quite remarkably, ASA and the salicylates have a built-in early warning system that can tell us when we are taking too much. A ringing in the ears (called *tinnitus*) signals impending salicylate overdosage. At that point, use must be suspended immediately.

Table 14.1 Recommended Adult Dosage Schedule for Nonstandard Aspirin Tablets

Amount of Aspirin Contained in One Tablet	*Number of Tablets That Can Be Taken Initially*	*Number and Timing of Tablets That Can Be Taken After Initial Dose*	*Total Number of Tablets That Can Be Taken in 24 Hours*
400 mg (6.15 grains)	1–2	1 after 3 hours	9
421 mg (6.48 grains)	1–2	1 after 3 hours	9
485 mg (7.46 grains)	1–2	1 after 4 hours or 2 after 6 hours	8
500 mg (7.69 grains)	1–2	1 after 3 hours or 2 after 6 hours	8
650 mg (10 grains)	1	1 after 4 hours	6

Note: The latest *Physicians' Desk Reference for Nonprescription Drugs* lists numerous OTC preparations that contain, variously, 81, 227, 250, 325, 500, 520 and 650 mg of aspirin, alone or in combination.

As mentioned earlier, aspirin is a most helpful drug for the reduction of fever. There is reason to believe that this result is produced, once again, by aspirin's inhibition of the synthesis of prostaglandins. Loss of body heat occurs as a result of increased peripheral circulation, sweating, and evaporation of water from the skin surfaces. Much the same effect can be achieved by means of an alcohol rub (body heat is lost in the evaporation of the volatile alcohol) or by bathing in tepid water.

Aspirin and the salicylates remain the cornerstone in the treatment of what TV ads call "arthritis and rheumatism." Because this is such a momentous subject financially as well as medically, we should take a closer look at the meaning of these terms.

Rheumatoid arthritis is a chronic inflammatory disease involving body joints and causing changes in joint fluid, cartilage, and skeletal muscle, often with marked deformities of the wrist, fingers, and feet. A prominent theory on the etiology of rheumatoid arthritis explains it as an autoimmune disease in which the immune system goes awry and attacks the patient's own cartilage. Women comprise 75% of the 2 million American rheumatoid arthritis patients, and in 80% of the cases, onset is before 40 years of age. Occurrence is more common in temperate climates. Rheumatoid arthritis is a severe, disabling, progressive disease requiring long and patient care. It must be differentiated from other inflammatory diseases such as gout, ostcoarthritis, rheumatic fever, and degenerative joint disease.

Rheumatism is a very general lay term used to describe numerous nonspecific illnesses characterized by pain, tenderness, and joint and muscle stiffness. The back is often involved, as are the neck, shoulders, thorax, and thighs. Rheumatism tends to disappear completely in a few days but may recur at frequent intervals.

Osteoarthritis (degenerative arthritis) is a noninflammatory degenerative joint disease seen mostly in older people. It causes alterations in synovial fluid, degeneration of the articular cartilage, and bone over-growth, accompanied by pain and stiffness.

Salicylates are very useful drugs in the therapy of rheumatoid arthritis. In fact, with most patients, this disease can be controlled by salicylates alone. The dose should be adjusted by the physician according to individual patient response. It is the anti-inflammatory action of the salicylates that is so valuable here, because with reduced inflammation there is less chance of bone deterioration, which means that crippling is delayed.

The FDA has a strong word of caution for people who attempt to treat their arthritis without medical supervision. If too low a dose of salicylate is taken, only arthritic pain may be relieved, masking the fact that the inflammation is proceeding and that bone and joint degeneration may follow. In most cases, larger doses of these drugs must be taken for longer periods of time, but a doctor's supervision is required because of the potentially serious side effects of salicylates and the possibility of drug interactions (see the following section). The FDA promulgates label restrictions on nonprescription analgesics prohibiting any claims relating to the treatment of rheumatoid diseases (such as "arthritis-strength" formula). Altogether, including salicylates, some 62 preparations are listed in the PDR as useful anti-rheumatoid arthritics. Other categories are: gold compounds, steroids, NSAIDs generally, and miscellaneous.

The use of salicylates by the general public to treat the symptoms of rheumatism is based upon the salicylates' ability to relieve minor pain. Oil of wintergreen (methyl salicylate USP) is an ingredient in preparations advertised for external use (liniments) in the treatment of rheumatism. Granted, salicylates can enter the bloodstream from skin absorption, but there is uncertainty about the total dose required by this route. Methyl salicylate must be kept out of children's reach; they perceive it as candy, sometimes with tragic results. See Appendix I for the chemical structures of ASA and acetaminophen.

14.4 Is Aspirin a Dangerous Drug?

When Americans swallow 20 tons of a medicine every day, it has to be one of the safest drugs. And in fact, the incidence of serious adverse reactions to aspirin is low, making the good it does all the more impressive.

With that commendation out of the way, let us look at what *can* go wrong with aspirin therapy. There are four major kinds of unpleasant results that can follow aspirin ingestion:

1. Nausea and vomiting (an "upset" stomach)
2. Loss of blood from the GI tract
3. A generalized, mild bleeding tendency
4. Hypersensitivity (allergy) to aspirin

Because aspirin is acetylsalicylic acid, it is capable of irritating the lining of the stomach (the gastric mucosa). Also, direct stimulation of the vomiting reflex center in the brain is possible. For most people taking small doses occasionally, this is of little or no consequence. But in sensitive people or those required to take heavy doses (arthritics), much stomach distress, even ulceration, is a real threat. Some of the irritating effects of aspirin on the stomach can be eliminated through the use of buffered products, but the FDA advises us not to rely on buffering to provide complete protection. According to the FDA, advertisements that indicate that buffering is the answer to all of the irritant effects of aspirin are misleading. So is the claim that buffered aspirin works twice as fast as plain aspirin. The FDA states that there is no evidence to show that the speed of onset of buffered aspirin in relieving pain is significantly increased over that of plain aspirin. Most of the published studies indicate that there is only little difference in the incidence of stomach upset after ingestion of buffered or plain aspirin.

In about 70% of normal subjects, ingestion of 1–3 g of aspirin a day will cause GI bleeding to the extent of about 5 mL; in some persons, much greater blood losses have been noted. Such GI bleeding becomes serious in habitual aspirin users, who lose so much blood that they become anemic. It is potentially very serious in ulcer patients and in alcoholics with gastritis, who may suffer massive hemorrhage induced by aspirin. In one recent year, 17,000 Americans died of gastric bleeding just from taking aspirin or other NSAIDs, says rheumatologist Sanford Roth of Arizona State University. Most had no warning. People taking an anticoagulant or people with a vitamin K deficiency or hemophilia should not take aspirin.

To help reduce the chance of bleeding and ulceration, it has been recommended that the user *not* swallow the tablet whole but break it up by chewing or by other means before swallowing. If you can stand the taste, this is a good idea, because it eliminates the concentration of drug on one part of the stomach lining. Usually, gastric distress is minimized by taking the aspirin with 8 ounces of water.

> ***Problem 14.1*** *Would it be a wise idea to chew or break up every drug tablet one takes? Can you think of two instances in which chewing is not indicated?*

Aspirin, even in small doses, prolongs the bleeding time in normal people; the effect lasts for days. It takes longer for the blood to clot. Again, for the average person this result is of little or no significance, but in the tonsillectomy patient, the hemophiliac, or someone taking a blood-thinning drug, there is every reason to consider aspirin a dangerous drug that must be avoided. Such high-risk people should switch to an analgesic such as acetaminophen (see, Section 14.6), which does not prolong the bleeding time or inhibit platelet aggregation (that is, sticking together), a crucial step in clot formation.

Examples of Other Medications That May Lengthen Bleeding Time

If you are scheduled for surgery and you have a prescription for any of the following medications, check with your doctor about whether you should continue taking it.

Dipyridamole (Persantine)	Phenylbutazone (Butazolidin)
Fenoprofen (Nalfon)	Piroxicam (Feldene)
Ibuprofen (Motrin)	Sulfinpyrazone (Anturane)
Indomethacin (Indocin)	Sulindac (Clinoril)
Naproxen (Naprosyn)	Tolmetin (Tolectin)

Pregnant women should avoid aspirin, especially near term. In fact, the FDA now requires that all OTC oral and rectal aspirin products include a label statement warning against use by pregnant women. Aspirin late in pregnancy can have adverse effects on fetal circulation and uterine contraction. This warning also extends to the use of ibuprofen in pregnancy.

Researchers at Bronx Veterans Affairs Medical Center found that aspirin significantly lowers the body's ability to break down alcohol in the stomach. Volunteers who took two extra-strength aspirin tablets before drinking alcohol had BACs 30% higher than when no aspirin was taken. Apparently, aspirin interferes with stomach enzymes that catalyze the breakdown of alcohol. We therefore conclude that it is a bad idea to take aspirin prophylactically before a party.

It is important to mention a few additional adverse reactions to aspirin, even though few people may ever experience them. A very small percentage of the population may be highly sensitive to aspirin (or allergic), suffering asthma attacks, pain, hives, shock, or even death from one dose. There is no way to tell who will experience such anaphylactic reaction to aspirin, although it's known that asthmatics are far more susceptible to an attack than nonasthmatics. Actually, about 20% of asthmatic children react to aspirin with breathing difficulty that begins about an hour after a dose is given. Salicylates can cause skin rashes, kidney damage, changes in blood cells, and ringing in the ears.

Drug interactions involving aspirin are not rare. Because of its extensive use, aspirin has an increased chance of interacting with other drugs currently administered. Three of the better known drug interactions of aspirin are:

Aspirin + An anticoagulant → Increased bleeding tendency, possible hemorrhage

Aspirin + Alcohol → Increased incidence of GI bleeding

and

Aspirin + Probenecid (used in gout) → Inhibition of probenecid

A statistical link has been established between aspirin and **Reye syndrome** (pronounced "rye"), a life-threatening condition that may follow influenza or chicken pox in children (including teenagers). Reye syndrome (RS) is characterized by sudden vomiting, violent headaches, and unusual behavior in children who appear to be recovering from an often mild viral illness. Fatal in about one-fourth of cases, RS was named after an Australian pathologist who in 1963 accurately described it as an edema (swelling) of the brain combined with liver malfunction. Data from four studies statistically link the use of salicylates during the antecedent viral illness with the development of Reye syndrome. The data are inconclusive (and are disputed by aspirin manufacturers), but nonetheless the FDA has urged parents not to use aspirin to treat their children's chicken pox flu-like symptoms and has ordered that all aspirin packages carry such a warning. Studies have found no statistical link between RS and the prior use of acetaminophen, and acetaminophen has now replaced aspirin in many conditions in children.

CAUTION: *Until we have more information about the pathogenesis of Reye syndrome and the role of salicylate in it, prudence dictates that aspirin in any form should not be given to children suffering from the flu or chicken pox—except on direct advice of the family physician. However, children and patients with certain conditions, such as rheumatic fever and Kawasaki disease, can benefit greatly from aspirin. In infants with severe birth defects of the heart, ASA is used to prevent blood clots after surgery.*

One of the primary causes of poisoning in children is aspirin overdosage. Children mistake orange-flavored aspirin for candy. The use of childproof containers limited to only 36 tablets has greatly reduced the incidence of poison episodes.

Among drugs used to commit suicide, aspirin is second only to sleeping pills. Sixty 325-mg aspirin tablets will kill the average 150-pound person; death results from respiratory failure after a period of unconsciousness. Before death, there is restlessness, incoherent speech, hallucinations, tremors, convulsions, and coma.

In doses of 1–2 g a day, salicylates reduce the quantity of uric acid excreted by the kidneys. As a result, urate plasma levels are elevated, increasing the chance of an acute attack of gout. For this reason, self-medication with salicylates by persons with a history of gout is ill-advised. Both aspirin and alcohol irritate the stomach lining, and it makes little sense to take ASA to relieve the inflamed stomach associated with a hangover.

14.5 Aspirin in the Prevention of Heart Attack and Stroke

There is now mounting evidence that low doses of ordinary aspirin, taken daily, can reduce the risk of strokes and fatal heart attacks in both healthy men and men with a history of stroke, heart attack, or angina pectoris. The reduction in risk can be from 18 to 47%. These conclusions are based on the results of many large, randomized, multicenter, placebo-controlled studies on hundreds of thousands of men in the United States and Europe. (Men have a greater risk of cardiovascular accident than premenopausal women.) ASA appears to be efficacious in patients who have undergone coronary artery bypass grafting or angioplasty. Aspirin's role in protecting women has been less certain, because far fewer women have been included in the research.[3] However, a study of 87,000 female nurses gave significant evidence that women too can be protected from heart attack, especially if they are over 50, have smoked, and have had hypertension or high cholesterol.

We have heard of older men who say they take one "baby aspirin" a day to prevent heart attack. However, the term "baby aspirin" is poor, because it is indefinite. A "children's" aspirin tablet containing 81 mg of ASA is sold with a recommended dose of 2 tablets for children 2–4 years of age who weigh 32–35 pounds, and more for heavier and older children. Most doctors and heart specialists prescribe a standard 325-mg ASA table for long term, heart-attack prevention. Scientists at Oxford University in England applied meta-analysis (see Section 5.7) to 287 published studies, involving some 200,000 people, on the effectiveness of ASA in preventing heart attack. The analysis showed that a 75–150 mg dosage works just as well as a standard U.S. 325-mg aspirin tablet, with less chance of internal bleeding; it also showed that people who haven't yet had a heart attack or stroke can reduce their risk of morbidity by as much as 25 percent. A new Italian study indicates that a daily 100-mg dose of aspirin protects men and women who have such risk factors as high blood pressure, diabetes, high cholesterol, or a familial history of heart trouble or stroke.

> ***CAUTION:*** *Before you decide to self-medicate with aspirin in the hope of preventing a cardiovascular accident, consult with your physician. Aspirin's role in all of this is still a little uncertain, getting exactly the right dose can be tricky, and aspirin can have well-known side effects if taken over a long period of time. And, aspirin does not have FDA approval for decreasing the risk of heart attack in healthy persons. A major aspirin maker has been fined $1 million for unsubstantiated advertising claims that regular aspirin could help the general population prevent heart attack and strokes.*

[3]Historically, women have been neglected in clinical trials of new drugs, devices, and biological products. However, through efforts of the federal Office of Women's Health (established in 1994), this situation is changing.

Aspirin is thought to reduce stroke and heart attack risk by blocking thromboxane A2, a substance that constricts blood vessels and powerfully induces platelets to stick together to form a clot. Heart attacks typically occur when clots form in the coronary arteries, cutting off the heart's own blood supply. Finding exactly the right dose of aspirin that will block thromboxane A2 without also blocking prostacyclin is tricky.

Other beneficial actions of daily low doses of aspirin appear to include protection of pregnant women from a risky type of high blood pressure that can induce preeclampsia; lowered risk of migraine headache; lowered risk of ischemic strokes; and possibly a reduction in incidence of colon and esophageal cancer.

Buyers beware. Mexican aspirin is not really aspirin at all but rather the potentially dangerous drug dipyrone (metamizole, methampyron) sold in Mexico as a cheap ASA substitute. Some children who have taken dipyrone have suffered agranulocytosis—an often fatal drop in the body's white blood cells. The U.S. FDA banned dipyrone from sale in the United States in 1977.

Acetaminophen has no effect on platelets in the blood; it offers none of the blood-thinning, clot-preventing benefits that can make aspirin effective against heart attacks or strokes. Neither does ibuprofen, except in risky amounts. Naproxyn sodium and ketoprofen have not been shown to have aspirin's beneficial action on cardiovascular health.

According to Bayer Market Research, more people now take aspirin for their hearts than for their aches. About 38% take it for prevention of heart disease, 23% for arthritis, and 14% for headache, with the rest for other body aches.

14.6 Acetaminophen

We have learned that aspirin is not always completely safe. It can and does cause loss of blood from the GI tract and an increased tendency to bleed, among other things. For these reasons, alternative antipyretic-analgesics have long been sought that could replace aspirin in high-risk individuals.

The most successful nonsalicylate competitor for aspirin's popularity is **acetaminophen** USP, in the public domain and marketed generically and under dozens of trade names such as Datril, Tempra, Tylenol, and Anacin-3. Acetaminophen, or *N*-acetyl-*p*-aminophenol, chemically is *not* a salicylate; it is derived from aniline, a coal tar chemical. It has gained popularity since 1949, when it was recognized as the major active metabolite in both acetanilide and phenacetin bio-transformation. In the FDA panel report referred to earlier in this chapter, acetaminophen was rated as "safe and effective" for the relief of pain and reduction of fever. It was found not to reduce inflammation and was not rated safe and effective for this purpose. The usual oral dose of acetaminophen is 325–650 mg every 4 hours for adults and older children, not to exceed 2600 mg a day.

Acetaminophen produces antipyresis (lowers fever) by a reaction on the brain's heat-regulating center in the hypothalamus, but current thought is that this mechanism does not involve inhibition of prostaglandin synthesis. Acetaminophen produces analgesia by elevating the pain threshold. Its advantages over aspirin include the following:

1. It does not cause the GI tract blood loss seen in aspirin use.
2. It does not increase the tendency to bleed. Therefore, it can be used in tonsillectomy patients and hemophiliacs.

3. As a nonsalicylate, it can be used in the presence of aspirin allergy.
4. It has less potential for causing nausea and vomiting.

Disadvantages to the use of acetaminophen include the following:

1. It is too weak an anti-inflammatory agent to be used for rheumatoid arthritis.
2. It occasionally causes skin rash.
3. In large doses, it is toxic to the liver. Acetaminophen shares this toxicity with all of the aniline derivatives. A single dose of 10 g may result in liver injury. Reversible liver damage was reported in a 16-year-old after ingestion of a single 5.85-g dose. A dose of 15–25 g is potentially fatal.
4. Reports suggest that even normal doses of acetaminophen may sometimes cause liver damage, especially if use is extended over a long period. For this reason, the maximum adult daily dosage of 4 g should not be exceeded, the duration of use should be limited, and extra-strength formulations should be used with great caution. This is doubly true in patients with preexisting liver disease. Young children are especially vulnerable to acetaminophen overdosage. The FDA's OTC drug advisory committee has recommended that manufacturers include a warning label on their products to advise people who drink large amounts of alcohol that acetaminophen in high doses can increase the risk of severe liver damage. A plaintiff in federal court was awarded $8.8 million in a suit that claimed his liver was destroyed by ordinary doses of a popular brand of acetaminophen.
5. Heavy doses of acetaminophen, such as one pill a day for at least a year, may double the risk of kidney failure, according to a study published in the *New England Journal of Medicine*. While such an effect is rare, researchers estimated that eliminating heavy use of acetaminophen could prevent 10% of all cases of kidney failure, reducing our medical bills by about $700 million annually.
6. It is of interest that in England acetaminophen rivals aspirin in frequency of use in attempted suicides. It is the leading suicide drug in young adults.
7. It may potentiate the effects of orally administered anticoagulants (a drug interaction).

There is preliminary evidence that long-time daily users of acetaminophen face three times the usual risk of kidney damage. The effect was not found when acetaminophen was taken as directed and emphasizes the importance of avoiding analgesic abuse, especially with the extra-strength products heavily advertised.

William Lee, director of the clinical center for liver diseases at the University of Texas Southwestern Medical Center in Dallas, warns that people who drink alcoholic beverages should limit their acetaminophen intake to no more than 2 grams a day instead of the 4 grams now recommended. Lee says acetaminophen is not to be considered a harmless OTC drug. On the DAWN list of emergency department drug episodes and mentions in 2000, acetaminophen ranked fourth (with 33,613), and aspirin and ibuprofen fifth and sixth, respectively. Acetaminophen with codeine ranked eleventh.

Case History *Extra-Strength Acetaminophen*

M., a 40-year-old electronics specialist, had taken a brand of extra-strength acetaminophen for years to control chronic ankle and wrist pain. He never exceeded the recommended dosage. During a bout with the flu, M. added another product to his regimen, a maximum-strength acetaminophen sinus medication. Each tablet contained 500 mg of acetaminophen, and he took at least two or three tablets of each daily, and some days as many as eight. (Note: Label directions permitted eight as the maximum daily dosage.) Thus on some days M. was ingesting 8 g of acetaminophen. In 5 days M. turned yellow and was admitted to a hospital, where doctors diagnosed acetaminophen-induced hepatitis. His condition very slowly and agonizingly worsened, and he died less than 4 years later. His wife sued the manufacturer, claiming her husband never exceeded the label dose, but she lost.

Case History *Acetaminophen Overdosage in a Child*

S., a 14-month-old toddler, had a fever. Her concerned mother began to dose S. with a major brand of grape-flavored infant acetaminophen, sold OTC as a safe alternative to aspirin. As the dosing continued, S. became pale, listless, diarrheic, glassy-eyed, and unresponsive. She had been given an accidental overdose of acetaminophen, and it destroyed most of her liver. The confusion was a common one: the infant-strength acetaminophen product used was concentrated, and three times stronger than the children's strength product. Further, there is but a small margin of error in children's acetaminophen dosage, despite the reassurances of safety proffered in the ads. S. was forced to undergo a partial liver transplant. The FDA reports that in 1991, 31,511 children under 6 suffered inappropriate exposure to pediatric acetaminophen.

Phenacetin, a close chemical relative of acetaminophen, was long sold in *aspirin-phenacetin-caffeine* (APC) formulations. Phenacetin, however, is also a *para*-aminophenol derivative and was banned by the FDA as too toxic to remain on the market. It can cause hemolytic anemia and damage to the liver. Daily users of phenacetin over long periods faced five times the usual risk of kidney damage. Phenacetin is metabolized in humans to acetaminophen.

14.7 Ibuprofen

One of the newer nonnarcotic, nonsteroidal pain relievers available for OTC sale has the generic name **ibuprofen**. It enjoyed large prescription-only sales for years under the brand names Motrin and Rufen. As an OTC drug, ibuprofen is sold under the trade names Advil, Medipren, Nuprin, Midol-200, and Trendar. Motrin is now sold OTC also. It is indicated for the temporary relief of minor aches and pains associated with the common cold, headache, toothache, backache, minor pain of arthritis, and menstrual cramps.

Ibuprofen also has antipyretic and anti-inflammatory activity. Its low dose (200 mg) makes it highly competitive with aspirin and acetaminophen. One negative factor is the frequency of cross-reactivity between allergy to aspirin and to ibuprofen. Labels on all ibuprofen products (Advil, Haltran, Medipren, Midol-200, Nuprin, Trendar) warn that those who are allergic to aspirin should not take ibuprofen. Use of ibuprofen during pregnancy is contraindicated.

Chemically, ibuprofen is racemic 2-(*p*-isobutylphenyl)propionic acid, which means that it is *not* chemically related to either the salicylates or the *para*-aminophenols. See Figure 14c in Appendix I for chemical structure. The Upjohn Company, which markets the Motrin brand of ibuprofen, says that its mode of action may be related to prostaglandin synthetase inhibition. See Section 14.3 for details.

One of the reasons ibuprofen was approved for OTC sales is its effectiveness in treating dysmenorrhea, where it appears to be clearly superior to aspirin. In other areas, ibuprofen in 200-mg doses appears to be at least as effective as 650 mg of aspirin. The recommended dose of ibuprofen is one or two 200-mg tablets every 4–6 hours as necessary, not to exceed 1,200 mg per day. Prescription-only Motrin contains ibuprofen in doses as high as 800 mg.

Side effects of ibuprofen, unfortunately, do occur. It can irritate the stomach and cause some bleeding. There are some indications that it can damage the kidney, especially in the elderly and in others predisposed to this action. Ibuprofen, like ASA, inhibits platelet formation and prolongs bleeding time, but the effect is not as long-lasting as that of aspirin.

Regarding drug interactions, ibuprofen can inhibit the antihypertensive and diuretic effects of diuretic drugs. It can also increase the risk of bleeding in patients taking oral anticoagulants or consuming alcohol. As with aspirin, the FDA now requires a label warning cautioning women not to take ibuprofen during pregnancy.

As noted in Section 7.15, caffeine enhances the painkilling effects of such nonprescription analgesics as ibuprofen, aspirin, and acetaminophen. A massive retrospective study, part of the Baltimore Longitudinal Study of Aging, conducted at Johns Hopkins Medical School, found that the risk of developing Alzheimer's disease can be reduced by as much as 60 percent by frequent use of ibuprofen over 2 years or longer. No similar effects for aspirin or acetaminophen were noted. Experts cautioned that there are substantial risks of kidney and stomach injury in people who take extended doses of ibuprofen.

14.8 Naproxen Sodium, Ketoprofen, and Fenoprofen

Another type of nonnarcotic analgesic now approved for OTC sale is **naproxen sodium**, sold under the brand name Aleve. (It was formerly sold on prescription only as Naprosyn and Anaprox.) This NSAID, in a 220-mg dose, is intended for the relief of minor pain associated with headache, toothache, backache, muscle ache, arthritis, menstrual cramps, and the common cold, and to reduce fever. **Allergy Warning**: Do not take naproxen sodium if you have had either hives or a severe allergic reaction after taking *any* pain reliever.

Advertisements for Aleve tout its longer half-life (see $t_{1/2}$, Section 4.5), approximately 12 hours in plasma. Hence, the manufacturer recommends dosing every 8–12 hours instead of the 6–8 hours or 4–6 hours recommended for other pain relievers.

Children younger than 12 years should never take naproxen sodium except under a doctor's supervision. For persons aged 12–65, the maximum daily dose is 3 tablets (with 8–12 hours between doses.) Persons older than 65 should not take more than 2 tablets a day, or 1 every 12 hours.

As with ibuprofen, naproxen sodium in excessive amounts can cause digestive problems such as heartburn and upset stomach. Vomiting has also been reported, as well as gastric bleeding, dizziness, fatigue, and depression.

On the DAWN list of drugs mentioned most frequently in emergency rooms in 2000, naproxen sodium ranked 33rd.

You will be hearing more about two other NSAIDs of the propionic acid type. **Ketoprofen** (sold as Orudis KT and Actron), is an analgesic, anti-inflammatory, and antipyretic drug, marketed in 12.5-mg tablets. It has been sold in other countries for some time and only recently became available OTC in the United States. It can cause the same adverse effects and allergic reactions associated with other nonsteroidal anti-inflammatory drugs of the propionic acid type. See Figure 14 in Appendix I for chemical structures of these drugs. **Fenoprofen** (Nalfon) is another NSAID of the propionic acid type.

Five other NSAIDs (of various chemical types) are: ketorolac tromethamine (Toradol), nabumetone (Relafen), mefenamic acid (Ponstel), etodolac (Lodine), and diclofenac (marketed as Voltaren and as Cetaflam).

A summary of four of the most popular pain relievers is given in Table 14.2.

14.9 Comparative Potencies

The ads will tell you "Hospitals prefer Tylenol" and "You can't buy a more potent pain reliever without a prescription." But a court decision labeled those claims "false and misleading." Truthfully, there is little difference in the potencies of aspirin, acetaminophen, and ibuprofen for ordinary aches and pains. Milligram for milligram, ASA and acetaminophen have the same antipyretic potency. But with $1.8 billion at stake in yearly sales, advertisers will stretch the truth as far as possible short of fraud to get you to buy their product. So, buyer beware. Know that ibuprofen can offer some advantages for dysmenorrhea, and acetaminophen is easier on the stomach than aspirin, but overall no one pain reliever offers significant pain relief advantages over the others. Also remember that aspirin, acetaminophen, and ibuprofen can be purchased generically at great savings.

14.10 COX-2 Inhibitors

Touted as a wonder drug, celecoxib (Celebrex, Searle) is the first in a class of selective painkillers termed COX-2 inhibitors, so named because at therapeutic doses

Table 14.2 OTC Pain Relief Primer

Type/Dosage	*Common Brands*	*What It Does*	*Possible Side Effects*
Aspirin, (325 mg, 500 mg)	Anacin, Ascriptin[b], Bayer, Bayer Plus[b], Bufferin[b], Ecotrin[c]	Relieves mild to moderate pain from headaches, sore muscles, menstrual cramps, and arthritis; reduces fever. (adults only)	Prolonged use may cause gastrointestinal bleeding, especially in heavy drinkers; may increase the risk of maternal and fetal bleeding and cause complications during delivery if taken in the last trimester; can cause Reye syndrome if given to children and teenagers who have the flu or chickenpox.
Acetaminophen, (325 mg, 500 mg)	Anacin-3, Excedrin[a], Pamprin[d], Midol[b], Tylenol	Relieves mild to moderate pain from headaches and sore muscles; reduces fever.	May cause liver damage in drinkers and those taking excessive amounts (more than 4000 mg daily) for several weeks.
Ibuprofen (200 mg)	Advil, Motrin-IB, Nuprin, Pamprin-IB	Relieves mild to moderate pain from headaches, backaches, and sore muscles; relieves minor pain of arthritis; provides good relief of menstrual cramps and toothaches; reduces fever.	Gastrointestinal bleeding; stomach ulcers; kidney damage in the elderly, people who have cirrhosis of the liver, and those taking diuretics.
Naproxen sodium (200 mg)	Aleve	Relieves mild to moderate pain from headaches, backaches, and sore muscles; relieves minor pain of arthritis; provides good relief of menstrual cramps and toothaches; reduces fever.	Gastrointestinal bleeding, stomach ulcers; kidney damage in the elderly, people who have cirrhosis of the liver, and those taking diuretics.

[a] Contains caffeine.
[b] Contains buffers.
[c] Enteric coated.
[d] Contains ingredients other than analgesics.

Source: *FDA Consumer*, January–February 1995, p. 13.

they block the inflammatory enzyme cyclooxygenase (see Section 14.3). Aspirin and similar anti-inflammatory agents do that, too, but they aren't selective, also blocking the COX-1 enzyme that acts to protect the stomach lining, leading sometimes to upset stomach and ulcers. The FDA has approved Celebrex as an Rx-only treatment for osteoarthritis and for rheumatoid arthritis (the most severe joint disease), but the agency does not recommend the drug for other types of pain. However, with exposure and examination in the marketplace, Celebrex's "Wonder Drug" appellation now appears unearned. Two studies, one by the FDA and the other reported in the *British Medical Journal*, concluded that Celebrex does not have a safety advantage over cheaper NSAIDs such as ibuprofen, nor is Celebrex a better analgesic or anti-inflammatory than ibuprofen, nor is it less likely to cause ulcers. Celebrex is a big money maker, costing up to 25 times more than the older, cheaper OTC pain killers. Celebrex is usually prescribed for osteo- and rheumatoid-arthritis patients; it is not for people who have asthma or are allergic to aspirin or other NSAIDs or to sulfa drugs. Celebrex, because it does not inhibit COX-1, is much less likely than older NSAIDs to cause bleeding. This is not to say that Celebrex is altogether free of adverse effects such as bleeding or stomach irritation. See your physician before taking any COX-2 inhibitor. The dose for celecoxib (for people 18 years and older) is 100–200 mg twice daily. Interestingly, Celebrex shows promise as a cancer weapon and is being investigated for prevention and treatment of Alzheimer's disease.

Vioxx (rofecoxib) is the second COX-2 inhibitor approved by the FDA. It is prescribed for osteoarthritis, menstrual pain, and acute pain. Its dosage, however, is much smaller than that of Celebrex. It has been shown that Vioxx has anti-inflammatory, analgesic, and antipyretic activity, and it does not interfere with platelet aggregation. The adverse effects of rofecoxib include possible headache, heartburn, upset stomach, and diarrhea. For additional information on Vioxx and Celebrex, browse **http://pharmacology.about.com**. For more information on arthritis, see **http://www.allaboutarthritis.com**.

Bextra (valdecoxib), a selective COX-2 inhibitor, has been approved by the FDA for relief of the signs and symptoms of osteoarthritis and adult rheumatoid arthritis. The manufacturer of Bextra claims that it shows a lower incidence of gastrointestinal side effects than its competitors. Common side effects of Bextra include headache, upset stomach, nausea, and abdominal pain; it should not be used by women in the third trimester of pregnancy or by anyone allergic to aspirin.

Web Sites You Can Browse on Related Topics

14.10 Aspirin
www.aspirin.org

Acetaminophen
http://www.emedicine.com/emerg/topic819.htm
Also: Search "acetaminophen"

NSAID in Arthritis
http://arthritis.about.com/health/arthritis/library
http://www.nsaid.net/
http://www.personalmd.com/drgdb/49.htm
Also: search "overdoses"
Also: search "NSAID"

Ibuprofen
http://www.rxlist.com/cgi/generic/ibup.htm

Study Questions

1. Complete the table by answering "Yes," "No," or "Probably."

	Aspirin	Acetaminophen	Ibuprofen
Useful analgesic	_____	_____	_____
Useful antipyretic	_____	_____	_____
Useful anti-inflammatory	_____	_____	_____
Classed as a salicylate	_____	_____	_____
Usual dose 325–650 mg every 4 hours	_____	_____	_____
Works by interfering with the synthesis of prostaglandins	_____	_____	_____
Indicated for use in tonsillectomy patients	_____	_____	_____
Contraindicated during pregnancy	_____	_____	_____
Has been shown to reduce risk of heart attack	_____	_____	_____

2. List five nonlethal disadvantages (or adverse reactions) of aspirin as a therapeutic agent in the general population. Which of these disadvantages is potentially the most serious and why?
3. Name three chemicals or drugs discussed in this chapter that are classified as salicylates.
4. Define (**a**) analgesia, (**b**) antipyresis, (**c**) tinnitus, (**d**) ASA.
5. Body temperature at any given moment is the result of a delicate balance between heat production and heat loss. Explain how aspirin works to reduce the body temperature of a febrile person.
6. A bottle of aspirin tablets was labeled "325 mg" in strength. When carefully examined, each tablet was found to have a weight of around 380 mg. Was the bottle mislabeled?
7. Explain what is meant by *buffered* aspirin.
8. The dose of aspirin for relief of minor pain is about the same as the dose used to treat fever. However, the dose of aspirin for treating rheumatoid arthritis may have to be much higher. Explain why this is so.
9. The FDA has proposed the establishment of a standard-strength tablet for use throughout the United States. What would be the advantage of such a tablet?
10. What do hemophiliacs, oral surgery patients, ulcer patients, asthmatics, and chronic alcoholics have in common in regard to aspirin use?
11. The FDA has stated that the combination of aspirin with an antacid for the treatment of heartburn, sour stomach, or acid indigestion is irrational. Explain the reasoning behind this statement.
12. On the basis of what you have read in this chapter, would you agree with those who suggest that aspirin should be restricted to sale by prescription only and not sold OTC?
13. What are the three most significant advantages acetaminophen has over aspirin?
14. Why is the FDA critical of advertising that promotes the use of extra-strength (e.g., 0.5 g) acetaminophen preparations?

15. What is Reye syndrome, and what is the relationship of salicylates to it?

16. Prostaglandins are required to help the stomach make the mucus barrier that protects the stomach lining and also to regulate overproduction of stomach acid. On this basis, explain aspirin's well-known ability to irritate the stomach and cause ulceration.

17. What do the letters NSAID stand for? Is morphine an NSAID?

CHAPTER 15 Additional OTC Drugs and Chemicals

Key Words in This Chapter

- Sleep aids
- Cold medicine
- Decongestant
- Cough remedy
- pH
- Antacids
- Antibiotics
- Laxatives
- Acne drugs
- Dandruff drugs
- Mouth rinses and gargles
- Diet aids

Learning Objectives

After you complete your study of this chapter, you should be able to do the following tasks:

- Name the active ingredient used in nearly all OTC sleep aids.
- Identify possible side effects of nasal decongestants.
- Write the definition of functional dependence.
- Write the definition of antitussive and give names of 2 important examples.
- Explain the pH scale and how it is used.
- Explain why the combination of aluminum hydroxide and magnesium hydroxide is a better antacid than sodium bicarbonate or calcium carbonate.
- Give the definition of drug resistance.
- Distinguish between laxatives and cathartics.
- State the four categories of acne treatments.
- Decide how valuable mouth-rinse products really are.
- Recognize possible serious side effects of phenylpropanolamine used as a diet aid.

15.1 Introduction

Check your library for a publication called *Physicians' Desk Reference for Nonprescription Drugs* (it is the same as the regular *PDR* but lists only OTC medicines). Look in the blue Product Category Index, and you will count approximately 144 different categories, starting with acne products and ending with wart removers. This *PDR* lists about 2,000 products that constitute the 144 categories, but this is only a fraction of the 300,000 OTC products that can be purchased in the United States. (Prescription drugs number about 65,000.)

Americans are daily using this vast, diverse number of OTC drugs and chemicals to relieve symptoms, preserve their health, and improve their appearance and their lives. Or at least that is what we believe. Prompted by never-ending television, radio, magazine, and newspaper advertising, we spend more than $13 billion a year on allergy relief pills, antacids, dandruff preparations, pregnancy tests, laxatives, antidiarrheals, pain relievers, sleep aids, contraceptives, vitamins, weight control products, gas relievers, and many more products. Surveys show that Americans suffer from aches, pains, and ills—about 12 self-treatable health problems a month for each of us. What is more, the use of nonprescription drugs is rapidly increasing—about 40% in one recent 5-year period. Three-quarters of Americans practice self-medication frequently, more than a third of them through the use of an OTC drug. The top OTC drugs (as you might have guessed from the heavy advertising) are pain relievers, cold-sinus-cough products, and vitamins.

Many drugs have been switched from prescription-only to OTC status, and this influx has greatly stimulated the OTC market. About 70 ingredients or dosages have been switched to OTC status, and 20 more are pending. Examples of important recent switches are antihistamines, painkillers (ibuprofen and naproxen), topical hydrocortisone, a decongestant, heartburn drugs, and antifungals.

What distinguishes a prescription-only drug from an OTC drug? The 1951 Durham-Humphrey Amendment to the federal Food, Drug, and Cosmetic Act provided the first statutory requirement that any drug be labeled for sale by prescription only. With the amendment, prescription drugs were defined primarily as those unsafe for use except under professional supervision. They include certain habit-forming drugs and any drug that is unsafe "because of its toxicity or other potentiality for harmful effect, or the method of its use, or the collateral measures necessary to its use." Medicines that can be used safely on the basis of product labeling alone, however, must (by law) be made available to Americans without a doctor's prescription.

In this chapter we examine 11 categories of OTC drugs and chemicals (i.e., preparations that may legally be purchased without a prescription). In a brief introductory way, we discuss the sources, pharmacological effects, and possible adverse effects of these substances. The reader is reminded that aspirin and other headache remedies are discussed in Chapter 14 and that antiobesity agents are included in Chapter 7 in the discussion of amphetamines.

We also look briefly at the importance of pH in drug and cosmetic advertising. One fact deserves emphasis: OTC drugs *are* drugs, with the same potential as prescription-only drugs for side effects, habituation, or drug interactions. Their easy availability should not lull us into a false sense of security about them. We should report OTC drug use to the doctor when we are asked, "What drugs are you currently taking?" Also, we should get into the habit of *reading labels*. The FDA recognizes that when the general public is making decisions about self-medication, it is imperative that adequate directions for safe and effective use and warnings against misuse be printed on the label of all OTC preparations. "Seven-point" labels for OTC drugs must include statements about their identity, manufacturer, net quantity of contents, ingredients, directions for safe and effective use, possible habit-forming nature, and cautions and warnings for protection of the user.

A note to parents: *To prevent possible overdosage in small children, parents should be sure to communicate with each other when a dose of medicine has been given. Also, parents must not confuse regular and extra-strength children's and infants' formulas.*

Before we begin our study of OTC drugs, here is a suggestion: regard any new drug on the market, especially one that claims to be a wonder drug, with considerable caution, if not skepticism. A case in point is **melatonin**, a human hormone produced by the brain's pineal gland. Physiologists have long known that melatonin inhibits gonad development and influences estrus in mammals, and that it is secreted at a rate inversely proportional to environmental lighting. Now, melatonin, with much hoopla, is being sold as a dietary supplement, touted to be a "natural wonder drug" that can combat aging, cure jet lag, help us sleep better, lower our blood pressure, improve our sex life, and reduce the risk of cancer and heart disease. Because it is termed a dietary supplement, and because no medical claims are made in its advertising, melatonin is not regulated by the FDA. Anybody can sell it, and nobody can legally question it or demand proof of purity, safety, or efficacy. Nobody knows how melatonin might interact with other drugs or what might be the consequences of long-term use. Remember, this substance is a human hormone (the chemical structure of which, by the way, is similar to that of the neurotransmitter serotonin), and hormones are powerful substances. Will melatonin prove itself over the years? Maybe, but why should Americans be the guinea pigs for the supplements companies? Should a human hormone be advertised as a dietary supplement? The wisdom of viewing melatonin—and other substances like it—with considerable caution seems obvious.

For melatonin's role in circadian rhythms, see Section 4.6.

15.2 OTC Sleep Aids

Chapter 8 was devoted to the barbiturates, the prescription-only sedatives used to induce sleep. In this section, we discuss agents that are advertised as sleep aids but that can be purchased without a prescription and used entirely at the discretion of the general public.

We spend more than $30 million a year on OTC drugs sold as sleep aids. This fact reflects the common occurrence of insomnia, real or imagined, in our population. Real insomnia may have its origin in thyroid disease, old age, disorders of the bladder or circulatory system, or tension or neurosis. Imagined insomnia may be the result of overestimating the number of hours of sleep we need. Not everyone needs 8 hours of sleep a night; some can function well on 5. Furthermore, almost everyone occasionally experiences sleepless nights caused by the previous day's excitement, by a stuffy nose, or perhaps by a caffeine-containing drink. Some advertisers would have us believe that this situation is sufficient reason to rush out and buy a chemical crutch to get us through the night.

On the other hand, no one denies that to some people insomnia is a real, distressing, and recurrent problem. It is often associated with psychological problems such as anxiety or depression. Getting at the cause of the insomnia is the best treatment, but that may entail a year's visits to a psychiatrist. The temptation is to turn to drugs, especially if they can be purchased without a prescription. Let's look at the ingredients in these OTC products and how they work.

The main ingredient in almost all OTC sleep aids is an antihistamine, mostly diphenhydramine (Benadryl, Parke-Davis). In fact, six OTC nighttime products are listed in the *Physicians' Desk Reference for Nonprescription Drugs* that list diphenhydramine as the active ingredient. Several others contain another antihistamine, doxylamine. Years ago, it was discovered that one side effect of antihistamines is drowsiness; this effect has become the rationale for their use as sleep aids. The PDR lists one product, Sleep-Tite, that contains ten herbs.

The user of an antihistamine sleep aid is warned, "Take with caution if ethyl alcohol has been consumed." This statement refers to the danger of combining two CNS-depressant drugs; the result can be overwhelming CNS depression in some people. The American Sleep Disorders Association cautions that for some types of insomnia, such as those caused by breathing disorders, OTC preparations might be dangerous. Alcohol in combination with diphenhydramine elicited 2,500 emergency department episodes in 1994. *Note*: One can purchase generic antihistamines OTC at a fraction of the cost of some of these very highly advertised brands.

Another ingredient in older OTC sleep aids was scopolamine (see skin patches, Section 4.2), an anticholinergic alkaloid obtained (along with atropine) from belladonna and henbane plants. In sufficient doses, scopolamine can cause drowsiness, amnesia, and dreamless sleep. Unpredictably, it may also induce excitement or hallucinations. Scopolamine has been used in obstetrics to cause a twilight sleep, a kind of amnesia in which the pain of delivery is felt but not remembered. It is not presently sold as a sleep aid but is found in small quantities in Elixir Donnatal.

The important question is, do the OTC sleep aids actually work? Users say emphatically, "Yes." Critics in the FDA and consumer advocate groups say, "At the doses recommended on the label, no." A dose of 25 mg of antihistamine or 0.5 mg of scopolamine is ineffective in sleep induction, they claim. For support, the critics cite a number of studies done in the last decade that cast much doubt on the advertising claims made for the OTC sleep aids. Other critics suggest that it is actually a placebo effect that helps the insomniac. In any event, studies show that these antihistamine products quickly lose their effectiveness.

With the wide variations in human need and in the response to drugs, however, it is possible that some insomniacs are benefiting from these patent medicines. Others may be wasting their money. And for those who have become habituated to sleep aids and who must take them for the rest of their lives, the drugs may be worse than nothing at all.

DAWN data for 2000 show that over 6,600 emergency room mentions were recorded for OTC sleep aids such as Sominex, Unisom, and Nytol, plus 6,270 episodes for diphenhydramine per se.

15.3 Cold Medicines and Decongestant Preparations

Most of the head colds we catch are caused by a rhinovirus, of which there are over 110 varieties. Influenza (the "flu") is also caused by a virus. Although we cannot say that there is a cure for the common cold, progress has been made in finding drugs that appear to halt replication of the cold virus and thus reduce duration and severity of cold symptoms. These drugs, based on the protease inhibitor concept (see Section 2.8), hold great hope but must be taken immediately upon your doctor's diagnosis that you are sick. These drugs are not yet fully evaluated. For more information, search the Internet for Picovir (pleconaril). The two new antiflu drugs that have won FDA approval, Relenza (zanamivir) and Tamiflu (oseltamivir), are neuraminidase inhibitors. Taken as soon as flu symptoms appear, these drugs can shorten the length of the flu attack and reduce symptoms. *However, do not use them if you are pregnant.*

Herbals and dietary supplements such as ginseng, echinacea, and zinc lozenges are promoted as cold preventives. Unfortunately, only a few have been tested and actually may have no scientific basis. The American Medical Association warns that large doses of zinc in zinc supplements can block absorption of needed copper and other minerals.

The status of vitamin C (ascorbic acid) in head cold treatment or prevention is still unsettled. There is no doubt that vitamin C helps maintain the integrity of the mucous membrane and is therefore a prerequisite for good health. The difference of opinion lies in *how much* ascorbic acid we should ingest to prevent colds or to treat them once they have started. Nobel Laureate Linus Pauling, who was an advocate of megavitamin therapy, suggested daily doses of thousands of milligrams. The federal government publishes an adult minimum daily requirement (MDR) for vitamin C of 30 mg and a *Recommended Daily Dietary Allowance* (RDA) of 60 mg (100 mg for smokers). However much you decide to take, you should take it daily, because this water-soluble vitamin is not stored well in body tissues and the body is not capable of synthesizing it. For those who are not able to obtain fresh fruit or vegetable sources of vitamin C, the inexpensive synthetic tablet form is the next best thing, because it is chemically identical to natural vitamin C. You should know, however, that in addition to vitamin C, fresh oranges and other fruits provide such things as bioflavonoids—factors that may help keep the capillaries healthy.

When a cold virus multiplies in the sinus area, inflammation and swelling of the tissues result in a stuffed-up nose, forcing the person to resort to uncomfortable mouth breathing. This condition will eventually correct itself without drugs, but drug advertisers constantly urge us to use their brand of nasal shrinker for symptomatic relief. The decongestants are sold as sprays, nose drops, inhalers, or in tablet form for oral use.

Examples of commonly used nasal mucous membrane shrinkers, or nasal decongestants, are phenylpropanolamine, pseudoephedrine, and phenylephrine.

> ***CAUTION:*** *all three of these drugs can be dangerous (see Section 5.6). Indeed, phenylpropanolamine is now banned by the FDA. Pseudoephedrine definitely should not be given to children under twelve or to babies in airplanes, who have pressure problems.*

If you reread parts of Chapter 5 you will see that these drugs are sympathomimetics, patterned after the catecholamines (such as epinephrine), and are therefore potent medications. Besides shrinking nasal membranes and blood vessels, sympathomimetics can excite the CNS and stimulate the heart and circulatory system. Although it is true that phenylpropanolamine and phenylephrine have been selected because their CNS and cardiac actions are minimal, these side effects still exist to some degree, and some users may discover their heart beating at a considerably faster rate after using one of these products (the drug enters the bloodstream via nasopharyngeal absorption). In addition, some users of nasal decongestants experience a **rebound congestive effect**; that is, their nasal congestion is even worse after the drug's effects wear off. This effect necessitates another dose of shrinker and then another, until the patient is sniffing the inhaler all day long, with no real long-lasting relief. This type of drug dependence is termed **functional dependence**.

Case Histories *Functional Dependence*

A 61-year-old male was examined who had been using decongestant nose drops regularly for 40 years. When seen, he was using the sympathomimetic product every 30 minutes for a sinus problem. He had developed audiovisual hallucinations in the form of threatening voices and frightening images at night. The nasal decongestant was gradually withdrawn, and his hallucinations disappeared. A major tranquilizer and an antidepressant helped him through the withdrawal syndrome.

A 50-year-old woman had been using OTC nose drops containing 1% ephedrine and 0.1% xylometazoline repeatedly every half hour to treat a nasal obstruction. She developed ischemic chest pain, tachycardia, and hypertension. When the drops were discontinued, all of her cardiovascular problems stopped.

You are placing yourself at risk if you use sympathomimetic nasal decongestants and if you:

- Have heart disease or high blood pressure
- Have hyperthyroidism
- Are taking an MAO inhibitor
- Are taking digitalis
- Have glaucoma or difficulty urinating

In each case here, the risky ingredient in the decongestant preparation is the sympathomimetic. It can stimulate the heart, affect metabolism, affect smooth muscle, and cause a crisis in the patient taking an MAO inhibitor. See Chapter 5 for a complete discussion of the effects of sympathomimetic drugs. See the discussion of diet aids for more on phenylpropanolamine.

The FDA deplores the "shotgun" approach of some cold products that include diverse combinations of an antihistamine, a vasoconstrictor decongestant, aspirin, caffeine, vitamin C, and/or a belladonna alkaloid as a drying agent. That is, in order to get one drug that might relieve symptoms, the person is forced to take four or five others that aren't needed and that may have side effects or allergic potential. One shotgun preparation is heavily advertised for the relief of coughs, colds, and flu. In each 20 mL, it contains 30 mg of dextromethorphan hydrobromide, 60 mg of pseudoephedrine hydrochloride, 4 mg of chlorpheniramine, and 650 mg of acetaminophen, plus 10% alcohol and 13 other supposedly inactive ingredients. The recommended dose for people 12 years and older (i.e., over 95 lb) is 15 mL, with 4 doses per day permitted. The question is, does one actually need all of these chemicals all at once? Is the product worth its high cost? How long would it take for the body to cure itself without any chemical assistance? How early, if at all, do parents want to start their children on the pill-popping bandwagon? Table 15.1 lists ingredients found in OTC cold products and the conditions they are supposed to relieve.

If you believe the medical experts who testified before Congress, antihistamines, found in most OTC cold medicines that cost Americans over $1 billion a year,

Table 15.1 Ingredients Found in Over-the-Counter Cold Products

Drug Category	*Condition Intended to Relieve*	*Examples of Drugs*	*Possible Adverse Effects*
Antihistamine	Rhinitis (see discussion in text)	Diphenhydramine, doxylamine, chlorpheniramine, pyrilamine	Drowsiness, impaired judgment
Decongestant	Stuffed-up nose or sinuses	Oxymetazoline, phenylpropanolamine, pseudoephedrine, phenylephrine	Increased blood pressure; excessive heart stimulation, insomnia
Expectorant	Accumulated sputum	Ammonium chloride, guaifenesin, terpin hydrate	(Little evidence they actually work)
Cough suppressant (antitussive)	Persistent, unproductive cough	Dextromethorphan, noscapine	Drowsiness, stomach upset
Analgesic-antipyretic	Body aches, fever	Aspirin, acetaminophen, salicylamide	See sections on aspirin and acetaminophen

are ineffective and pose health and safety risks to the user, especially to children. Contrary to what the commercials tell us, antihistamines do not provide any relief from the sneezing and runny nose caused by colds. Studies show that antihistamines work no better than placebos—in fact, no better than giving nothing at all. And the drowsiness and impaired judgment they cause make driving a car or operating machinery hazardous. Antihistamine overdosage is common—nearly 28,000 cases a year reported to poison control centers nationwide, more than half involving youths under 17.

In summary, **there is nothing you can take to cure a cold or even to hasten recovery significantly**. There are OTC preparations that can help relieve symptoms, but there can be clear-cut disadvantages to their use. If aches and fever accompany a cold, aspirin may be the drug of choice. But why pay top price for a highly advertised, shotgun cold product? Buy an inexpensive brand of plain or buffered aspirin and take it as directed, with lots of fluids such as orange juice.

15.4 Cough Remedies and Expectorants

Among the OTC cough remedies and expectorants, two types of pharmacological actions must be discussed:

1. Suppression of the cough reflex
2. Mobilization of thick mucus through an expectorant process

The cough reflex is a highly desirable, protective mechanism initiated by the presence of mucus or a foreign irritant in the air passages. (In serious cases, a cough may have its origin in heart disease, bronchitis, or emphysema.) A cough is an explosive expulsion of air from the lungs for the purpose of expelling an obstruction, but if the cough reflex is triggered repeatedly without the irritant being removed, it becomes unproductive and possibly exasperating.

Treatment of such a persistent cough consists of administering a drug that will suppress part of the network of nerves that make up the cough reflex arc. Actually, this suppression occurs in the so-called cough center in the brain, where the drug (termed an **antitussive**) makes nerve transmission through the center more difficult. In other words, it elevates the threshold for coughing. The coughing usually is not stopped, but it is reduced in frequency.

Dextromethorphan is a synthetic antitussive having no narcotic action or addiction liability. In 15–30-mg doses, it acts to suppress the brain's cough center. It can be purchased OTC in syrups and lozenges under many trade names; check the label to make sure. Dextromethorphan has become a popular drug in place of codeine for treating coughs. As a point of interest, heroin is an excellent antitussive, an action noted when heroin was first introduced (legally, then) into medicine in Germany in the 1890s.

As with all OTC preparations, it is important to read the label for the list of ingredients in the cough product you are about to purchase. Some cough remedies may contain up to 20% alcohol. Some products contain large amounts of sugar or antihistamines that can make one drowsy.

Cough drops on the market may look and smell like potent medicines, but there is no clinical evidence that they are better than hard candy. Menthol, eucalyptol, and topical anesthetics may work no better than an old-fashioned lemon drop, says the University of California at Berkeley *Wellness Letter*.

An expectorant is a drug that acts to loosen thick mucus, sometimes called *phlegm*, associated with a chest cold and causes the mucus to be propelled upward in the bronchial tree so that coughing can remove it entirely. Tiny hair-like projections called *cilia* line the air passages, and it is through their wave-like motion that the phlegm is propelled upward. For smokers, a word of admonition: cigarette smoke damages the cilia, rendering them incapable of normal functioning. The problem with discussing expectorants is that there is little scientific evidence that they actually work. The FDA has concluded that terpin hydrate, ammonium chloride, horehound, pine tar, and spirits of turpentine are all ineffective as expectorants. Their use in OTC expectorant products was banned in 1990. Only one expectorant, guaifenesin, retains FDA approval for use in nonprescription OTC cough and cold medications, and it is no miracle drug. In fact, one can get as much relief by inhaling steam from a vaporizer using plain water.

A persistent cough is a warning sign that something is wrong. It is better to get medical advice than to try to suppress the cough for an extended period of time.

15.5 pH in Advertising

Even if we have never taken a single course in chemistry, most of us know that certain substances are acidic and others alkaline. We have heard of acid stomach, battery acid, pool acid, carbonic acid, and acid soil for azalea growth. And we have heard of alkaline (or basic) substances such as lye, soap, Drano, and ammonia water. In order to compare *how* acidic or alkaline something is, chemists have created a pH scale:

pH = 0 1 2 3 4 5 6 7 8 9 10 11 12 13 14

← increasing acidity — Neutral — increasing alkalinity →

Note that if a solution has a pH of 7, it is neither acidic nor basic; it is neutral. A solution or substance with a pH of 1 would be very acidic. Each unit change in the pH scale represents a 10-fold change in the strength of the acid or base. For example, a solution of pH 5 would be 10 times as acidic as a solution of pH 6. A pH 4 solution is 100 times as acidic as a pH 6 solution. Carbonated beverages have a pH of about 5 or 6; ammonia water has a pH of about 11.

The human body is remarkably capable of maintaining a physiologically proper pH in its tissues and fluids. Blood's pH, for example, is ordinarily maintained strictly between 7.3 and 7.4. Cells in the lining of the stomach are able to secrete *hydrochloric acid* (HCl), giving stomach contents a very acidic pH of 1 or 2. This pH is normal and desirable for the digestion of food. In contrast, the contents of the small intestine are distinctly alkaline.

Advertising copywriters have used the pH concept to sell shampoos, hand lotions, acne treatments, and other products. A preparation for the skin, pHisoderm, claims that it will protect the natural pH 5.5 balance of the skin. This claim is in reference to what has been called the *acid mantle* of the skin, meaning that the skin's surface is slightly acidic under normal conditions.

Most of the advertising claims about controlled product pH overemphasize the importance of this aspect of tissue chemistry. Human skin takes care of itself quite well, and most people get through an entire lifetime without once worrying about the pH of their skin or scalp.

15.6 Antacids

Substances that are capable of neutralizing or destroying acids are given the general name *antacids*. In the previous section, we learned that acidic solutions have a pH of less than 7.0 (e.g., a carbonated drink at pH 5.8). If such an acidic solution is treated with an antacid, all of the excess acid may be destroyed and the pH changed to 7.0 (neutral) or higher, depending on how much antacid is used. Americans currently spend close to $1 billion yearly on OTC antacids.

The public turns to antacids in order to treat indigestion, acid stomach, or ulcers. The last-named condition (peptic or duodenal ulcers) is worsened by the release of hydrochloric acid from the lining of the stomach. However, hydrochloric acid is a physiologically natural and desirable chemical needed in the process of food digestion, and it should not be neutralized unless its release is aggravating a condition like an ulcer. In fact, some few people suffer from too little hydrochloric acid (a condition known as *achlorhydria*) and must take some of the acid by mouth.

Indigestion and *sour stomach* are imprecise terms for conditions vaguely described as an uncomfortable feeling in the abdomen, a burning pain in the lower chest, gas, belching, or nausea. Although it is most likely that the problem does lie in the stomach (irritated by alcohol, spicy foods, or tension) and that an antacid will help relieve the symptoms, it is also possible that a far more serious condition such as heart disease or gallstones is the cause of the symptoms (in which case an antacid is highly unlikely to alleviate the discomfort). This fact makes self-medication with antacids potentially dangerous and has made the FDA critical of the advertising blitz for antacid products. Advertisers would like us to believe that if we feel bloated after a large meal we should take an antacid. But the FDA has found that no product on the market shows evidence of being safe and effective for this condition. It is inappropriate, therefore, to take an antacid or gas reliever just because you ate a big meal.

A common household antacid is "soda bicarb" (sodium bicarbonate, or baking soda, $NaHCO_3$). When mixed with a water solution of an acid (whether it is in the stomach or on a car battery), sodium bicarbonate quickly neutralizes the acid, releasing bubbles of carbon dioxide gas. Occasional use of sodium bicarbonate as a stomach antacid in an otherwise healthy person is safe. But repeated or continuous use is definitely to be avoided, since it can result in *acid rebound*, a condition in which the stomach is stimulated to secrete even greater amounts of hydrochloric acid in a futile attempt to overcome the alkalizing effects of the antacid. A vicious cycle can develop, with possible habituation to the bicarbonate. Heart or kidney patients on a low-sodium diet must avoid the use of sodium bicarbonate.

Acid rebound is also possible after repeated ingestion of the antacid *calcium carbonate*, found in one of the heavily advertised roll products. Occasional use of calcium carbonate is harmless, but it is a sad comment on the American way of life that many people are seduced into believing that they need to consume a roll a day of this antacid, week after week. That is abnormal use and could result in problems such as constipation and acid rebound. A recent ad for an antacid suggested its use as a source of calcium for teenagers. Adequate calcium intake is ensured by a proper diet of fruits and vegetables. To depend regularly on an antacid for one's supply of calcium is wrong because of the highly abnormal systemic alkalizing effect from constant use of the antacid.

Antacid products with the word *seltzer* in their name are highly advertised for relief of indigestion. These products typically contain sodium bicarbonate and an acid such a citric acid, so that when they are mixed with water they fizz through the release of carbon dioxide gas. Some also contain aspirin, and this is what gives the people at the FDA heartburn, for aspirin can be a stomach irritant, and it makes no sense to include it in a preparation advertised for *relief* of indigestion.

Aluminum hydroxide, $Al(OH)_3$, is an effective, inexpensive antacid that does not cause acid rebound and can be used by low-sodium-diet patients. Unfortunately, it tends to constipate the user. After this fact was recognized, aluminum hydroxide was combined with magnesium hydroxide (milk of magnesia), a known laxative and itself an antacid. The combination has been successful, and numerous combination products are marketed with the approval of medical authorities and consumer groups (see Maalox and Mylanta in Table 15.2).

Table 15.2 OTC Antacid Products

Brand Name	*Dosage Form*	*Ingredient(s)*	*Comments*
Alka-Seltzer	Effervescent tablet	Sodium bicarbonate, citric acid	High sodium
Axid AR	Tablet	Nizatidine	Heartburn prevention
Amphogel	Liquid	Aluminum hydroxide gel	Low to medium sodium
Arm & Hammer	Powder	Sodium bicarbonate	High sodium; for acid indigestion only
Maalox	Liquid	Aluminum and magnesium hydroxides	Low sodium
Mylanta AR	Tablet	Famotidine (Pepcid), simethicone	Low sodium
Rolaids	Tablet	Magnesium hydroxide, calcium carbonate	Low sodium
Tagamet HB 200	Tablet	Cimetidine	Acid reducer
Tums	Tablet	Calcium carbonate	Low to medium sodium

***CAUTION:** The aluminum–magnesium-hydroxide combination product should not be used by patients with chronic kidney disease, because a healthy kidney is needed to handle the challenge to the body's acid-base balance posed by the ingestion of the antacid combination.*

Case History *Antacid Poisoning in an Infant*

A 25-day-old baby was admitted to the hospital, limp and unresponsive. Her parents, anxious about what they perceived as constipation or difficulty in defecation, had appealed to their physician. Despite the fact that the baby's stools were normal in all respects, the doctor instructed the mother to add 1 teaspoonful of magnesium hydroxide (Phillips' Milk of Magnesia) to each feeding. Thus the infant had received 8 teaspoonfuls a day for 3 days. On admission, the infant's serum magnesium was about four times normal. Excess magnesium can act to paralyze skeletal muscle, slow the heart, lower the blood pressure, and depress the brain. The magnesium was discontinued immediately, and potassium chloride and calcium gluconate were given intravenously. In 8 hours, the baby had recovered most of her functions.

All things considered, if you can stand the taste of the aluminum–magnesium hydroxide combination product, it is probably your best bet as an antacid when used appropriately.

By way of summary, the FDA has the following advice on the use of antacids:

1. Unless your doctor approves, don't use any antacid for more than 2 weeks.
2. For occasional use, sodium bicarbonate and calcium bicarbonate antacids are acceptable. For repeated and frequent use, rely on the aluminum and magnesium hydroxide combination type.
3. Low-sodium-diet patients must read the label of the antacid product to determine the sodium ion content.
4. Liquid or suspension dosage forms are more efficacious than tablets, because liquids can more easily coat a larger surface and usually have more buffering capacity. If you use a tablet, chew it up before swallowing.
5. It is wiser to eliminate the cause of the indigestion than to mask the symptoms with an antacid.
6. If stomach problems persist, give up self-medication and get medical help.

Some antacids are formulated with simethicone, supposedly an antiflatulent (gas reliever). The FDA has concluded that simethicone lacks evidence of effectiveness as an antiflatulent.

Misuse Potential. High. It is all too easy to listen to the misleading advertising and overrely on antacids. Overreliance leads to habituation and compulsive use. This result is dangerous, because antacids can change the normal acid-base balance of our body, possibly leading to kidney stones in susceptible people. Some antacids are constipating.

Adverse Effects. Aluminum hydroxide antacids can cause constipation, while sodium bicarbonate antacids can cause acid rebound. If your doctor tells you to avoid sodium, then you should avoid sodium bicarbonate antacids. The use of antacids for days or weeks can lead to undesirable changes in the acid-base balances in your body.

Drug Interactions. There are two ways antacids can interfere with other drugs: by changing gastrointestinal (GI) absorption and by affecting kidney elimination. Antacids of the aluminum, calcium, and magnesium types can bind to (or absorb) many prescription drugs, thus decreasing their absorption from the GI tract. Examples of drugs that can be bound are tetracycline antibiotics, digitalis drugs, indomethacin, and chlorpromazine. For other drugs, such as quinidine, the use of antacids can increase blood concentrations or, potentially, their effects. Consult your physician before concurrent use of any prescription drug and an antacid. Proper spacing of doses of antacids and prescription drugs can eliminate possible interactions. Because antacids tend to make the GI tract alkaline, they can cause enteric-coated tablets to dissolve at a faster rate.

H2-Blockers. Suppression of stomach acid before it can cause heartburn is the mechanism of action of a class of acid control drugs known as H2-blockers. Four drugs of this type are now sold OTC: Pepcid AC (active ingredient famotidine), Tagamet (cimetidine), AxidAR (nizatidine), and the antiulcer drug Zantac (ranitidine). Since the H2 histamine receptor mediates the promotion of stomach acid secretion, blocking it will diminish stomach acid and presumably prevent heartburn. Critics of this approach cite the misleading advertising, the fact that allowed drug dosage is probably too low to be really effective, the need to take the product before you expect an attack of heartburn, and the wide variation in response seen with these products. It would be far better to eat wisely and selectively and to be sure the pain you are experiencing is really heartburn and is not something far more serious. The manufacturers of H2-blockers strongly urge that users read the label directions and warnings before taking these drugs.

In the prescription-only category, older H2 antagonists are being supplanted by more effective proton pump inhibitors such as Prilosec and Prevacid. Astra Merck spent $72 million in one recent year promoting Prilosec, much of it in *direct-to-consumer* (DTC) ads.

15.7 Antibiotics

When Alexander Fleming discovered penicillin over 70 years ago, he opened the door to a vitally important class of agents we call *antibiotics,* that is, drugs produced by molds or bacteria that are capable of inhibiting germ growth (*bacteriostatic* action) or even of killing germs outright (*bactericidal* action).

Although it is true that most antibiotics are available only on prescription, some are sold OTC; it is for this reason that we discuss the topic in this chapter. Then, too, all of us are tempted to self-medicate with antibiotics left over from a previous illness, a situation much like OTC drug use.

Some widely used antibiotics are the penicillins (such as penicillins G and V and ampicillin), tetracycline, erythromycin, zithromycin, streptomycin, bacitracin, neomycin, nystatin, and polymixin B (all generic names).

Antibiotics available to the physician can differ from each other on the basis of:

- Effectiveness by the oral route
- The kind of invading germ they will attack (broad-spectrum versus narrow-spectrum action)
- Whether or not they are resistant to enzymes causing the destruction of antibiotics
- Whether or not they are absorbed well from the GI tract

It seems wise to limit the purchase of major antibiotics such as penicillin and tetracycline to **prescription-only**, because indiscriminate and casual use of antibiotics by the general public for every real or imagined illness could well lead to the development of adaptation or **resistance** in the invading organisms. Such resistance develops when insufficient doses of the antibiotic are taken repeatedly or when a sufficient dose is not continued until the conclusion of the infection. These situations can give the germ time to make a supply of enzymes to help destroy the antibiotic or can permit mutant forms of the bacterial genes to replace the susceptible forms. Either way, resistance can be a big problem—as seen, for example, in the current difficulty in treating penicillin-resistant strains of the organism that causes gonorrhea. This venereal disease has already reached epidemic proportions, and a crisis could be imminent if resistance continues to develop. In 1995, health officials in Seattle reported the first documented U.S. cases of a drug-resistant form of gonorrhea, found in prostitutes in Seattle and in a Denver man who went on a "dating tour" of the Philippines where he had sexual contact with seven or eight prostitutes. The gonorrhea strain detected showed high-level resistance to treatment with fluoroquinolones, one of two drug types recommended for gonorrhea. Besides gonorrhea, resistance has been found in salmonella-induced diarrhea, in infections due to staphylococcus, streptococcus, and pneumococcus organisms, and in typhoid fever, tuberculosis, and even malaria. According to Dr. Donald Guiney, a nationally recognized authority on infectious disease at the University of California, San Diego, the biggest contributors to the problem of antibiotic resistance are the countries that fail to regulate the drugs. Antibiotics can be purchased OTC in Mexico, certain countries in Central America, South America, the Far East, and Europe. Travelers from these countries can carry antibiotic-resistant germs to other countries. Dr. Guiney says that unrestricted antibiotic use is the reason penicillin-resistant gonorrhea first emerged in the Philippines, where even the most potent antibiotics are sold without prescription and where prostitutes regularly take penicillin as a prophylaxis against venereal disease. Also, according to Dr. Jeffrey Kaplan, director of the Center for Disease Control and Prevention, more than 50 million antibiotic prescriptions every year are unnecessary—too often written for people with viral colds or other minor ailments.

Another reason for restricting antibiotics to prescription only is that they can cause serious side effects, and a doctor should be involved if this situation happens. Some individuals are allergic to penicillin (including ampicillin, a member of the penicillin group). Life-threatening allergic episodes (called **anaphylactic shock**) can unexpectedly follow its administration. Then, too, antibiotics can upset the normal balance of microorganisms in the human body, such as occurs in the intestine or vagina, which may lead to serious problems. In women taking broad-spectrum antibiotics, growth of the fungus *Candida albicans* can result in a distressing vaginal itch for which treatment may be difficult. Some of the milder side effects observed occasionally with penicillin use are skin rashes, upset stomach, irritation of the mouth, and diarrhea.

Viral infections such as the common cold and infectious hepatitis are not curable with antibiotics. It is unwise to try to treat a viral sore throat with an antibiotic. This is the reason your doctor may want to take a culture of your throat before antibiotic therapy is begun. If the culture shows that the infecting germ is not susceptible to any antibiotic, he or she will not want to risk drug side effects or the development of resistance. If the culture identifies a susceptible germ, then the doctor can select just the right antibiotic and in the dose and drug regimen that will completely knock out the infection. The patient must remember to take *all* of the dose prescribed and at the correct time. The aim is not to give the bug a second chance.

It is unwise to self-medicate an infection with leftover supplies of antibiotic, because specific illnesses require specific types of antibiotics. Indiscriminate use of a leftover antibiotic may only mask symptoms, thus greatly confusing the situation or even prolonging the morbidity.

The only types of antibiotics that the general public can purchase OTC are the ointment or cream types used to treat infections on external surfaces of the body, plus the nasal spray or lozenge types. It has been questioned whether one can expect any value from topical (surface) use of antibiotics, especially in the nose and throat.

Neomycin and bacitracin are the antibiotics most commonly selected for effective use on the skin.

Drug-drug and **drug-food interactions** in the antibiotic category are potentially significant. One of the most famous is the negative effect of milk and other high-calcium foods on the antibiotic effect of tetracycline. Calcium ions in the milk or milk product combine chemically with hydroxyl groups in the tetracycline, impairing absorption from the GI tract. Tetracycline is one drug that should *not* be taken at mealtime.

Besides milk products, antacids such as calcium carbonate, milk of magnesia, and aluminum hydroxide interfere with the absorption of tetracycline. Consequently, such antacids should not be taken to offset possible stomach upset caused by the tetracycline.

Use of tetracycline antibiotics during tooth development may cause permanent discoloration of the teeth. Tetracycline may induce photosensitivity, as indicated by an exaggerated sunburn reaction. Pregnant women should avoid using tetracycline, since damage to the embryo may occur. Tetracycline antibiotics should not be combined with penicillin, because these two types of antibiotics have different mechanisms of action, and combining them reduces their therapeutic effectiveness.

The simultaneous use of an anticoagulant drug with many of the antibiotics may lead to an increased bleeding tendency and even hemorrhage. The reason is that many antibiotics reduce the growth of bacteria occurring normally in the intestine. Fortunately, these bacteria synthesize vitamin K, an important factor in rapid blood coagulation. Their suppression leads to decreased vitamin K synthesis, with consequent increased tendency toward bleeding.

If you discover an old antibiotic prescription in your medicine cabinet and there is no expiration date on the label, it's best to throw it out. Many antibiotics, especially in liquid or suspension form, rapidly decompose at room temperature, losing their effectiveness. If you are uncertain, call your pharmacist. He or she is trained to answer questions about the shelf life of drugs.

15.8 Laxatives

Americans are a super-bowel-conscious society. We spend more than $300 million yearly on the purchase of more than 700 OTC laxative products. Apparently we believe the ads that suggest that we must be regular if we are to be healthy and happy, and that one day without a bowel movement is tantamount to chronic constipation. Well, there is a true, documented story of a man who did not have a bowel movement for one whole year and lived to tell about it. The truth is that many healthy people defecate promptly whenever they feel the urge, without resorting to a rigid, daily schedule. Furthermore, our eating habits and diet can change from day to day, and the presence or absence of bulk (cellulose), bran, prunes, or other foods can influence emptying of the colon. The degree of body hydration is reflected in stool consistency. Astronauts spending time in space travel know that a diet can be followed that leaves practically no residue.

In actual fact, then, normal elimination may be anywhere from twice daily to twice weekly. However, it is difficult to convince some people of this fact. Through family training or by association with health faddists, they have come to believe that daily defecation is important if "toxic wastes" are to be eliminated. Thus, they resort to laxatives (or even daily enemas). These individuals are especially receptive to advertising that suggests that their sluggishness is due to their "sluggish" bowels. What they may fail to perceive is that left alone, the healthy rectum will take care of itself, and the more we stimulate it artificially to produce, the less likely it will to do so on its own.

The FDA also warns that prolonged laxative use can also deplete the body of fluids, salts, vitamins, and essential minerals and inhibit the absorption and effectiveness of other drugs. What is more, it can cause dizziness, confusion, fatigue, skin irritation, diarrhea, irregular heartbeat, belching, and other side effects, depending on the laxative used.

Other individuals do indeed suffer from chronic constipation, and they benefit from laxative administration. Genuine constipation may be tied to emotional stress, may be postsurgical, or may be due to a serious disease such as cancer. Antacid drugs can induce constipation, as can the opioid narcotics, certain antispasmodics, clonidine, heavy metals, and iron.

Since laxatives are with us and are likely to stay, let us examine the various types and some nomenclature. The term **laxative** should be reserved for milder agents that promote elimination by softening the stool. **Cathartics** are purgatives; that is, they induce a more fluid elimination. It is possible, however, to give smaller doses of a cathartic drug to produce a typical laxative effect. The more commonly used agents are classified in Table 15.3.

Mineral oil is obtained from petroleum (crude oil) and is a clear liquid consisting of hydrocarbons 18–24 carbons long. It has long been believed that mineral oil is chemically and physiologically inert, passing through the body unchanged, and that it is safe to use it as a laxative. In fact, it is widely used as an intestinal lubricant and stool

Table 15.3 Classification of Laxatives[a]

Emollients or Fecal-Softening Agents. Make defecation easier, especially when hard fecal masses are present. Require 1–3 days to work.
Examples: dioctyl sodium sulfosuccinate (DDS, docusate, Colace, Correctol, Feena-Mint, Surfak, Doxidan).

Bulk Formers. Swell up in contact with water inside bowel. Best choice for simple constipation because they closely imitate normal evacuation. Require 1–3 days to work.
Examples: bran (from foods), methyl cellulose (Citrucel), polycarbophil (FiberCon), psyllium seed (Metamucil (Perdiem)).

Lubricants. These oils coat feces, reducing absorption of water by bowel. This makes feces soft and easy to eliminate.
Example: mineral oil; no other products on market.

Saline agents. Attract water inside the intestine, resulting in mechanical stimulation. Also work by other mechanisms. Act in 1–3 hours. Not intended for long-term management of constipation.
Examples: hydrated magnesium sulfate (Epsom salts), magnesium hydroxide (Phillips' Milk of Magnesia), sodium phosphates.

Stimulants. Local irritants to bowel mucosa; increase propulsive peristalsis in 6–8 hours. Not recommended for routine use; can produce dangerous side effects. Contraindicated in presence of abdominal pain, nausea, or vomiting.
Examples: senna glycosides (Senokot, Ex-Lax), cascara sagrada (Nature's Remedy), bisacodyl (Correctol, Dulcolax).

[a] CAUTION: See text for contraindications to use. Some laxatives can be dangerous to use in certain conditions.

softener. However, there is now enough evidence to conclude that mineral oil is not safe to use. The greatest hazard in its use is the danger of producing lipid pneumonia. This is a well-known condition in which some of the oil gets into the lungs either from oral ingestion or through use of oily nose drops and causes fluid accumulation in the lungs, with consequent impairment of breathing. Lying down (as at bedtime) after an oral dose of mineral oil can result in oil getting into the lungs. Mineral oil also is absorbed to a limited but significant extent from the intestines and gets into the lymph system, liver, and spleen. Mineral oil taken at mealtime or shortly thereafter can dissolve vitamins A, D, E, and K from foods and carry them out of the body in the stool. Serious loss of vitamin A can cause night blindness and damage to the eye. Serious loss of vitamin D can reduce the body's ability to absorb and store calcium. The consequences of loss of vitamin E in humans are not clear. Loss of vitamin K reduces the clotting ability of the blood. For the hemophiliac or the patient taking an oral anticoagulant, this situation could have serious consequences. If you use mineral oil as a laxative, you may experience seepage of the oil from the anus and possibly an irritated anal sphincter. If you feel that you must use mineral oil, take only a small dose on an empty stomach, and do not lie down for at least a half hour after the dose.

CAUTION: *Do not use this or any other laxative if you have nausea, abdominal pain, or any sign of appendicitis.*

Mineral oil is another of the chemical crutches some people rely upon to support their neuroses, in this case the unrealistic anxiety about constipation. Consume a diet with a reasonable amount of fiber in it, drink plenty of water, and promptly respond to nature's call when it comes, and you will not require any laxative.

Bulk-forming laxatives act through their ability to absorb water and swell to a large, soft bolus of material inside the lumen of the intestine. This action promotes peristalsis and maintains the feces in a hydrated, soft condition. Defecation is thereby promoted. It seems clear that Americans should include more natural bulk-forming foods in their diet: bran, vegetables, cereals, fresh fruit, and celery. These foods contain a high percentage of cellulose, the nondigestible polysaccharide that is so abundant in nature and that has been called nature's laxative. It is even more clear that popular refined types of foods (sugar, white bread, white rice, cakes, ice cream) offer calories but little or no bulk and should not comprise our main dietary intake.

Saline cathartics work because they remain inside the intestine and attract large volumes of water from surrounding tissues. This influx of water promotes peristalsis and rapid evacuation of the bowels. Magnesium sulfate (Epsom salts) is the most well-known saline cathartic.

Problem 15.1 *How does the concept of osmosis explain the action of saline cathartics?*

Stimulant cathartics work owing to their irritant action on the lining of the intestine and are to be considered the harshest of the various cathartic types. Painful cramping is a possible adverse effect when using contact cathartics. Senna has replaced phenolphthalein, because the latter was found to cause cancer in animals.

Although the makers of some laxatives have promoted the ingredients in their products as "natural" enhancers of bowel movements, an FDA advisory panel found such claims unacceptable and recommended against including them on labels as indicators of safety or effectiveness. The FDA also disapproves of claims for effectiveness based on gender, age, or a pleasing taste of the product. Recently, the FDA announced a ban on 23 laxative ingredients due to lack of proof of effectiveness.

The experts who deal with constipation problems, real or imagined, have some words of advice: do not be anxious about bowel movements; do not promote anxiety in children about regularity; use laxatives as a last resort and then only infrequently; eat wisely, and allow regularity to take care of itself; haste does not make waste.

15.9 Acne Treatments

To a teenager, acne can be one of the catastrophes of life. Just at the point of development when relationships are becoming important, when the teenager wants to blend into the crowd and look most attractive, those ugly blotches appear on the face. And then they develop into "zits"—pustules or pus-filled pimples that never seem to go away. With this condition, the teenager can suffer intense mental anguish.

Ordinary mild to moderate acne is termed **acne vulgaris**. The more severe form of the condition, in which inflammation, pustules, cysts, and scarring can occur, is termed **cystic acne**. The incidence of acne is greatest in adolescents, of whom nearly 100% are afflicted. The peak incidence in girls is reached at 14 years and in boys at 16 years. Most, but not all, acne ends by age 25 or 30. Severe, disfiguring acne afflicts an estimated 500,000 Americans annually. Acne products represented a market exceeding $130 million in one recent year.

The unfortunate acne patient buys expensive soap and scrubs the face six times daily, applies one sure-cure, expensive OTC product after another, alters his or her diet, and eventually winds up in the dermatologist's office in the hope that this expert will prescribe the cure.

Dr. James Fulton, Jr., founder of the Acne Research Institute of Newport Beach, California, says that acne is not caused by chocolate, dirty skin, or sexual frustration. He believes that the cause is mainly genetic. It is believed that common acne originates in the spurt of hormone production at puberty, and androgen (the male sex hormone) appears to be the precipitating factor. (Recall that females, too, normally have small amounts of the opposite-sex hormone in their bodies.) Androgens stimulate oil glands in the skin to secrete oily *sebum* into and around the hair follicles. In the acne sufferer, however, the follicles or adjacent pores become plugged, then red and inflamed, and finally filled with sebum, bacteria, decomposed cellular matter, and pus. This process can go on for years.

Treatment of acne vulgaris is aimed at relieving symptoms. So many OTC remedies (more than 150) are offered because manufacturers know that the acne sufferer is highly motivated to do *something* about a pimply face and is pitifully susceptible to ads that promise relief. Drugs used to treat acne may be categorized as OTC or prescription only.

1. *OTC drugs:* benzoyl (pronounced *BENZ-oh-eel*) peroxide, sulfur, salicylic acid, and resorcinol. These chemicals are used because they kill off cells in the top layer of skin, causing the skin to scale and peel. If the OTC product has a high enough concentration of active ingredient, prolonged use can result in a kind of leathery skin, with relief of pimples and other symptoms. I

emphasize, however, that these chemicals are destructive to protein; after their initial use, the skin becomes red and inflamed and later peels off.

2. *Estrogen.* Since the male sex hormone appears to be part of the cause of acne, it seems that the female sex hormone (estrogen) would afford relief. Indeed, treatment of acne vulgaris with estrogen is effective in both sexes, and this is the reason oral contraceptives are prescribed in this dermatological condition. The Pill is administered just as it is in suppressing ovulation. However, readers of Chapter 13 will recall the danger signals: Estrogen use is statistically implicated in blood clot formation and in other serious side effects. An acne patient must be informed of the risks that accompany the benefits of estrogen use. Furthermore, estrogen use in males, especially in pubescent males, can have serious consequences. Testicle growth can be suppressed, breast growth can be stimulated, and personality changes are possible. It seems ironic that marijuana is condemned because it appears to lower testosterone levels in young males, while at the same time estrogen use is considered for boys with acne vulgaris.
3. *Antibiotics.* Ordinary acne is not an infection, and antibiotics have no place in its therapy. However, inflammatory (cystic) acne requires treatment by a physician, who may prescribe an oral antibiotic, usually tetracycline (250–2000 mg a day). Most acne patients remain on the drug for long periods. The American Academy of Dermatology recognizes tetracycline as safe and effective in the treatment of acne.
4. **Isotretinoin (Accutane).** Vitamin A is the essential food factor that, among other things, keeps the skin healthy and prevents it from drying out. Isotretinoin (13-*cis*-retinoic acid) is a chemical cousin of vitamin A that has been shown to be highly effective in the treatment of acne, even to the point of being termed a cure. Introduced in 1982, isotretinoin reduces sebum production up to 90%, producing an almost complete loss of facial oiliness. It accomplishes this by shrinking the oil (sebaceous) glands in the skin. Prescribed orally in doses of 1–2 mg/kg daily for 15–20 weeks, isotretinoin is remarkably effective in patients with severe inflammatory or cystic acne that has resisted other treatments. What is more, this drug's helpful effects last for months or even years. Ninety percent of the nation's dermatologists have prescribed isotretinoin at least once since its release. Accutane's annual sales are $60 million. Research at the National Cancer Institute has shown that isotretinoin can also prevent skin cancers, although long-term use gave serious adverse effects such as liver abnormalities and high blood fat levels. A chemically related drug, **Tegison (etretinate)** is a prescription-only drug used to treat the chronic skin disease psoriasis. It works by making cells mature and stop dividing and hence is effective against skin diseases that are characterized by abnormally rapid cell division.

FDA's Accutane Website is **`www.fda.gov/cder/drug/infopage/accutane`**.

CAUTION: *Isotretinoin and Etretinate are now recognized as powerful teratogens. Accutane is known to cause severe birth defects in about 25% of babies exposed to it in utero. Defects include a malformed face, missing or misplaced ears, heart defects, or* *mental retardation. Tegison can cause malformed face, flipper-like vestigial arms and legs (see phocomelia, Section 3.2), missing ears, ears below the chin, heart defects, and mental retardation. Many prescriptions for isotretinoin are written for women of child-bearing age, and perhaps as many as 600 defective babies have already been born after their mothers took this drug. Etretinate can remain dormant in body tissues for years after the last dose and can cause birth defects during this time. One researcher reported that it can be detected in patients' blood more than 2 years after they stopped taking it.*

A unique warning label is now required by the FDA for the patient package insert for Accutane (see Figure 15.1). In addition, users of Accutane receive a pregnancy prevention program kit from the prescribing physician. This kit contains information for the doctor and patient, including a pretreatment checklist, a consent for treatment, and a referral form for contraceptive counseling. Women should be told that their new prescription will be filled each month only after the start of a new menstrual period to make sure there is no pregnancy. Physicians who prescribe Accutane should have special competence in the diagnosis and treatment of severe, recalcitrant cystic acne, for which the drug should be exclusively prescribed. Physicians, by the way, are not legally bound to follow FDA regulations, although most comply.

In a study of 50 young men who were taking isotretinoin for cystic acne, no changes in the subjects' sperm count, sperm motility, or sperm morphology were found, nor were there any significant changes in ejaculate volume or seminal plasma fructose concentration.

Although no other acne medicine works as well as Accutane for severe acne, its adverse effects can be just as dramatic. Accutane is statistically related to miscarriage, teratogenicity, and damage to the liver, skeletal system, ear, eye, brain, and intestine. Psychiatric adverse effects include depression, suicidal behavior, and (rarely) suicide. The acne patient has alternatives to isotretinoin: Tazorac has been FDA approved; topical retinoids such as Retin-A Micro and Differin are available, and in Europe, laser and mild skin-radiation techniques are used.

FDA CAUTION: Because Accutane is a high-risk drug, it should be reserved for cases of "severe recalcitrant nodular acne."

Figure 15.1 Warning label to alert the user that a drug can have teratogenic effects.

Ralph Nader's consumer group argues that Accutane is being overprescribed and should be taken only by patients with the most severe form of acne and only after other drugs have failed. People taking Accutane should not donate blood.

A point of clarification: **Retin-A** (trans-retinoic acid, tretinoin), a stereoisomer of isotretinoin, is being highly touted as an antiwrinkle agent. Although one study showed "mild to moderate" wrinkle improvement in those using Retin-A cream, it must be emphasized that this agent does not affect deep wrinkles and has not been properly evaluated for long-term use. We do know that it can cause severe swelling and peeling of skin; severe allergies to it are possible. Retin-A should never be used during pregnancy, immediately after washing, or during the day (it increases susceptibility to sunlight). There are good reasons for being cautious in the use of this drug.

Relief of acne symptoms is possible in some cases by the use of ultraviolet (UV) radiation from either sunlight or sunlamps. Here again, the skin toughens and tends to peel, reducing redness and pimple formation. Wise use of UV light is a must! Never allow this high-energy radiation to enter your eyes; wear protective goggles when sitting under a sunlamp. Do not expose your skin for too long. Remember, true UV light is invisible to the human eye; we may not see it, but it is certainly there.

15.10 Dandruff Treatments, Shampoos, and Conditioners

Lucky is the person who needs not worry about dandruff. Most people are forced to pay some attention to the scales on their scalp or the flakes on their clothing. Dandruff is a complex and poorly understood condition. The experts cite too much oil secretion as a cause, or too dry a scalp, or dead skin cells that are shed faster than normally. In the past, microorganisms (specifically the common yeast *Pityrosporon ovale*) were implicated in the cause of dandruff, a theory supported by the current success of antimicrobial treatments such as pyrithione and selenium sulfide.

Frankly speaking, no one knows the cause of dandruff, and treatment is not aimed at a cure but at alleviation of symptoms. Some degree of drying and scaling is to be expected even on the healthiest scalps; this condition can be controlled by shampooing several times a week with a detergent product or soap. In severe dandruff, which is common, scales build up on the scalp and can be scratched off in large chunks. Dermatologists call chronic, inflammatory, oily scaling of the skin **seborrheic dermatitis**; it can spread to the eyebrows, nose, and skin folds. It is difficult to distinguish between a bad case of dandruff and seborrheic dermatitis.

One ingredient long used and still approved for the control of dandruff is salicylic acid. Its mechanism of action is to loosen the scales by a keratolytic process, followed by washing away this debris.

Problem 15.2 *Explain the meaning of the word keratolytic. Hint: keratin is dead outer skin protein, as in animal nail or horn; lysis means a cleaving or splitting.*

Zinc pyrithione is the ingredient now used in most of the popular products. Check the label if you are uncertain of ingredients. Selenium sulfide lotion USP is a therapeutic shampoo used to treat severe cases of dandruff, seborrheic dermatitis, and other dermatoses. Selenium in the sulfide form is insoluble and is therefore safe for external use. Soluble forms of selenium such as the selenate, selenite, and organo-selenium compounds are about 20 times more toxic. Selenium-containing products should not be used on damaged or inflamed skin, because these conditions promote increased systemic absorption. This is a general rule to follow in using all externally applied drugs.

In a final ruling, the FDA has announced that only five active ingredients are safe and effective for treating dandruff. These are coal tar preparations, pyrithione zinc, salicylic acid, sulfur, and selenium sulfide. Although most popular products already contain one of these ingredients, some treatments for cradle cap, such as Diaparene Cradol, will be removed from the market because their active ingredient, methyl benzethonium chloride, does not work. Only four ingredients are approved for seborrheic dermatitis: coal tar, pyrithione zinc, salicylic acid, and selenium sulfide. For psoriasis, only two are approved: coal tar and salicylic acid. Dandruff products are not to be used on children under 2 years of age. The FDA has banned 27 other ingredients in dandruff shampoos, because they were not safe and effective.

At this point, let us discuss the difference between synthetic **detergents** (simply called *detergents*) and **soaps**. Both types of cleaners work to emulsify oils and greases, thereby permitting the rinse water to remove the oily matter and the dirt it contains. Both detergents and soaps alter the "wettability" of fabrics by changing the surface tension of water, hence their general name: *surface-active agents* or *surfactants*. However, there are major chemical and functional differences between the two cleaners. Soaps are sodium or potassium salts of fatty acids (e.g., sodium stearate) and as such can be precipitated into an insoluble scum by calcium and magnesium salts and other ions found in hard water. Hence, soaps make relatively poor shampoos in hard-water areas. Synthetic detergents, on the other hand, are not salts of fatty acids and are not sensitive to ions found in hard water. Shampoos containing detergents give a good lather (i.e., emulsify oil well) and do not leave a scummy residue upon rinsing. Detergents on the market today generally are biodegradable, breaking down chemically after use into nonactive fragments that do not pollute the environment. The disadvantages of detergents are their higher cost and greater tendency to induce allergic reactions or eye irritation.

The FDA points out that shampoos or other cosmetics advertised as hypoallergenic are not necessarily foolproof against allergies. Hypoallergenic simply means that there is less likelihood of an allergic reaction; there is no way to formulate a product to which no one is allergic. Unfortunately, the FDA lost a court case on a regulation demanding that cosmetic manufacturers prove by studies on humans that their so-called hypoallergenic product did indeed induce fewer allergies than other nonhypoallergenic cosmetics. Because the FDA lost the court decision, manufacturers can continue to claim that their cosmetics are hypoallergenic without presenting any supporting evidence.

Some modern shampoos and rinse conditioners contain hair-building agents such as lauric acid amide, glycol stearate, and imidazolidinyl urea, which act to thicken the hair shaft of what the ads call fine, limp, or damaged hair. These products also contain hydrolyzed animal protein, that is, protein that has been chemically broken down into small peptide fragments or free amino acids. It remains to be seen how much, if any, of the hydrolyzed protein actually enters and repairs the damaged hair shaft, which is dead tissue. In some products, antistatic agents are included to help eliminate "flyaway" during brushing.

15.11 Mouth Rinse–Gargle Products

A visit to a large drugstore reveals the amazing variety of mouth rinse–gargle products offered for sale. Most of the nonprescription mouth rinses are simply breath fresheners, having little effect on gingivitis or plaque (the thin film on teeth harboring bacteria). Some popular brands contain cetylpyridinium chloride, and one has 27% alcohol. Realistically, it's better to eliminate the cause of bad breath than try to mask it. Brush teeth thoroughly and often, and use dental floss to free food particles from between teeth. Odors can come from infections of the throat and nasopharyngeal region; such infections, if serious, require medical advice. Mouth rinses can reduce bacteria to some extent, but reducing bacteria temporarily is of little significance. When selecting a mouth rinse, you may wish to look for the *American Dental Association* (ADA) Seal of Acceptance that states the product reduces plaque, gingivitis, and bacterial growth. Physicians recommend a saltwater gargle (compare costs) for certain upper respiratory infections.

Numerous "active" ingredients can be found in oral rinses, including sanguinaria, enzymes, detergents, aromatic oils, hydrogen peroxide, and baking soda (not to mention the pretty artificial colors). Some of these ingredients have never been tested in mouth rinses; others have failed convincingly to show effectiveness against plaque or gingivitis. Remember, no mouth rinse can ever substitute for diligent brushing and flossing.

Replacing alcohol with saltwater seems sensible in view of a single government report of a study of 866 oral cancer patients that states that people who use mouthwash containing more than 25% alcohol for many years may have a 40–60% increased risk of developing oral cancer.

Some mouth rinses contain fluoride, for people who don't have fluoridated water.

15.12 Diet Aids (Appetite-Suppressant Drugs)

There are factors in our society that seem geared, consciously or unconsciously, to help us gain weight. Sugar is used by food processors to sell their products. Fast food outlets smother us with their carbohydrate-rich, lipid-laced offerings. Soft drink and beer distributors entice us with irresistible advertising. Even dear old Mom, with love as her only motivation, saddles us with 3,000-calorie dinners and delicious desserts.

And Americans cannot or will not resist. Our population is overweight by an estimated 1 billion pounds. At the time of the Civil War, the average weight of a 5-foot 8-inch man between 30 and 34 years was 137 pounds; today he is likely to weigh 170 pounds. To lose weight, many Americans have turned to stimulant and laxative drugs. It has been known since the 1930s that amphetamines and other sympathomimetics depress the appetite by stimulating the brain and putting the user into a semiexcited state much like that of a flight-or-fight situation. When we are angry, excited, or fearful, we don't have an appetite, and that is the kind of state these drugs mimic. Hence, they are widely used as diet aids, with a yearly U.S. market of more than $200 million.

However, of all the OTC and prescription drugs we should take with caution, diet pills stand out. Phenylpropanolamine (PPA), formerly a key ingredient in 35 brand-name appetite suppressants, is now banned by the FDA because of its link to hemorrhagic strokes. The infamous, now banned fen-phen combination (fenfluramine + phenteramine), once widely promoted for weight loss, has been implicated in the development of a rare and dangerous heart-valve defect. Dexfenfluramine (sold as Redux) can increase the risk of primary hypertension. Even pseudoephedrine, touted as the replacement of PPA in diet pills, has the potential to overstress the cardiovascular system (see Section 5.6).

Table 15.4 lists three types of antiobesity (or anorectic) drugs that are sold by prescription only. Type I drugs act in a manner similar to the amphetamines; they stimulate the CNS and jack up the body's metabolism. Type II drugs get people to eat less through a release of serotonin, a brain chemical that induces feelings of fullness and satisfaction (Section 5.7). Type III drugs are lipase inhibitors; that is, they inhibit fat digestion and absorption.

CONCLUSION: *Be wary of drug fads, hard sell advertisements, and the everyone-is-taking-it-so-it-must-be-safe deception. You may wind up being one of the guinea pigs who prove how dangerous some drugs can be.*

Diet drugs of the stimulant type have a high abuse potential. It is easy to get hooked on them, because after all, stimulation of the brain and spinal cord is a pleasurable sensation, and when it is ended our natural desire is to take more drug to re-experience it. What is more, tolerance to these drugs can easily develop, requiring that we take ever larger doses to achieve the same effect. If we suddenly stop taking a stimulant drug, depression can occur. It is very tempting to relieve this depression by taking more of the drug.

Type II drugs get people to eat less probably through a mechanism that involves serotonin, a brain chemical that induces feelings of fullness and satisfaction (see Section 5.7). For example, the makers of Meridia (sibutramine) describe it as a "*serotonin norepinephrine reuptake inhibitor* (SNRI) which decreases appetite by preventing the reuptake of serotonin and norepinephrine." Physicians caution that type II drugs are intended for use by people 20% or more over recommended weight. They are not magic pills, and probably will not result in weight loss of more

Table 15.4 Antiobesity Drugs (Prescription Only)

Generic Name	*Trade Name*
I. Sympathomimetic-amine type	
Benzphetamine hydrochloride	Didrex
Diethylpropion hydrochloride NF	Tenuate
Methamphetamine	Desoxyn
Phendimetrazine tartrate	Adipost, Bontril PDM, Plegine, Prelu-2
Phenteramine hydrochloride	Adipex-P, Fastin, Ionamin
II. Serotoxin-release type	
Sibutramine	Meridia
III. Lipase-blocker type	
Orlistat	Xenical

than 10% on average. They should be used as part of a regimen that includes nutritional advice and exercise.

Stimulant laxatives—sold in health food stores and through mail-order catalogs—are employed for weight loss in some users' belief that increased bowel movements will prevent absorption of calories. They will not, says the FDA, which found no significantly reduced absorption of calories. This is because the laxatives do not work on the small intestine where calories are absorbed, but rather on the colon. The products that most concern the FDA contain one or more of the substances senna, rhubarb, aloe, buckthorn, cascara, and castor oil. Some teas and fruit drinks also contain stimulant laxatives. The advertising of these and like products can border on the mendacious, and the consumer must be careful.

Drug Interactions. Taken concurrently with other sympathomimetics (such as epinephrine) or with ergot alkaloids, appetite-suppressant drugs can cause excessive stimulation of the cardiovascular system. These drugs counteract the actions of anti-high blood pressure drugs. Irregular heartbeat can occur if phenylpropanolamine and digitalis are combined.

CAUTION: *The use of any monoamine oxidase inhibitor drug simultaneously with PPA or other sympathomimetics is strongly contraindicated.*

Besides stimulants, there are the following approaches to weight control:

- Bulking agents such as methyl cellulose, psyllium seed, and agar. These drugs supposedly swell up in the water of the GI tract and thus provide satiety; their success has not been established.
- Benzocaine, a local anesthetic that dulls the taste buds
- Limitation of salt intake and therefore of water retention
- Psychiatric counseling
- "Tummy tuck" surgery

Web Sites You Can Browse on Related Topics

OTC General Information
http://www.health.org/nongovpubs/prescription/
Also: Search "over the counter drugs"

Accutane, Retin-A
http://www.druginfonet.com/faq/faqaccut.htm
http://www.fda.gov/cder/drug/infopage/accutane/default.htm
http://www.capederm.com/info_retin_a.htm

Sleep Aids—Insomnia
http://womenshealth.miningco.com/maub87.htm

Cold Medicines
http://www.biomedcentral.com/news/20001108/03
http://www.kidshealth.org/research/cold_medicines.html

Cimetidine (Antacid)
http://www.e-antacid.com/
http://www.rxlist/com/cgi/generic/cimet.htm

Laxatives
http://www.eating-disorder.org/laxatives.html
Also: Search "laxatives"

Obesity
Search "obesity drugs"

Study Questions

1. Our text says, "OTC drugs *are* drugs, having the same potential as prescription-only drugs for side effects, habituation, and addiction." Why, then, does the FDA allow these drugs to be sold without a prescription?
2. One reason Americans spend billions of dollars yearly on patent medicines is to avoid the expense of consulting a doctor. Give three other reasons we rely so heavily on OTC medications.
3. The drug route from physician to pharmacist to patient offers certain advantages not found in the route from sales clerk to patient. What are these advantages?
4. Match the chemical in the left-hand column with its OTC drug product category in the right-hand column. A product category may match up with more than one chemical.

a. Aluminum hydroxide	**1.** sleep aid
b. Bacitracin ointment	**2.** decongestant
c. An antihistamine	**3.** cough suppressant
d. Methylcellulose	**4.** expectorant
e. Benzoyl peroxide	**5.** antacid
f. Phenylephrine	**6.** antibiotic
g. Zinc pyrithione	**7.** laxative/cathartic
h. Potassium iodide	**8.** acne treatment
i. Dextromethorphan	**9.** dandruff treatment
j. Calcium carbonate	
k. Phenolphthalein	

5. Some health conditions can be effectively treated with OTC products, and others are quite resistant. Select the condition among the following that you feel is least likely to respond to any OTC medication, and defend your answer: indigestion, dandruff, viral head cold, insomnia, constipation.
6. Define (**a**) rebound effect, (**b**) functional dependence, (**c**) antitussive drug, (**d**) anaphylactic shock, (**e**) sebum.
7. Why is it not wise for the average person to take antacids every day, keeping the stomach acid constantly neutralized?
8. A pH of 7.0 means that a solution is neutral. State whether each of the following solutions is acidic or alkaline: (**a**) blood, pH 7.4; (**b**) urine, pH 5.0; (**c**) tears, pH 7.2; (**d**) orange juice, pH 4.9; (**e**) rainwater, pH 6.6.
9. True or false:
   - **a.** There are over a quarter of a million OTC products for sale in America.
   - **b.** The health problem Americans are most likely to treat with a nonprescription medicine is dandruff.
   - **c.** The main ingredient in most OTC sleep aids is an antihistamine.
   - **d.** Scopolamine is an alkaloid that can induce amnesia.
   - **e.** Vitamin C is well stored in body fat tissues.
   - **f.** Vitamin C is a cure for the common viral head cold.
   - **g.** Heroin is an effective antitussive.
   - **h.** Hydrochloric acid is a normal constituent of stomach contents.
   - **i.** It is wise to avoid milk and dairy products when taking tetracycline.
   - **j.** Laxatives generally have more powerful actions than cathartics.
   - **k.** Isotretinoin is a diet aid.

l. For cleaning in hard water, detergents are generally more effective than soaps.
m. Cetylpyridinium chloride is an ingredient in some mouthwash products.

10. A recent ad for an OTC laxative urged the listener to make the product "a member of your family." What is so very wrong with this type of advertising? What kind of dependence does it foster?

11. From how many self-treatable health problems do you or your family suffer each month? List them. Are home remedies or OTC medicines used to treat any of the problems? List and categorize these medicines.

12. **a.** What danger lies in stopping the dose of an antibiotic as soon as the symptoms of the illness begin to disappear, although half of the dose remains?
b. What danger lies in taking only half of the prescribed dose, leaving some medicine for the next episode of illness?

13. Give an example of a laxative and cathartic. What is the difference in action between these two types of drugs?

14. What do PPA, fenfluramine, caffeine, and pseudoephedrine have in common?

15. *Advanced study question*: Assume that you have the authority to demand that a cosmetic manufacturer perform tests to *prove* that his product is hypoallergenic. What test protocol on what kind of subject would satisfy your requirement for proof?

16. What drug is in Afrin nasal spray that might be responsible for a dependency on this decongestant product?

CHAPTER 16

Personal Drug Testing

Key Words in This Chapter

- Urinalysis
- Screening test
- Confirmatory test
- Thin-layer chromatography
- EMIT
- Abuscreen
- False positive
- False negative
- Detection cutoff level
- Persistence period
- Workplace testing
- Employee assistance program

Learning Objectives

After you complete your study of this chapter you should be able to do the following tasks:

- State the protocol for urine collection and storage.
- Distinguish between screening tests and confirmatory tests.
- Explain the use of TLC and immunoassay in drug screening.
- State why a confirmatory test is important.
- Discuss the nature and occurrence of false positives and false negatives.
- List the eight drugs of abuse (or their categories) and give their detection periods and metabolites.
- Cite the limits of hair analysis for drugs.
- Take a stance on the legality or illegality of personal drug testing in America.

16.1 Introduction

As the use of drugs increases in America, and as their impact is felt in almost every aspect of our society, tests for the personal use of drugs are increasingly demanded. A railway accident is blamed on the use of drugs, and the engineer is tested.[1] A tragic bus accident occurs, and the driver is found to be under the influence of drugs. An Olympic weight lifter is stripped of his gold medal when his urine tests positive for anabolic steroids. An employer has reason to suspect that absenteeism and poor productivity can be explained by pervasive drug use and demands plant-wide personal drug testing. Similarly, the military, athletic teams, and correctional facility officials

[1]Since 1975, about 50 train wrecks and accidents, which resulted in 37 deaths, 80 injuries, and $34 million in damage, have been attributable to drug or alcohol abuse.

demand that the persons under their charge prove that they are drug-free by submitting to urinalysis.

More than 500 U.S. school districts have screening programs in place. An Ohio senior housing community requires applicants to submit to a drug test for membership eligibility. And today nearly 9 out of 10 top corporations demand personal drug testing as an integral part of the hiring process.

It is unfortunate that personal drug testing must be imposed on our society. Some say it is not only unfortunate but is also an illegal infringement of privacy and civil rights. Lawyers have argued that random testing violates the constitutional rights contained in the Fourth and Fourteenth Amendments against unreasonable search and seizure. Others believe that when one's use of drugs imperils others or constitutes an unfair advantage, society has a right to intervene. Of course, authorities have been testing for blood alcohol levels (see Chapter 9) for decades, and the use of such data in a punitive way is accepted by society. No matter what the area, the reliability of test procedures becomes a critical factor.

Because personal drug testing is becoming a way of life in America, you will need to know how it can be used, the possible pitfalls, costs, and how to protect yourself. You will want to know what a false positive is and when passive inhalation can give positive results.

In this chapter we examine the procedures used to screen for and confirm the presence of drugs, how long some drugs can be detected, proposed tricks to circumvent detection, and the legality of it all.

16.2 Methods of Drug Detection

The five largest drug-testing laboratories in the country are now processing more than a million specimens per month. Most of their clients are corporations screening employees for the common drugs of abuse listed in Table 16.1. The best body fluid to use for drug testing is the blood, but it is also the most painful and expensive to sample. Urine, breath, and saliva are also used for drug detection, with urinalysis being the most commonly used because it is simple to sample and its contents are usually a direct reflection of what is, or has been, in the blood.

In urinalysis, tests are made for the drug itself and/or for drug metabolites, that is, the chemical breakdown products resulting from enzymatic activity in the liver or other organ. Metabolites and detection-cutoff levels (see Table 16.1) are important when we are dealing with cocaine, marijuana, and benzodiazepines such as Valium.

To eliminate errors and deception in urinalysis, the person must be observed urinating, that is, the container or specimen bottle must be within sight before and after the individual has urinated. Each container must be tightly capped, correctly sealed, and labeled. In addition, for legal purposes, the *chain of custody* for the specimen must be established by the execution of custody forms for each transfer of possession from point of collection to final disposition in the laboratory. Substitution of an old sample of urine for one's own can be detected by temperature differences (freshly voided urine measures 38°C, body temperature); attempts at gross adulteration with alkali or acid can be detected by measuring the pH of the specimen. Ammonia can be detected by its odor. However, drinking a large volume of fluid can

Table 16.1 Detection Cutoff Levels for Drugs of Abuse

		Detection Cutoff Levels (ng/mL)			
Drug	*Drug or Metabolite Detected*	*Military*[a] *IA*	*GC-MS*	*Civilian*[b] *IA*	*Civilian*[c] *IA*
Amphetamines	Methamphetamine,[d] amphetamine	500	500	1000	1000
Barbiturates	Secobarbital, pentobarbital, talbutal butabarbital, and others	200	200	200, 300[e, f]	200
Benzodiazepines	Oxazepam, diazepam, temazepam, chlorazepate, others		—	300[e]	—
Cannabinoids (marijuana)	11-Nor-Δ^9-tetrahydrocannabinolcarboxylic acid and other THC metabolites	50	15	50, 100[f]	25, 50, 100[f]
Cocaine	Benzoylecgonine	150	100	300	300
LSD	*N*-desmethyllysergide	0.5	0.2[g]	—	0.5[h]
Methaqualone	Methaqualone, hydroxy metabolites		—	300[e]	300[e]
Opiates	Heroin, morphine, codeine, morphine-3-glucuronide	2000	4000[i]	2000	300
Phencyclidine	PCP, 4-phenyl-4-piperidinocyclohexanol	25	25	25	25
Propoxyphene	Norpropoxyphene, dinorpropoxyphene		—	300[e]	300[e]

IA = Immunoassay screen; GC-MS = gas chromatography-mass spectrometry confirmation; ng = nanogram.

[a]Data courtesy Naval Medical Center, San Diego, California.

[b]Data courtesy Syva Corp, Palo Alto, California.

[c]Data courtesy Roche Diagnostic Systems, Somerville, New Jersey.

[d]For methamphetamine to be reported positive, its metabolite (amphetamine) must also be present at a minimum of 200 ng/mL. Thus is to rule out false positives.

[e]Cutoff levels not mandated under current federal guidelines.

[f]The end user may select the cutoff to suit local requirements.

[g]That is, 500 picograms/mL or 200 pg/mL.

[h]Using RIA.

[i]The new, higher cutoff levels are necessary to account for opiates from possible poppy-seed ingestion.

dilute the concentration of any drug in the urine, possibly below that of the cutoff level (see the following discussion), and for this reason, the specific gravity of pale-looking urine should be checked. Urine with a specific gravity of less than 1.010 should be considered suspicious for dilution. It has been learned that salt added to a urine sample can "defeat" the EMIT (enzyme-multiplied immunoassay test); since added salt is very difficult to detect, it is imperative that the donor be observed while urinating.

Other attempts at deception that have been tried include a plastic bag of "clean" urine hidden under the arm (where it is kept warm); carrying the urine in a thermos bottle; use of colored liquids such as diluted tea, coffee, beer, Listerine, apple juice, pear juice, and cola; use of a stand-in (more than a driver's license or credit card ID must be demanded); and voiding over hands or fingernails dipped into a supposed drug-destroying chemical such as bleach or baking soda. None of these attempts at deception will work if the person is required to wash hands prior to urinating, to strip, is observed voiding, and if the pH, color, temperature, and specific gravity of the urine sample are checked. One of the most dangerous techniques for deception has been practiced by some prison inmates who use a catheter to empty their bladder and refill it with someone else's drug-free urine. Serious infections can result from this practice, not to mention the possible pain involved.

Case History *An Applicant for an Industrial Position*

Job applicant F. arrived at a testing center that contracts with employers to carry out preplacement medical examinations. F. put on a hospital gown and was examined by a physician, who checked to see that no plastic bag of urine was taped to his body. F. was then directed to provide his urine sample privately in a "dry bathroom" in which the hot water had been turned off, soap removed, and the toilet water dyed blue. After the applicant came out of the bathroom with his urine sample in a jar, he was accompanied at all times by a nurse who checked the color and temperature of his specimen and divided it into two tamper-proof, sealed containers, one to be frozen and stored for a year, and the other to be analyzed. The samples were given nonsequential identification numbers. At this point, F. and the nurse signed a form stating that the numbered containers were properly sealed and indeed represented the applicant's urine. A courier delivered one container to an outside laboratory to be analyzed. The results of F.'s urinalysis by immunoassay were positive for marijuana; confirmation was made using GC-MS (see a description of GC-MS later in this chapter). F. was denied the job, but he was not told the exact reason that he failed his physical examination. (The company is under no legal obligation to tell the applicant that he failed the drug test.)

Cheating on urine tests is an ongoing challenge to valid results.[2] The Internet is replete with ads for products "guaranteed" to protect the user who must submit to drug testing. Typically, such products contain diuretics that will "wash out" the drug. "Carbo" drinks are supposed to provide a "one-hour flush." The use of diuretics or hydration by drinking large volumes of fluid the night before is termed *in vitro adulteration*. Users are told to spike their urine specimen with bleach, Drano crystals, or hydrochloric acid. Cheating on hair drug analysis is attempted using "clean shampoos." See Section 16.10 for more on cheating on urine tests.

Test procedures for drug detection fall into two categories: **screening** and **confirmation**. A **screening test** must be applicable to thousands of samples a week, reasonably accurate, and low in cost. It is now agreed by every expert in personal drug testing that if a sample tests positive in a screen, it should be subjected to a confirmatory test. If an agency is unwilling to spend the money to confirm a screening test, it should not be testing for drugs in the first place.

Today, by far the most-used drug screening tests are of the immunoassay type. This type is designed to detect very small quantities of drugs, typically in the range of 25–1000 nanograms[3] per milliliter (ng/mL). Immunoassays use antibodies to detect the presence of drugs. The antibody is prepared to be specific for a drug (such as an opiate) or a drug metabolite. If the drug is present in the urine, the antibodies attach themselves. This binding can cause a measurable color change or the release of a low level of radiation, depending on the test method. Today, to a great extent, radioimmunoassay has been supplanted by other, easier-to-use immunoassay screening procedures that do not involve the disposal of radioactive waste.

Two widely used immunoassay screens are **EMIT**, the enzyme-multiplied immunoassay test sold by Syva Company, Palo Alto, California, and **Abuscreen**, marketed by Roche Diagnostics, Somerville, New Jersey. They are designed to detect eight major abused drugs or drug classes: amphetamines (speed), barbiturates, benzodiazepines (such as Valium), cannabinoids (marijuana), cocaine, methaqualone ("ludes"), opiates (morphine, codeine, heroin, synthetics), and phencyclidine (PCP). Table 16.1 shows cutoffs (the lowest concentration of drug that can reliably be detected by the test procedure) for popular immunoassay screens.

Seeking quicker, cheaper alternatives, manufacturers have developed simple, rapid assay devices that require no laboratory processing and can be used on-site. Figure 16.1 shows Roche's TESTCUP, a rapid assay kit for simultaneous detection of cocaine, morphine, and THC, which is based on the principle of competitive microparticle capture inhibition. The test relies on the competition between drugs, which may be present in the urine being tested, and drug conjugate immobilized on a membrane in the test chamber. Urine is collected directly in the cup. When sufficient drug is present in the specimen, it binds to antibody-coated microparticles, thereby inhibiting the particles from binding to the drug conjugate, and no blue band is formed. A positive drug sample results in the membrane remaining white.

---

[2]Contact PHARMCHEM Laboratories, 1505A O'Brien Dr., Menlo Park, CA 94025-1435.

[3]A nanogram is a billionth of a gram.

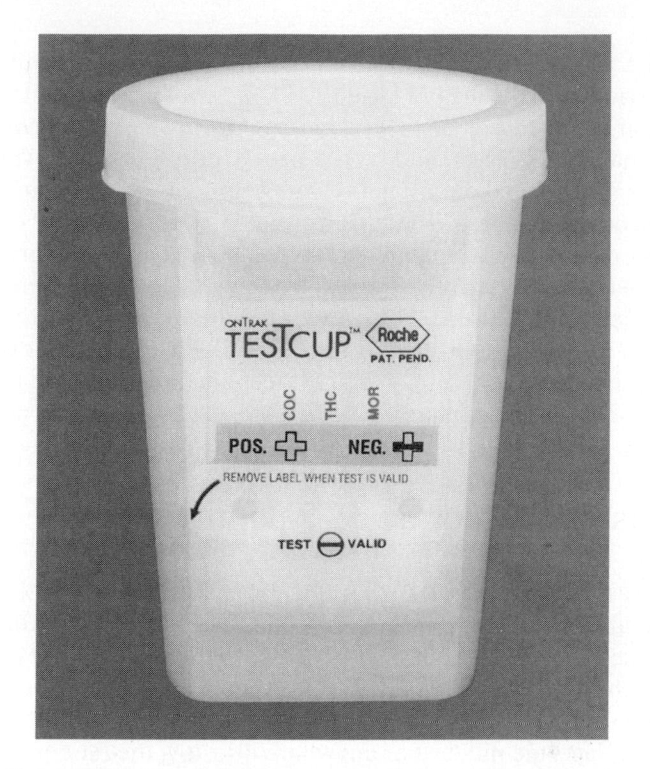

Figure 16.1 Roche's TESTCUP simultaneously analyzes for cocaine, morphine, and THC in under three minutes. (Photo courtesy of Roche Laboratories, Nutley, New Jersey.)

The TESTCUP profile (cutoff) consists of cocaine (300 ng/mL), morphine (300 ng/mL), and THC (50 ng/mL), all tested for in one urine sample. Roche's Abuscreen ON-TRAK is a rapid assay device for six different drugs of abuse that is based on latex agglutination inhibition. Drug metabolites in the urine bind to the limited amount of antibody present, preventing the binding of a latex–drug compound to the antibody. If agglutination occurs, visible as a white clumping in the window of the test kit, this means that no drug was present in the urine. If the solution in the window keeps its smooth appearance, this shows that the drug being sought is present in the urine. An ON-TRAK test takes about 3 minutes and is designed for use in clinical situations such as obstetrics and the emergency room and nonclinical settings such as prisons, parole and probations checks, and work-release programs. Where permitted by law, employers can apply the test. Another rapid-assay device for on-site use, Abbott Laboratories' TDx and ADx system, is based on a fluorescence polarization immunoassay. *Note*: Results from a rapid-assay kit used on-site *must* be confirmed by a more powerful method such as GC-MS, described in the following paragraphs.

Screening tests, which cost \$10–\$60 per sample, are highly accurate when used by expert technicians in laboratory environments, but they are not meant to be the final answer. They are meant to eliminate the negatives so that more definitive tests can be performed on the positives. We call these **confirmatory tests**, and they can cost up to \$100 or more per sample. Confirmation of a positive screen is especially desirable if the results are to stand up in court. Although confirmation tests are sometimes carried out by repeating the EMIT or Abuscreen test, it is much better to use a separate method such as *high-performance liquid chromatography* (HPLC). For marijuana, this would be the procedure of choice. Overall, however, confirmation is best carried out using the *gas chromatography—mass spectroscopy* (GC-MS) procedure, because it gives absolute proof of identity of the drug. In GC-MS, the sample components are separated by gas chromatography and then are broken down into electrically charged fragments (ions), with each drug giving a unique fragmentation pattern. The apparatus then matches up the pattern with that of a known drug, and identification is made. GC-MS requires expensive, highly complex equipment and a highly trained technician to operate it; it can detect how much of a drug or its metabolite is present, as well as what drug is present.

16.3 False Positives and False Negatives

A false-positive test result occurs when the laboratory reports the presence of a drug or its metabolite, whereas in actuality the specimen does not contain that substance. In a false-negative, the lab fails to detect and report the presence of a drug. Fortunately, with methods such as EMIT and Abuscreen, false results are very infrequent, and when they occur, they are almost always false negatives.

In the workplace, the military, or anywhere else, it is the false positive that causes the most concern, because a person can be subjected to dismissal, loss of reputation, or unjustified punitive action. A false positive can result, of course, from a laboratory error, but that is not common. It is more likely to be the result of other chemical substances in the person's blood—for example, a false positive for amphetamines could be given by a stimulant drug such as ephedrine or methylphenidate (Ritalin), a drug containing *phenylpropanolamine* (PPA) (examples: Acutrim, Allerest, Contac, Dimetapp, Robitussin-CF, Dexatrim, Permathene, or Prolamine), or a decongestant product containing a sympathomimetic (examples: Alka-Seltzer Plus, Allerest, Actifed, Sinarest, Vicks 44, Contac, NyQuil, Sinutab, Sudafed, and Triaminic). There has been criticism of the marijuana test because it allegedly discriminates against blacks, who have more melanin metabolites supposedly free-floating in their urine, where they can be mistaken for *tetrahydrocannabinol* (THC) metabolites. There is no scientific evidence to support this criticism.

Additional false positives have been reported. In tests for barbiturates, Doriden, Nalton, and Dilantin have reported positive on screening tests. For opiates, Thorazine and dextromethorphan have reported positive. For *phencyclidine* (PCP), Benadryl, Demerol, chlorpromazine, Mellaril, dextromethorphan, and Unisom have reported positive. For cocaine, coca leaf tea has reported positive.

In the case of false positives, confirmation tests can clear up the error. Indeed, confirmation should be routine for any positive test obtained in an initial screen.

There is yet another kind of problem in drug testing, which is not a false positive but a false pitfall. Even your dear old sainted grandmother would have shown up positive for opiates if she had taken paregoric for diarrhea, Demerol or codeine for pain, Lomotil, or Vicks Formula 44. All of these substances are legally available OTC or by prescription and contain an opiate or opioid (an opium-like compound). You may have every right to be taking them, but your urine is going to show up positive, and you will either have to stop taking them a month in advance of the test or announce the fact of use to the testing agency. It has now been confirmed that a single poppy seed muffin, eaten 1–5 hours before the urine test, can give a false positive for opiates.[4] People who eat poppy seed items can test positive for opiates (mainly morphine) for up to 72 hours.

Case Histories *Tetrahydrocannabinols and Opiates in Urine*

B., a Marine lance corporal and a bodybuilder, used a dietary supplement containing hemp seed oil (a legal byproduct of marijuana plants), because the oil had been recommended by nutrition promoters, in the dose of 1–2 tablespoonsful a day. Under the military's tough policy of random drug-abuse testing, B.'s urine tested positive for marijuana metabolites, and he was court-martialed. At the trial B.'s defense attorney argued that THC can be found in some brands of hemp oil, and that the amounts were high enough to show up in humans. The jury voted to acquit B. As more people use hemp seed oil, it is likely there will be more cases of false positives. We note that none of the commercial hemp oil products will give users a high.

C., a 22-year-old cyclist, 6′3″, 178 pounds, ate one poppy-seed bagel one-half hour before an 85-mile road race. Placing second among 130 riders, he was required to submit a urine sample. Several days later he was informed by mail that his urine tested positive for opiate. With the aid of witnesses, and knowing that there was no other possible source of opiate, he was able to convince the cycling federation that his positive test was due to poppy seed ingestion.

PharmChem Laboratories, Menlo Park, California, now tests for glucose in all specimens that confirm positive for ethyl alcohol because of the naturally occurring fermentation of sugar in the urine of diabetics.

Regarding testing accuracy, the *American Association of Clinical Chemistry* (AACC) believes that labs can achieve nearly 100% accuracy in testing for illegal drugs in the urine. The AACC has conducted a number of studies of testing accuracy. In the most recent, 31 out of the estimated 150 commercial laboratories that carry out testing for U.S. employers were sent blind samples of urine containing cocaine, marijuana, opiates, amphetamines, or PCP. The labs correctly identified samples containing drugs 97% of the time, with no false positives. Screening was done by immunoassay and confirmation using GC-MS. In another test, 47 hospital and independent laboratories across the country volunteered to participate. The overall accuracy rate was 99.3%, with 39 of the 47 labs achieving 100% accuracy. It should

---

[4] A. Belson, *Journal of the American Medical Association*, Vol. 266, 1991, p. 3130.

be noted that all of these labs were staffed by professionally trained scientists, were involved in external quality control programs, and conducted in-service continuing training programs for their staff. The study demonstrates that high accuracy can be achieved by our country's best labs. Other studies have been conducted that show a much poorer level of accuracy.[5]

It should be noted that urinalysis results for drugs tell us only that the person had ingested the drug and approximately when. The results do not tell exactly how recently the drug was ingested or how stoned, intoxicated, or impaired the user was, if at all. Remember, it is not necessarily illegal to have a controlled substance in your urine or blood, a point that will be discussed further in Section 16.6.

16.4 Which Drugs Are Detected and for How Long

Table 16.2 lists drugs of abuse and their detection periods (i.e., retention times), but some words of caution are necessary. The detection periods given are approximate, because there can be considerable variation according to the test method used, how long and how much of the drug was used, its route of ingestion, and possible dilution in body fluids. The laboratory cutoff level selected is also important. Nevertheless, Table 16.2 provides reasonable guidelines.

Drug testing for marijuana deserves some special comments. Δ^9-THC, the active ingredient in marijuana, is highly fat-soluble and tends to dissolve in body-fat reservoirs (lungs, brain, ovaries, testes, and body fat in general), from which it is slowly released to the blood and liver; it is then metabolized to the products we test for in the urine. In man, the primary metabolite of THC is the pharmacologically inactive 9-carboxy compound excreted as the glucuronide (see Appendix I, Figure 13). Researchers at PharmChem laboratories in Menlo Park, California, state that the smoking of a single high-potency marijuana cigarette can be detected for at least 3 days and possibly as many as 10 days after use. Furthermore, they believe that heavy use of pot can be detected by the EMIT 100-ng cutoff method for 3–4 weeks or even longer. One person admitted to smoking 7–10 joints a day for 7 years; he tested positive 11 weeks after last use (based on the EMIT 20 ng/mL cutoff). Thus in order to pass a urine drug test as a condition for employment, one should avoid smoking pot for at least a month before submitting the urine specimen.

Passive inhalation of pot smoke can lead to detectable blood levels of THC. Nonsmokers sitting in a small, closed room and passively inhaling heavy pot smoke achieved urine levels as high as 50–100 ng/mL. However, for a person to claim that he or she inhaled enough pot smoke at a concert or a party to test urine positive is more than likely stretching the facts.

There is now a blood test that can help identify alcohol abuse. Called the **GGT test**, after gamma-glutamyl transferase, the test is some 70–80% accurate in identifying alcohol abuse. When a person drinks four to six or more drinks a day, the liver

---

[5]The *Clinical Laboratories Improvement Amendments Act* (CLIA) of 1988 requires all laboratories conducting tests "for the assessment of health" to be certified by the Health Care Financing Administration, a part of the Department of Health and Human Services.

Table 16.2 Retention Times of Drugs in Urine

Drug	*Approximate Retention Time*
Amphetamines	48 hours
Barbiturates	Short acting (e.g., secobarbital); 24 hours
	Long acting (e.g., phenobarbital); 2–3 weeks
Benzodiazepines	3 days if therapeutic dose ingested; up to 4–6 weeks after extended dosage (i.e., years)
Cocaine metabolite	2–4 days
Ethanol	2–14 hours
LSD	12–36 hours
Methadone	Approximately 3 days
Opiates	2 days
Propoxyphene	6–48 hours
Cannabinoids	Moderate smoker (4 times/week)—5 days
	Heavy smoker (daily)—10 days
	Retention time for chronic smokers may be 20 days
Methaqualone	2 weeks
Phencyclidine	Approximately 8 days
	Up to 30 days (mean value = 14 days) in chronic users

Note: Interpretation of retention time must take into account variability of urine specimens, drug metabolism and half-life, patient's physical condition, fluid intake, and method and frequency of ingestion. These are general guidelines *only*.

Source: Used by permission of DADE BEHRING, NEWARK, DE.

responds by increasing GGT blood levels. When the person stops drinking, the enzyme levels soon drop back to normal. Thus the GGT test can be a clinical tool to help identify alcoholism and monitor abstinence. Because many other factors such as disease can elevate GGT levels, the diagnostician must consider the patient's history, general health, possible concurrent use of other drugs, and other factors before a final diagnosis can be made.

16.5 Drug Testing in the Workplace

The federal government has estimated that reduced productivity due to alcohol and drug abuse costs the United States $100 billion annually. Drug use on the job can result in accidents, absenteeism, health problems, memory loss, turnovers, and loss of coordination skills. The recognition of this huge problem with drugs has led segments of government and private industry to institute drug testing in the workplace, and among 230 of our biggest companies, one-third have now started worker drug tests to hire employees or to monitor them. Among Fortune 500 companies, 50% are estimated to use drug screening programs. A growing number of police agencies now require drug tests for their officers, applicants, and recruits. Metropolitan police departments in Chicago, Honolulu, Miami, and Louisville conduct "reasonable suspicion" testing. The metropolitan police department of the District of Columbia operates its own in-house drug screening program, testing applicants and probationary officers more than once during the first year and undercover officers and veteran officers after leaves of absence. During the first seven years the U.S. Navy enforced a drug-testing program, the positive rate decreased from more than 30% to less than 5%. The Navy operates five drug analysis laboratories and tests 1.8 million samples yearly. Chicago-based Discovery Zone, a chain of indoor playlands for kids, places signs to assure parents that all of its 15,000 employees have been tested for drugs.

Alcohol and marijuana are drugs commonly abused in the workplace, along with amphetamines, barbiturates, benzodiazepines, cocaine, methaqualone, opiates, and PCP. Federal government testing guidelines, established by the Department of Health and Human Services, state that at a minimum, marijuana and cocaine must be tested for. Furthermore, the initial test should use an immunoassay procedure that meets the requirements of the FDA for commercial distribution; cutoff levels to determine a positive result should be 100 ng/mL for cannabinoids, 300 ng/mL for cocaine metabolites, 300 ng/mL for opiates, 25 ng/mL for PCP, and 300 ng/mL for amphetamines.

Is employee drug testing legal? The Fourth Amendment to our Constitution states that "the right of the people to be secure in their persons ... against unreasonable searches and seizures, shall not be violated ...." Searches are permitted only "upon probable cause, supported by Oath or affirmation, and particularly describing the place to be searched, and the person or thing to be seized." The United States Supreme Court has ruled that extracting body fluids constitutes a search within the meaning of the Fourth Amendment. Therefore, when the government plans to test the urine of one of its employees, it must comply with due process and first provide probable cause (plausible evidence of illegal activity.) Mass random testing is a different matter. Most, but not all, courts reviewing proposals to conduct mass urine tests on government employees have found these proposals to be illegal. Notwithstanding this fact, the military continues to conduct routine testing in its various branches.

The **Americans with Disabilities Act** (ADA) of 1990 prohibits employers with 25 or more employees from discriminating against qualified persons with disabilities with respect to all terms, conditions, and privileges of employment. The ADA does not consider users of illegal drugs disabled, but it protects "from discrimination on the basis of past drug addiction" persons who are no longer using drugs illegally and are receiving treatment for drug addiction or who have been rehabilitated. With some limitations, the ADA permits drug and alcohol testing. Tests for illegal drugs are not considered medical examinations. Such tests can be required of job applicants before a job offer is made. However, a blood alcohol test is considered a medical examination under the ADA. An employer may not ask a job applicant what prescription drugs he or she is taking. Employers must also comply with all provisions of the Drug-Free Workplace Act of 1988.

The employer must keep urine drug test results private, lest he be sued for libel or slander. This requirement is in effect because under federal law, the mere presence of a controlled substance in one's blood without an authorizing prescription is not illegal, and a positive test is not a basis for forensic action. (In 37 states, however, it is illegal merely to have a controlled substance in one's body without an authorizing prescription.)

16.6 The Problem of the Employee Who Abuses Drugs; Employee Assistance Programs (EAPs)

Anyone who doubts there is a serious problem regarding the abuse of alcohol and other drugs in the workplace needs only to reflect on the following statistics from the federal government as shown in Table 16.3:

- Among full-time workers aged 18–49 in 2000, 8.1% reported past-month heavy alcohol use, and 7.8% reported past-month illicit drug use.
- In the past year, 7.4% of these workers were dependent or abusing alcohol, and 1.9% were dependent or abusing illicit drugs.

What resources are available to the employer who identifies an employee with a drug problem? If an employee screens positive for a drug and the test is confirmed, he or she can be fired, referred to outside professional help, referred to Alcoholics

Case History *One Company's Program to Control Drug Abuse*

Acme Trucking Company (not its real name), one of our largest interstate truck lines, previously never tested its workers for anything but recently discovered that 17% of its work force was abusing alcohol and other drugs. The company operates a fleet of 1,400 vehicles, including trucks, automobiles, and tractor trailers; it logs more than 11 million miles a year. The company decided to put a comprehensive drug-screening program into effect, coupled with a new company policy prohibiting the use, possession, distribution, or sale of alcohol, controlled substances, or illegal drugs. The Teamsters Union agreed to a new contract provision authorizing drug screening and specifying an exact laboratory procedure for urine and blood analysis. Before any testing was begun, all employees were given details of the program, including a list of drugs in the screen, cutoff levels, and the consequences of testing positive. If a worker tested positive, he was immediately informed in the presence of a union representative and was told that he or she would be retested in 30 days. If the employee tested positive a second time, he or she was considered drug dependent, and disciplinary action was taken. The State Department of Motor Vehicles was also notified. Alcohol and marijuana were the two drugs most often found, while cocaine and amphetamines were seldom discovered. Alcohol abusers were given the chance to enroll in the Teamsters' Alcohol Rehabilitation Program on a 30-day leave of absence without pay. Job applicants who tested positive were not given a second chance. Results: Four months into the program, the number of employees who tested positive for drug abuse was down by almost 50%. The company continues to test everyone who operates company vehicles or machinery, including salespeople, supervisors, and managers; only clerical workers are excluded.

Anonymous, offered no assistance, or offered a combination of resources. Company-operated **employee assistance programs** (EAPs) are now widely available in U.S. industry. Unions such as the International Brotherhood of Teamsters have adopted collective-bargaining agreements that provide for the handling of drug abuse cases, including strict laboratory-testing procedures. The following is an example of one company's successful attempt to deal with drug abuse among its employees.

The concept of the EAP began to take hold in the United States in the late 1960s. As public awareness of alcohol and other drug problems increased, the workplace as a site for alcohol and drug abuse received increasing attention. At about the same time, research in the field of organizational psychology was demonstrating that firing employees with personal problems rather than attempting to ameliorate those problems was cost ineffective and cost inefficient for businesses and organizations. The idea of the "impaired employee" emerged. This is the person with personal problems that affect work performance who could be "saved" from termination and returned to the work force a more productive and efficient employee. Termination, hiring, and training costs were demonstrated to be much higher than the costs of implementing programs to identify, refer, and treat impaired employees. On September 15, 1986. President Ronald Reagan signed Executive Order 12564, which established the goal of a Drug-Free Workplace. While focusing specifically on federal workers, this Executive Order spurred the development and implementation of EAPs in all sectors of American society.

Blum and Roman[6] have provided an overview of workplace problems that can be addressed by EAPs. These include the following:

- Drug use reduces workers' productivity in terms of both the quality and quantity of performance on the job.
- Drug use creates unpredictable and disruptive behavior in the workplace, affecting the behavior and productivity of coworkers.
- Drug use is a threat to safety in the workplace.
- The presence of illicit drug use is typically an indicator of illicit drug dealing. Not only does this imply the presence of serious criminal behavior in the workplace, but it also suggests the increased likelihood that pushing will occur, which may induce nonusing employees to use drugs.
- Drug-using habits are expensive and encourage theft from both the employer and fellow employees.

EAPs are unique in that the best of them can become self-contained "continuums of care" for any business or organization. A continuum of care is a range and flow of services that are available to address a health or drug problem. The basic structure of a continuum of care involves at least three levels. *Primary prevention* consists of programs and services that attempt to keep people from developing substance abuse problems in the first place. *Secondary prevention* (also known as *intervention*) is designed to identify and refer persons with existing abuse or

---

[6]Blum, T. C. and Roman, P. M., Employee assistance and human resources management. In K. Rowland and G. Ferris (Eds.), *Research in Personnel and Human Resources Management*, Vol. 7, Greenwich, CT, JAI Press, pp. 258–312, 1984.

Table 16.3 Prevalence of Substance Use, Abuse, or Dependence Among Full-time Employed Workers Aged 18 to 49: 2000 NHSDA

		Rates of use (%)			
	Estimated population (000s)	*Past month heavy alcohol use*	*Past month any illicit drug use*	*Past year dependence or abuse of alcohol*	*Pastyear dependence or abuse of illicit drugs*
TOTAL	87,672	8.1	7.8	7.4	1.9
Male	50,466	11.4	9.2	9.9	2.4
Female	37,206	3.6	5.9	4.0	1.2
Age groups					
18–25	15,190	13.5	14.9	13.5	5.3
26–34	24,464	8.7	7.9	8.2	1.8
35–49	48,017	6.0	5.5	5.1	1.0
By type of occupation					
Executive, administrative, and managerial	14,822	6.5	6.5	6.9	1.1
Professional specialty	13,222	4.9	4.7	5.3	1.4
Technical and sales support	13,239	8.9	8.0	8.2	1.8
Administrative support	10,714	4.9	6.9	5.5	1.9
Services	10,047	7.7	9.7	8.0	2.3
Precision production, craft & repair	10,786	12.6	11.2	9.2	2.5
Operators, fabricators, and laborers	12,428	11.2	8.6	9.3	3.0
By type of industry					
Construction & mining	8,267	15.7	12.3	10.9	3.6
Manufacturing	14,610	9.4	6.7	6.7	1.7
Transportation, communications, and other public utilities	6,541	7.6	7.2	8.2	1.4
Wholesale and retail	15,881	9.2	10.9	10.5	2.9
Service—business & repairs	7,883	9.4	9.0	8.7	1.9
Finance, insurance, real estate, and other services (personal and recreation)	8,320	5.9	7.7	7.4	1.7
Service—professional	19,125	4.0	5.0	4.4	1.3
Government	4,252	6.3	3.7	3.3	0.6

(Source: National Household Survey on Drug Abuse. SAMHSA, September 2002.)

dependence disorders. *Tertiary prevention* involves actual substance abuse treatment for severe problem cases and aftercare services to avert relapse.

An effective and comprehensive Employee Assistance Program will involve a complete continuum of care for employees. Typical components of such a program are:

1. *Drug policy promulgation*. Addressing substance abuse in the workplace begins with an institutional drug policy which sets forth the parameters and defines a drug-free workplace. Having a drug policy is the first step in primary prevention because it serves an educational and awareness-promoting function for all members of the company or organization. Perhaps the most important element of an effective policy is the guarantee of absolute confidentiality for employees utilizing the EAP and the dissemination of information that employees will not be fired or demoted for voluntary self-referral to the EAP.
2. *Formal training for supervisory/management personnel*. Training on work performance indicators of personal problems of employees, and training on referral strategies to the company/organization EAP.
3. *Drug screening testing*. Either preemployment screening or screening for cause when there is reasonable suspicion that an employee's work performance is impaired due to substance abuse. Drug testing is a secondary prevention (intervention) strategy.
4. *Provision of counseling services*. Some businesses and organizations provide internal (meaning they have their own counselors on staff) counseling services for short-term outpatient counseling. Other businesses and organizations have external programs (meaning they contract for the provision of counseling services by counseling providers). The provision of counseling services is tertiary prevention as is the provision of aftercare support, usually in the form of attendance at self-help groups such as Alcoholics Anonymous or Narcotics Anonymous and/or Aftercare support group therapy.
5. Provision for referral to long-term outpatient treatment or inpatient hospitalization for persons requiring more intensive treatment services.

Finally, the most effective programs utilize what is known as a "broad brush" approach to the provision of employee assistance programming. Broad brush is the term for an employee assistance model that covers a number of different personal problems that can affect an employee's job performance. Problems such as mental health and anxiety disorders, marital and family discord, stress, financial difficulties, legal problems, job performance problems, eating disorders, termination counseling, and retirement counseling, among others, can be dealt with in a broad brush employee assistance program. These programs have the additional benefit of not stigmatizing or singling out the substance-abusing employee for using the EAP, because the employee could be utilizing the EAP for any number of types of problems.

EAPs have proven to be an effective strategy in the fight against drug abuse in the workplace. The basic concept has lately been expanded to other institutions of society, most noticeably the school, where student assistance programs are proliferating in our nation's middle and high schools.

Health maintenance organizations (HMOs) currently are playing an increasingly important role in EAP. More companies are turning over the administration of their EAP to an HMO.

16.7 Drug Testing in the Home

A new home drug-testing kit is now on the market. SherTest Corp. sells DrugAlert, a product designed to detect even trace quantities of residual PCP, marijuana, hashish, and crack cocaine on any doorknob, article of furniture, desk top, sink, floor, or book—in short, anything that has been touched by a child whose parents suspect of having used drugs in the home. To perform a test, the parent uses a white paper towel or tissue to wipe a surface that may have been contaminated. The paper is then sprayed with DrugAlert (supplied in an aerosol can). If the paper turns a color that is on the can label, the drug test is positive. The manufacturer recommends performing the test in the child's presence in order to break down barriers of denial between child and parent.

Some have expressed their belief that this type of in-home test is counterproductive, violating the child's constitutional rights and invading his or her privacy. Others commend such testing, citing awareness as a parent's best defense against drug use.

The first **sweat-patch** test system for speed, cocaine, and opiates has received FDA approval for use by trained testing professionals in clinical centers. A waterproof, adhesive pad about the size of a playing card is worn up to 7 days on the back, arm, or chest; it will collect perspiration that might contain drugs of abuse during that time. The patch has a tamper-proof feature.

Another kit contains a moist pad that parents can use to wipe any surface their child may have touched. It tests for contact with any of 30 drugs, and parents can use an identification number instead of a name for secrecy when they return the pad in a special envelope. Alco-Screen, supplied by Chematics, Inc., North Webster, Indiana, is a two-minute test for blood alcohol levels that uses a highly specific enzyme to detect alcohol in saliva directly in the mouth.

16.8 Impairment Testing for Drugs

Impairment testing has been proposed as an alternative to urine testing for drugs. One form of the impairment test is a hand-eye coordination examination. An employee tracks a moving object on a computer screen while the computer compares his performance with past standards, and records the results. In another approach, cognition skills are challenged by analyzing images or sequences of numbers and letters, or answering questions. Promoters of impairment testing claim these tests are cheaper than urinalyses and do not invade the person's privacy. They admit that impairment testing identifies not only the drug-impaired but also those impaired by fatigue, illness, or other nondrug factors, but they claim this factor is actually an advantage. For workers who repeatedly fail impairment tests or whose behavior otherwise suggests drug use, urinalysis may be used. Promoters claim that impairment tests are better than urine tests, and should largely replace them.

16.9 Analysis of Hair for Drugs of Abuse

The hair follicle, from which the shaft grows, is nourished by the blood. Consequently, the hair follicle is exposed to any chemical or drug that might be in the blood. It has been shown that small amounts of the chemical or drug will be deposited in the core of the hair shaft and remain there as the hair grows. Typically, the hair sample is 1.5-inch-long strands cut near the scalp; they give a drug history over the past 90 days. Proposals have been made that hair can thus be used to detect drug use that might have occurred weeks or months before.

For a bibliography of publications relating to all aspects of hair testing (from SAMHSA), browse: **http://www.health.org/workplace/autry723c.aspx**.

Actually, hair has been used as a specimen in forensic (used in courts or legal proceedings) investigations for decades. Hair tests for drugs are currently performed in some areas of the United States as part of preemployment, parole, and probation procedures. Authorities who have reviewed the present state of hair analysis, however, have serious doubts about its value in routine drug-abuse testing. They say that too many critical questions remain unanswered for test results to be accurately interpreted. Some of the questions are:

1. What acceptable reference material is available to standardize the analytical method?
2. What is the ratio of drug to drug metabolite in hair?
3. Is the amount of drug (especially cannabinoid) sufficient for detection by routine GC-MS procedures?
4. Although we presently have anecdotal results, where are the rigorous scientific studies?
5. How are drugs and drug metabolites distributed in hair?
6. How long after drug use does the drug appear in the hair shaft?
7. What is the relationship between dose of drug and its concentration and detection in hair?
8. What is the influence of age, sex, and race on drug uptake?
9. How does washing the hair affect its drug content?
10. If a drug has a high affinity for the pigment melanin, won't dark-haired people test higher than blonds?

A panel of experts convened by the Society of Forensic Toxicology and NIDA reviewed this subject and concluded that because of the unanswered questions, hair analysis alone is not a sufficient technology for workplace drug testing at this time. The FDA agrees, at least in one area. It has concluded that the radioimmunoassay test procedure for hair analysis is unreliable, unproven, and unsupported by the scientific literature or well-controlled studies and clinical trials.

Despite misgivings, hair analysis is being used. A Chicago company took a lock of hair from each of its 15,000 employees for testing by Psychemedics Inc., of Los Angeles, which claims the test allows detection of drugs taken within the prior 90 days. As more research is done, the analysis of hair may in time become an effective tool in controlling the abuse of drugs in the workplace, but presently it is not defensible in court.

16.10 Are Teens Too Smart for Drug Testing?

Today, many schools expect their teenage students who engage in optional activities such as competitive sports to submit to **random (suspicionless)** drug testing. Some administrators plan to extend the testing to all extracurricular activities, including chess, drama, and chorus.

In a 2002, 5-4 decision, the U.S. Supreme Court approved suspicionless drug testing in public high schools, ruling that students' rights to privacy must give way to their school's desire to rid the campus of drugs. The ruling applies not only to student athletes but to all who engage in competitive after-school activities. It does not apply to random testing for any student. When a student signs up for an activity, he or she is considered to have agreed to obey the school's policies on drug abuse. It's much like the motorist driving on public highways, who is assumed to have given his consent for drug testing. Tests for drugs are usually performed on urine samples, but breath, saliva, hair, and perspiration (via multiple skin patches) are also used. Avitar's Oral-screen (**www.avitarine.com**) is a saliva-based test for THC, cocaine, and opiates.

Probably most observers of teenagers and their drugs would posit that random drug testing in schools would never work. Teenagers, they say, are too smart. From the Internet and teen publications, they know about "chain of custody," "no test" criteria, and drug-retention times. They know about products such as "KLEAR," "Pass It," "Urine Luck," "Testclean," "Carboclear," and "4 Clean P." Further, if they know they are going to be tested, they expect to find ways to get around it. Schools cannot be expected to establish International Olympics–type screening programs. Unfortunately, the use of drug tests on students in extracurricular activities could turn potential participants away from such activities.

On the other hand, there are drug-free students on every campus for whom testing is likely not a problem. A program of drug testing might be just the impetus for a student considering trying drugs to opt for abstinence. In strenuous sports, the use of stimulant drugs can be disastrous. The side effects of anabolic steroids can be serious in both sexes, especially in teenagers, and as the teenager matures into adulthood, he or she is very likely to recall drug use in school as "the dumbest thing I ever did."

For more insights on this topic, see the following Web sites:
http://www.drugs.indiana.edu/issues/suspicionless.html, **http://www.sportsafe.com/**, and **http://www.drugfreeschools.com/**. Also, search for "schools.drugs.testing"; "teen.drugs.testing"; and "teens.drugs".

Web Sites You Can Browse on Related Topics

Personal Drug Testing

http://www.drugfreeworkplace/org
http://www.Pharmchem.com/

Workplace Testing

http://www.drugs.indiana.edu.publications/ncadi/primer/drug test.htm
http://lectlaw.com/files/emp02.htm

DEA Drugs of Abuse

http://www.usdoj.gov/dea/pubs/abuse/content.htm

Suspicionless Drug Testing

http://www.drugs.indiana.edu/issues/suspicionless.html

Hair Testing

http://www.drugtestnow.com/pdt90.html

Home Test for Alcohol
Search "test for alcohol"

Study Questions

1. What is the difference, if any, between testing government employees and testing private company employees for the presence of drugs in the workplace?
2. A certain private company appears to be having no drug problems, and its productivity is considered normal. One day, without warning, all employees are required to open their personal lockers in a search for drugs. They refuse and take the matter to court. Why is it possible that the court will hold in favor of the employees?
3. True or false:
   - **a.** In urinalysis, the test is always made for the drug itself.
   - **b.** In urinalysis, the "chain of custody" refers to the heavy metal linkage used to secure the box of specimen.
   - **c.** Salt, added to a urine sample to defeat a test, would *increase* the specific gravity of the sample.
   - **d.** A nanogram is one one-millionth of a gram.
   - **e.** The "cutoff level" of a drug in urinalysis refers to its half-life.
   - **f.** Ammonia added to urine would tend to make the urine alkaline.
   - **g.** If a urine sample tests positive in a screen, it should be confirmed only if the donor requests it.
   - **h.** EMIT is an example of an immunoassay test.
   - **i.** Abuscreen is an example of an RIA test.
   - **j.** The most valuable procedure used today to confirm a positive urinalysis is GC-MS.
   - **k.** In actual laboratory practice, false negatives occur more commonly than false positives.
   - **l.** Urinalysis for a drug is usually able to tell us how much drug was taken and for how long.
   - **m.** Water-soluble drugs tend to remain in the body longer than fat-soluble drugs.
   - **n.** It is possible for a person to smoke one joint and still test positive for marijuana 3 days later.
   - **o.** In the urine of an Olympic athlete, gene-spliced human growth hormone could be detected because the gene-spliced form is different from natural hGH.
4. **a.** A person takes Dexatrim to control weight. What drug of abuse could show up as a false positive?
   - **b.** An epileptic takes Dilantin. What drug of abuse could show up as a false positive?
   - **c.** What procedure could be followed to clear up these false positives?
5. Benzoylecgonine is a major metabolite of what drug?
6. Of THC, cocaine, and methadone, which drug typically persists for the longest time in the urine? Which for the shortest time?
7. You work as a bookkeeper for a private company that makes hairnets. Your employment and personal records are clean. One day your employer demands that everybody in the company submit to urine drug testing, including you. Do you comply, or do you challenge him in the courts? Explain your position.
8. Would you support the stand that alcohol testing must be made mandatory for *all* locomotive engineers? Explain your position.
9. Test your forensic skill. A newspaper columnist reports that a 20-year-old Ohio man was arrested for stealing Christmas gifts off doorsteps. The thief received a year in jail instead of the usual probation after the court ordered that he submit to a urine test, which came back positive, but for pregnancy! Propose an explanation for the unexpected urinalysis result.

10. Regarding testing in the workplace, Harvey Grossman of the American Civil Liberties Union has said: "Who made the decision that the employer in our society is the guardian of public morality? ... They don't use lie detector tests for adultery. Should they test for scofflaws, or for gambling?" We could add, should engaged couples be tested for AIDS? Your views, please.
11. In the recent Olympic games, six weightlifters were banned after testing positive for, variously, furosemide, testosterone, and amphetamine. Describe what each of these drugs does in the body. Why would an athlete take any one of these drugs?
12. Critics of drug testing cite the occurrence of false positives and false negatives. What is another valid criticism of the usefulness of drug testing in the workplace?
13. You are on a jury deciding the propriety of firing two city employees who regularly worked around high-voltage electric wires and whose urines tested positive for drugs of abuse. The workers claimed Fourth and Fourteenth Amendment protection against unreasonable search and loss of jobs. The prosecutor claimed that the search was not unreasonable because of the hazardous work. How would you vote, and why?
14. Regarding parents using kits to test their children for drug use, a 16-year-old sophomore in Chicago says, "Drugs mess up your body. Parents can check up on a kid's safety by using the kits." But an 18-year-old senior at a nearby school says, "It's an invasion of privacy and shows lack of respect." How do you feel?
15. The Dade County (Florida) School District has approved random drug testing of all high school students whose parents give approval. If signed, consent forms will allow pulling students out of class at random and testing them for marijuana, cocaine, barbiturates, and amphetamines. As a student in this district, how would you handle this situation?
16. The Spanish Chess Foundation has banned the use of certain drugs in championship tournaments. Predict the kinds of drugs banned by the Foundation.

CHAPTER 17 Drugs in Sports

Key Words in this Chapter

- Doping
- Psychomotor stimulant
- Anabolic steroid
- Human growth hormone (hGH)
- Erythropoietin (EPO)
- Beta-blocker
- Masking agent
- Testosterone
- Masculinizing agent

Learning Objectives

After you complete your study of this chapter, you should be able to do the following tasks:

- Define doping, stacking, and masking.
- Define psychomotor stimulant and give an example.
- Explain the difference between an anabolic protein and an anabolic steroid, and give examples.
- Cite the potentially serious side effects in the use of anabolic steroids.
- Discuss androstenedione: source, actions, legality.
- Take a personal position on the propriety of forced personal drug testing in school sports.

17.1 Introduction

One headline put it this way: "A Pumped-Up World. Illegal Doping Is Everywhere Now, and the Culprits Are Rarely Caught." It's true. Drugs, the "dirty little secrets of sports," are indeed now out in the open—widely discussed and practically accepted as a normal, even necessary part of competition. Performance-enhancing drugs are used by football linemen, cleanup hitters, cyclists, sprinters, archers, wrestlers, and female gymnasts. It is now routine for steroids to be uncovered at all levels of competition. Examples: An international competition cycle team support person was arrested in possession of 400 vials of steroids and performance-enhancing *erythropoietin* (EPO), cycling's favorite drug. A routine search of the luggage of an international swim team revealed 13 vials of Somatropin, packed in ice in a Thermos flask. The team was suspect to begin with because of their massive shoulders, deep voices, odd metabolic profile of urine samples, and the ready availability of steroids in their home country. One of baseball's greatest stars claims that 85 percent of major-league baseball players are using steroids. NIDA states that more than half a million eighth- and tenth-grade students are now using these "dangerous drugs," and increasing numbers of high school seniors say they don't believe the drugs are risky. A U.S. Olympic bobsledder was banned for two years after he tested positive for the anabolic steroid methandienone. And in the small-potatoes department, five English professional snooker players were fined and stripped of tournament points after failing drug tests.

The DEA estimates there are 1 million steroid abusers in the United States alone, the majority believed to be bodybuilders or adolescents trying to impress others with their bulging biceps. According to Dr. Linn Goldberg of Oregon Health Sciences University, national surveys indicate that girls account for about one-third of the high school students who abuse steroids. And a Penn State University study published in the *Archives of Pediatric and Adolescent Medicine* shows that as many as 175,000 high school girls have used steroids. Some take the drugs to become leaner, while others use them to build muscle.

17.2 Doping and Examples of Drugs

The use of drugs to enhance performance is termed **doping**. The *International Olympic Committee* (IOC) has defined doping as the use of either substances foreign to the body or physiological substances in abnormal amounts—or drugs used by abnormal methods with the goal of obtaining an unfair advantage during competition.

Types of drugs that are or have been used in sports training or competition are:

1. CNS (psychomotor) stimulants, such as ephedrine, amphetamines, or caffeine
2. Anabolic steroids
3. Pain relievers, such as narcotic analgesics
4. Diuretics
5. Peptide hormones, of which the most popular are *human growth hormone* (hGH), *erythropoietin* (EPO), and ACTH
6. Beta-blockers such as propranolol (14 of which are IOC banned)
7. Masking agents, which hide the presence of other drugs (See HES in the following discussion)
8. Antiasthmatics, such as albuterol

Amphetamines, cocaine, caffeine (in excessive amounts), pemoline, methylphenidate, strychnine, and phenmetrazine are examples of stimulant drugs banned from use in the Olympic Games. So are the amphetamine-like drugs that society usually thinks of as nasal shrinkers, decongestants, or diet pills. These drugs have significant CNS stimulant actions.

Stimulant drugs excite the CNS, increase the heart rate and blood pressure, and generally make the user more aggressive, more alert, and better prepared to meet a stressful situation. Physiologists recognize, however, that adrenaline is already flowing during a tense sports contest, and to add to the catecholamine blood levels with exogenous drugs is inviting overexertion with dangerous consequences. Cycling, professional football, and soccer are games in which psychomotor stimulants are likely to be used.

Beta-blockers are used in archery and shooting events. Presumably, the beta-blocker will slow the heart, making it easier to shoot between beats (which is what the shooter is striving to do).

Case Histories *Inadvertent Ingestion of a Banned Substance*

At the Pan American Games in Argentina, a cold capsule cost a Canadian rower her gold medal in the quadruple sculls. The 30-year-old woman had taken 1 Benadryl-D decongestant capsule the night before the race, "in order not to cough and wake her roommates." She did not know that the product contained the banned substance pseudoephedrine in addition to Benadryl, an antihistamine. Games officials were sympathetic to her mistake, but they felt they "had to issue sanctions according to the rules."

The first British skier to win an Olympic medal was stripped of his award when he tested positive for methamphetamine (desoxyephedrine) contained in a Vicks inhaler. The IUC rule is, "the mere presence in the body constitutes a doping offense."

Masking agents are supposed to hide the presence of banned drugs in the user's system. A diuretic, for example, might "hide" a drug by flushing it out in the urine. Probenecid can block the release of drugs *into* the urine. Fruit pectin is believed by some to hide drugs. Hydroxyethylstarch (HES) is a plasma expander; that is, in the blood it attracts water and thus increases total body blood volume. HES, now banned by the IOC, is taken by athletes who take EPO (see following) to increase their red blood cells (RBCs) in endurance contests. HES masks this increase by expanding blood volume and giving the appearance of normal concentrations of RBCs. HES is administered by slow intravenous drip. It has potentially lethal side effects. HES is not a steroid.

Anabolic proteins and **anabolic steroids** constitute additional types of drugs used in sports. (Anabolism means "building up.") **Erythropoietin** (EPO), an anabolic protein, is made naturally in the human kidney and is now available commercially as Epogen by gene splicing (see Section 2.6). Used medically to combat anemia associated with kidney failure, EPO works by increasing the blood's ability to supply oxygen. EPO is now widely abused by athletes who are seeking increased stamina and performance in endurance contests such as cycling, long-distance running, and cross-country skiing. However, in athletes who already have high levels of natural EPO, injecting more of it can thicken the blood, increasing susceptibility to dangerous clotting in vital organs. EPO abuse is suspected in the deaths of 18 cyclists in Holland and Belgium. The drug is obtained from black market sources in Europe and Mexico; the cost is about $40 a dose.

Steroids (see Figures 6c and 6h in Appendix I) are chemicals found naturally in our bodies or made synthetically that have a special kind of polycyclic chemical structure. Many hormones are steroids. Anabolic steroids are hormonal drugs that induce the buildup of protein tissue and other body tissue associated with nitrogen utilization.

Anabolic steroids such as nandrolone and oxymetholone are prescribed by physicians for certain anemic and postsurgical patients, for persons suffering from osteoporosis, or in general for patients who need to increase their appetite and body tissue-building processes (i.e., anabolism). One can easily see, therefore, how weightlifters, football linemen, and others would be motivated to use anabolic steroids to build body mass and gain a competitive edge. And indeed, this practice is now open and widespread. But do anabolic steroids really work to build bigger and better bodies, or is it motivation, placebo effect, and the accompanying training that produces results? Most studies suggest that anabolic steroids do increase body weight and muscle size, but whether they do so consistently is debatable.

Examples of anabolic steroids that have been or could be used in sports are: clostebol, boldenone (Equipoise), ethylestrenol (Maxibolin), fluoxymesterone (Halotestin), kabolin, methandriol, methandrostenolone (Dianabol), methenolone, methyltestosterone, nandrolone (Durabolin, Deca-Durabolin), oxandrolone (Anavar), oxymesterone, oxymetholone (Anadrol), stanozolol (Winstrol), testosterone, and trebolone (Finaject). (Actually, stanozolol was found in the urine of an Olympic athlete and cost him his gold medal.) Typically, anabolic steroids are assigned to Schedule III under the Controlled Substances Act of 1990. The federal Anabolic Steroids Act makes the unauthorized manufacture, distribution, possession, or use of anabolic steroids criminal offenses, with penalties ranging from 1 to 10 years in prison.

Synthetic anabolic steroids are chemical relatives of testosterone, the primary male androgen (sex hormone). Consequently, all of the anabolic steroids used by athletes or young men or women have the potential of masculinizing the user. This characteristic can manifest itself in males in testicular atrophy (with exogenous androgen, the testes do not need to produce much of their own), difficulty in urination because of changes in the prostate or seminal vesicles, accentuation of male pattern baldness, decreased sperm production, impotence, inability to ejaculate, a too-high blood cholesterol, fluid retention, insulin resistance, high blood pressure, heart attack, and increased risk of kidney disease. Masculinization in women is characterized by loss of body curves, deepening of the voice, enlargement of the clitoris,[1] menstrual problems, and irreversible pattern baldness. Increased aggressiveness has been seen. There is evidence that anabolic steroids can cause stunted growth and permanent short stature in children and teenagers. The violent mood swings, termed *'roid rages*, that have been associated with heavy steroid use were not substantiated in one study reported in the *New England Journal of Medicine.* Actually, there is a dearth of scientific information on the adverse effects of anabolic steroids, especially in long-term use. There is evidence, however, that these drugs can cause liver and prostate cancer if taken in large doses over a long period of time. For answers to many questions you may have about anabolic steroids, consult this Internet address: **`http://www.cesar.umd.edu/metnet/`**. NIDA maintains a useful website: **`www.steroidabuse.org`**.

Athletes do take large doses of anabolic steroids over lengthy periods. T.H., a 37-year-old weightlifter, readily admits to taking 100 mg of oral and injectable anabolics each day while in training (6 mg a day is the recommended dose for aplastic anemia patients). Experienced steroid users employ a "**stacking**" or "pyramiding" dose regimen,

[1]Laura Dayton, editor, *Female Body Building* magazine, has known women who have had to tape down their steroid-enlarged clitoris to keep the bulge from showing in competition. "What Price Glory," *Women's Sports and Fitness*, March 1990, pp. 52–55.

which involves taking progressively increasing doses of one or more products beginning 3–4 months before competition, reaching a peak dose at mid-cycle, and gradually reducing the dose toward the end of the regimen. The abuser often mixes tablets and injectable forms and can build up doses 10–40 times greater than ordinarily prescribed by a physician. The steroid user believes that stacking will maximize effects and detection avoidance while minimizing health risks, but there is no scientific evidence that stacking achieves these goals.

Oral anabolic steroids can be detected in body fluids for several weeks after the last dose. After injection into muscle tissue, anabolics can be detected months after the last dose. Injected forms are better tolerated by the human body, especially by the liver, which is subjected to all-at-once greater concentrations after oral administration.

Problem 17.1 *In the controversy that has arisen over the use of anabolic steroids in sports, proponents cite the accepted use of birth control pills in female athletes to delay menses until after a meet, and the practice of blood packing ("blood doping"), a procedure in which blood is taken from an athlete and then infused back into his body just prior to competition. Cite your reasons for taking a stand, for or against any or all of these procedures. (Note: To counter blood doping, The International Cycling Union has passed a rule requiring cyclists to have their volume of red blood cells—the hematocrit—tested before competition. This process caught 4 riders at a recent Giro d'Italia.)*

Despite their prescription-only status, anabolic steroids are easy to obtain. They are smuggled from Mexico or Europe and sold via the mail or through health clubs and gyms. California's Department of Justice Bureau of Narcotics Enforcement says that 85–90 percent of all illicit steroids on the U.S. market come from Mexico. Most are produced in pharmaceutical factories that ring Mexico City. According to the Department of Justice, Mexico believes the steroid problem is in the demand, not the supply. (The then-president of Colombia said the same about cocaine.)

Problem 17.2 *Based on your perceptions of our societies, do the roots of the problems such as anabolic steroid abuse and trafficking lie in supply or demand?*

More than half of the black market steroids are counterfeit (see Figure 17.1). Although of uncertain purity they usually are pharmacologically active. Some smuggled products have been found to contain high amounts of cortisone. BEWARE! If you purchase anabolic steroids in Mexico, be aware that bringing them back into the United States or possessing them without a prescription is a federal offense, punishable by up to 5 years in prison under the Anabolic Steroids Act of 1990. The FDA says that anabolic steroids are now a $300–$400 million-a-year business.

For more photos of counterfeit steroids, browse: **`http://www.steroids101.com/counterfeits.html`**.

Human growth hormone (hGH; see also Section 2.5) is now widely injected by athletes who believe that it builds more competitive bodies. There is no scientific evidence, however, that hGH can improve athletic performance. A study reported in the *American Journal of Physiology* concluded that hGH does not help the muscles of young men grow faster or work more efficiently than exercise alone. It is true that

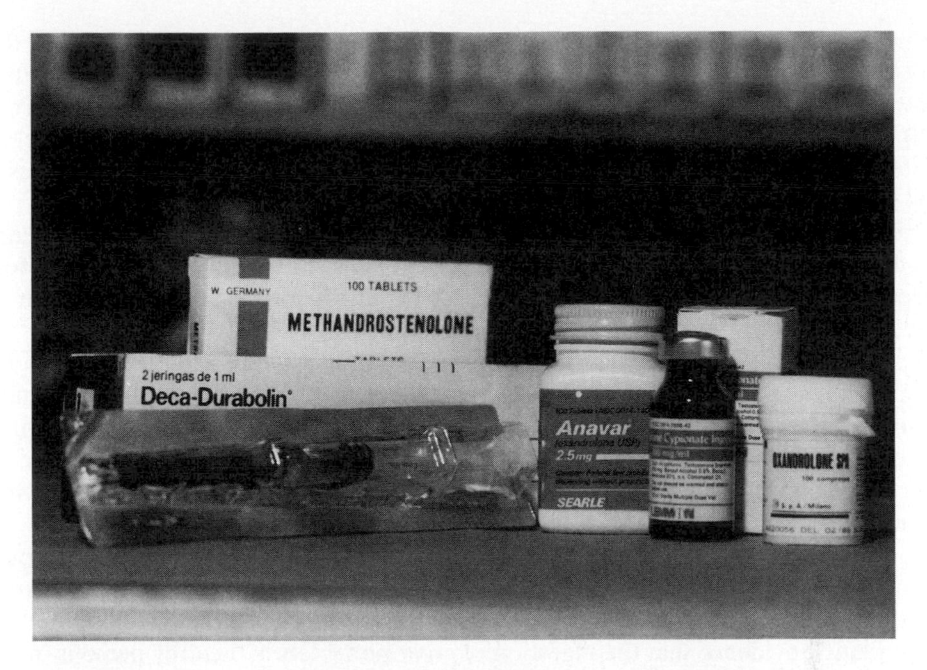

Figure 17.1 Purported anabolic steroids purchasable in Tijuana, Mexico. According to U.S. Customs, at least one of the drugs in this photo, Anavar, is counterfeit. Other purported steroids purchased from foreign sources have been found to contain ingredients other than those indicated on the label. (Source: Union Tribune Publishing Co., San Diego, CA. Used with permission. Reproduction prohibited.)

hGH has been shown to cause adults to lose fat and gain muscle—*but only if they were deficient in this hormone to begin with*. It's a myth that otherwise healthy young men can use hGH to "remodel" their bodies. Serious adverse effects of hGH can include carpal tunnel compression (numbness or tingling in the fingers), giantism (jaw, joint enlargement), and elevated blood sugar, possibly evolving into diabetes later in life.

Problem 17.3 *In sports competition, why would the detection and abuse of testosterone be especially difficult, if not impossible, to prove?*

Androstenedione, as well as **testosterone** and hGH, are produced in the human body. A normal adult male's daily *in vivo* production of testosterone is 2.5–11 mg. In females the corresponding quantity is 0.25 mg. Our livers make testosterone from another steroid, androstenedione, and also metabolize testosterone to androstenedione, the latter secreted by the testis, adrenal cortex, and ovary. Thus all these substances occur naturally in the adult human body, as contrasted to drugs such as amphetamines, synthetic anabolic steroids, or the stimulants phenylpropanolamine or pseudoephedrine.

Androstenedione is prohibited by the IOC, NFL, and NCAA, but it is not a federally controlled substance, and its use is widespread in sports. Androstenedione is sold legally in nutrition stores as a sterol "dietary supplement" of plant origin (its chemical precursor is obtained from yams). That makes it a food in the eyes of the FDA, and no data need to be submitted to prove bodybuilding claims or identify short- or long-term adverse effects, if any. A bottle of 60 100-mg tablets costs from $30 to $40. Androstenedione ingestion results in a short burst in the blood level of testosterone, but the latter is rapidly metabolized, and the chemical fingerprints usually vanish within 24 hours.

Recently, Iowa State researchers tracked 20 young men using OTC androstenedione. After 8 weeks of a daily 300-mg dose, subjects showed no increase in testosterone, but did show increased levels of estrogen, which can cause young men to develop breasts.

An Olympic gold medalist failed an International Amateur Athletic Foundation urine test for androstenedione when the inspecting team showed up unannounced at his home base. One of the premier home run hitters in professional baseball has readily admitted to frequent use of androstenedione. He says use of this drug in baseball is common. Major league baseball has not banned androstenedione.

Creatine, made naturally in the kidney from glycine and arginine, has been touted as a muscle builder and is legally used by professional athletes. While it is true that creatine phosphate plays an important role in muscle during exercise, there is no scientific evidence that the ingestion of creatine tablets by healthy persons on an adequate diet will result in increased strength or muscle mass. In spite of this, large numbers of children as young as twelve are using creatine; usage rises to 44% among high school seniors. In the human, creatine is converted into creatinine and is excreted in the urine.

In summary, among today's elite athletes, the popular performance-enhancing drugs are hGH, EPO, and designer steroids (see Section 12.8 for the discussion of designer steroids). Currently, there are no tests that can detect abuse of hGH, EPO, or most designer steroids. In cyberspace, there are Web sites that offer all the abuser needs to know to select and acquire what he or she believes to be performance-enhancing drugs. Muscle-media magazines are sold nationally, as are anabolic reference guides.

17.3 Testing for Drugs in Sports Competition

Performance-enhancing drugs continue to be used by athletes in American colleges, and programs are in force to detect drug abusers. The *National Collegiate Athletic Association* (NCAA) tests athletes for street drugs and steroids at all championship events and bowl games. Altogether, the NCAA conducts between 10,000 and 12,000 drug tests a year on athletes. The NCAA also has implemented a random year-round program for Divisions 1 and 2 football and for Division 1 track and field, testing for steroids, diuretics, and urine manipulators (masking agents). Colleges—on 48-hour notice—can be selected for testing anytime during the year.

Reacting to criticism based on Fourth Amendment protection of privacy, NCAA and school officials contend that students waive their privacy rights when they consent to drug testing as a condition of their athletic scholarship. "If you don't want to be tested, don't sign," they say. Court decisions in Fourth Amendment cases have been conflicting and confusing. But recently, the United States Supreme Court ruled (in a 6–3 vote) that public schools can require drug tests for athletes, whether or not they are suspected users. The corollary ruling was that privacy rights must sometimes yield to the fight against drugs. "School sports are not for the bashful," said the majority. The minority concluded that under this ruling, millions of students, "an overwhelming majority of whom have given school officials no reason whatsoever to suspect they use drugs at school, are open to an intrusive bodily search." The case began when a seventh-grader was banned from his junior high football team for refusing to submit to a urinalysis.

Ninety percent of schools and colleges that play Division 1 football conduct their own drug-testing programs, albeit with varying protocols and penalties. One school tests for five street drugs—marijuana, amphetamines, cocaine, opiates, and PCP—and alcohol.

Olympic Games drug testing began in 1968 when amphetamine became the first drug to be scrutinized. But it was not until the summer of 1976, when the IOC began to use high-tech, state-of-the-art instrumentation—mainly GC-MS—that athletes could be certain that any drug use prior to the games would be discovered. Today, the first four finishers plus at least one competitor chosen at random must provide a urine sample within 1 hour of the end of competition. Wide publicity is given to the testing in the hopes of deterring athletes in the first place. In 1992 the International Tennis Federation approved a comprehensive random drug testing program that applies to all events under its jurisdiction. The IOC has approved a protocol for EPO that uses both the traditional urine test and a blood test.

Prohibition does not mean that athletes have stopped taking drugs; many still use them but terminate use far enough in advance of testing (they expect) to avoid detection. Not all escape detection, as shown by the following case history.

Case History *Young Anabolic Steroid User*

> A 14-year-old South African runner competed at a national junior championship meet. She was required under meet rules to submit a urine sample. The sample tested positive for traces of the anabolic steroid nandrolone and a stimulant, fencamfamine. Her suspension from competition makes her the youngest track and field competitor anywhere in the world to be accused of using performance-enhancing drugs.

The classes of drugs that are IOC-banned include amphetamines and other stimulants, narcotics, analgesics, anabolic steroids, diuretics, and beta-blockers. Gas chromatography combined with mass spectroscopy is used to screen for anabolic steroids and beta-blockers. Some injectable forms of anabolics can be detected months after the last dose; oral forms have a much briefer half-life and disappear more quickly. Radioimmunoassay is used to screen for cocaine. Caffeine is detected using *liquid chromatography* (LC); its concentration in the urine is permitted up to 15 ng/mL, the normal consequence of drinking coffee, tea, or colas. Athletes must be careful not to use cold medicines that contain ephedrine, phenylpropanolamine, pseudoephedrine, or related sympathomimetic amines, because these drugs could give false positives for amphetamines.

A *National Football League* (NFL) player who fails a random urinalysis may enter a three-stage program. Stage 1 consists of evaluation and treatment, and failure to cooperate carries a fine equal to three games' pay. In Stage 2, as many as 10 urine tests a month may be required. Any positive test results in a fine equal to four games' pay and a possible four-week suspension. Another positive drug test results in as much as a six-week suspension, and Stage 3 results in three years of urine tests up to 10 times a month and possible banishment from the NFL for at least a year.

The *National Basketball Association* (NBA) tests its players for marijuana and steroid use. The NBA tests veteran players only once, but rookie players are tested throughout the season.

The sports-loving public generally agrees that winning a contest with the help of a drug is unfair and that drug use in sports is to be condemned. One opponent of drug use put it this way: The basic nature of a sports contest is to test the natural abilities of athletes against each other. Drugs change that basis. Therefore, drugs have no place in sports. That advice may fall on deaf ears if the athlete is of the elite international caliber or is striving for professional contracts or college scholarships. The rewards in certain sports are high, competition is fierce, and the competitors believe that drugs will make a difference.

Athletes continue to find novel drugs and techniques to circumvent the system, especially at the international level and at the Olympic level. In a recent Olympic Games, a U.S. woman shotputter, a U.S. male hammer thrower, and several British weightlifters tested positive for Clenbuterol, an asthma and veterinary drug banned by the IOC because it has CNS stimulant and possibly some anabolic activity. Clenbuterol is not approved in the United States and is not an anabolic steroid, but is used in some European countries by animal trainers to build strength in exhibition livestock. It can have serious, immediate adverse effects in humans.

Web Sites You Can Browse on Related Topics

Drugs in Sports
http://www.olympic.org/uk/index_uk.asp
http://www.raceagainstdrugs.org/
Also: Search "IOC drugs"

Doping
Search "drug doping"

Androstenedione
http://www.usatoday.com/life/health/mem/ihmen033.htm
Also: Search "androgen"

Erythropoietin
http://www.druginfonet.com/faq/faqerypo.htm

Beta-Blocker Drug
http://www.rxlist.com/bblock.htm

General
Search "steroid abuse"

Study Questions

1. Define: **(a)** psychomotor stimulant, **(b)** diuretic, **(c)** anabolic steroid, **(d)** doping, **(e)** blood doping, **(f)** masking agent, **(g)** "pyramiding" doses, **(h)** sterol, **(i)** Schedule III drug, **(j)** GC-MS, **(k)** anemia.
2. Identify each substance and specify its category:

Substance	Occurs naturally in humans?	Pharmacological category
hGH	________	________
Caffeine	________	________
Pseudoephedrine	________	________
EPO	________	________
Androstenedione	________	________
Propranolol	________	________
Testosterone	________	________

3. Give an example of **(a)** a peptide hormone, **(b)** an amphetamine, **(c)** a masking agent, **(d)** a Fourth Amendment right.
4. What chemical relationship explains the fact that anabolic steroids such as nandrolone have masculinizing effects in humans?
5. Explain why an oral anabolic steroid would have a shorter half-life in the human than an injectable.
6. From these three—nandrolone, methamphetamine, morphine—select the drug most *acutely* toxic to an athlete in actual competition, and the *least* acutely toxic.
7. Explain what an anabolic steroid is expected to do when:
   a. A physician prescribes it for an anemic patient.
   b. An athlete takes it during training.
8. True or false:
   a. All anabolic steroids used in sports have some degree of male sex hormone activity.
   b. The IOC uses blood tests to determine the possible presence of banned drugs in Olympic competition.
   c. Because anabolic steroids can lower HDL levels, they could be implicated in hardening of the arteries.
   d. Anabolic steroids can act to make a woman less feminine.
   e. Anabolic steroids can act to stunt the growth of children and teenagers.
9. Athletes employ different kinds of performance-enhancing drugs for different kinds of competition. What type of drug might **(a)** a cyclist be inclined to use?, **(b)** a weightlifter?, **(c)** a soccer player?, **(d)** a football lineman?
10. It has been said that women athletes who dope their bodies with testosterone have more to gain than male athletes who do the same. Explain why this could be so.
11. Who is more sensitive to possible adverse effects of steroids, youth or adults? Why?
12. Is a commercial product a drug if it is intended to **(a)** cause weight loss, **(b)** enhance metabolism, **(c)** increase energy, or **(d)** act as a bodybuilder? Note: Which of these are health claims?
13. You are the parent of a high school football player. One day your son tells you his coach is offering to sell him creatine. What is your reaction?

APPENDIX Structure–Activity Relationships (SARs)

A.1 Introduction to SARs

In addition to knowing the chemical structure of thousands of synthetic drugs, medicinal chemists have been able to determine the structures of a large number of naturally occurring, biologically active compounds. All of this knowledge permits researchers to relate pharmacological action to chemical structure in a series of like-acting drugs. For example, we know the chemical structure of the barbiturates (see Figure 1a,b), and we know that substitution of various chemical groups on carbon number 5 (Figure 2) can modify the sedative-hypnotic action in kind, degree, and duration. Substitution of an aromatic ring ($-C_6H_5$), as in phenobarbital, gives a barbiturate that is less water-soluble and more resistant to biodegradation—and thus is longer acting. With this kind of SAR knowledge, medicinal chemists were able to plan the synthesis of several thousand different barbiturates and to select those that showed the greatest promise in clinical application.

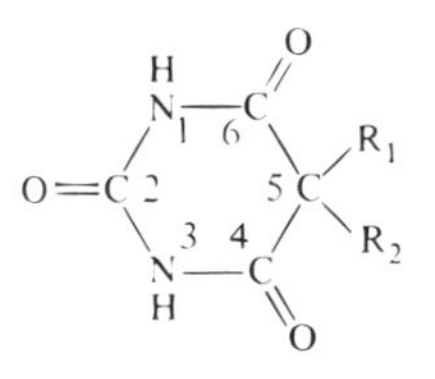

Figure 1a Barbiturate nucleus

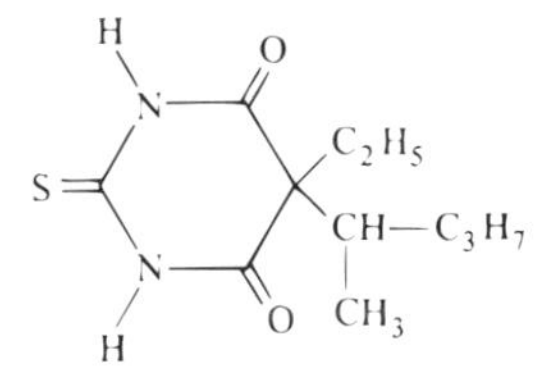

Figure 1b Pentothal sodium

Substituent groups in phenobarbital:

R_1 = ethyl ($-C_2H_5$), R_2 = phenyl ($-C_6H_5$);

substituent groups in seconal:

R_1 = allyl, R_2 = 1-methylbutyl.

Figure 2 Chemistry of barbiturate derivatives

With SARs, researchers have been able to predict and synthesize useful new chemical structures. They have also been able to perceive relationships between chemical compounds occurring in the human body and others occurring in nature. These chemical relationships help us to relate pharmacological actions and even to postulate theories of the cause of certain illnesses. An interesting case in point is a theory of the etiology of schizophrenia that developed from the recognition of a chemical structure common to LSD, various hallucinogens, and the natural brain amine *serotonin* (5-hydroxytryptamine). The heavy-lined segment of the LSD molecule in Figure 3a represents a chemical structure known as indolethylamine. Some investigators believe it is significant that this same structure had been found in other potent hallucinogens such as psilocin (Figure 3b) from Mexican mushrooms, in the hallucinogenic harmala alkaloids (Figure 3c), in DMT, bufotenin, and (especially) in serotonin (Figure 3d). Since LSD can produce model psychotic behavior in microgram doses, and since its chemical cousin serotonin is known to have a role in normal brain function, the possibility arises that schizophrenia (psychotic behavior) may be caused by the abnormal production of an indolethylamine type of chemical in the brain of the mentally ill person. The chemical may develop from an inborn error of metabolism or a genetic defect, and because it is chemically similar to serotonin, it may influence serotonin receptor sites. One great difficulty in the pursuit of this theory is the nearly impossible task of locating a few micrograms of such a chemical in human brain tissue. Nevertheless, SAR considerations have suggested fascinating new approaches to the understanding of mental illness.

A newer theory of the etiology of mental illness also uses SARs. It examines the relationship between the normal brain amine *norepinephrine* (NE) (Figure 4a) and dopamine (Figure 4b). Dopamine can arise biogenetically from dihydroxyphenylalanine (dopa, Figure 4c). It is postulated that abnormal brain production of dopamine results in overstimulation of adrenergic receptor sites, with consequent psychotic behavior. SAR examination shows the close similarity of dopamine to NE and the likelihood that both will fit the same receptors.

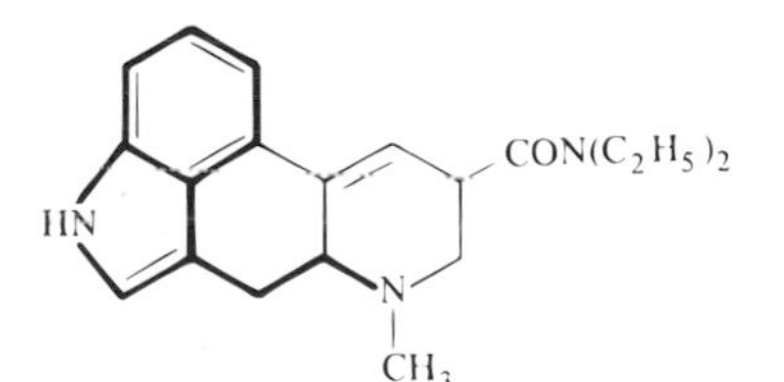

Figure 3a LSD

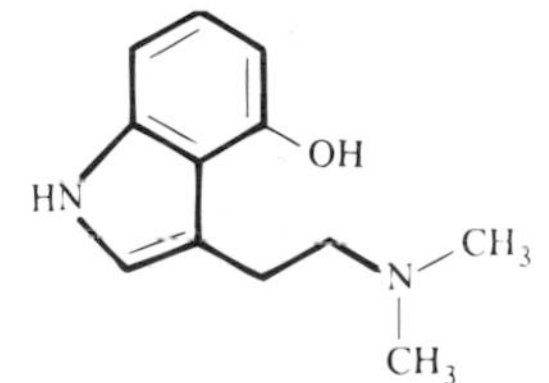

Figure 3b Psilocin

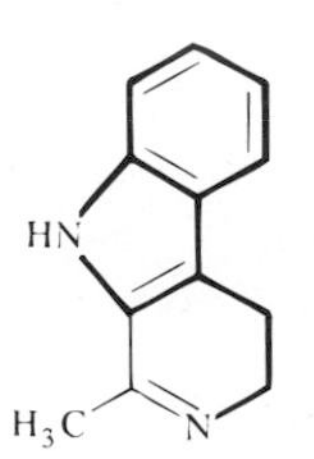

Figure 3c Harmala-type alkaloid

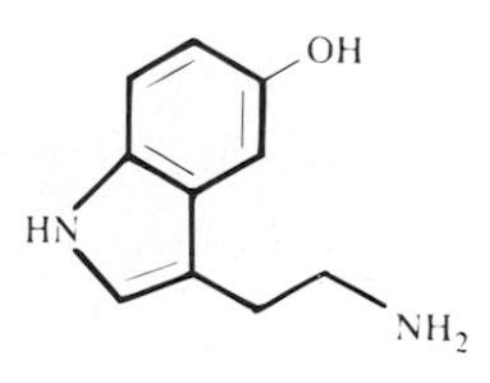

Figure 3d Serotonin

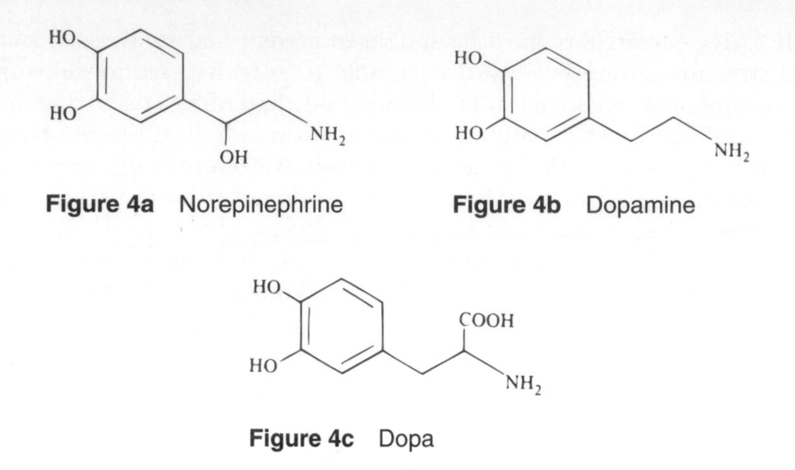

Figure 4a Norepinephrine

Figure 4b Dopamine

Figure 4c Dopa

A.2 Antihistamines and Antinauseants

Histamine (Figure 5a) is released in human tissue as the result of contact with pollen, dust, and other allergens, from cell damage, and from contact with chemicals to which the individual has become sensitized. In the human body, histamine's powerful effects include dilation of capillaries, increased capillary permeability, bronchoconstriction, and increased gastric secretion.

Synthetic antihistamine agents have been known since 1937. Two famous examples are diphenhydramine USP (Benadryl, Figure 5b) and chlorpheniramine (Chlortrimeton, Figure 5c).

The mode of action of the antihistamines appears to be competitive antagonism of histamine for its receptor sites. Saturation of these sites by antihistamine molecules prevents the pharmacological action of histamine. Regarding SARs, note in Figures 5a–c the common occurrence of the ethanamine chain ($-CH_2CH_2-N=$) attached to either a heterocyclic ring, an electron-rich atom, or an atom carrying aromatic ring systems.

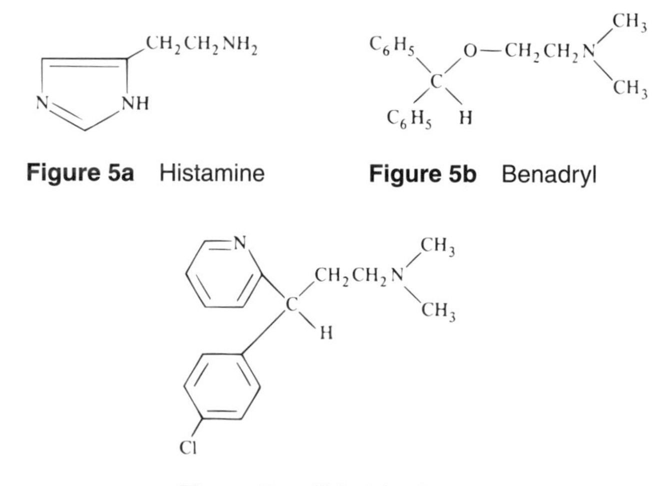

Figure 5a Histamine

Figure 5b Benadryl

Figure 5c Chlortrimeton

Dimenhydrinate USP (Dramamine), used to control motion sickness and nausea in pregnancy, is actually identical to Benadryl except that the chlorotheophylline salt is used instead of the hydrochloride. This fact established the CNS action of the antihistamines; that is, the blocking in the brain of the reflex arc that controls vomiting. CNS depression by antihistamines accounts for their use in sleep induction.

A.3 Steroid Hormones

Medicinal chemists recognize a large, diverse group of biologically active agents possessing the fused tetracyclic steroid nucleus (Figure 6a). The numbering of the carbon atoms is standard and helps to identify the location of substituent groups. Variously classed as steroids are (1) the female sex hormone estradiol (Figure 6b), (2) the male sex hormone testosterone (Figure 6c), (3) oral contraceptive agents, or progestins, one example of which is norethindrone (Figure 6d), (4) anti-inflammatory agents such as prednisone USP (Figure 6e), and (5) cardiac stimulant drugs such as digitoxin (Figure 6f). The structure for diethylstilbestrol (DES, Figure 6g) shows that although it is not a steroid estrogen per se, it does appear to be capable of acting as a steroid precursor in the human body upon ring closure to a tetracyclic structure. Indeed, it is now accepted that this result is what happens to DES after ingestion.

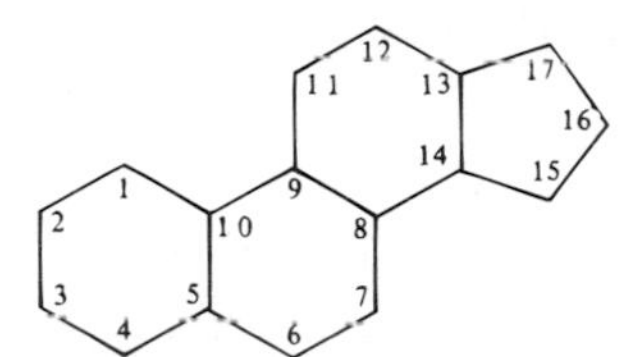

Figure 6a The steroid nucleus

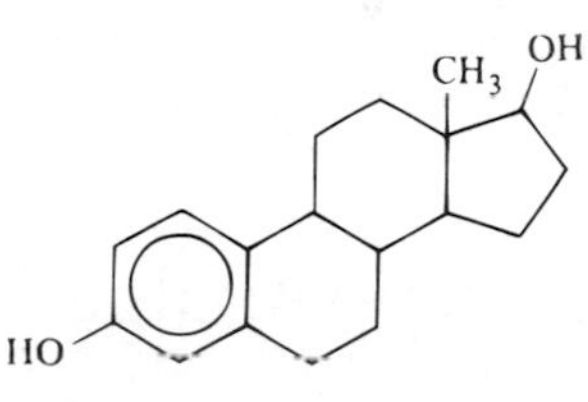

Figure 6b Estradiol

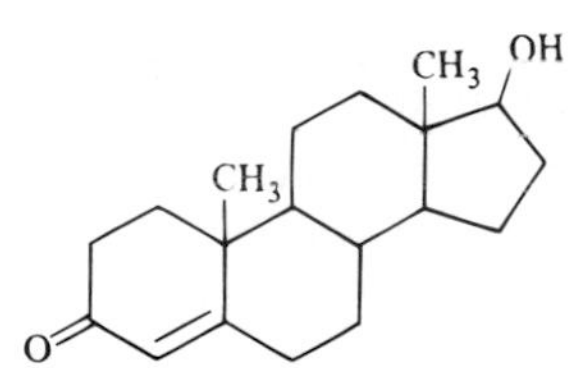

Figure 6c Testosterone

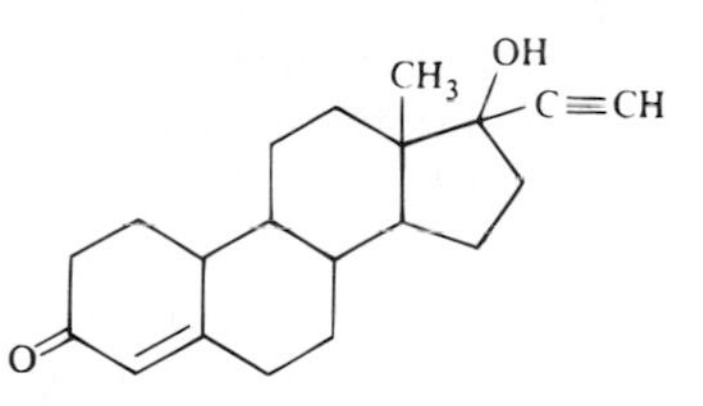

Figure 6d Norethindrone

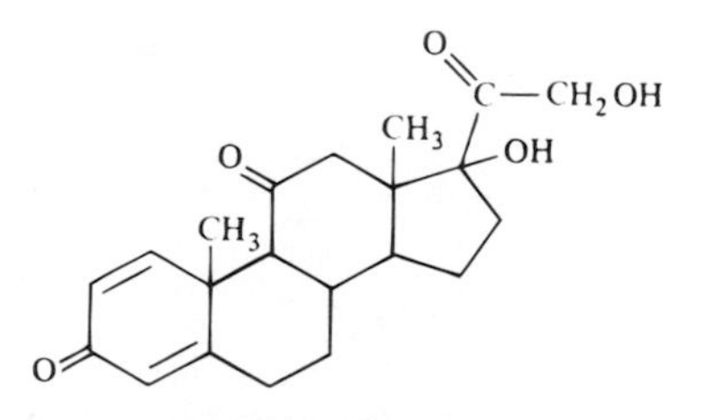

Figure 6e Prednisone

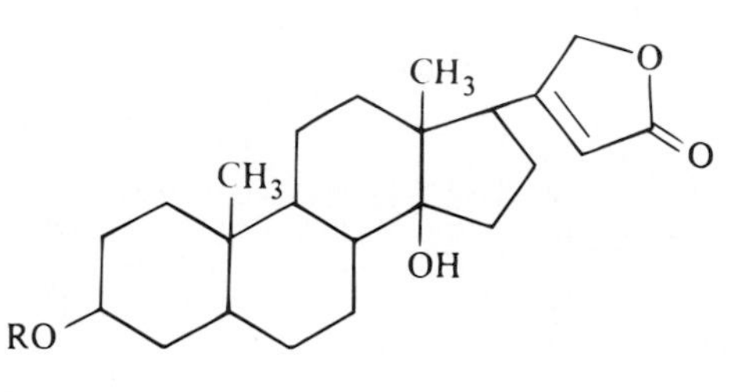

Figure 6f Digitoxin

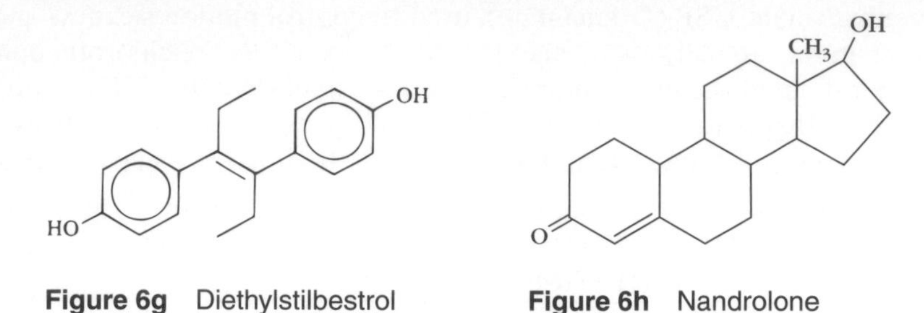

Figure 6g Diethylstilbestrol **Figure 6h** Nandrolone

The anabolic steroid nandrolone (Figure 6h) is very similar structurally to testosterone (Figure 6c) but lacks the 10-methyl group of the latter. Both of course are potent androgens. Androstenedione, discussed in Section 17.2, is also very similar to testosterone, but the former has a ketone group at C-17 instead of a secondary alcohol. When androstenedione is ingested, the liver converts the 17-ketone to a 17-OH (i.e., testosterone). We don't have the space here to discuss stereochemistry of these compounds, but that would have to be considered also in comparing structures.

The steroid structures we have just examined teach us much about SARs and show how relatively minor chemical modifications can cause tremendous differences in types of biological activity. A good example of this is the less than profound change on carbons 1–6 in *estradiol*, which converts it to the male sex hormone testosterone!

A.4 Narcotic Analgesics

The humble poppy plant *Papaver somniferum* is a veritable chemical laboratory capable of synthesizing large, complex alkaloids with potent biological activity. From the opium poppy we obtain morphine (Figure 7a), a T-shaped molecule that fits human brain receptor sites, producing analgesia. Morphine's hydroxyl, ether, and amine groups undoubtedly help bind it to the receptor site. Pharmacologists know that the illicit substance heroin (Figure 7b) is easily converted in the human body to morphine by the enzyme-catalyzed hydrolysis of the two ester (-$OOCCH_3$) linkages. Figure 7c shows the chemical structure of the potent narcotic antagonist naloxone. Note that by changing a hydroxyl group to a ketone, a methyl group to an allyl, introducing another -OH group, and reducing a double bond, chemists obtain a drug that still fits the opiate receptors but has lost all of its narcotic and respiratory depressant activity.

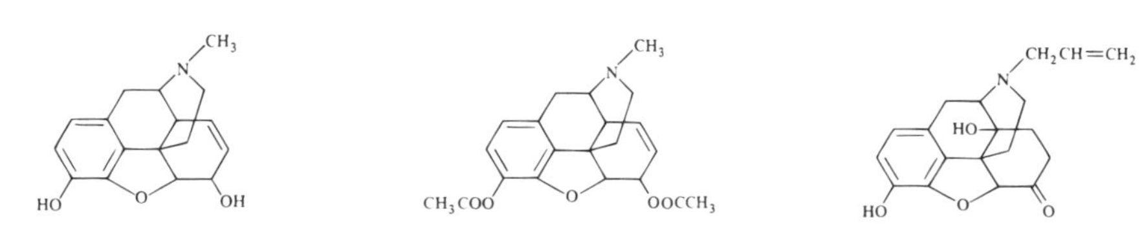

Figure 7a Morphine

Figure 7b Heroin

Figure 7c Naloxone

A.5 Tranquilizers

The structure of the minor tranquilizer diazepam (Valium) is shown in Figure 8a. Comparison of its structure with that of Figure 1a clearly reveals that Valium is not a barbiturate, although there are some similarities. The ring structure in Valium that encompasses the two N atoms is called the *benzodiazepine nucleus*. Librium is also a benzodiazepine derivative, but Miltown (Figure 8b) obviously belongs to a different class (in this case, the propanediol carbamates).

Minor tranquilizers such as Valium and Miltown are quickly distinguished from a major tranquilizer like Thorazine (Figure 8c) by the differences in their respective chemical structures. Such major differences strongly suggest that different brain receptor sites are involved when major tranquilizers exert their pharmacological activity.

By examining the structures of the three tranquilizers shown, scientists can predict the mechanism by which the human liver detoxifies (metabolizes) these drugs. Thus SAR considerations offer clues to the biodegradation products of these drugs.

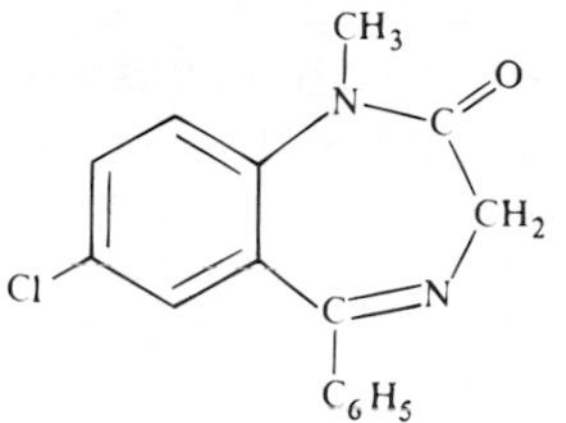
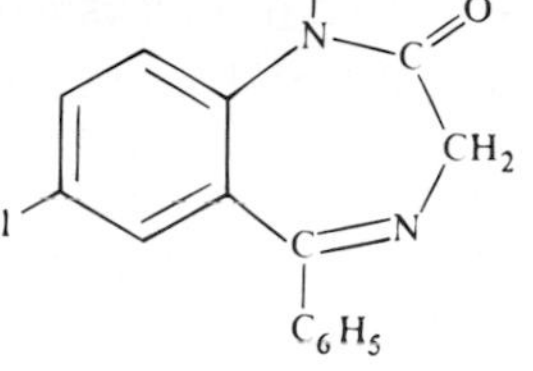

Figure 8a Valium

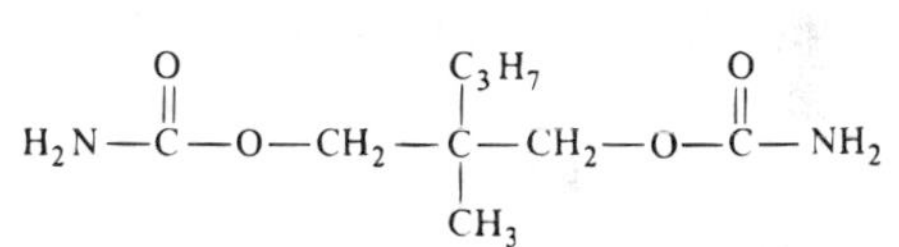
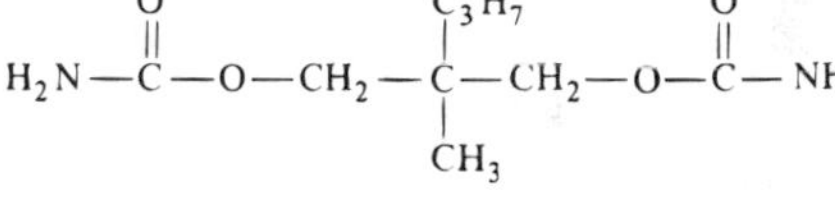

Figure 8b Miltown

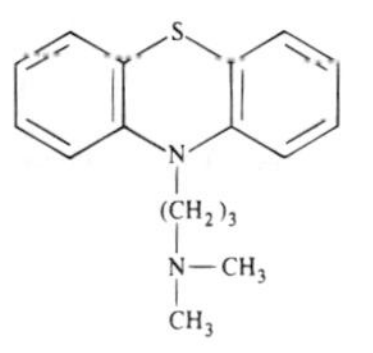

Figure 8c Thorazine

A.6 Cocaine and Synthetic Local Anesthetics

Cocaine is an alkaloid. That means it is an alkaline, nitrogen-containing compound obtained from a plant. In addition to being a CNS stimulant, cocaine is an effective local anesthetic, and this property has stimulated the preparation of *synthetic* local anesthetics patterned after it. When we examine the structure of cocaine (Figure 9), we find an aminoid nitrogen atom connected to an ester grouping by a three-carbon chain (heavy lines). This same system is found in the synthetic substances Surfacaine and benzocaine, although in the latter an aromatic ring separates the amine group

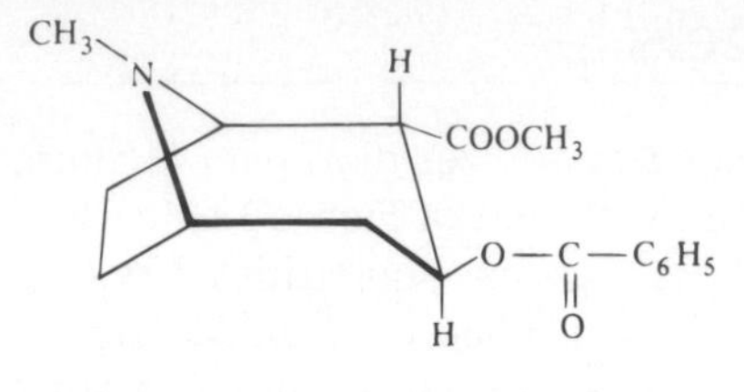

Figure 9 Cocaine

from the ester group. Here again, SAR studies have opened the door to new, active analogs based on clues from nature.

A.7 Antimetabolites in the Treatment of Cancer

Nowhere have SARs been applied more elegantly than in the synthesis of "fraudulent" compounds designed to fool a cancer cell into accepting them as the real metabolite. Consider mercaptopurine (Figure 10a). This bogus molecule is so similar to the genuine metabolite adenine (Figure 10b) that rapidly growing cancer cells can mistake it for adenine with consequent disruption of cancer cell proliferation. The use of S in place of N is possible because of their similar electron pair structures—and the process is similar to building an automobile with an engine that has no cylinders. The car will not function; neither will the cancer cell. The antimetabolite idea is only one approach to cancer chemotherapy, but it has become very important. Another example of an antimetabolite is 5-fluorouracil (Figure 10c), in which a foreign *fluorine* atom has been introduced. Here again, the cancer cells incorporate the antimetabolite into their enzyme systems as though it were a useful substrate, but it only serves to impede cell growth. 5-Fluorouracil is effective and is used in treating cancer of the breast, colon, stomach, ovary, bladder, and skin.

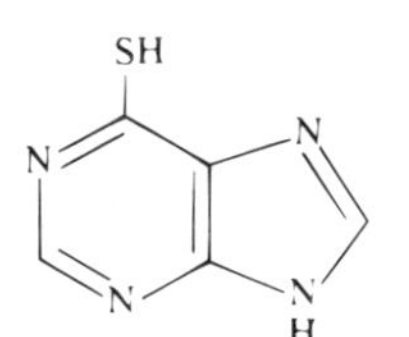

Figure 10a Mercaptopurine

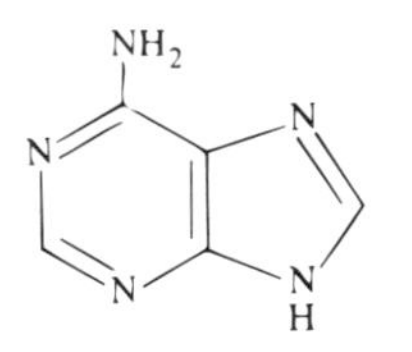

Figure 10b Adenine

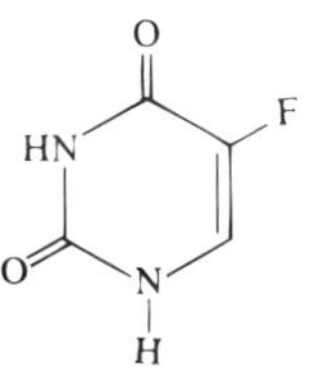

Figure 10c 5-Fluorouracil

A limitation to the antimetabolite approach is the fact that *normal* cells can also mistakenly use the fraudulent molecules, with toxic consequences to the host organism.

A.8 Amphetamines and Related Compounds

The synthetic substance amphetamine (Figure 11a) is pharmacologically classified as a **sympathomimetic** because its actions mimic stimulation of the sympathetic division of the autonomic nervous system (see the discussion of pharmacology in Chapter 7). Amphetamine and its N-methyl analog, methamphetamine (Desoxyn, Figure 11b), have a strong stimulatory effect on the brain and the entire CNS. The amphetamines are patterned after norepinephrine (Figure 4a) and contain the beta-phenethylamine moiety, $C_6H_5CH_2CH_2NH_2$.

The CNS stimulatory effects of the amphetamine-type compounds have long been recognized by drug abusers seeking chemical highs, and consequently many synthetic analogs (chemical relatives) have sprung up from illicit laboratories, including MDA, MMDA, and STP (see Table 12.2). Figures 11c–e show the close chemical relationship of these hallucinogens to amphetamine per se and to the naturally occurring hallucinogen, mescaline (Figure 11f).

All of these compounds, plus epinephrine, NE, and dopamine, possess the beta-phenethylamine moiety ($C_6H_5CH_2CH_2N$), modified or unmodified. From all the

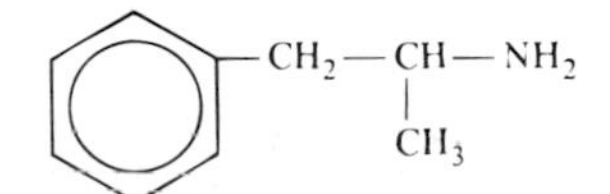

Figure 11a Amphetamine

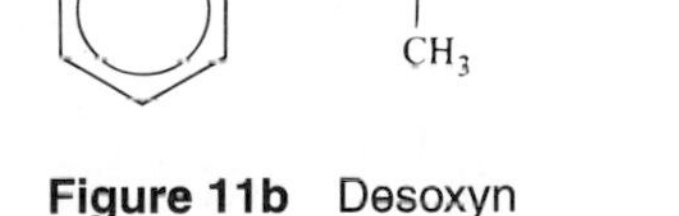

Figure 11b Desoxyn

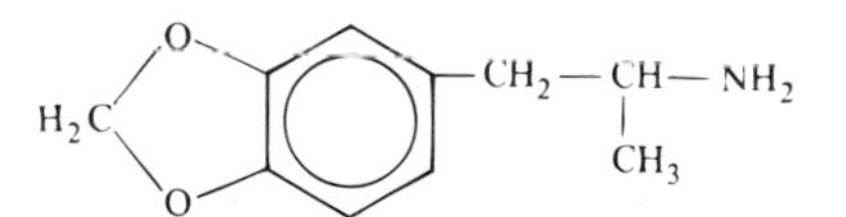

Figure 11c MDA

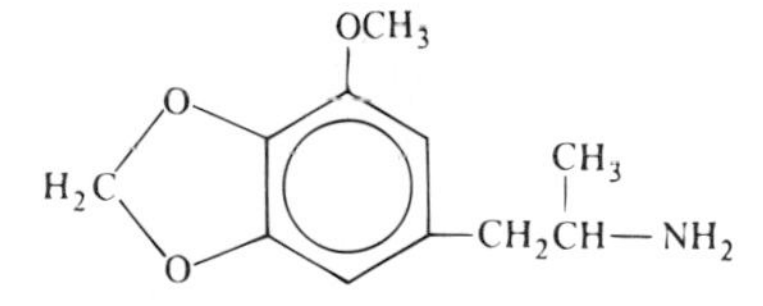

Figure 11d MMDA

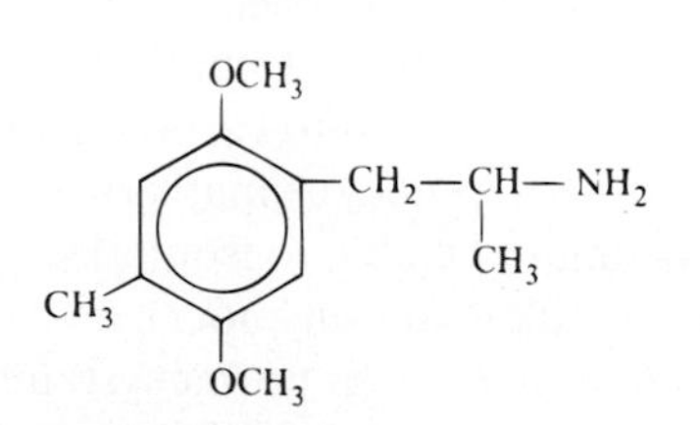

Figure 11e STP

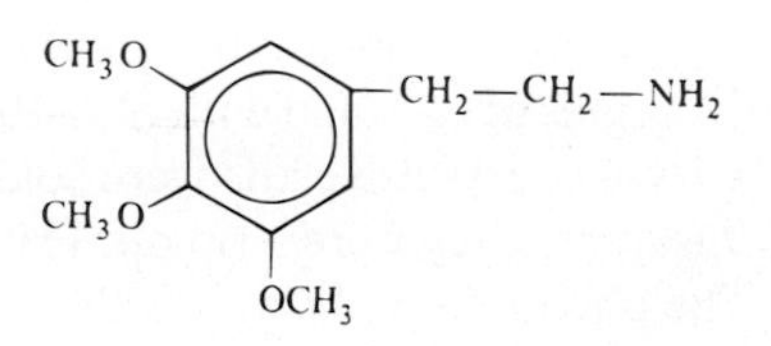

Figure 11f Mescaline

research on amphetamines, scientists have accumulated a vast knowledge of the SARs in these compounds. For example, introduction of an alcoholic OH group on the carbon next to the aromatic ring (as in epinephrine) is known to reduce central stimulatory activity. Reducing the aromatic ring or replacing it with an alkyl group produces compounds with little CNS stimulatory activity but with vasoconstrictive action useful in nasal decongestants. STP is about 100 times as potent a hallucinogen as mescaline but only about 1/30 as potent as LSD.

A.9 Designer Drugs: Fentanyl Analogs

As discussed in Section 12.8, designer drugs are contrived in illicit laboratories to be chemical analogs of a well-known controlled substance. Their slight difference in chemical structure makes them different substances and therefore supposedly outside the jurisdiction of the law. Figure 12 shows the structural relationship between the legitimate narcotic analgesic fentanyl and two analogs of it designed to circumvent the law. The relatively minor structural changes do not alter the fundamental narcotic analgesic nature. Actually, in these cases, much more powerfully active narcotics were obtained, possibly because of a better fit to receptor sites in the CNS. It is of interest to note that many of today's designer drugs were synthesized years ago by legitimate medical researchers but rejected because of serious adverse effects.

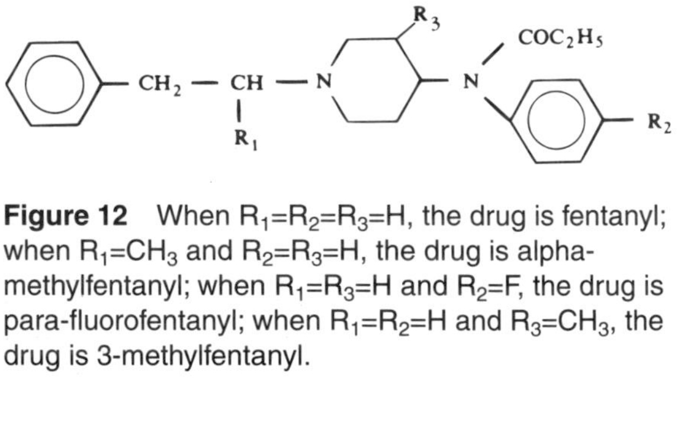

R_3 COC_2H_5 CH_2 — CH — N N R_2 R_1

Figure 12 When R_1=R_2=R_3=H, the drug is fentanyl; when R_1=CH_3 and R_2=R_3=H, the drug is alpha-methylfentanyl; when R_1=R_3=H and R_2=F, the drug is para-fluorofentanyl; when R_1=R_2=H and R_3=CH_3, the drug is 3-methylfentanyl.

A.10 Tetrahydrocannabinol (THC)

As discussed in Section 2.4 and 11.1, the most important psychoactive ingredient in marijuana is delta-9-tetrahydrocannabinol. In naming, sometimes the Greek capital letter delta (Δ) is used; this gives Δ^9-THC. Note the system of numbering shown in Figure 13. We term this isomer delta-9 because the carbon-to-carbon double bond (C=C) begins at C-9 (and ends at C-10). Another less potent cannabinol is the Δ^8 isomer. Years ago, expensive kits were sold purporting to be able to isomerize the C-8 to the C-9 isomer. They were a rip off. THC is not an alkaloid (why not?); it is a phenolic ether. The phen*ol* (at C-1) gives it the name cannabin*ol*. Other newer systems of numbering the atoms are also in use. Hence we also see the names Δ^1THC and $\Delta^{3,4}$THC.

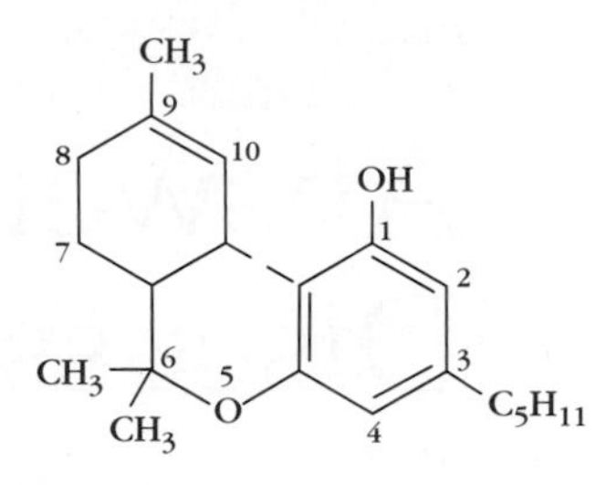

Figure 13 Δ^9-Tetrahydrocannabinol (THC)

A.11 Nonsteroidal Anti-inflammatory Drugs (NSAIDs)

Chapter 14 introduced us to NSAIDs, and we noted that there are distinct chemical differences between them. Aspirin (Figure 14a) is a salicylate (the salicylic acid core structure is emphasized in **bold lines**). Acetaminophen (Figure 14b) is an aniline derivative (core structure in **bold lines**). All of the remaining are derivatives of propionic acid (core structure in **bold**): ibuprofen (Figure 14c), naproxen sodium (Figure 14d), ketoprofen (Figure 14e), and fenprofen (Figure 14f).

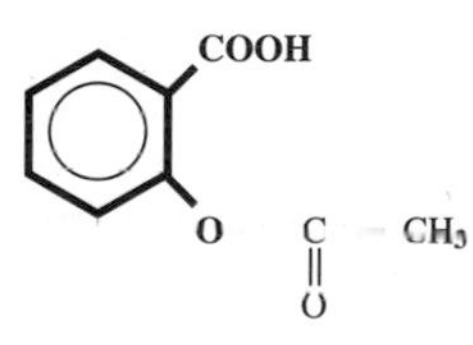

Figure 14a Aspirin

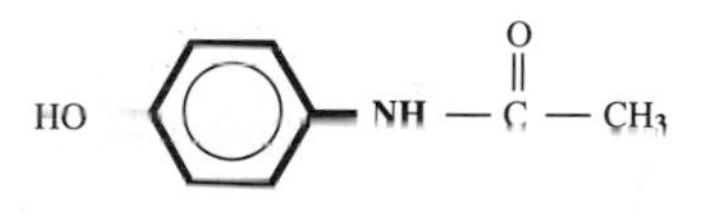

Figure 14b Acetaminophen

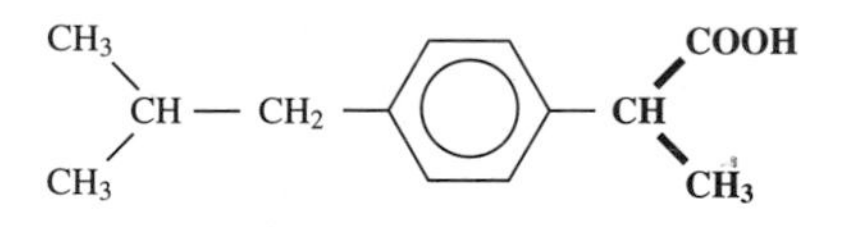

Figure 14c Ibuprofen

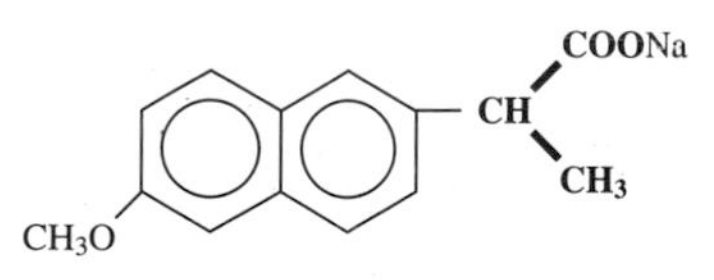

Figure 14d Naproxen sodium

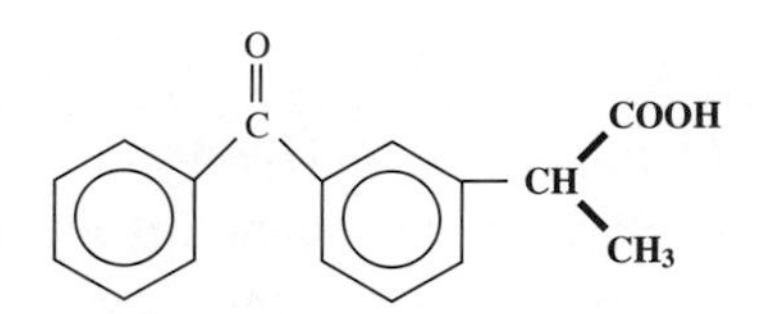

Figure 14e Ketoprofen

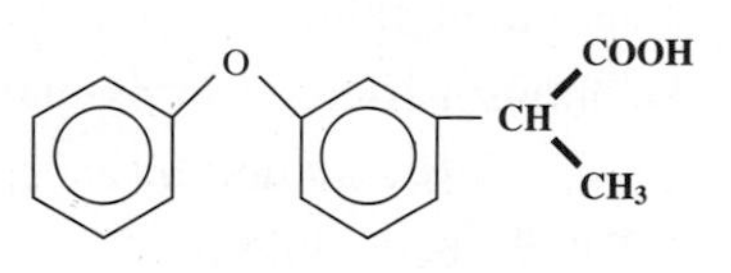

Figure 14f Fenoprofen

APPENDIX II

Answers and Discussion: Understanding Alcoholism

The following are answers to the self-test on alcoholism, Section 9.12:

1. Most acceptable definitions of alcoholism include a reference to the approximate quantity of alcohol consumed per unit of time.

   Answer: False. Alcoholism is most accurately seen as pathological dependence upon the drug ethyl alcohol, which is used excessively and inappropriately to the extent that harmful consequences occur to the individual or to society. It has little to do with absolute quantities of alcohol consumed. In fact, some social drinkers consume more alcohol without serious consequences than do advanced alcoholics. In the later stages of alcoholism, due to a decreased physiological tolerance, the alcoholic may require relatively little alcohol to achieve intoxication.

2. As well as suffering adverse consequences from his drinking, the alcoholic usually drinks according to different patterns than does the normal drinker.

   Answer: True. The alcoholic uses alcohol as a drug or medicine, is preoccupied with its use, often conceals his or her supply, gulps drinks, suffers blackouts, drinks alone, and may calm morning shakes with a morning drink. These patterns of drinking are distinctly *nonsocial* and help distinguish pathological from normal drinking.

3. Many health professionals consider alcoholism our leading health problem.

   Answer: True. Dr. Roger Egeberg, among other prominent physicians, has so termed this disorder. It certainly ranks among the top four health problems in our country, along with heart disease, cancer, and mental illness. The prevalence of alcoholism in general hospital populations is staggering. Studies have shown the overall incidence to be 10–29% of the hospital population, with a male/female ratio of approximately 4 to 1.

4. Alcoholism affects approximately 1% of our adult population.

   Answer: False. Generally acceptable data place the prevalence of alcoholism at 5% of the adult population, or about 13 million.

5. Alcoholism should be considered a symptom of an underlying personality or mental disorder, as opposed to a disease per se.

Answer: False. If drinking begins as a symptom of an emotional problem, by the time it has reached the stage of true alcoholism it has taken on the properties and dynamics of a true illness. While other disorders often coexist, the alcoholism is usually by far the most serious in its impact. It is recognized as a disease by the American Medical Association as well as the World Health Organization. Attempts to treat alcoholism simply by treating the presumed underlying disorder have been notably unsuccessful.

6. Approximately one-fourth of all alcoholics are on skid row.

Answer: False. The proportion of alcoholics on skid row is less than 5% of the total alcoholic population. Ninety-five percent of alcoholics are responsible persons who are usually employed, married, and not distinguishable from the general population by any socioeconomic factors.

7. Research has failed to establish any specific genetic, environmental, social, or personality factors as the cause of alcoholism.

Answer: True. Despite a rapidly increasing body of knowledge, no specific etiologic factor has been demonstrated. There is evidence, however, that each of the factors cited may contribute to the cause of alcoholism much like other chronic and relapsing illnesses (e.g., coronary heart disease) have been found to have multiple causes.

8. Becoming unconscious or passing out from excessive drinking is known as an alcoholic blackout period.

Answer: False. A blackout is a period of alcohol-induced amnesia, usually a period of hours, during which time the person is totally conscious and often functioning in what appears to be a normal manner, frequently carrying out somewhat complex activities that he or she is later unable to remember.

9. A drinker who never consumes anything stronger than beer is probably not an alcoholic.

Answer: False. This myth is common and is one often exploited by beer drinkers in their defense system. In fact, there is as much ethyl alcohol in one 12-ounce bottle of beer as in an ounce of 86-proof whiskey. A case of beer is equivalent to approximately one-fifth of such whiskey.

10. A brief drinking history should be obtained from every new patient.

Answer: True. Because of the high incidence of associated medical and nervous disorders, the alcoholic often consults a physician. Drinking history and alcohol-related problems are usually avoided in the medical interview because of the patient's defense system, and a significant alcoholism problem may be missed by the physician who does not review the drinking history (patterns and consequences of drinking) with each patient.

11. One may be a thoroughly reliable worker on the job and still be an alcoholic.

Answer: True. Although many alcoholics have job problems, studies show that the majority stay reliable on the job for many years, often being the subject of disciplinary action or being fired only at a late stage of the illness. Job longevity is often used as a handy alibi in the denial system of the alcoholic and should not be considered a contraindication for the diagnosis of alcoholism.

12. The ability to confine drinking to weekends suggests that a person is probably not alcoholic.

Answer: False. This statement is another myth commonly exploited by alcoholics. A large number of alcoholics spend all week counting the hours until the closing whistle on Friday so that the weekend bender can officially commence. The rationalization "I never drink during the week" is used to try to hide the damaging effects of extensive weekend drinking which often extend into Monday morning. The test of alcoholism is not whether one can stop drinking, but whether one can drink without continuing adverse consequences.

13. An alcoholic must hit bottom before he can begin the recovery process.

Answer: False. This is another common belief that has been shown by experience to be false. While it is true that motivating for change usually requires some crisis or other serious consequences of drinking, these occasions will occur many times before any extreme bottom, in the sense of skid row deterioration or organic psychosis, is reached. The implication is, of course, that treatment should be instigated as early as possible in the illness, with known success rates being better than those achieved after gross deterioration has occurred.

14. Alcoholics are prone to abuse any other chemical substance given to them that also produces a sedative effect.

Answer: True. Alcoholism is best considered a particular form of chemical or drug dependence. This dependence will usually manifest itself whether the sedative is in the form of an alcoholic beverage or a medication. Any chemicals affecting the mood, such as stimulants and analgesics, pose an equally dangerous risk of abuse in the drug-dependent person.

15. It is usually wise to conceal liquor when entertaining a recovering alcoholic in your home and to advise relatives of alcoholic persons to do so.

Answer: False. Hiding liquor is both a naive and condescending act toward the abstinent alcoholic. It is naive in its assumption that physical proximity alone causes the alcoholic to drink and is condescending in that it fails to show respect for the fact that the person is constantly exercising choice to abstain although liquor is easily available.

16. The suicide rate among alcoholics is markedly higher than that for the general population.

Answer: True. Alcoholism has been shown by studies to be one of the most potent factors in predicting the risk of suicidal behavior. Depressive affect and social isolation, other important factors in suicidal acts, are obviously correlated with the progression of alcoholism. As a CNS depressant, ethyl alcohol may also produce depressing effects on a chemical basis. Furthermore, the alcoholic person is usually remorseful and guilt-ridden after a bout of drinking and often feels worthless and deserving of death. Finally, the state of intoxication may impair judgment and produce suicide even when the intention to die is not strong.

17. A person's real character emerges when under the influence of alcohol.

Answer: False. Although alcoholic persons will often reveal strong feelings while intoxicated that they find difficult to express otherwise, there is little to suggest that this is their "real" personality. Drug-related behavior, whether produced by the influence of LSD, barbiturates, heroin, amphetamines, or alcohol, is a product of the effects of the drug upon the individual in that particular situation.

18. Many people who say alcoholism is an "illness" often behave toward the alcoholic as though he had a moral weakness.

Answer: True. While public education aimed at changing attitudes has led many to pay lip service to the disease concept of alcoholism, the majority of lay and professional people show by rejecting behavior and avoidance that they really believe the alcoholic person is not sick but simply weak-willed, self-indulgent, and morally wrong.

19. Tranquilizing drugs such as Librium or Valium are often valuable in maintaining the recovering alcoholic through his first year or so of sobriety.

Answer: False. These drugs are too often prescribed to alcoholics under the false premise that pathological drinking has been the result of an anxiety state. In fact, the commonly prescribed minor tranquilizers tend to alter the mood in such a way as to pose a grave risk either that drug dependence will be transferred from alcohol to the prescribed drug or that the person will return to alcohol use, resulting in a dual problem. These tranquilizers, however, do have a valuable place in the medical armamentarium as agents to control potentially dangerous withdrawal symptoms when the alcoholic is being detoxified.

20. An alcoholic with over 10 years sobriety may safely take an occasional social drink.

Answer: False. There has been a brief flurry of interest in "controlled drinking" studies with alcoholics who are taught to limit their drinking to a specified quantity in experimental situations. Although some studies have shown that this approach may be promising with certain individuals, widespread experience shows that a high risk of returning to alcoholic drinking is present regardless of how long a period of abstinence precedes the attempt to return to social drinking. Reports of individuals successfully engaging in controlled drinking are rare.

21. Coming from a family background of teetotalism is relative insurance that one will not develop alcoholism.

Answer: False. In fact, evidence suggests that persons raised in an atmosphere of teetotalism have a higher risk of developing alcoholism than those raised in families where alcohol is used normally. This result probably stems in part from the ambivalent attitudes about alcohol use that develop in a situation where family values consider drinking bad, and society as a whole considers it good.

22. The first step in psychotherapy with an alcoholic person is determining the underlying reasons for drinking.

Answer: False. The initial step in therapy is to attempt to help the person recognize and accept the condition of alcoholism, thereby motivating him toward sobriety. Only when the person has been abstinent from alcohol for a period of time can any associated emotional problems be identified or successfully handled.

23. Alcoholics Anonymous has been more effective than psychiatric treatment in helping alcoholics to recover.

Answer: True. Almost everyone agrees with this observation. This fact, however, does not negate the effectiveness of psychiatric treatment as much as it does the usual psychiatric approach. All too often, psychiatrists have approached alcoholism as a symptom of an underlying disorder and have attempted to uncover the cause of the disorder (much as one does in treating the neurotic) rather than tackling the alcoholism as an entity in itself.

24. The spouse of the alcoholic is often a primary cause of the alcoholism.

Answer: False. No person can make anyone else an alcoholic. Nor, for that matter, can a person make someone else a drug addict, obese, or habituated to cigarettes. At worst, another person's behavior may provide a convenient and often accepted alibi for pathological drinking. Too often the spouse of the alcoholic has become convinced that he or she is to blame, and occasionally physicians reinforce this misunderstanding.

25. An alcoholic who has "fallen off the wagon" (relapsed into drinking) more than four times can usually be regarded as untreatable.

Answer: False. Attendance at a meeting of Alcoholics Anonymous (AA) or experience with any population of alcoholics demonstrates that it is unsafe to classify anyone as a "lost cause," because a significant number of those previously so labeled are now leading sober, productive lives. The probability of recovery may decrease with repeated treatment failures, but one never knows which individual will ultimately recover.

26. The strong resistance among alcoholics to admitting their problem is in large part due to society's attitude toward alcoholism.

Answer: True. Although several factors contribute to the denial and defense system of the alcoholic person, the continuing enormous moral stigma attached to alcoholism may be regarded as a major component of resistance.

27. Involuntary treatment of an unmotivated alcoholic has been shown to be effective in many cases.

Answer: True. Contrary to popular belief, motivation prior to treatment is not essential. There is much evidence that many if not most alcoholics become motivated during treatment rather than before. In fact, many "voluntary" patients submit to treatment initially only as a way of coping with strong social pressure or external coercion. Alcoholism almost by definition is an insightless illness whose victims are usually unaware of their disorder. To wait until the alcoholic person says, "I need to be treated" usually results in so much delay as to make death or significant physical and mental deterioration the likely alternative.

28. Professionals are often wise to advise the spouse of an alcoholic to consider precipitating a crisis, often by separation from the unmotivated alcoholic, after lesser measures have failed.

Answer: True. Related to the previous question, some form of pressure or coercion is often necessary to initiate treatment after simple confrontation and persuasion have repeatedly failed. Depending on individual circumstances, a separation may be the most reasonable way of producing the critical situation, and the physician may well advise this alternative to the ambivalent spouse who is suffering deeply along with the rest of the family.

29. Most wives of alcoholics tend to become more emotionally disturbed when their husbands are maintaining sobriety.

Answer: False. Another prevalent myth is that the wife of an alcoholic is invariably seriously neurotic and becomes worse when her husband is sober because she "needs" an alcoholic husband. Certainly such women may exist. However, the majority of wives of alcoholics are normal people who develop emotional problems as a result of the severe family disorganization of alcoholism, and these problems usually diminish with sobriety. Most wives will become noticeably less disturbed as the process of sobriety and recovery continues.

30. Treatment (versus no treatment) improves the alcoholic's chances for recovery to a greater extent than the neurotic's.

Answer: True. The rate of recovery for the average alcoholic can be improved from an estimated 10–15% with no treatment to 50–75% with optimal treatment. Comparable estimates for neurotics are 60–70% without treatment to 70–90% with optimal treatment. Thus, although a higher percentage of neurotics improve with treatment, treatment plays a relatively more useful rule with alcoholic persons.

31. A spouse or other informant should be interviewed if possible whenever a drinking problem is suspected.

Answer: True. The denial system of the alcoholic is such that the extent of the alcohol problem (consequences and patterns of drinking) may come to light only when information is obtained from another source. Also, alcoholism deeply involves all members of the family, and it is desirable to include key people from the beginning of treatment.

32. Alanon is the companion group to AA for female alcoholics.

Answer: False. Alanon is an independent organization parallel to AA but is exclusively for the family of the alcoholic. The spouse or other family members may receive immense help in Alanon through information, support, and the guiding philosophy available, whether or not the alcoholic is involved in helping himself.

33. Alcoholics often seek help for emotional or family problems without ever mentioning a drinking problem to the interviewer.

Answer: True. Symptoms such as depression, anxiety, insomnia, and psychosomatic ailments, as well as marital or family conflicts, may be directly related to alcoholic problems that are never spontaneously revealed. It is not uncommon even for the spouse of the alcoholic to ignore or minimize the role of drinking because of the moral stigma of alcoholism and to see the drinking behavior as a voluntary act or merely a symptom. In practice, many or all of the problems may be the result of the alcoholism process itself.

34. The drug disulfiram (Antabuse) has proven to be a dangerous and unsatisfactory treatment modality for alcoholic persons.

Answer: False. Disulfiram used in proper doses (250 mg per day maximum) causes few adverse side effects except when alcohol is ingested. Disulfiram then interrupts the metabolic process, creating a buildup of acetaldehyde that is associated with an unpleasant physiological reaction, thereby acting as a deterrent to alcoholic drinking. This drug has been shown to be of great benefit to many alcoholic persons who are also involved in a reliable treatment program such as AA or group therapy. The drug virtually eliminates the common hazards of reactive or impulsive drinking. Used alone, however, without an ongoing counseling or treatment program, disulfiram is usually an unsatisfactory modality.

35. Education about alcoholism often helps the alcoholic reduce his resistance to accepting the facts about his condition.

Answer: True. Almost invariably, alcoholics in their denial system will believe that an alcoholic is someone more severely affected than themselves. When objectively presented with education showing the facts about the illness, its nature, symptoms, progression, and treatability, they will often eventually recognize and accept this problem and be motivated to change.

36. The alcoholic who is maintaining sobriety has no greater number of serious emotional problems than the population in general.

Answer: True. Sober alcoholics reflect the entire spectrum of psychological adjustment from the normal healthy personality to the psychotically disturbed. Alcoholism has been found to exist in persons suffering from all known mental illnesses as well as in persons who have no other detectable mental or psychiatric disorder. The latter situation is usual in that, apart from mild personality problems, the typical alcoholic does not have significant associated psychiatric illness.

37. The physician can best help the alcoholic by adding his pleas to that of the family in urging the alcoholic to quit drinking.

Answer: False. A plea, whether given by a physician, another professional person, or anyone else, is almost always ineffective, as are threats of death or dire physical complications. Education about alcoholism plus the identification of adverse effects of drinking, done objectively by the physician in conjunction with other family members, offers a much greater probability of creating a desire to recover.

38. Alcoholism can be seen as a type of drug addiction.

Answer: True. The World Health Organization defines drug addiction as a state of chronic or periodic intoxication detrimental to the individual or society characterized by (1) an overpowering desire to take the drug, no matter how it must be obtained, (2) a tendency to increase the dose, and (3) psychological and sometimes physical dependency upon the effects of the drug. Alcoholism certainly fits this definition.

39. A significant emotional problem or disorder generally precedes the development of alcoholism.

Answer: False. Alcoholism generally develops over a period of years, not after a particular problem. Furthermore, most alcoholics do not suffer from significant psychiatric illness apart from alcoholism. Although the evidence is not yet in, animal data suggest that it may be true that anyone who drinks enough over a sufficiently long period will develop the signs of alcoholism.

40. Physicians frequently misdiagnose psychiatric problems in alcoholic persons.

Answer: True. The use of alcohol by an alcoholic person can produce anxiety and depression and may also reveal psychological mechanisms and traits that are not apparent without alcohol. For this reason, assessment of the psychiatric status of the alcoholic, when done during active drinking or in the early withdrawal stages, can be grossly misleading. Psychiatric assessment should properly be done both during active drinking and in the state of sobriety.

41. Cross-dependence (or "cross-addiction") on other sedative or tranquilizing drugs in the alcoholic may begin iatrogenically.

Answer: True. Probably the most common criticism of the way physicians manage alcoholic persons is that they have a tendency to prescribe medications that, in fact, may result in additional drug dependence problems. At times, alcoholics attempt to manipulate their physicians into prescribing these drugs for their known euphoric and mood-altering effects. Such cross-dependence is so common that it must be investigated in any person suspected of alcoholism.

42. The incidence of alcoholism and drug dependence is lower among physicians than in the general population.

Answer: False. Estimates of physician addiction rates range from 12 to 20 times that of the general population. In a Mayo Clinic study, Duffy found alcoholism or drug dependence to be the major disorder in 50% of all physicians admitted to the psychiatric unit. State boards of medical examiners in Arizona and Oregon report that 2% of practicing physicians were subject to disciplinary action for drug dependence over a period of 10 years.

APPENDIX C

DAWN Data Summary

Since the early 1970s the *Drug Abuse Warning Network* (DAWN) has collected information on patients seeking hospital *emergency department* (ED) treatment related to their use of an illegal drug or the non-medical use of a legal drug. The survey provides data that describe the effect of drug use on EDs in the United States. To be included in DAWN, the person presenting to the ED must be at least 6 years of age, must have been treated in the hospital's ED, must have had a problem induced by a drug (either an illegal drug or non-medical use of a legal drug), and must have taken the drug (a) because of dependence, (b) in a suicide attempt or gesture, or (c) for psychic effects.

Hospitals eligible for DAWN are non-federal, short-stay general hospitals that have a 24-hour emergency department and are located throughout the coterminous United States, including 21 metropolitan areas.

The DAWN system also collects data on drug-related deaths, sampling medical examiners nonrandomly.

C.1 Highlights of the Updated 2001 Drug Abuse Warning Network Report: Estimated Numbers of Hospital Emergency Department Episodes Directly Related to Use of an Illegal Drug or Non-Medical Use of a Legal Drug

TOTAL DRUG-RELATED ED EPISODES

- In 2001, there were 638,484 drug-related ED episodes in the coterminous U.S. a rate of 252 ED episodes per 100,000 population. On average, 1.8 drugs were reported per episode, for a total of 1,165,367 drug mentions. ED drug mentions and ED drug episodes each increased 6% from 2000 to 2001. Total ED visits (that is, ED visits for any reason) increased 5% (from 96.1 million to 100.5 million) during this period.
- Eight out of every 10 ED drug mentions (82%) come from only seven categories: alcohol-in-combination, cocaine, heroin, marijuana, benzodiazepines, antidepressants, and analgesics. In 2001, alcohol-in-combination was a factor in 34% of ED drug episodes (218,005 mentions); cocaine in 30% (193,034), marijuana in 17% (110,512), and heroin in 15% (93,064). Taken together, the benzodiazepines, antidepressants, and analgesics constituted 339,484 ED mentions in 2001, or nearly 30% of total ED drug mentions.

- From 2000 to 2001, significant increases in drug episodes were found in 5 of the 21 metropolitan areas oversampled in DAWN: Atlanta (30%, from 11,112 to 14,456), Minneapolis (26%, from 5,197 to 6,521), Boston (13%, from 14,902 to 16,853), Denver (11%, from 4,944 to 5,468), and San Francisco (9%, from 7,857 to 8,575). From 2000 to 2001, significant *decreases* in drug episodes were found in two metropolitan areas: New Orleans (20%, from 4,664 to 3,729) and San Diego (2%, from 7,094 to 6,962).
- Adjusting for population differences, the highest rates of ED drug episodes in 2001 were apparent in: Philadelphia (573 ED drug episodes per 100,000 population), Chicago (558), San Francisco (546), Seattle (538), and Baltimore (505). Among the 21 metropolitan areas in DAWN, Dallas had the lowest rate of ED drug episodes (210 per 100,000 population) in 2001.

C.2 Metropolitan Area Estimates of 2001 ED Drug Abuse Mentions; Rates per 100,000 Population

From ED mentions, one can derive an idea of the extent of drug use in various metropolitan areas. Exhibit 1 shows the extent of drug use in 20 metropolitan areas.

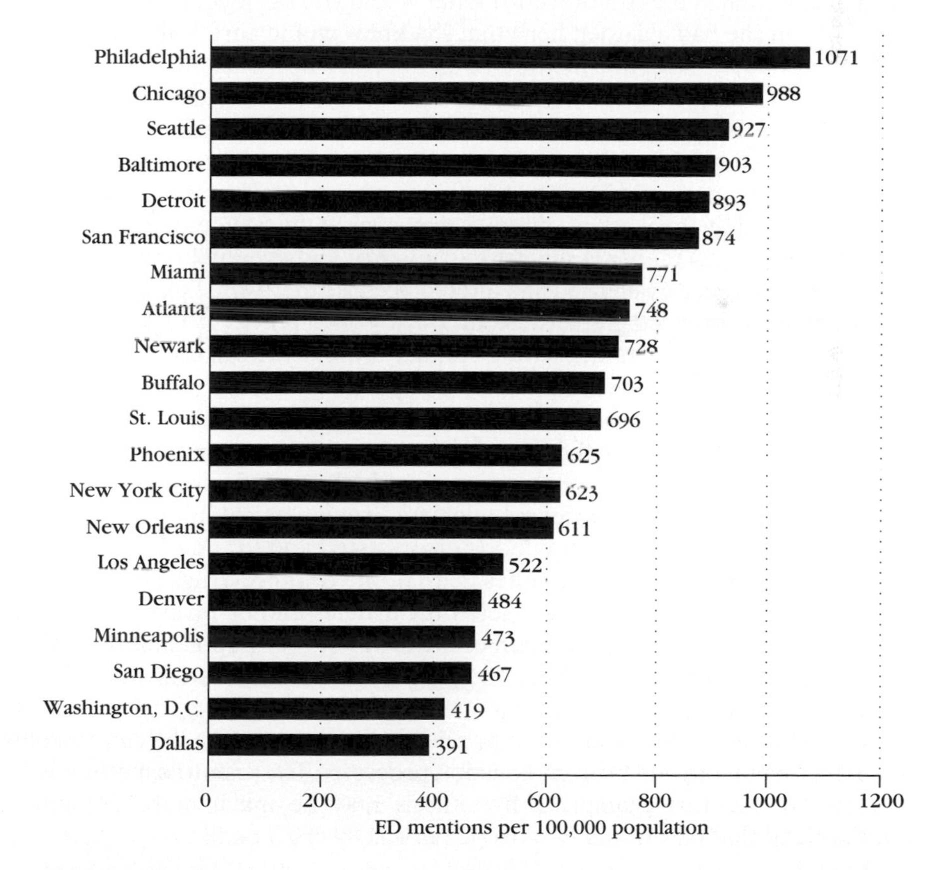

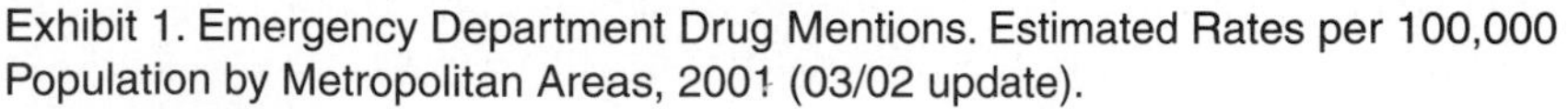
Exhibit 1. Emergency Department Drug Mentions. Estimated Rates per 100,000 Population by Metropolitan Areas, 2001 (03/02 update).

APPENDIX IV

The Scientific Method

The **scientific method** is a tool for problem solving and theory making. Scientists use the scientific method

(a) to find an explanation for an observation in nature or in the laboratory, or
(b) for solving a problem or understanding a societal dilemma, or
(c) for gathering a pattern of facts for a specific set of conditions, leading to a *natural law*.

Scientists also use the scientific method to rule out opinion based solely on prejudice or personal ambition at the cost of truth. (What would you say if your dear old grandmother told you she had a garden herb that she knew would cure diabetes?)

We can define the scientific method as an unbiased approach to an observation or to a problem, with the expectation of finding an explanation that withstands testing by other scientists. The core of the scientific method is observation of a phenomenon or identification of a problem, postulation of an explanation or plan of action (**hypothesis**), and then experimentation to confirm or deny the hypothesis. If your conclusion is validated by the work of others, it may eventually be accepted as a **theory**. A theory is not a law; a theory is subject to challenge and eventual reversal. In fact, many theories have been challenged and altered because they were limited or because new scientific data provided greater insight.

Researchers use the scientific method continually, and you can use it in your daily work. Consider the following examples:

A. You are aware that some athletes competing at your school complain of health symptoms. Working with the athletic department, you interview all of the athletes, including those with no symptoms, and identify the specific problems being encountered. You also collect pharmacological and toxicological information on all drugs, food supplements, and energy boosters ingested and for how long. You analyze data from all athletes, and hypothesize that one or more drugs is the cause of the health problems. Using published scientific articles and by investigation of individual cases, you conclude that certain of the drugs or food supplements are damaging the health of the athletes. Your theory is substantiated by further research reported nationally.

B. An earth warming trend is noted by scientists, and from data the culprit is suspected to be carbon dioxide from combustion processes. The scientific method is applied. Note that in this example a hypothesis may be much more difficult to substantiate and that bias would have to be rigorously checked for.

C. A national organization examines published data and speculates that violence depicted on television and in the movies is associated with violence in real life. To support their hypothesis the organization gathers additional data from 1,000 households across the nation and from law-enforcement agencies. Note to the reader: Do you see areas in this example in which the scientific method would have to be applied with extra rigor?

APPENDIX V Glossary

Note: Definitions of over 2,400 street drug terms can be found at **http://www.urban75.com/Drugs/drugterm.html**.

Abortifacient A drug or substance capable of inducing abortion
Abstinence syndrome *See* Withdrawal syndrome.
Acetylcholine A neurotransmitter in the human nervous system
Acid head A regular user of LSD or another psychedelic substance
Active ingredient The specific chemical in a plant or drug responsible for the drug action ascribed to the entire plant
Acute effect The immediate, short-term response to one or a few doses of a drug. Opposite of chronic effect.
Addiction (physical) A drug-induced change in a person such that he or she requires the continued presence of the drug to function normally and to prevent the occurrence of a withdrawal syndrome
Additive effect The combined effect of two or more drugs having similar pharmacological action; it is equal to the sum of their separate effects.
Adrenergic Refers to nerves, receptors, or actions that involve the release of epinephrine or NE. Most postganglionic sympathetic nerve fibers are adrenergic.
Adverse effect *See* footnote 3 of Chapter 1.
Affect Feeling experienced in association with an emotion
Agonist A drug that stimulates a receptor to produce a pharmacological effect
Akathisia A condition of motor restlessness with an urge to move about constantly and an inability to sit still, or, at worst, self-destructive outbursts; a common adverse effect of neuroleptic drugs
Alcohol Without qualification, means ethyl alcohol (ethanol); the active ingredient in alcoholic beverages
Alcohol dehydrogenase A liver enzyme that catalyzes (speeds up) the conversion of ethyl alcohol to acetaldehyde
Aldehyde dehydrogenase An enzyme that catalyzes the biotransformation of acetaldehyde to acetic acid. This enzyme can be blocked by the drug Antabuse.
Alkaloid An alkaline, nitrogen-containing chemical obtained from a plant. Many alkaloids are useful drugs.
Alpha receptor Adrenergic drug receptor site that responds to norepinephrine and to blocking agents such as phentolamine
Ambulatory Able to walk, that is, not confined to a bed or wheelchair
Anabolism Metabolism (body functioning) in which tissues or chemicals are built up from simpler substances
Analgesic Pain-relieving
Anaphylactic shock (anaphylaxis) An unexpected, exaggerated allergic reaction to a drug or allergen; potentially fatal

Androgens Male sex hormones or any masculinizing substance

Anesthesia The condition in which pain or other sensation is partially or totally lost, but consciousness is not

Angina pectoris Pain in the chest due to an insufficient oxygen supply to the heart muscle

Anorectics Appetite suppressant drugs used in weight control; also termed *anorexics*

Antagonist A drug that binds to a receptor, producing no action of its own and blocking the action of the agonist

Antibiotic A drug produced by a microorganism that has the ability to inhibit the growth of or kill other microorganisms

Antibodies Specific protein molecules synthesized by the immune response system as a defense against invading antigens

Anticholinergic A drug that blocks the action of acetylcholine; examples are atropine and nicotine

Antiemetic A drug used to prevent or control vomiting

Antigen A toxin, foreign protein, or microorganism that invades the body and induces the production of defense antibodies

Antihypertensive A drug used to lower high blood pressure

Antimetabolite A drug with a structure so similar to that of a natural metabolite that it can inhibit or block the utilization of the natural metabolite by preferentially occupying the receptor site

Antipsychotic A drug for the treatment of schizophrenia and other psychoses

Antipyretic A drug used to reduce fever

Aphrodisiac Any substance that arouses the sexual instinct

ASA Aspirin

Atrophy To shrink, shrivel up, or waste away

Autism A mentally introverted, self-centered condition in which reality is excluded. Autistic children have little affect; they show either no emotional response or a very inappropriate response.

Barbiturate A unique chemical class of drugs used for their sedative and sleep-inducing properties

Benign Not harmful

Benzodiazepines Minor tranquilizers such as Valium that have a characteristic seven-membered ring with two nitrogen atoms

Beta blocker A drug capable of competing with beta-adrenergic receptor-stimulating agents for available receptor sites

Beta receptor Adrenergic drug receptor that responds to epinephrine and to blocking agents such as propranolol. Classified as beta^1 or beta^2.

Bioavailability The degree to which a drug becomes available to its target tissue after administration

Biotransformation The chemical alteration of a drug by enzymatic actions in the body

Black-box warning Used by the FDA, this is a special warning to physicians when prescribing a drug that can result in significant, sometimes life-threatening, risks

Blood sugar Glucose, synonymous with dextrose

Blunts Hollowed-out cigars, refilled with marijuana and smoked; equivalent to about six marijuana joints

BMR Basal metabolic rate; a measure of the extent of body metabolism at complete rest and under the influence of no drugs

Bong A water pipe used in smoking marijuana

Bradycardia An abnormally low heart rate

Brompton Cocktail A combination of drugs (e.g., morphine plus cocaine) used to relieve pain in terminal cancer patients without drugging them into unconsciousness

Buccal cavity The space between the inside of the cheek and the teeth; a good area for drug absorption

Bufotenin A hallucinogenic substance obtained from the skin glands of a toad
Bummer A bad trip (experience) with a psychedelic drug such as LSD
Calorie A unit used to measure the heat energy available in a food
cAMP Cyclic adenosine monophosphate
Carcinogen A drug or factor that can cause cancer
Catatonia In psychiatry, a syndrome characterized by muscular rigidity and mental stupor
Catecholamines Neurotransmitters in the body having the dihydroxyphenethylamine chemical structure; examples are epinephrine, norepinephrine, and dopamine
Cellular target *See* Receptor site.
Central nervous system *See* CNS.
Cholinergic Refers to nerves that use acetylcholine as their synaptic transmitter
Chronic effect Experienced over a long time, as in weeks to years; constant, habitual, and inveterate
Chronic toxicity The harmful effect of a drug or chemical experienced over a long period of time
Circadian rhythms Physiological processes (or neural clocks) that operate on a 24-hour cycle, influenced by daily cycles of dark and light, temperature, meal times, exercise, and stress
Clone (verb) To create a genetically identical duplicate of an organism, that is, one having the same DNA
Clone (noun) The duplicate organism produced by cloning
CNS The central nervous system; that is, the brain and spinal cord
Cognitive Pertaining to all actions of the mind involved in thinking, perceiving, and remembering
Controlled substance A drug or chemical regulated under the federal Controlled Substances Act of 1970 as amended. Its manufacture, distribution, and sale are subject to federal control or punishment. The key criterion for controlling a substance is its potential for abuse and dependence. Also referred to as a *scheduled substance*.
Corpus luteum A yellow mass in an ovary found at the site of expulsion of the egg. In pregnancy, it secretes estrogen and progesterone.
Cross (or crossed) dependence A psychophysical state in which one drug can prevent the withdrawal syndrome associated with dependence on a related but different drug
Cross (or crossed) tolerance The diminished response to one drug caused by development of tolerance to another drug
CSF Cerebrospinal fluid; the fluid that bathes and cushions nerves in the CNS. In a spinal tap, CSF is withdrawn.
DAWN Drug Abuse Warning Network, operated by the federal government. See Appendix III.
DEA Drug Enforcement Administration, a division of the U.S. Department of Justice
Dependence The physical or psychological state in which the usual or increasing dose of a drug is required to prevent the occurrence of withdrawal symptoms
Designer drug A chemical, patterned after a real, pharmacologically active drug, but altered just enough to let it slip outside the law while retaining potent pharmacological activity
Detoxification (1) The process by which the body chemically alters a drug or other substance to make it less poisonous or more easily excretable from the body (*see* Biotransformation); (2) a treatment designed to free an addict of his or her drug and drug dependence
Diabetes mellitus A serious disorder in which carbohydrates are improperly metabolized, leading typically to excessively high blood sugar levels
Dilation An increase in the inner diameter of a blood vessel or organ
Diuretic A drug or substance that increases the formation of urine
Downer A CNS-depressant drug
Drug Any absorbed substance that alters or enhances any physiological or psychological function in the body or aids in the diagnosis, treatment, or prevention of disease
Drug interaction The action of one drug upon the effectiveness or toxicity of another
Drug screen A test designed to detect the presence of a drug or its metabolite in a body

fluid applicable to a large number of specimens as part of an initial program to detect abuse; typically followed up by confirmatory tests

Dysphoria Mental uneasiness, restlessness, or anxiety

EEG Electroencephalogram; a recording of the electric potentials on the head (i.e., "brain waves")

Embryo The developing creature in the uterus during the first trimester of pregnancy (cf. *fetus*). Embryonic tissue is more susceptible to teratogenic effects than is fetal tissue.

Endocrine system All of the glands that secrete hormones

Endogenous Coming from within

Endometrium The inner membrane (or lining) of the uterus

Enteric coated Covered with a substance that prevents the medication from dissolving until it reaches the small intestine

Enzyme A biological chemical, a protein, produced by living cells, that can influence the rate of body processes. Enzymes can act independently of the cells that produce them.

Estrogen A female sex hormone

Ethical drug A drug sold only on prescription

Euphoria A mental high characterized by a sense of unusual well-being and buoyancy, in which cares disappear

Exogenous drug One that has its source outside the body

FDA Food and Drug Administration, a branch of the U.S. Department of Health and Human Services

Fetus The developing creature in the uterus during the second and third trimesters of pregnancy (cf. *embryo*)

Flashback The spontaneous reexperiencing of a hallucinogenic drug's effects long after the last dose of the drug has presumably left the body

Free-basing Use of the nonsalt form of an alkaloidal drug. For example, cocaine can exist as the hydrochloride salt or as the nitrogenous base freed of its salt.

FSH Follicle-stimulating hormone

GABA Gamma-amino butyric acid

GAD Generalized anxiety disorder. *See* Table 5.4.

GI tract The gastrointestinal tract, consisting of the stomach and the intestines

Glucose Blood sugar, also called *dextrose*

Glycoside A plant product consisting of an organic molecule combined with a sugar. Some very important drugs are glycosides (e.g., cardiac glycosides).

Gonad A sex gland, that is, an ovary or testis

Gonadotropin A hormone secreted by the pituitary gland that stimulates the function of the gonads

Goofball Classically the combination of a downer drug with an upper

GRAS substance A drug or food additive *Generally Regarded As Safe* because of its long use by the population without apparent deleterious effects

Half-life The time (usually in hours) required for one-half of the dose of a drug to be excreted from the body

Hallucination The perception of an object, sound, or other external stimulus in its absence

HDL High-density lipoprotein, the "good" cholesterol. Currently considered by many authorities as a key substance in the prevention of hardening of the arteries.

Head Heavy abuser of drugs

Head shop A store specializing in the paraphernalia of drug use

hGH Human growth hormone

Hormone A potent chemical secreted by a ductless gland directly into the bloodstream and having a specific action on a specific target organ

Hypercholesterolemia A too high blood level of cholesterol

Hypnotic A drug or substance that induces sleep

Hypodermic Under or inserted under the skin, as in hypodermic injection

Iatrogenic Pertaining to any adverse condition in a patient caused by the physician's treatment

Inflammation A defensive body process characterized by redness, heat, pain, and swelling and caused by injury to tissue

Ionizing radiation High-energy rays such as X-rays or gamma rays that can damage the DNA of a cell

Ischemia Stoppage of flow of blood to organ or vessel

JAMA *Journal of the American Medical Association*

Joint A marijuana cigarette

Lacrimation Secretion and discharge of tears

Larynx The voice box

LDL Low-density lipoprotein, the "bad" cholesterol implicated as a prime cause of arterial blockage leading to heart attack

Legend drug One that bears on its label the warning "Caution: Federal law prohibits dispensing without a prescription"

Lente Slow or delayed onset of action

Libido Sexual desire or drive

Lipid A fat or fat-like substance

Mainline (a drug) To inject intravenously

Menopause The time in life when menstruation permanently ceases. Both pre- and postmenopausal times are recognized.

Meta-analysis A newer technique of evaluating a drug's effectiveness by statistically analyzing data obtained by combining results of *many* studies previously done on the drug singly. A meta-analysis of Prozac, for example, combined data from all 13 studies of the drug in the literature.

Metabolism (of drugs) All of the chemical and physical reactions the body carries out to prepare a drug for excretion

Metabolite Any substance produced by the body as a result of normal functioning, including breakdown products of drugs

Miosis Contraction of the pupil of the eye; opposite of mydriasis

Modal Occurring most frequently in a series

Monoamine oxidase (MAO) One of a group of body enzymes that speed up the body's biotransformation of amines such as norepinephrine, dopamine, serotonin, and drugs containing aminoid nitrogen

Morbidity Characterized by disease; state of being physically or mentally diseased

Morning-after pill A drug taken shortly after sexual intercourse to prevent implantation of any fertilized ovum and thus prevent pregnancy

Mutagen An agent that causes mutations, that is, permanent changes in genetic material

Mydriasis Unusual dilation of the pupil of the eye

Narcotic A substance that produces insensibility or stupor

NDA *New Drug Application,* required by the FDA

NE Norepinephrine

Neuroleptics Major tranquilizing drugs used to treat psychotic episodes; *see also* Antipsychotic.

Neurotransmitter A chemical released from the end of one nerve to carry the nerve impulse across the synapse to the next nerve

NHSDA National Household Survey on Drug Abuse

NIAAA National Institute on Alcohol Abuse and Alcoholism. One of the federal National Institutes of Health.

NIDA National Institute on Drug Abuse

NIH National Institutes of Health

NIMH National Institute of Mental Health

NNICC National Narcotics Intelligence Consumers Committee; a report published by the DEA

NSAID Nonsteroidal anti-inflammatory drug such as aspirin

Opiate Any remedy derived from opium

Opioid Any synthetic narcotic analgesic that has opiate-like pharmacology but is not derived directly from opium

Orphan drugs Those drugs used in the treatment of rare diseases. Companies are reluctant to produce such drugs because they are not profitable.

OTC Over-The-Counter; legally purchasable without a prescription

Outpatient A patient who comes to a hospital or clinic for treatment but does not occupy a bed

Oxytocic A drug that stimulates the uterus to contract

Pap smear A test used to detect and diagnose conditions and diseases of the female genitalia, especially cancer of the cervix

Parenteral Administered by any route of injection

PDR *Physicians' Desk Reference*

Peristalsis The wave-like contractions of the GI tract by means of which the contents are propelled along

pH An artificial scale from 1 to 14 used by chemists to express the acidity or alkalinity of aqueous solutions

Pharmacodynamics The study of what drugs do in the body and how they do it

Pharmacokinetics The study of the fate of drugs in the body, including absorption, distribution, biotransformation, and excretion

Pharmacology The science that deals with the sources, body distribution, effects, and detoxification of drugs

Pharynx The portion of the alimentary canal that connects the mouth and nasal passages with the esophagus

Pituitary gland The hypophysis, the master gland of the body; situated at the base of the brain, it secretes many important hormones

Placebo A pharmacologically inactive substance given to satisfy a patient's demand or used as a control to test the true effectiveness of an experimental drug

Plasma Blood from which the formed elements (red blood cells, white blood cells, and platelets) have been removed

Platelets Normal constituents of the blood, smaller than red cells and essential to the process of blood clotting

Polycyclic hydrocarbon a carcinogen found in partially burned hydrocarbons, as in cigarette and marijuana smoke

Polydrug abuse Simultaneous abuse of more than one psychoactive drug or other substance

PPA Phenylpropanolamine, an adrenergic drug used as a decongestant, CNS stimulant, and anorexic

Precursor A chemical ancestor; a chemical the body uses to convert into a needed product

Pressor agent One that tends to increase blood pressure

Progestin A synthetic substance having actions similar to those of progesterone

Prophylactic Warding off disease

Prospective study A study that is planned ahead of time to collect information from people as they live

Prostaglandins A large and diverse group of fatty acids made in the body that contain 20 carbon atoms and have an effect variously on blood pressure, uterine contraction, peristalsis, body temperature, inflammation, perception of pain, and many other body processes. Aspirin apparently works by inhibiting the body's synthesis of prostaglandins.

Proteins A class of substances found in all areas of nature that are composed of amino acids bound together in peptide linkages

Psychedelic A drug that produces visual hallucinations or intensified perception

Psychoactive Having an effect on the mind or behavior

Psychotomimetic Having an effect resembling that seen in psychoses, that is, hallucinations and distortions of perception

Psychotropic Affecting the mind in some way

Radioactive tag A chemical or drug, used in research, bearing a radioactive atom (tag) that permits a researcher to follow its behavior and ultimate fate in the body

Rave An all-night dance, held in an unusual place (such as a warehouse), where drugs (such as Ecstasy) are a factor

Rebound effect A response sometimes seen when a drug is discontinued, in which the suppressed body function returns as strong as or stronger than before, necessitating another round of drug administration

Receptor site Specialized cells in a body tissue to which a drug or chemical attaches to exert its pharmacological effect

Reservoir (drug) Any system, fluid, organ, or part of the body that can bind and hold a drug for an extended period of time, thus preventing its rapid elimination from the body

Retrospective study A study that collects information from people *after* they have contracted a disease in an attempt to discover what caused it

Rhinitis Inflammation of the mucous membranes of the nose

Risk/benefit ratio Comparison of the value of a drug in treating or preventing morbidity with the hazards of its use, leading to a conclusion to use the drug or find an alternative

SAMHSA Substance Abuse and Mental Health Services Administration, a federal agency

Scheduled substance *See* Controlled substance.

Screen *See* Drug screen.

Serotonin A neurotransmitter that causes a broad range of effects on perception, movement, and our emotions by modulating the actions of other neurotransmitters in most parts of the brain

Shelf life The period of storage time during which a drug is expected to remain stable and effective. "Shelf" can refer to the drugstore, home medicine cabinet, or other storage place.

Side effect An unexpected or unwanted drug effect that occurs in addition to the anticipated action. Side effects can range from relatively innocuous to life-threatening.

Sinsemilla The flowering tops of unfertilized female marijuana plants, cultivated for its high THC content

Skin popping Injecting a drug like heroin under the skin

Spliff A marijuana cigarette

Steroid A fat-like substance containing a unique fused four-ring carbon system (termed *cyclopentanoperhydrophenanthrene*) like that found in cholesterol

Stoned Intoxicated or under the influence of a drug (e.g., marijuana)

Subcutaneous Under the skin

Sublingual Under the tongue

Sympathomimetic A drug whose actions are similar to those seen upon stimulation of postsynaptic nerves of the sympathetic division of the autonomic nervous system. *See* Adrenergic.

Syndrome All of the signs and symptoms associated with a disease

Synergism The mutual potentiation of two or more drugs such that their joint effect is greater than the sum of their effects when the drugs are taken singly

Tachycardia An abnormally high heart rate

Target tissue The glands or cells in the body to which a drug binds and thus exerts its pharmacological effect

Teratogen A drug or agent capable of causing birth defects by its action on the embryo or fetus

THC Tetrahydrocannabinol; the active, mind-altering chemical in marijuana

Therapeutic Curative or healing

Tincture A liquid medication in which the solvent is predominantly alcohol

Tolerance The need for increasing doses of a drug to produce the same effect; results of the resistance built up with chronic use

Topical Pertaining to a surface, as on the skin

Toxicity The poisonous character of a substance. Acute and chronic aspects of toxicity are recognized.

Uppers CNS stimulants

USAN United States Adopted Name

USP *United States Pharmacopoeia,* the official drug compendium

Uterus The womb

Vasoconstrictor A drug or agent that can cause a blood vessel to reduce its inner diameter (i.e., constrict its opening)

Vasodilator Opposite of vasoconstrictor

Vasopressor agent *See* Pressor agent.

Volatile Low-boiling; easily passing to gaseous state

Withdrawal syndrome A crisis, with varying degrees of physical and emotional severity, that can accompany the abrupt removal of a drug on which the person has become dependent

Woman-years A measure used to rate effectiveness of contraceptives; the number of women using the method multiplied by the number of years used

Index

Note: ch = case history;
f = figure; ff = following;
fn = footnote; t = table